Quantum Mechanics

Springer-Verlag Berlin Heidelberg GmbH

JULIAN SCHWINGER

Quantum Mechanics

Symbolism of Atomic Measurements

Edited by
Berthold-Georg Englert

Springer

Julian Schwinger (1918–1994)

Dr. Berthold-Georg Englert
Gleissenweg 23
85737 Ismaning, Germany

Clarice Schwinger
10727 Stradella Court
Los Angeles, CA 90077, USA

With 78 Drawings and Figures,
and 351 Problems

Library of Congress Cataloging-in-Publication Data.
Die Deutsche Bibliothek – CIP-Einheitsaufnahme
Schwinger, Julian Seymour:
Quantum mechanics : symbolism of atomic measurements /
Julian Schwinger. Ed. by Bertold-Georg Englert. – Berlin; Heidelberg;
New York; Barcelona; Hong Kong; London; Milan; Paris; Singapore;
Tokyo: Springer, 2001
(Physics and astronomy online library)

ISBN 978-3-642-07467-7 ISBN 978-3-662-04589-3 (eBook)
DOI 10.1007/978-3-662-04589-3

Originally published by Springer-Verlag Berlin Heidelberg New York in 2001.
Softcover reprint of the hardcover 1st edition 2001

Typesetting: Camera ready copy by the editor using a Springer TEX macropackage
Cover design: Erich Kirchner, Heidelberg

Printed on acid-free paper 55/3111 5 4 3 2 1 SPIN: 10980460

To Clarice and Ola

Julian SCHWINGER (1918–1994)

Preface

Julian Schwinger had plans to write a textbook on quantum mechanics since the 1950s when he was teaching the subject at Harvard University regularly.* Roger Newton remembers:†

> [A] group of us (Stanley Deser, Dick Arnowitt, Chuck Zemach, Paul Martin and I forgot who else) wrote up lecture notes on his Quantum Mechanics course but he never wanted them published because he "had not yet found the perfect way to do quantum mechanics."

The only text of those days that got published eventually – following a suggestion by, and with the help of, Robert Kohler‡ – were the notes to the lectures that Schwinger presented at Les Houches in 1955. The book was reissued in 1991, with this Special Preface by Schwinger [3]:

> The first two chapters of this book are devoted to Quantum Kinematics. In 1985 I had the opportunity to review that development in connection with the celebration of the 100th anniversary of Hermann Weyl's birthday. [...] In presenting my lecture [4] I felt the need to alter only one thing: the notation. Lest one think this rather trivial, recall that the ultimate abandonment, early in the 19th century, of Newton's method of fluxions in favor of the Leibnizian calculus, stemmed from the greater flexibility of the latter's notation.
>
> Instead of the symbol of measurement: $M(a',b')$, I now write: $|a'b'|$, combining reference to what is selected and what is produced, with an indication that the act of measurement has a beginning and an end. Then, with the conceptual analysis of $|a'b'|$ into two stages, one of annihilation and one of creation, as symbolized by
>
> $$|a'b'| = |a'\rangle\langle b'|\,,$$
>
> the fictitious null state, and the symbols Ψ and Φ, can be discarded.
>
> As for Quantum Dynamics, I have long regretted that these chapters did not contain numerous examples of the practical use of the Quantum Action Principle in solving physical problems. Perhaps that can be remedied in another book, on Quantum Mechanics. [...]

*See chapter 10 in the recent biography by Mehra and Milton [1]. †As quoted by Schweber in section 7.11 of [2]. ‡See the preface to [3].

The change in notation mentioned here was systematically incorporated in the set of lecture notes that Schwinger wrote up for the students of the three-quarter course on quantum mechanics that he taught twice in the mid-1980s at the University of California, Los Angeles (UCLA). I had the great luck to still be at UCLA during much of the first round, taking care, in fact, of office hours and problem sessions, and I continued to receive Schwinger's handwritten lecture notes after I had left.

Indeed, these notes were meant to be the basis of the intended book for which Schwinger had selected the natural title *Quantum Mechanics* and the less obvious – to others, that is, not to him – subtitle *Symbolism of Atomic Measurements*. This choice is the succinct pronouncement of his philosophy, which is spelt out in the Prologue. The quote from page 10:

> [P]hysics is an experimental science; it is concerned only with those statements which in some sense can be verified by an experiment. The purpose of the theory is to provide a unification, a codification, or however you want to say it, of those results which can be tested by means of some experiment. Therefore, what is fundamental to any theory of a specific department of nature is the theory of measurement within that domain.

is to the point.

Schwinger's continuing interest in frontier physics was a permanent and, in hindsight, unfortunate distraction from the book-writing enterprise. Eventually, his untimely death put an end to all plans, and so his quantum mechanics book is not to be.

Yet there are those UCLA lecture notes. Although they are certainly not identical to the book Schwinger would have written, they do represent a first draft and are the closest thing to the unwritten book that is available. From many conversations I had with him, I know that Schwinger was quite happy with the way he induces the general structure of quantum kinematics and establishes the dynamical principle, his quantum action principle. I think that he had finally "found the perfect way to do quantum mechanics."

I always thought that the notes should be put into a form that makes them accessible to the broad public, but it needed the encouragement of a few friends to actually go about it. Particularly decisive was the gentle push by Robert Finkelstein who, in response to my remark – during a lunch session at the UCLA faculty club (the Chatham had disappeared years earlier) – that somebody should put the notes into print, just said: "*You* should do it." Thank you, Bob.

And then, of course, there was the consistent support by Clarice Schwinger who gave me the feeling – very calmly and, I'm sure, very consciously – that she couldn't think of anyone else to do it. Thank you, Clarice.

My dear wife Ola had to be content with a much too small share of my attention while I was working on this project. Knowing well how much the book means to me, and why it does, she never complained. Dziękuję Ci, Olu.

The lecture notes of Schwinger's UCLA course consist of three parts corresponding to the three quarters of teaching. Here is a brief summary of the contents.

Part A, the material of the fall quarter, begins with an analysis of experiments of the Stern–Gerlach type that accomplishes "a self-contained physical and mathematical development of the general structure of quantum kinematics" [4]. Much technical material is delivered in passing. In particular, unitary transformations are studied from various angles, and the algebra of angular momentum is treated in depth. Then, an analysis of Galilean invariance yields the non-relativistic Hamilton operator.

The winter quarter, Part B, proceeds from there. The response to infinitesimal time displacements establishes the equations of motion. Then the Quantum Action Principle is derived, and accepted as a fundamental principle. In a sense, the rest of Part B and all of Part C consist of instructive applications of the action principle – the "numerous examples" referred to above. Part B contains treatments of, among others, the (driven) harmonic oscillator, bound-state properties of hydrogenic atoms, and Rutherford scattering.

Part C (spring quarter) begins with the two-particle Coulomb problem, including the modifications for two identical particles. The treatment of systems with many identical particles follows, where the notion of second quantization eventually leads to the concept of the quantized field. As a first application, the Hartree–Fock and Thomas–Fermi approaches to many-electron atoms are presented, the latter in considerable detail.§ The second and final application is the quantum theory of electromagnetic radiation, which is developed to the extent necessary for an understanding of (the non-relativistic aspects of) the Lamb shift.

During his oral lectures, and in the handed-out notes, Schwinger never took any credit for his own very substantial and highly original contributions. But, of course, he mentioned the names of others whenever appropriate. I decided to stick to this practice when preparing the notes for print.

Distributing notes to the students that attend your lectures is one thing, writing a book for the anonymous reader is quite another. So, some editing was unavoidable in the course of turning Schwinger's lecture notes into book form, but I tried to change as little as possible. In addition to the UCLA notes of the mid-1980s, Chapter 12 contains some material from lectures that Schwinger gave at the University of New Mexico, Albuquerque in 1987. The Prologue is based on the transcript of the audio record of a public lecture that he delivered in the early 1960s.¶ Most of the problems are as formulated by Schwinger; in addition to the ones that came with the lecture notes, I discovered many good problems in the Schwinger Papers [5] that are archived at the UCLA Research Library. In fact, all the raw material that I used can

§This is an example that teaching and research were closely related activities for Schwinger. At the time of these lectures, he was working on refinements of the Thomas–Fermi method. ¶Section 7.10 in [2] comments on this lecture.

be found there. Charlotte Brown, Curator of the UCLA Special Collections, has been very helpful in my search of the Schwinger Papers. I thank her sincerely.

I wish to thank Herbert Walther for the splendid hospitality extended to me over the years at the Max-Planck-Institut für Quantenoptik in Garching; the institute's infrastructure was of great help while I was working on this book. During the past year, the crucial stage of this undertaking, I was supported by the Universität Ulm; I thank Wolfgang Schleich for the generous invitation to join his Abteilung Quantenphysik temporarily.

I acknowledge with gratitude the support by the editorial staff of Springer-Verlag; in particular, the help of Wolf Beiglböck, Christian Caron, and Brigitte Reichel-Mayer was invaluable. And I thank Jens Schneider, who turned the handwritten notes into electronic files that I could then work on.

Ismaning, September 2000 *BG Englert*

References

1. Jagdish MEHRA and Kimball A. MILTON: *Climbing the Mountain. The Scientific Biography of Julian Schwinger* (Oxford University Press, Oxford and New York 2000)
2. Silvan S. SCHWEBER: *QED and the Men Who Made It: Dyson, Feynman, Schwinger, and Tomonaga* (Princeton University Press, Princeton 1994)
3. Julian SCHWINGER: *Quantum Kinematics and Dynamics* (W.A. Benjamin, New York 1970; reprinted by Addison–Wesley, Redwood City 1991)
4. Julian SCHWINGER: 'Hermann Weyl and Quantum Kinematics'. In: *Exact Sciences and their Philosophical Foundations. Proceedings of the International Hermann Weyl Congress, Kiel, Germany, 1985*, edited by Wolfgang DEPPERT et al. (Verlag Peter Lang, Frankfurt/Main and New York 1988) pp. 107–129
5. Julian Schwinger Papers (Collection 371), Department of Special Collections, University Research Library, University of California, Los Angeles

Contents

Prologue

It seems to me that there are deep philosophical lessons to be learned in the way in which the practicing theoretical physicist thinks about the foundations of the subject, i. e., the manner in which he approaches the problems, the general criteria he brings to bear on what is a reasonable solution. So, the important thing then is to display the general world view, the world picture that the theoretical physicists has.

This is particularly significant in connection with the philosophical implications of quantum physics, because quantum physics or quantum mechanics – by which I think we mean finally the rational mode of understanding of microscopic or atomic phenomena – has perhaps had the greatest impact of any of the developments of physics upon the mode of thinking or the world picture of the physicist and thereby, indirectly, of the general citizen.

Now, if we want to understand specifically the origins of quantum physics, we must go back to see how the stage has been set through the developments of what is called classical physics, and then compare with quantum physics. By classical physics, we mean the precise formulation of all of the properties of matter as they were finally expressed in their essentially perfect form at the beginning of the 20th century. Classical physics is characterized by the fact that, whereas the underlying conceptions are idealizations – it was no easy job to be a Galileo or to be a Newton – nevertheless, these conceptions still strike very close to common, ordinary everyday affairs. To understand the principles of physics, as they are expressed within these great generalizations of classical physics, is not very difficult. Our school children manage it all the time. But quantum physics is something different. In quantum physics, you go far beyond the ordinary situations of everyday life. We strike at a level of idealization that is hard to appreciate until you have seen how this historical line of development has come about.

The first grand physical theory, of course, was that of Newtonian physics. This is a theory of massive point particles which interact by means of actions at a distance. The traditional theory of gravitation is the classic example of this. And to characterize this theory in a general way in terms of its philosophical foundations, let me say that Newtonian physics, or Newtonian mechanics is a causal, deterministic theory.

By *causal*, one means essentially that when the state of the system is given at a particular time – and we must return to precisely what we mean by "state" – then the state is completely determined at any other time; this is what we mean by causality. Causality is inference in time: given the state of affairs at one time, the state of affairs is uniquely determined at another time. What makes it *deterministic* is that the knowledge of the state also determines all possible physical phenomena precisely.

This distinction between causal and deterministic may not seem very important until we come up to the rather different situation that appears in quantum mechanics. I have spoken of the state within the framework of Newtonian physics. It is familiar that when we specify the state of the system of particles which interact with each other by means of instantaneous forces as the gravitational force was conceived at that time, that the full specification of state is the indicating of precisely where the particles are at a given instant of time and how they are moving. In more technical language, it is the specification of the positions and the momenta of the particles. If these are known at a given instant of time, and it is known precisely under what law of force the particles are moving, such as the grand statement of the inverse square law of gravitation, then the physical situation is specified completely. The state at any other time can be predicted and also, since this is a deterministic theory, the knowledge of the state is the origin of the full knowledge of the answer to all possible physical questions that can be asked.

Now, to indicate that Newtonian physics is not something that lies completely behind us in the history of physics, let me perhaps remind you of the fact that the triumph of Newtonian physics is indicated every time we have an announcement of a successful space mission. That is, Newtonian physics, as any general theory of physics must, remains perfectly valid in its own domain. Here, of course, the domain is the motion of material bodies under the action of known laws of force which are instantaneous interactions and cover, therefore, fully the motion of artificial satellites in the perfectly well-known field of force that is provided by the gravitational attraction of the Earth, or the Moon, the planets, and so on.

Physics introduces new theories not because the theories in a particular domain are found to be unsatisfactory, although they may be so also, if the technique of experiment becomes finer and finer and new phenomena are found which pass beyond the level of accuracy in the earlier theories, but primarily because the domains of physics which come into question are ever extended. To follow the line of historical development: the introduction in the 17th century of the Newtonian concepts led to a steady development of these ideas, their applications being primarily to astronomical phenomena, which lasted for essentially a full 200 years, while the technical means for the inference, in a precise sense, of the implications of these laws were developed until one could fully carry out the calculations necessary to follow the paths of the planets in full detail and so on. It was, however, during the 19th century that new areas of physical experience began to be met, in particular in the domain of electromagnetism. And then we came finally, towards the second half of the 19th century, to essentially a new physical theory going beyond anything that had been, so to speak, conceived within the framework of Newtonian physics. This is Maxwell's field theory of electromagnetism.

As far as the broad categorization of these theories is concerned, this was also a causal, deterministic theory. But what made it so very different was what was involved in the specification of state. Recall that the Newtonian theories were concerned with point particles and the specification of state was the indication of where these particles were and how they were moving at any particular time. It is a discrete description: a finite number of particles, and a finite number of quantities is needed to characterize everything about this physical system. By contrast, in a field theory – let's continue to have in mind the very specific example of the electromagnetic field – the specification of state requires not a finite number of things (where the particles are and how they are moving) but an infinite number. We must, in principle at least, specify what the electromagnetic field is doing; how the electric field is pointed; and how the magnetic field is pointed at every point of space – and this at a given time. And what makes this a causal theory is that if we know the state, if we know the distribution throughout space of the electric and magnetic field at a given time, then we can predict at any later time what the distribution of the electric and magnetic field will be. Given the state at one time, the state is uniquely implied at another time. That makes it a *causal* theory.

Again, what makes it deterministic is that the knowledge of the state, the knowledge of the electric and magnetic fields, completely determines the answers to all questions that can be asked about the electromagnetic field. And, we should note again that – while we are talking about a domain of classical physics, with the inference that this is not the final word – nevertheless, the quantitative success of Maxwell's theory is demonstrated every day. We just have to look at the ever-expanding development of radio communication systems, and microwaves of radar or of television, not to mention the ubiquitous cellular phones. This is not a past history in the development of physics, it is something whose validity is confirmed all the time. The point, however, is that it refers to a limited domain of experience. It is not all of the physical world.

So, here then we have two very different kinds of physical theories, both causal and both deterministic, but widely different in the nature of what characterizes the specification of a state. One lies, so to speak, at one limiting domain of the spectrum, it is a discrete description; a finite number of quantities is specified. The other, at the other end, is a continuous description; fields are involved, distributed throughout all space. These then are the models of the two limits of classical behavior: the discrete, the continuous. And it is particularly interesting to see how these two entirely different classical modes of description have in a sense become unified or, perhaps better, transcended in the further development of quantum physics.

Still continuing with our historical development, I recall what is all familiar to you, that at the beginning of the 20th century there were further very important developments, associated with Einstein's name – the developments of the special and general theory of relativity. But yet, these were not radically new developments in the sense that I mean quantum mechanics to be; they were rounding out the framework of classical physics. They were the recognition that – once we had placed the field phenomena, the electromagnetic field specifically, on the same foundation as the theories of particles – there was a modification in the strict Newtonian point of view. While we are still dealing, within this framework, with point particles, they no longer interact via instantaneous forces. We now recognize, as is particularly emphasized by the relativity theory, that the interactions between particles are propagated through space by means of the intermediary of the field. Incidentally, I should also emphasize this difference between the two classical modes: the strict Newtonian viewpoint is one of the instantaneous interaction at a distance; the field point of view is one of local interaction propagated from one point of space to the contiguous points. Within the field concept, there is no longer any idea of instantaneous propagation. It is propagation through space and time by means of a mechanism which is, in fact, intrinsically limited in speed. This is, of course, the famous constancy of the speed of light, the starting point for all the investigations on the special theory of relativity.

And so, what finally emerges from all of this is a theory with a dualistic point of view, in which there are particles and fields, standing side by side, neither explained in terms of the other – a fundamental duality that is the culmination of classical physics: the strict Newtonian, *discrete*, point of view of the particle has been modified because we now recognize that the interactions between particles are not instantaneous but are propagated through the mechanism of the field with its *continuous* point of view. The field is there to supply the dynamical agency by which particles interact.

It was to be the purpose of further developments of quantum mechanics that these two distinct classical concepts are merged and become transcended in something that has no classical counterpart – the quantized field that is a new conception on its own, a unity that replaces the classical duality. We must try to trace the development of quantum mechanics, starting from this classical background, up to this much deeper quantum mechanical foundation and explanation.

So much then for a summary which can hardly do justice to several hundred years of hard work by many physicists in attempting to lay the foundations of these laws of what I will now call macroscopic phenomena because it was, of course, in the investigation of the microcosmos – of atomic phenomena – that an entirely new world and a new system of order was opened up. It was here that it was found that the laws, which served so very well to range from ordinary experience on the Earth to extraterrestrial experience in terms of the motions of the planets, are not in accordance with the observed phenomena. When we turn not outward, but inward, we find new laws of motions, new laws of physics, new ways of thinking, new philosophical conceptions.

Now, how did this come about? Of course, in just a few sentences I cannot give a fair account of the tremendous development of physics which occurred during the last years of the 19th century and the early years of the 20th, but let me remind you that these developments began in what may seem to be a paradox. As pointed out above, we had two distinct laws of behavior in classical physics, and one never trampled on the other. We either had particles, and they were discrete objects, or we had fields, and they were continuous objects distributed throughout space. The fields could be attenuated as much as you want – a radio wave, as we travel out through space, becomes weaker and weaker and weaker in a perfectly continuous way. A particle, on the other hand, has discrete properties which it carries with it. And so the remarkable thing was the discovery, in investigations of various atomic phenomena, of an apparent paradox. Light, for example, was known, from various interference phenomena, to possess properties typical of waves, which are field phenomena; and light is spread out through space in a characteristic field way. When performing suitable experiments, one now found that these light waves appeared to acquire particle-like properties under certain experimental conditions. Unlike the classical notion of a light wave, which suggests that its energy is simply distributed continuously throughout the area that it occupies, light exhibited the ability to transfer definite and discrete amounts of energy, acting then as if it were a particle – first simply called light quantum, but now more commonly *photon* – whose characteristic, of course, is to have associated with it, in a certain state of motion, a definite energy, a definite momentum. This was the paradox found in the early days of these developments: that light waves exhibit, under certain circumstances, definite particle-like behavior. The two quite distinct classical notions, mutually exclusive as they are, are nevertheless in some sense realized jointly within the microscopic domain. An example is the classic experiment of the photoelectric effect in which light waves, falling upon metals, would transfer energy to electrons and liberate them, whereby a definite amount of energy was absorbed every time, despite the picture of a classical field distributed throughout space in which you might absorb more or less energy, depending upon the accidental circumstances of that particular electron.

Here then was light, a characteristic example of a wave or a field phenomenon acting in a particle-like manner. The converse was also true, although the experimental proof of this would have to wait for some 27 years. But at this distance in time, I think we can lump all these things together and say that experimentally, as an important aspect of this same development, the converse was true. Electrons were the characteristic example of microscopic particles. Electron beams could be produced in evacuated tubes and they would move in straight lines. And when exposed to electric and magnetic fields, they would change their direction just as material bodies were supposed to do. But, nevertheless, under appropriate circumstances, namely when electron beams are scattered from crystalline bodies, one found interference rings which would be typical for the type of wave phenomena that is characteristically associated with a distributed field. In other words, instead of being scattered as material objects would be, electrons moving through a crystal would be scattered in the way that would produce a characteristic interference pattern much as light of a certain wavelength would do. Here then were objects originally thought to be essentially classical particles which, under new experimental conditions, would exhibit continuum or wave-like phenomena.

We had then a remarkable duality in which apparently the same objects could, under some circumstances, act as classical particles; under some other circumstances, they could act as classical waves. And, of course, this was something for which there was no preparation in any other phenomena of physics as they had been known.

Meanwhile, the detailed investigation of the properties of atoms, as they were revealed by spectroscopic experiments, had proceeded. The possibility of producing atomic spectra in the suitable circumstances of very thin gases had made it possible to study the behavior of individual atoms. There were attempts to understand the observed spectra in terms of the motion of electrons within the atoms, which ended with a complete failure of classical physics to account for these phenomena. Indeed, the mere existence of atoms and their ability to radiate precise spectral lines is a conflict with classical physics. If we'd accept any picture of an atom as electrons moving around some central nucleus, as Rutherford discovered to be the situation in 1911, then, according to the classical laws of electromagnetism, the accelerated charges in their motions around the nucleus would always continue to radiate until finally they had exhausted all possible energy and would fall into the nucleus. First of all, this meant that atoms were not stable, a quite flagrant violation of simple experience; and more than that, in the course of this process the electrons would radiate spectral lines whose frequencies would change as they got closer and closer to the nucleus, and you would have nothing analogous to the empirical situation of sharp spectral lines, of definite frequencies, characteristic of individual atoms. Clearly, the laws of macroscopic physics failed completely within the microscopic world.

In the detailed analysis as it was carried out, primarily in the hands of Niels Bohr, and others of this important Copenhagen School of Physics, it was found that the only successful attempt to understand the properties of atomic spectra consisted of introducing a bold hypothesis: that physical quantities – such as energy and angular momentum, the two most important examples – can only have certain definite values. In classical mechanics, they would assume any possible values; classically they are continuous objects. A particle in ordinary life can be given any energy one wishes by simply providing the appropriate amount of energy; and if you set a body into rotation, the angular momentum that this will have can be given any value. There is no particular selected set of values that are natural. But, nevertheless, the analysis of the facts of atomic spectroscopy indicated that the energy values that atoms, or electrons in atoms, could have were not continuous but assumed definite values. This was the only explanation that could be given of the discreteness of spectral lines. And all of the certain definite values of angular momenta are simple multiples of a new natural constant, known as Planck's constant of action, which was first discovered in connection with other attempts to understand particularly significant characteristics of atomic phenomena.

So here, then, we had first of all the major break with the phenomena of classical physics: quantities which classically would be given continuous values, now had discrete values. This, in other words, is the general observation of the microscopic world: that the phenomenon of atomicity is all-prevailing. We must account for the very existence of atoms, which after all is not a classical conception. Classically, there should be no limit to the extent to which you could subdivide matter. The fact that this subdivision cannot be carried out indefinitely, but ceases when we reach the atomic scale is, of course, the most fundamental statement that something new is involved. Here is the phenomenon of atomicity, not only in the mere existence of atoms but also in the laws of mechanical motion: an atomicity of angular momentum, an atomicity of action, to put it in the most general way, was a basic phenomenon of microscopic physics. And we simply had to accept – I say "we" but, of course, I was not involved at the time; there is, nevertheless, the feeling of kinship here – that the physical properties of atoms must be understood in terms of new laws which transcended anything that was familiar before. This was a new world.

Beyond this phenomenon of atomicity, which I mark as the one basic fact, *the* new fact of microscopic physics, there is another one which appears at the same time. This is the essential statistical nature of microscopic phenomena – another fundamental feature which must be accepted as the way that the microscopic world operates. That the microscopic world is necessarily statistical can be understood in the example mentioned above of the diffraction phenomenon that an electron beam, interacting with a crystal, exhibits. Now, let's think about how this diffraction phenomenon would come about. A crystal is a regular arrangement of atoms with a characteristic distance separating them and, by virtue of this characteristic dimension, for any wave phenomenon that falls upon the crystal, there will be certain preferred definite directions of scattering, selected by the relation of the wavelength to the interatomic spacing of the crystal. (This phenomenon is, for example, well known in the case of X-rays – in fact, its demonstration represented one of the experimental proofs that X-rays are wave phenomena.) So, when we carry out an experiment in which a beam of electrons falls upon this crystal, and then moves in various directions toward a screen, it will produce a characteristic interference pattern there; instead of the electrons arriving more or less at random, with a uniform intensity all over this screen, you will find preferred places.

Now, if I make the beam so weak that one electron within a perfectly definite time interval moves through this crystal in some way, it will finally be detected by landing in a perfectly definite place on this screen. This may be a scintillation screen, for example, and if you have a weak beam of electrons and you look at the screen, you'll suddenly see a flash of light, not all over the screen but in one place. The electron exhibits its particle-like characteristics when it is finally detected, when it finally exhibits its position by producing an appropriate chemical process, the result of which is a flash of light. And we observe that the electron is here. For one electron, there is certainly no interference pattern. The interference pattern does not appear all at once. You have simply an individual electron. Now, a second electron arrives in the course of this very weak beam. What happens to it? Does it land at the same spot? No, it lands at random, at some other point that has no relation to the first. But as we continue, and more and more electrons land upon this scintillation screen – all coming under the same experimental conditions, but each arriving independently of the others – eventually the pattern of this interference behavior emerges. Many electrons land here, none land there, and so on, until the final picture of intensity is one which gives the overall pattern. But nevertheless, this has come about as the result of the random landing of the electrons at various points on the screen.

To emphasize another aspect, now suppose we had carried through such an experiment and ten billion electrons had finally produced a certain interference pattern, a picture of intensity of relative number. Then I repeat the experiment. I prepare exactly the same circumstances, and I turn on the electron gun and see what happens. Again, the first electron moves through the crystal. Will it land at the same place as the first electron of the previous run? Certainly not. We begin all over, but the pattern *as a whole* repeats itself in a random way. The first electron of the second experiment will land somewhere; the second electron of the repeat experiment will land somewhere else – with no relation between them or the electrons of the first experiment. In other words, the individual particles arrive in a perfectly random statistical way. There is no possibility of controlling this. This is, of course, a generalization from half a century of attempts to do so: the picture which we must accept is that, within the domain of microscopic phenomena, we are unable to control how the individual particles will behave. But what is perfectly determined, and will be reproduced every time you repeat the experiment, is the interference pattern; the characteristic features of the pattern are predictable and reproducible. Once we have this apparent duality of entities – electrons that behave under some circumstances like discrete entities, particles, landing at definite places on that screen, but in other respects act as waves, producing in their overall intensity characteristics the wave phenomenon of interference – we must accept that the interference pattern is not going to be repeated in miniature every time an electron lands; and that, therefore, there must be an aspect of randomness about where the electrons do land. The interference pattern is merely finally the statement of relative probabilities: with endless repetitions, you will find many more particles there than here – and this in a perfectly regular way.

It is then, in a sense, almost an automatic inference of everything that we have said, but I would prefer to take it as really the more fundamental thing: it is a basic characteristic of the laws of microscopic phenomena that they are statistical. It is not possible to predict, in general, the outcome of a specific event. But what one can predict is the average result, the statistical result, the net situation for the repetition, a sufficiently large number of times, of the same experiment – and it is the purpose of microscopic physics or quantum mechanics to make such predictions.

So here, then, we have the great challenge of microscopic physics – that there are these two new basic aspects which we must incorporate into a world picture: the fact that phenomena are atomistic and that they are statistical. But that the new physics is statistical and therefore fundamentally different from the fully deterministic classical physics does not mean that we have failed. We simply recognize what the nature of this new microscopic physics must be. It is not to predict the outcome of each individual event. It is to predict rather what the outcome must be on average; what the probable outcome must be. And as the electron-scattering experiment – or rather this simple-minded description of the experiment – indicates, this is something that we necessarily must put up with.

In fact, as we should perhaps mention here, if we attempt in any way to control precisely where the first electron shall land, we can indeed do that; we can produce a new experimental situation in which, with essential certainty, the electrons will land at the pre-assigned spot – all of them. But then we shall have no interference pattern. In other words, there are now two situations that we're talking about. The first, in which an interference pattern does appear, is one definite experimental situation, and in that it is not possible to predict or control in any way where the individual electrons will appear. Then there is the second experimental situation, in which we can control and predict where the electrons will appear in the course of moving through some apparatus. It will be a different apparatus and that apparatus could never produce an interference pattern. We are dealing, so to speak, with two distinct aspects of the microscopic world and it requires a different experimental situation to display one or the other.

Let's think now of how – once having recognized that we have two basic features of the microscopic world, the aspects of atomicity and of the statistical nature of microscopic events – we could proceed to construct a theory that would incorporate this very bizarre situation.

We must have a mathematical theory which, in some way, will represent a suitable mathematical model or idealization and enable us to predict in a coherent way – in much the same manner as physics has always done – what the outcome of experiments will be if we are given correctly all the conditions that fully characterize the nature of the experiment. To see what we have to do, I think we must go back and think a little more consciously of some of the fundamental principles (call them philosophical, if you like) which underlie classical physics – or shall I say macroscopic physics, because that's now the distinction.

I think specifically of the theory of measurement. And here, of course, we have to recognize the fundamental philosophical conception that physics is an experimental science; it is concerned only with those statements which in some sense can be verified by an experiment. The purpose of the theory is to provide a unification, a codification, or however you want to say it, of those results which can be tested by means of some experiment. Therefore, what is fundamental to any theory of a specific department of nature is the theory of measurement within that domain.

Now, what was characteristic of the theory of measurement in the macroscopic classical physics? Well, the essential thing that was basic to it was the conception of a non-disturbing measurement. It is, of course, perfectly obvious to anyone who has ever come near a physics laboratory that in the process of a measurement, aimed at gaining information about a particular object, we must interact with it physically in some way. But nevertheless we would like to be able to idealize that interaction in such a way that we could meaningfully state what that property would be as though the interaction did not occur. (You may take as the simplest model the insertion of a thermometer into a body of water with the objective of determining the temperature of the water – ideally as it would be without any disturbance by means of the thermometer. Without the presence of the thermometer, however, there is no means to determine what that temperature is.) As you know, whenever the interaction occurs there must be some disturbance as a net effect of that interaction with the object in question. (The insertion of a thermometer into a pail of water changes the mass of the water. It will change the temperature that is to be measured in some way.) But what is characteristic of classical physics is that we can state, and correctly so, that it is meaningful to talk of an idealization in which that interaction can be made as small as we please without, however, it becoming zero; because if it is zero, we have no means of gaining information.

That the interaction is so small that it does not disturb the object of interest, is in fact not always necessary, nor is it always possible. For example, some measurements might represent chemical changes, which are large alterations in the nature of the substance, and these are certainly not negligible disturbances. It is here that a second aspect of macroscopic physics comes into play. Since classical physics is causal and deterministic, we can calculate as accurately as we please and correct for the effect of these unavoidable disturbances. It is indeed familiar that in classical physics any measurement has a theoretical description associated with it, which represents the recognition of what the disturbance has been and the calculation of how to correct for it in order, therefore, to come back to what an idealized non-disturbing measurement would be. So, for the record: the two basic features of measurements in classical physics are that we can either make the interaction so small that there is a negligible disturbance; or, in a particular experimental circumstance, by the nature of the experiment we wish to perform, if we cannot make the disturbance arbitrarily small, we can calculate the effect of that disturbance and compensate for it with arbitrary precision.

This is, simply stated, the theory of classical measurement; and associated with it is the idea that there is then no limit to the accuracy with which we could make measurements simultaneously of any number of physical properties, as in the very statement of the concept of state, for example, in Newtonian physics. When I assert that the state is the specification of the positions and the momenta of all the particles in question, then implicit in that is the assumption, consistent with the whole scheme, that in fact I can carry out the measurements necessary to give the numerical values that those quantities have at every time. And similar remarks apply to measuring the distribution of electromagnetic fields throughout all of space.

In summary then, the classical theory of measurement says that there is no limit to the accuracy with which we can assign numerical values to all the quantities that are needed to specify the state, and since all of these are deterministic theories, that means to all physical properties at once. Since physical properties can be assigned numerical values consistently, one has never in classical physics drawn any distinction between the physical properties and the numerical values which they have at any particular time. In classical physics, we are always able to assign to the physical properties, considered as an abstract thing, a very concrete representation by means of numerical values which a non-disturbing measurement would find for them at a particular time.

This restates the foundations of classical physics: the idealization of non-disturbing measurements and the corresponding foundations of the mathematical representation, the consequent identification of physical properties with numbers because nothing stands in the way of the continual assignment of numerical values to these physical properties.

Now, by contrast, what is the situation in microscopic physics? Drawing upon the vast body of experimental data, accumulated over the course of several decades, I have summarized above the properties of microscopic physics – or, if you like, of microscopic measurements – under these two basic headings: atomicity and the statistical nature of the phenomena.

What does this mean? First of all, *atomicity*: atomicity means that the microscopic entities have many of their properties carried in certain basic units. There is no half of an electron. The electron has a definite mass; it has a definite charge. If the interactions that I am concerned with are electrostatic in nature, I cannot reduce them arbitrarily in strength because there is no half of a unit of charge. This indicates to you immediately, I think, the basic difference between the laws of microscopic measurement and macroscopic measurement. I must take into account the fact that the strength of the interaction – which must be present if I am to talk of measurement at all and, therefore, talk meaningfully of physical phenomena – cannot in general be made arbitrarily small because the physical objects that interact (the atoms, the electrons) in general have relevant physical properties which come in certain units – *quanta*, the origin of the name of the subject that we are discussing: quantum mechanics.

Now, this might seem as though it were not an unsurmountable difficulty. We recognized, even in classical physics, that there might be circumstances in which the act of measurement produced definite disturbances that we could not minimize because of the particular kind of measurement we carried out. In classical physics, we said the situation may be such that the measurement interaction is very strong and cannot be made arbitrarily weak, but this does not upset the underlying philosophy of measurement because I can calculate with arbitrary precision what the effect of that interaction was and compensate for it, correct for it.

Can I still do that now, in the realm of atomic measurements? The answer is *no*, because this is where the second fundamental aspect of microscopic measurement comes into play; namely, the phenomenon of statistics, the fact that we cannot predict in detail what each individual event will do but only make predictions on an average or statistical scale. The measurement act involves a strong interaction – I repeat: on the microscopic scale it is necessarily strong because we cannot cut the strengths of the charges in half; we cannot change the properties of these fundamental particles; we must accept them as they are – and so the measurement unavoidably produces a large disturbance, which we cannot correct for in each individual instance, for we cannot control what happens in each individual event in any detail. We can only predict or control what happens *on the average*, never in any *individual instance*. Therefore, the program of computing what the effect of the disturbance was and correcting for it is, in general, impossible. Accordingly, the two basic tenets of the theory of macroscopic measurement are both violated. Either the interactions cannot be made arbitrarily weak because of the phenomenon of atomicity, or if we wish to accept this and correct for it, we cannot do so because we do not have a detailed, deterministic theory of each individual event; we have only the ability to anticipate or control the statistical average.

So, here then is the general implication from the mass of experimental data that for microscopic physics, if we are to construct a theory, we need a whole new theory of microscopic measurement, and to go with this we need a whole new scheme of mathematics, which is to say that we can no longer speak meaningfully of the numerical values that physical properties have at a given time. Put differently, I wish to point out that the failure of these fundamental assumptions means equally well a failure of the ability to represent physical phenomena in the microscopic realm by numbers which change in time as we do in the macroscopic or classical domain. Something of an entirely different mathematical nature is needed, such that it represents, or mimics, the basic facts of microscopic measurement.

To emphasize the relevant point, I may say this. Macroscopically, we can measure one physical property; we assign a number to it. We measure a second physical property; we assign a number to it. We can then speak of this pair of numbers as the values of this pair of physical properties at a given time. There is no contradiction here. We can perfectly well go back and check that the first property has still the same value it had before if we could, in an idealized way, carry out these measurements rapidly enough (or regenerate the physical circumstances in such a way that we could repeat the first measurement).

By contrast, suppose we have indeed succeeded in measuring in some way one physical property of an atomic system. Now we go on to make a measurement of the second physical property. That measurement necessarily will involve an interaction, the strength of which is not arbitrarily weak and the effect of which is not controllable, in such a way that it will, in general, produce changes in the physical circumstances that specify the conditions of the first measurement. In other words, the system that is being measured is disturbed in an uncontrollable way in such a manner that if we now went back and asked for the value of the first physical property, checking to see that it still had the same value as before, we would now find not at all the same value but a random assortment of all the possible values that it could assume, with various probabilities that depend in detail upon precisely what we have done. This is so because the second measurement has introduced a new physical situation; the interaction of the second measurement has disturbed the physical system of interest so violently that we have no way of knowing, except under very special circumstances, whether the system has been left in precisely the same physical situation that would enable us to say that the first physical property still has the same value it had before the second measurement. In other words, once we recognize that the *act* of measurement introduces in the *object* of measurement changes which are not arbitrarily small, and which cannot be precisely controlled, then we must acknowledge that every time we make a measurement we introduce a new physical situation that is essentially different from the situation before the measurement.

So, if you measure two physical properties in one order, and then the other – which classically, of course, would make absolutely no difference – these are simply two different experiments in the microscopic realm. You have two different physical situations which come about depending upon whether you first measure property A and then property B, two successions of disturbances which have this microscopic character, or do it in the reverse order, first B and then A, which involves entirely different disturbances. And since the final physical situation depends crucially upon the order in which the microscopic measurements are performed, it is, in general, no longer possible to say that property A and property B have these values. That would only have meaning if you could get the same numerical values no matter in which order the measurement was carried out. Exceptional situations aside, in the "first A then B" order of measurements only B emerges with a known value, and only A in the "first B then A" order.

Therefore, the mathematical scheme for microscopic measurements can certainly not be the representation of physical properties by numbers, because numbers do not have this property of depending upon the order in which the measurements are carried out. The assignment of a pair of numbers to two physical properties introduces no sense of order, no sense of sequence. We must instead look for a new mathematical scheme in which the order of performing physical operations is represented by an order of performance of mathematical operations. The mathematical scheme that was finally found to be necessary and successful is the representation, in a very abstract way, of physical properties not by numbers but by elements of an algebra for which the sense of multiplication matters. In other words, the multiplication of these algebraic symbols was found to be the proper counterpart of the successive performance of measurements: that the order of measurements is significant, as a consequence of the unavoidable disturbances, is reflected in that, correspondingly, the sense of multiplication of these symbols must be significant.

And so we are led to a much more sophisticated and deep mathematical scheme in which physical properties are set into correspondence with the non-commutative elements of an algebra or, as they often are referred to, non-commutative *operators* as compared to the very elementary representation of physical properties by *numbers*. And with every physical state – the idea of state reoccurs – we associate a *vector* in a suitable abstract space on which these operators act.

As a result of all this, a very beautiful mathematical scheme has emerged which gives a wonderful account of all of these seemingly bizarre and incomprehensible facts of microscopic physics. This symbolization of atomic measurements is quantum mechanics, developed by Heisenberg, Born, Schrödinger and others, essentially in the years 1925 to 1927, still very distant from our present point of view.

Let me now describe, within the same general framework, what the nature of quantum mechanics is. It is still a causal theory. Given the state at one time, the state at any other time is uniquely determined, but what makes it different is that it is not a deterministic theory. It is a causal, *statistically* deterministic theory. The knowledge of the state at one time fixes the state at another time, but what information is obtainable from this knowledge of the state? Recall that classically, if you knew the state, you knew everything; if you knew where the particles were and how they were moving, you could predict any other physical property you happened to be interested in with arbitrary precision. But, as we have just said, arbitrary precision of individual predictions cannot exist in the microscopic world. Nevertheless, in a science of observation we must be able to make precise predictions – and we are, only that in quantum mechanics these precise predictions are of a statistical nature. The knowledge of the state enables you to predict the statistical, the average outcome of the measurement of any physical property, but never the result of any specific event. In other words, if you know the state, you can then predict what the result of repeated trials of measurement of a particular physical property will be. You will have perfectly determinate, statistical predictions but no longer individual predictions.

I repeat: the causal connections between states at different times is still present. (This seems to be fundamental in any branch of physics as we know it.) But what has changed drastically is that the knowledge of the state does not imply a detailed knowledge of every physical property but merely, in general, of what the average or statistical behavior of physical properties may be. This, in a sense, is the final understanding of these remarkable apparent paradoxes in the earlier developments of the theory. They are now resolved in terms of this statistical determinate rather than individually determinate theory.

I have spoken of states but have in no way indicated how a state is to be defined. The answer to this can be given if we think of a model of a physical system which still comes very close to classical models. For example, we began by thinking that atoms were to be understood simply by electrons (small material bodies, each carrying a unit of negative electric charge) moving in a certain definite field of force, the Coulomb force field of attraction of the positively charged nucleus. Here is a situation which seems to fit the Newtonian mold; a definite law of force, a finite definite number of material bodies. What failed was not that the dynamical picture was not correct but that the laws of microscopic physics were different. They were not such as to permit a detailed, deterministic prediction but they have this character of the statistically deterministic theory.

We have a physical model which is classical in picture. When we describe an atom, we say how many electrons there are and what the nuclear charge is and the picture is still classical, at least as far as our minds are concerned. What is very different is how we go about calculating. For simplicity, think of a hydrogen atom where there is only one electron. So, here is one electron and the electron has associated with it physical properties of position and physical properties of momenta, and classically we would say that there is no limit to the accuracy with which we could measure those positions and momenta simultaneously. But the distillation of what physicists have learned throughout this line of development that culminated eventually in the mathematical scheme of quantum mechanics, is that this is not the proper definition of a state of an electron in an atom. The best we can do to specify a state is not to assign numerical values simultaneously to *all* of these classical properties of position and momenta but to only *half* of them.

We can, in fact, produce experimental situations in which we know precisely where an electron is. It lands on a scintillation screen, and the flash of light reveals essentially the position of the electron, or we can produce experimental situations in which we know precisely what the momentum of the particle is. That, in fact, is what I had in mind when I described this beam experiment. Here is the particle, moving in a definite direction with a definite speed. And, having a definite mass, that means I know the momentum; but when I know the momentum, in a sense I cannot know where the position is, and the appearance of the interference pattern formed by the random falling of the particles on the screen is the sign of that fact.

On the other hand, I could produce a very different experimental situation in which I arrange matters so that the electrons always land at a pre-chosen site. Then I'd have a position measurement, and I can predict precisely what the result of such a measurement would be. We will then never have the interference pattern that is characteristic of the very different physical situation in which the momenta are perfectly definite.

What has changed basically is the nature of a state. If we have a certain number of particles – electrons, for example – the specification of the quantum state amounts to telling where the particles are at a given time *or*, alternatively, how they are moving at a given time, but never both together. So, indeed, by comparison with what would be a full specification of state in classical physics – where the electrons are *and* how they are moving – in quantum mechanics the state is specified by telling, with arbitrary precision, what the result of the measurement would be of half of those properties. But then we are completely incapable of predicting the individual values of the other half. They will then simply have random probability distributions.

If you make measurements again and again on this state, about which you know precisely the positions, you will always find that position, of course, comes out precisely as it should. But if, on that state, you make momentum measurements, you will never find a definite value, you will find a random statistical distribution. And the more precisely you specify the position, the wider will be this momentum distribution.

This simple situation is essentially a statement of what is perhaps the widest philosophical principle that has emerged from these studies of microscopic physics. This is what is known as Bohr's principle of complementarity. We have used, and shall continue to use, the example of the so-called wave–particle duality, tied to the names of Einstein and de Broglie, for the illustration of complementarity. The general development, however, establishes that wave–particle duality is just one consequence of the fundamental complementarity.

So, by the idea of complementarity we mean the final unification, within these general principles, of what began by seeming to be a paradox. Electrons, under certain perfectly definite experimental situations, act as particles would, and under other experimental situations, they act as waves – this is, so to speak, what we have expressed now in a more precise way using the insight that the definition of state never refers to all of these physical properties but to only half of them. You have the privilege of designing experiments in which different choices are made as to which will be the physical properties whose values are precisely known. Waves, for instance, represent the option, the choice on the part of the experimenter, to produce an experimental situation in which the momenta are selected to have definite values, the positions can then not be controlled.

Bohr's principle of complementarity is the statement that we have in microscopic physics first of all a new world (that's the important thing to recognize), in which classical analogies fail. But, nevertheless, there are certain situations in which analogies of a classical nature do hold – situations, for example, in which it is possible to speak meaningfully of particles with regard to certain circumstances and certain measurements; and other situations where electrons, or what have you, can be spoken of as waves. Two distinct classical pictures can hold under different physical situations, never simultaneously, and the applicability of one picture prevents the applicability of the other – the two classical analogies are mutually exclusive. But both pictures are on the same footing. We can produce experimental situations in which either classical picture can be applied, and the other is then inapplicable. The idea of complementarity is that a full understanding of this microscopic world comes only from the possibility of applying both pictures; neither in itself is complete. Both must be present, but when one is applied, the other is excluded.

This is, in essence, the entirely new situation which has no counterpart in any of the classical philosophical modes of thought. It is something that simply must be accepted. At least all physicists have accepted it – it is essentially the way in which the laws of microscopic physics have been understood, and as the result of which the enormously successful development of quantum mechanics has arisen. Within the space of a few years, the application of these understandings in microscopic phenomena has completely swept away what was traditionally regarded as the great classical problems of physics. At least in principle, if not in practice, a reduction of chemistry to physics has been brought about. The understanding of all of the various diverse properties of matter in all its forms under all ordinary circumstances is reduced to a few simple facts. The laws of quantum mechanics and the specification of which particular configuration you happen to be talking about is all you need – in principle at least, and with great success in practice – to understand the properties of material bodies, as indicated by the enormous developments of the underlying theory of the solid state and many other applications.

All of these are in large measure the expression of the understanding of the laws of atomic physics that have been codified – and unified – in quantum mechanics.

In a fundamental sense, this was completed in 1927. It is by no means the end, however, of physicists' investigation of the physical world. The development has gone on and gone on in the direction of looking for entirely new realms of physical experience within the domain of higher energies, smaller distances. We have spoken again and again of the atom, but in the center of the atom is the nucleus, and within the nucleus are the nucleons, and as we now understand, the nucleons are made up of other still more fundamental primordial entities. The search goes on.

To understand a little bit in what language this development is continued, I must come back to an idea that I mentioned earlier: the notion of the quantized field, because here we have perhaps the deepest expression of what has been learned within the framework of these microscopic phenomena. Let me introduce this in terms of another basic philosophical idea which is given an entirely new turn within the phenomena of microscopic physics. This is the concept of identity or indistinguishability. It is, of course, perfectly clear to you that when we speak even in classical physics of electrons that mere terminology indicates that we understand that one electron is just like another. If we measure any of the fundamental non-accidental properties of this electron, or that electron – its mass, its charge, and whatever more sophisticated properties you may be concerned with – these are invariably the same. This is, if you like, the fundamental conception of the uniformity of nature without which physics could never begin to operate. We must assume that one sample of a particular substance is like any other sample if no relevant circumstances are involved.

So, if we take any two electrons, describing them classically as we might have if we had two beams of electrons moving in some evacuated chamber, then we understand that one electron is just like another, and *in classical physics* there are no particular implications in this, because despite the identity, the indiscernability in principle of the two electrons, in classical physics we are never in any difficulty of being able to specifically distinguish them. We can say this electron originated from this region of space; the second electron from that region of space. Classically, I am able to follow in detail the trajectories of these particles, and I can at every instance say precisely where this particular electron came from and trace its path continuously to the point of origin. No matter how these electrons may interact in some complicated way within a vacuum chamber or a radio tube, for example, I could always identify – in principle at least, thereby using nothing but the classical laws of physics – at any stage precisely which each electron is which.

Now, I think you may see immediately that the situation must be very different when we recognize the reality of the microscopic world which is not governed by those classical laws. Consider, for example, a collision experiment. Suppose I have two beams of electrons. (You could make them protons just as well. It is, of course, important to realize that the laws of nature which I am speaking about govern all of the various manifestations of matter. If I speak of electrons, that is historical convention. Those could be protons, neutrons, hyperons, π mesons, the laws of physics are the same.) If the electrons are to collide, they must come into intimate interaction. Of course, what they do *not* do now is carry out, as they would do classically, a detailed trajectory because for that to be meaningful, I must be able to always, without disturbing the nature of this experiment, check precisely what a particular electron is doing at every instant of time. That calls for a degree of control, and thus of determinism, that is impossible in the microscopic world. If I have produced an experimental situation in which the particles head toward each other, then they have rather definite momenta. Then, as I have tried to suggest, the complementary physical measurements no longer can be specified in detail. I have no way of knowing precisely where these electrons are, and can give no meaning to such statements about their positions.

And when finally these beams separate, so I no longer have any doubt about which is which, I have no right or ability to tell whether this electron is that one – or that one – because I have not been able to follow in detail precisely what has happened. In other words, these basic physical phenomena, the atomicity, the statistical nature of things, the inability to control in detail individual events, imply correspondingly the absence of an ability in the fundamental experimental sense to tell which particle is which at every stage of the interaction. This requires, therefore, that my description must take into account this fundamental failure of being able to place a tag on every particle. I repeat: there is no experiment I can perform that gives reality to that label because to do so would represent an intervention into this experiment, the performance of a detailed microscopic localization experiment, which would change completely the nature of the collision experiment. It would no longer be a simple collision, there would be something else there, equally important, interacting with these particles. It is a *different* experiment. And this is the whole point, of course: the recognition that I must indicate in detail precisely every experiment I have in mind because every measurement I wish to perform changes the conditions of the experiment, produces a strong, non-controllable interference in the other things that are going on. The net result of all of this is the recognition that the description of the states of several particles necessarily can only be done in a way which incorporates from the very beginning the fact that they are indistinguishable, that particular labels have no significance. This simply means that when we have several indistinguishable particles, the states can only be described in a way that is perfectly symmetrical among all the particles that contribute to it.

Now the word "symmetrical" is to be understood in a more general sense. The traditional manner in which states are specified is in terms of numerical quantities that are known as wave functions. I spoke of vectors, state vectors, which are abstract entities, and wave functions are particular numerical representations of these abstract vectors. I have several identical particles and the wave function is a function of their positions, for example. The only description that is tenable is one in which this wave function is completely symmetrical among all these positions, so that all the particles are on exactly the same footing. The arbitrary labels – which could say this particle is the first, that particle is the second, and so on – must be deprived of any distinguishing significance by insisting that, no matter how the labels are assigned, the wave function of the state is the same. The states themselves must remain unaltered if I decide to distribute the arbitrary labels differently among the identical particles. But this can be ensured either by making the states completely symmetrical or completely antisymmetrical – the more general sense of "symmetrical" referred to above. In this way we recognize the existence of two very different types of systems of identical particles. In fact, of the two basic examples we have used, the photon and the electron, the detailed properties of physical phenomena show that the photon belongs to this class of symmetrical states, and the electron falls into the framework of antisymmetrical states.

Let's take a closer look at those wave functions, beginning with just a single particle. Its wave function is simply a function of its position – or, if you like, of the three coordinates that specify the position – and of time. Here you might say that I have, in a sense, a field; a field is after all physically a distributed object in space and time. This is very much analogous, as far as this space–time structure is concerned, to the electromagnetic field of Maxwell's theory; and indeed it might seem that a wave function is just a field of some other kind. But, as soon as we deal with several particles, the wave function cannot be a field in the conventional sense, because it is now the function of two, three, or more positions (and time). In other words, here I have rather a much more abstract, multidimensional configuration space, and the multiparticle wave function is a way of indicating what several particles are doing at once in a highly hypothetical mathematical space, which is very remote from ordinary three-dimensional space because there are many three-dimensional spaces now being considered at the same time.

So, we have states of single particles, two particles, three particles, and so on, and clearly the picture is getting ever more complicated. An enormously fruitful change of perspective begins by recognizing that there is also a no-particle state – the *vacuum*. States with just one particle can then be thought of as brought about by creating a particle. First you have the vacuum, then you create a particle and you have a one-particle state. You create another particle and have a two-particle state – then a third, a fourth particle, and so on. At the moment this is nothing but imagination, idealized, abstract creation processes that, in a sense, just make book-keeping easier. What makes it useful and important is that I can now deal with experimental situations with varying numbers of particles – situations in which particles are physically created and annihilated.

In the mathematical scheme, then, we describe the creation by the application of a creation operator – apply it, for instance, to a state with five particles and the result is a state of six particles. It is an operator, rather than a numerical quantity, because it symbolizes a physical property that is something beyond what we are accustomed to thinking of, and also because it acts on a state to produce another one with one more particle.

So, I will imagine, for example, that I apply the creation operator to the vacuum state and create particle 1 at the time t; I will then create particle 2 at the time t (or any number of additional particles by the repeated action of this creation operator, but let's have just two right now). In this manner, I produce something which has, essentially, the nature of a two-particle wave function which involves both positions and the time t. And the important thing is that these requirements of symmetry (or antisymmetry) on this wave function can now be converted into algebraic statements on these operators. If the situation is one of symmetry, as in the case of photons, then I assert that whether the creation operators are multiplied in one order or the other, the result is the same; this independence of the multiplication order produces symmetry. If it is antisymmetry, as in the case of electrons, I assert that if they are multiplied in one order, the result is the negative of what occurs when they are multiplied in the reverse order. In other words, the properties of these two classes of identical particles – or statistics as it usually is referred to – becomes replaced by an algebraic property of these operators. In short, we are led, instead of talking about a system of a definite number of particles, to think of physical systems with an indefinite number of particles because we can produce whatever number we are interested in by the application of this creation operator. We now then, so to speak, transfer our attention to this operator as the basic physical object. And this is what I mean by the *quantized field*: it is on the one hand a field, a mathematical quantity which varies continuously in time and space; on the other hand, it is certainly not a classical field because these operators are not physical quantities which can be measured simultaneously. In the operator character, in the fact that the sense of multiplication is significant (in a much deeper way than I can describe it here) we have the elements of discontinuity which is the essence of the particle concept.

The two entirely unrelated classical conceptions of discreteness of particles, of continuity of the field, are now unified in this entirely new conception of the quantized field – more than unified: transcended because the new conception is beyond either. The two are, after all, incompatible in the classical sense because there is nothing that is both discrete and continuous. We have arrived at something which has neither of those properties but which, in limited domains, can be characterized in terms of either of these conventional concepts.

So, here is the fundamental unification, brought about by this idea of the quantized field. I have emphasized how the quantized field enables us to speak meaningfully of creation acts, but it should be obvious that, correspondingly, the inverse processes of annihilation, in which particles get destroyed at various points in space and time, are equally accounted for by the quantized field.

As it arose historically, the quantized field was merely a convenient way of summarizing the mathematical properties of indistinguishable particles, but soon, through the ever-broadening developments of experimental science, what was conceived of simply as a mathematical idealization became reality. I am referring to the ability, as enough energy was available, to supply the rest energy of particles according to the famous Einstein relation $E = mc^2$: given the energy E that corresponds to its mass m, a physical particle can be produced. (It may be necessary for other reasons, such as the conservation of electric charge, to produce them in pairs, as is in fact the case for the negatively charged electron and its positively charged counterpart, the positron.)

By the early 1930s, these experiments had been performed. In later years, pairs of protons and antiprotons, of neutrons and antineutrons were produced for which vastly much greater amounts of energy are required, and by now experiments have created (pairs of) most of the other particles known to be the building blocks of nature. These experiments alone then give physical reality to the quantized field, in its interpretation of symbolizing acts of creation and destruction.

But now, in the course of the development, it soon became realized that the moment you have introduced the quantized field in direct association with particles as we know them, inevitably this situation could not persist. A new level of abstraction had to be reached and was reached. It occurred essentially during the late 1940s and the 1950s while attempting, as demanded by the refinement of experimental data, to understand better the properties of atomic phenomena that were successfully accounted for in the first flush of the development of quantum mechanics. The experiments went on; more and more refined properties became known; further and sharper applications of the theory were required. For example, in the case of electrons in atoms, from the field point of view, we are really concerned with the dynamics of two fields. There is the quantized field which is associated with the electrons and also their counterpart, the positrons; there is the quantized field which is associated with the electromagnetic field, to use its classical name, the field of photons. The photon field and the electron field are in interaction. And as a consequence, the identification of each field by these physical names is only an approximate one. Only if the interaction between the two fields is weak, as to a large extent it is in this example, can we use physical names in relation to these mathematical objects. But in a more refined theory in which the interaction between them must now be taken into account, we have to recognize that what we call physically an electron is only partially to be associated with that electron field alone. It is also partially to be associated with the photon field, because the two are in interaction. Physically, an electron can sometimes radiate a photon; it can then also reabsorb it. In other words, what we physically call an electron would, at a deeper level, be described as sometimes the action of the electron field only, but another fraction of the total history also involves the action of the photon field. And conversely, what we call a photon, propagating through empty space, is not merely the result of the creation act of an analogous photon field, because the photon can occasionally materialize itself in space and become replaced by an electron and a positron which then, in the course of time, recombine to reform the photon.

In other words, the physical object we call the photon is not what is created all the time by applying the mathematical photon-creation operator to the vacuum state. The other operators, the quantities that represent the creation of electrons and positrons, also come into play. Once you recognize this, we draw a distinction between two levels of physical description. There is the phenomenological level at which we recognize the properties of electrons and photons as we see them, out of, of course, the enormously detailed analysis of microscopic experiments. Then there is the attempt to deepen the understanding in terms of more primitive objects which are these quantized fields, which are no longer placed in immediate correspondence with the phenomenological particles but through a chain of dynamical development. And, in fact, this program of renormalization, to mention the technical term, as it was applied specifically to the case of electrons and photons, through the development of what is called quantum electrodynamics, led to a description on a more fundamental level of some of the finer features of electron and photon behavior. What were once considered to be anomalies, things that were unexpected, became the predicted outcome of this deepened understanding. The lesson was learned that our most profound understanding is not to be found in terms of what we actually see but at a more fundamental level.

So it has always gone throughout the history of physics. We begin with atoms as fundamental objects and then we attempt to understand the properties of atoms in terms of electrons and a nucleus, which is taken as unanalyzable. Then we move down to the level of the nucleus, and analyze it in terms of the properties of nucleons and so on and on. In very simple terms, this is the conception of how we go about it in terms of smaller and smaller particles, smaller and smaller regions of space.

The analysis of particles as we know them and as we associate them with fields is an attempt at understanding, at a deeper level, that strives for a simplification in terms of yet more fundamental quantized fields which have fewer properties, that strives for deeper, more symbolic laws with fewer arbitrary constants. For example, unlike the experimental situation in which the charge of the electron, the mass of the electron, the magnetic moment of the electron, are all unrelated constants, the deeper understanding attempts to explain one or more of these in terms of a fewer number of fundamental things.

So it has gone in the case of quantum electrodynamics. This has been a very successful application of this idea that it is the quantized field conception which is the statement of our deepest level of understanding of microscopic phenomena. But as I mentioned, and it must be familiar to you to some extent, in the course of the development of higher and higher energy machines, more and more particles have become known and these have appeared in a bewildering array of properties. Some of them are stable; some of them are unstable. They decay into each other in all possible conceivable ways. As sufficient energy is available, they are produced very copiously as the result of obviously strong interactions. They then proceed very slowly to die successively, moving down to the final stable particles that we know, which are still the electrons, protons, plus a few others.

Now, the interactions which are involved here in these basic studies of nuclear phenomena are of an entirely different kind than the electromagnetic ones. The electromagnetic forces are essentially rather weak. And on the basis of this, one has been able to develop technical methods of handling these interactions. But when one comes to the very much stronger bonds that not only hold the nucleus together, but hold together the particles that compose the nucleons, we are at a much more difficult level in the sense that not only are the phenomena bewilderingly complicated, but we also lack the mathematical means to draw the implications of any particular hypothesis about what is going on. And, as a result of this, there is doubt about what should be the fundamental nature of an explanation at this level.

Should it be the continuation of this point of view of the searching for deeper understanding in terms of ideally a very small number of fundamental fields, which in their dynamic interplay, and as a result of the complexity of that dynamics, finally bring about the manifold nature of the world as we see it? Or must we really abandon this attempt completely, replacing the difficulties by the anticipation of a fundamental impossibility, and simply describe nature in terms of what happens when we take various microscopic particles and perform experiments on them? We send electrons, protons into the various kinds of nucleons, where we perform experiments in which these particles enter a certain very small region. We make no attempt to describe what goes on there and simply try to finally characterize what emerges when the particles are separated again. Is the purpose of theoretical physics to be no more than a cataloging of all the things that can happen when particles interact with each other and separate? Or is it to be an understanding at a deeper level in which there are things that are not directly observable (as the underlying quantized fields are) but in terms of which we shall have a more fundamental understanding? Well, this question – idealized, frankly, beyond all recognition – is in a sense the deep philosophical problem that confronts theoretical physics at the frontier of high-energy physics, where we attempt to understand the structure of matter as it is revealed to us, in all of its complexity, using the ever-rising level of energy that has become available to study the basic building blocks of matter and, in the course of this, to create new kinds of matter.

Part A

Fall Quarter: Quantum Kinematics

1. Measurement Algebra

1.1 Stern–Gerlach experiment

I presume that all of you have already been exposed to some undergraduate course in Quantum Mechanics, one that leans heavily on de Broglie* waves and the Schrödinger† equation. I have never thought that this simple wave approach was acceptable as a general basis for the whole subject, and I intend to move immediately to replace it in your minds by a foundation that *is* perfectly general.

In checking out my impression of undergraduate courses, I happened to glance through a particular elementary textbook and found this statement:

> The laws of quantum mechanics cannot be *derived*, any more than can Newton's‡ laws or Maxwell's§ equations. Ideally, however, one might hope that these laws could be deduced, more or less directly, as the simplest logical consequence of some well-selected set of experiments. Unfortunately, the quantum mechanical description of nature is too abstract to make this possible.

Despite that last pessimistic assertion, I propose to present just such an ideal *in*duction (the more accurate term) of the general laws of quantum mechanics from a well-selected set of experiments – indeed, from a *single* type of experiment.

The experiments are atomic beam measurements of magnetic properties, developments of the original experiment carried out by Stern¶ and Gerlach‖ in 1922. With the exception of ferromagnets, matter in bulk is unmagnetized in the absence of a magnetic field. A paramagnetic substance acquires a magnetization proportional to an applied magnetic field, which is understood as the field lining up the individual atomic magnetic moments against the disorganizing effect of thermal agitation. This, then, is an indirect measurement of an atomic magnetic moment. The Stern–Gerlach experiment attempts a direct measurement of the atomic moment through the *mechanical* effect of the field.

---

*Prince Louis–Victor DE BROGLIE (1892–1987) †Erwin SCHRÖDINGER (1889–1961) ‡Sir Isaac NEWTON (1643–1727) §James Clerk MAXWELL (1831–1879) ¶Otto STERN (1888–1969) ‖Walther GERLACH (1889–1979)

A magnetic dipole moment $\boldsymbol{\mu}$ in the magnetic field $\boldsymbol{B}$ has the energy $-\boldsymbol{\mu}\cdot\boldsymbol{B}$, so that the force on it is

$$\boldsymbol{F} = -\boldsymbol{\nabla}(-\boldsymbol{\mu}\cdot\boldsymbol{B}(\boldsymbol{r})) = \boldsymbol{\nabla}\boldsymbol{\mu}\cdot\boldsymbol{B}(\boldsymbol{r}) . \tag{1.1.1}$$

In the Stern–Gerlach experiment we meet a situation where, at a particular point, $\boldsymbol{B}$ has a single component, B_z, which varies strongly in the z direction. So, the z component of force is

$$F_z = \mu_z\left(\frac{\partial}{\partial z}B_z\right) . \tag{1.1.2}$$

Here is the experimental setup:

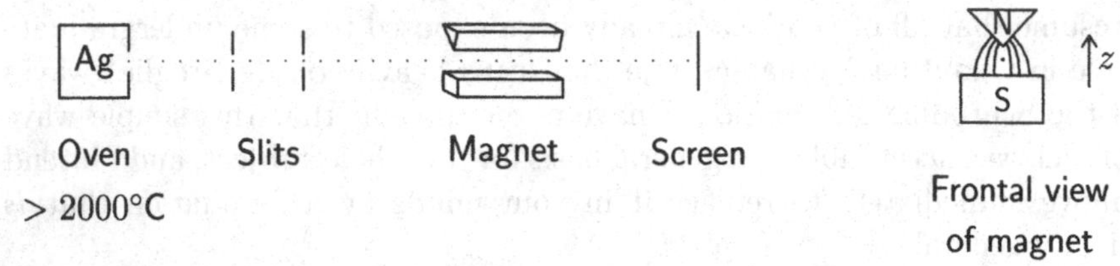

Silver atoms evaporate from the oven and pass through the slits, traversing the magnet at the center point where there is a strong variation of B_z in the z direction. Depending on the value of μ_z for a particular atom, the force on it will deflect the atom up or down and correspondingly deposit it on the screen. Thanks to the high temperature, the Ag atoms certainly have their magnetic moments (m.m.) distributed uniformly in all directions, which is to say that the distribution of μ_z should be uniform between the limits μ and $-\mu$. So, after the beam has run for a while, the distribution of atoms on the screen should be

What did Stern and Gerlach find? This:

The atoms were deflected up or down; nothing in between! It is as though the atoms emerging from the oven have already sensed the direction of the field in the magnet and have lined up accordingly. Of course, if you believe *that*, there's nothing I can do for you. No, we must accept this outcome as an irreducible fact of life and learn to live with it!

As a first step, we refine the Stern–Gerlach experiment by dealing with single beams. (In the following we speak of a +m.m. or a −m.m. according as the atom is deflected up or down.) Suppose we stop the −m.m. beam:

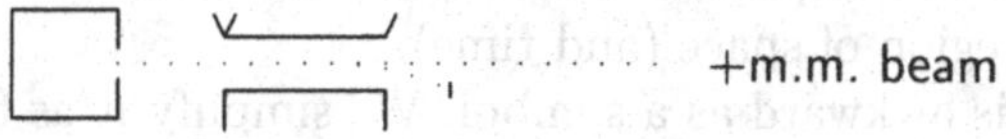

$$\text{+m.m. beam} \tag{1.1.3}$$

or the +m.m. beam:

$$\text{−m.m. beam} \tag{1.1.4}$$

How can we be sure that we have produced a pure +m.m. beam or a pure −m.m. beam? By repeating these selective measurements:

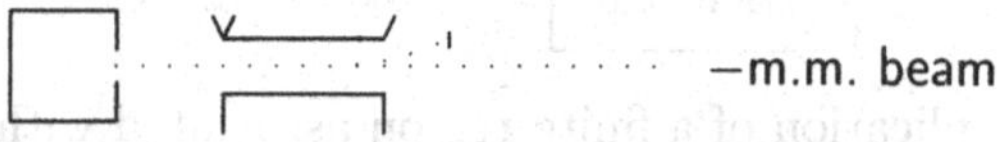

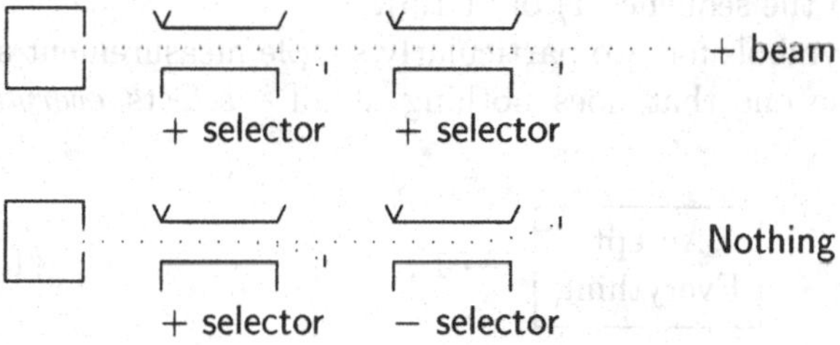

1)   + beam   Yes, a + beam.

     + selector   + selector

2)   Nothing   Not a − beam.

     + selector   − selector

$$\overline{\qquad\qquad\qquad}$$

Pure + beam.

$$\tag{1.1.5}$$

In sequence 1) we look for a +m.m. beam and find it; in sequence 2) we look for a −m.m. beam and do not find it. Together that establishes that the initial +m.m. selection indeed selects +m.m. atoms. Similar pictures apply to an initial −m.m. selection.

1.2 Measurement symbols

The Stern–Gerlach experiment using silver atoms is the measurement of a physical quantity, μ_z, that happens to have only two possible values, say $+\mu$ and $-\mu$. We now generalize by regarding μ_z as just an example of a physical quantity A that has the possible values $a_1, a_2, \ldots a_n$; a typical value will be designated as a' or a''. The specific physical apparatus that measures μ_z and selects a particular outcome (say +m.m.), as suggested by

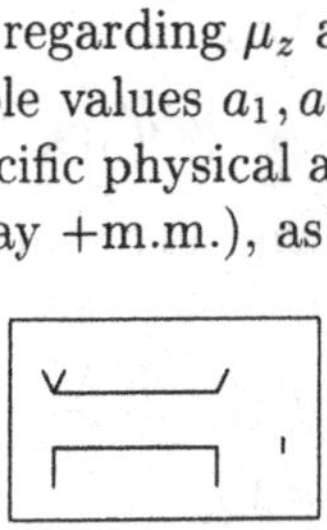

then becomes an unspecific apparatus, suggested by

$$\boxed{\begin{array}{r}\text{Measure } A\\ \text{Select } a'\end{array}}$$

all of which carries the implication that a measurement is a physical act occupying a finite region of space (and time).

The above box is awkward as a symbol. We simplify it as follows

$$\boxed{\begin{array}{r}\text{Measure } A\\ \text{Select } a'\end{array}} \longrightarrow |a'a'| \,, \tag{1.2.1}$$

which retains the implication of a finite region associated with the measurement act. Physical property A is implicit, adequately implied by a'. But why the repetition of a'? First, it prepares the way for a generalization; second, it is a reminder that a selective measurement involves an initial act followed by its verification, as in the sequence 1) of (1.1.5).

We also introduce symbols for two particularly simple measurement acts: the unit symbol for the one that does nothing at all – selects *everything* without bias,

$$\boxed{\begin{array}{c}\text{Accept}\\ \text{Everything}\end{array}} \longrightarrow 1 \,, \tag{1.2.2}$$

and the null symbol for the one that rejects everything, accepts nothing:

$$\boxed{\begin{array}{c}\text{Accept}\\ \text{Nothing}\end{array}} \longrightarrow 0 \,. \tag{1.2.3}$$

A first step toward constructing an algebra for these symbols is made on representing *successive* acts of measurement – displaced in time – by sequential *multiplication* of the respective symbols. Thus the generalization of 1) in (1.1.5), which says that the repetition of a selective measurement confirms the measurement is symbolized by

$$|a'a'||a'a'| = |a'a'| \,. \tag{1.2.4}$$

The generalization of 2), that two distinct selection acts end up by selecting nothing, is

$$a' \neq a'' : \qquad |a'a'||a''a''| = 0. \tag{1.2.5}$$

Then, as reasonably obvious statements about multiplication of the measurement symbols 1 and 0, we have

$$\begin{aligned} |a'a'|1 = 1|a'a'| &= |a'a'| \,, \\ 11 &= 1 \,, \\ 10 = 01 &= 0 \,, \end{aligned} \tag{1.2.6}$$

and

$$\big|a'a'\big|0 = 0\big|a'a'\big| = 0\,,$$
$$00 = 0\,. \tag{1.2.7}$$

We regress temporarily to note the equivalence

$$\boxed{\begin{array}{c}\text{Measure } A\\ \text{Select } a'\end{array}} \equiv \boxed{\begin{array}{l}\text{Measure } A - a'\\ \quad\text{Select } 0\end{array}}\,. \tag{1.2.8}$$

What then do we mean by $(a' \neq a'')$

$$\boxed{\begin{array}{l}\text{Measure } (A - a')\,(A - a'')\\ \quad\text{Select } 0\end{array}}\;? \tag{1.2.9}$$

This is an A measurement in which either the outcome a', or a'', is accepted without distinction. It is a less selective measurement, which I propose to represent by the *addition* of the respective symbols:

$$a' \neq a'': \qquad \boxed{\begin{array}{l}\text{Measure } (A - a')\,(A - a'')\\ \quad\text{Select } 0\end{array}} \equiv \big|a'a'\big| + \big|a''a''\big|$$
$$= \big|a''a''\big| + \big|a'a'\big|\,, \tag{1.2.10}$$

which incorporates the complete symmetry between a' and a''. Continuing in this way,

$$a' \neq a'' \neq a''' \neq a':$$
$$\boxed{\begin{array}{l}\text{Measure } (A - a')\,(A - a'')\,(A - a''')\\ \quad\text{Select } 0\end{array}} \equiv \big|a'a'\big| + \big|a''a''\big| + \big|a'''a'''\big|\,, \tag{1.2.11}$$

we end with

$$\boxed{\begin{array}{l}\text{Measure } (A - a_1)\cdots(A - a_n)\\ \quad\text{Select } 0\end{array}} \equiv \big|a_1a_1\big| + \cdots + \big|a_na_n\big| = \sum_{a'}\big|a'a'\big|\,. \tag{1.2.12}$$

Just as there are $2 = 2!$ equivalent sums in (1.2.10), permutations give $3! = 6$ equivalent forms in (1.2.11) and $n!$ ones in (1.2.12).

The measurement that accepts ***all*** possible outcomes without distinction is symbolized by 1. So the sum in (1.2.12) must be equal to the unit symbol,

$$\sum_{a'}\big|a'a'\big| = 1\,, \tag{1.2.13}$$

which states the *completeness* of the measurement symbols $\big|a'a'\big|$.

Before continuing, notice the simple properties of 0 with respect to addition. Given the option of accepting either something or nothing one ends up with something:

$$
\begin{aligned}
&|a'a'| + 0 = 0 + |a'a'| = |a'a'| \,,\\
&1 + 0 = 0 + 1 = 1 \,,\\
&0 + 0 = 0 \,.
\end{aligned}
\tag{1.2.14}
$$

Is it consistent with the known properties of the $|a'a'|$ that $\sum_{a'} |a'a'|$ acts like 1, the unit symbol? If so, we must have

$$
\left(\sum_{a'} |a'a'|\right) |a''a''| = |a''a''| \,. \tag{1.2.15}
$$

This is indeed true if a product with a sum is the sum of the products (distributive law of multiplication):

$$
\begin{aligned}
\left(\sum_{a'} |a'a'|\right) |a''a''| &= \sum_{a'} |a'a'||a''a''| \\
&= |a''a''||a''a''| + \sum_{a'(\neq a'')} |a'a'||a''a''| \\
&= |a''a''| + 0 + \cdots + 0 = |a''a''| \,.
\end{aligned}
\tag{1.2.16}
$$

We therefore accept the distributive law of multiplication.

The notation $|a'a'|$ is an invitation to generalization: $|a'a''|$, $a' \neq a''$. What can this mean? Return to the m.m. example and consider arrangement 2) of (1.1.5):

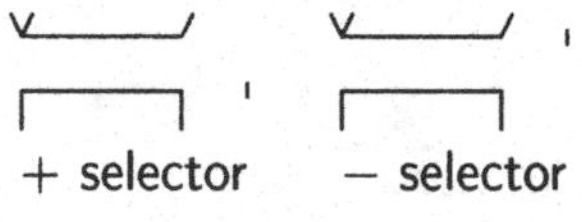

As it stands it stops everything. But suppose in the region between the + selector and the − selector we reversed the direction of the m.m.? A homogeneous magnetic field can do this; a dipole precesses around the direction of the field:

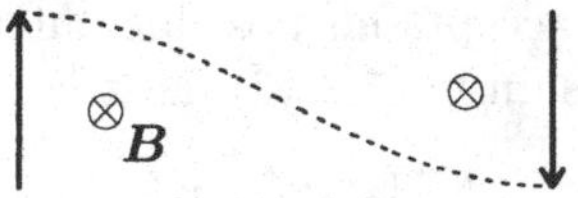

It is only necessary to control the time in the field. The outcome is a selective measurement in which only +m.m.'s enter, and only −m.m.'s leave: $|+-|$.

For successive measurements of this general type we have

$$\left|a'a''\right|\left|a''a'''\right| = \left|a'a'''\right| \tag{1.2.17}$$

and

$$\left|a'a''\right|\left|a'''a^{\mathrm{iv}}\right| = 0 \qquad \text{if} \quad a'' \neq a''' \,. \tag{1.2.18}$$

Again, we have the proper behavior for the unit symbol 1,

$$\begin{aligned}\left(\sum_{a'}\left|a'a'\right|\right)\left|a''a'''\right| &= \sum_{a'}\left|a'a'\right|\left|a''a'''\right| \\ &= \left|a''a''\right|\left|a''a'''\right| + 0 + \cdots \quad = \left|a''a'''\right| .\end{aligned} \tag{1.2.19}$$

Now notice this:

$$\begin{aligned}\left|a'a''\right|\left|a''a'\right| &= \left|a'a'\right| , \\ \left|a''a'\right|\left|a'a''\right| &= \left|a''a''\right| .\end{aligned} \tag{1.2.20}$$

The products on the left side differ only in the order of multiplication; the right sides are different if $a' \neq a''$. The order of multiplication can be significant! This evolving algebra is non-commutative for multiplication. And, as we could have noticed before, it is not a division algebra,

$$a' \neq a'' : \qquad \left|a'a'\right|\left|a''a''\right| = 0 \tag{1.2.21}$$

does not imply that either $\left|a'a'\right|$ or $\left|a''a''\right|$ is 0; similarly

$$a' \neq a'' : \qquad \left|a'a''\right|\left|a'a''\right| = 0 \tag{1.2.22}$$

does not imply $\left|a'a''\right| = 0$.

The *commutator*

$$[X, Y] \equiv XY - YX \tag{1.2.23}$$

of two measurement symbols X and Y vanishes if the order of their multiplication does not matter; if it does, one has $[X, Y] \neq 0$. A related quantity is the *anticommutator* $\{X, Y\}$, defined by

$$\{X, Y\} \equiv XY + YX \,. \tag{1.2.24}$$

It equals $2XY$ or $2YX$ if $[X, Y] = 0$; otherwise one can regard $\frac{1}{2}\{X, Y\}$ as a symmetrized product of X and Y. Note the identity

$$XY = \tfrac{1}{2}\{X, Y\} + \tfrac{1}{2}[X, Y] , \tag{1.2.25}$$

an immediate consequence of these definitions.

The outcome of a measurement is a number. We must have numbers as well as abstract symbols of measurement in our algebra. The obvious definitions of the basic numbers one and zero are

$$\underset{\substack{|\\ \text{number}}}{1}\,\big|a'a''\big| = \big|a'a''\big| , \qquad \underset{\substack{|\\ \text{number}}}{0}\,\big|a'a''\big| = \underset{\substack{|\\ \text{symbol}}}{0} . \tag{1.2.26}$$

They are convenient in synthesizing the products

$$\begin{aligned} \big|a'a''\big|\big|a'''a^{\text{v}}\big| &= \left\{ \begin{array}{r} \big|a'a^{\text{v}}\big| = 1\,\big|a'a^{\text{v}}\big| \text{ if } a'' = a''' \\ 0 = 0\,\big|a'a^{\text{v}}\big| \text{ if } a'' \neq a''' \end{array} \right\} \\ &= \delta(a'',a''')\big|a'a^{\text{v}}\big| , \end{aligned} \tag{1.2.27}$$

where

$$\delta(a'',a''') = \begin{cases} 1 \text{ if } a'' = a''' , \\ 0 \text{ if } a'' \neq a''' , \end{cases} \tag{1.2.28}$$

is Kronecker's* delta symbol.

Notice something else. What meaning shall we give to $\big|a'a''\big| + \big|a'a''\big|$? Well, if we accept the distributive law generally, this is

$$1\big|a'a''\big| + 1\big|a'a''\big| = (1+1)\big|a'a''\big| . \tag{1.2.29}$$

And what is one plus one? Two, naturally!

$$1\big|a'a''\big| + 1\big|a'a''\big| = 2\big|a'a''\big| . \tag{1.2.30}$$

1.3 State vectors

Let's think a little more about the meaning of $\big|a'a''\big|$. Only an atom having the value a' of property A, an a' atom for short, can enter (left to right reading, indicated by L $\rightarrow$ R where necessary) and what leaves is an a'' atom. It is *as though* the entering a' atom is destroyed and in its place an a'' atom is created. This is a mental two-step process that is indistinguishable from the actual one. We symbolize the composite viewpoint by introducing little brackets

*Leopold KRONECKER (1823–1891)

$$|a'a''| \equiv |a'\rangle\langle a''| \, .$$

[L → R] (1.3.1)

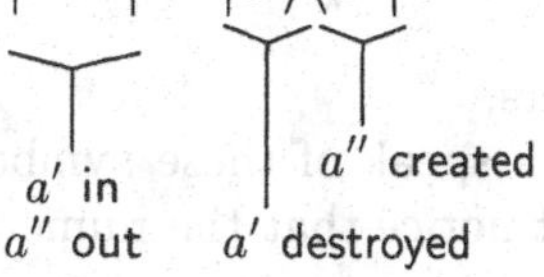

So far this is innocuous. But we take a giant step forward by viewing this as the *product* of two symbols of a new type! But is it compatible with the known algebraic properties of the symbol $|a'a''|$? We must have

$$|a'\rangle \underbrace{\langle a''||a'''\rangle}_{=\langle a''|a'''\rangle} \langle a^{\text{iv}}| = \delta(a'',a''')|a'\rangle\langle a^{\text{iv}}| \tag{1.3.2}$$

where we simplify the notation by writing $\langle a''|a'''\rangle$ rather than $\langle a''||a'''\rangle$ here and in all subsequent products of this kind. Observe that (1.3.2) is satisfied if

$$\langle a''|a'''\rangle = \delta\,(a'',a''') \, . \tag{1.3.3}$$

Note the consistent physical meaning:

$$\langle a''|a'''\rangle = \begin{cases} a'' = a''' : \text{ Yes, represented by number 1,} \\ a'' \neq a''' : \text{ No, represented by number 0,} \end{cases}$$

[L → R] (1.3.4)

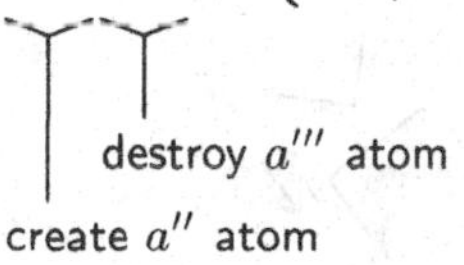

if this creation–annihilation act is considered in isolation, so that the destruction of an a''' atom, where only an a'' atom, $a'' \neq a'''$, is available, is not possible.

So now we have n symbols $\langle a'|$ and n symbols $|a'\rangle$ such that the product $\langle \; | \; \rangle$ of those with the same label equals one; the product of those with different labels equals zero.

This has the ring of familiarity, most obviously in the three unit vectors $\boldsymbol{i}$, $\boldsymbol{j}$, $\boldsymbol{k}$ of a spatial coordinate system, or more systematically, $\boldsymbol{e}_k$, $k = 1, 2, 3$. Indeed,

$$\boldsymbol{e}_k \cdot \boldsymbol{e}_l = \begin{Bmatrix} 1 \text{ if } k = l \\ 0 \text{ if } k \neq l \end{Bmatrix} = \delta(k,l) \equiv \delta_{kl} \tag{1.3.5}$$

characterizes unit orthogonal vectors. This statement about orthonormality is supplemented by the completeness relation

$$\sum_k e_k e_k = 1\,, \tag{1.3.6}$$

a sum of dyadic products.

Accordingly, we shall speak of these symbols $\langle a'|$ and $|a'\rangle$ as *vectors*, n component vectors. But notice that the numerically-valued product

$$\langle a'|a''\rangle = \delta(a', a'') \tag{1.3.7}$$

involves *two* distinct types of vectors; certainly $\langle a'|$, symbolizing a creation act, cannot be equated to $|a'\rangle$, representing an act of destruction (reading L → R). The kind of geometrical space in which these vectors lie must be somewhat more general than a Euclidean* space. We shall speak of $\langle a'|$ as a *left*-vector and $|a'\rangle$ as a *right*-vector, from the respective positions in the numerical product. [Dirac† calls $\langle a'|a''\rangle$ a bracket, and the vectors: bra and ket, respectively.]

1.4 Successive measurements. Probabilities

The measurement of the magnetic moment in the z direction, or of property A, is only one of a myriad of possible measurements, of properties $B, C, \ldots$. This is most obvious in the m.m. example through the possibility of measuring the m.m. in any other direction:

In the rotated apparatus atoms are also deflected up or down, in the direction set by the rotated magnetic field. Considered in itself, this experiment is indistinguishable from the original one. But what happens when a beam from one apparatus is sent through the other one, as suggest by

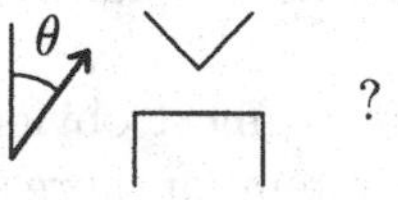

There are two situations where we already know the answer: When the angle θ is 0° or 180°.

---

*Euclid of Alexandria (fl. B.C. 300) †Paul Adrien Maurice Dirac (1902–1984)

$\theta = 0°$ 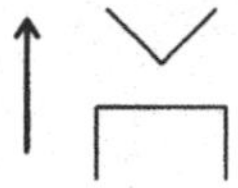 The atoms in the approaching beam are all +m.m. The transmitted beam is entirely bent *up*.

$\theta = 180°$ 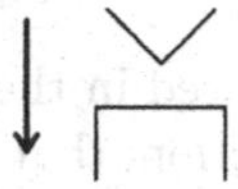 The atoms in the approaching beam are −m.m. The transmitted beam is entirely bent *down*.

Then, what about

$\theta = 90°$ 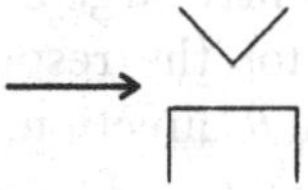 ?

It helps to think of the first apparatus as being gradually rotated from $\theta = 0°$ to $\theta = 180°$. As one does so, the initial situation of the transmitted beam, entirely up, no atoms in the down beam, must change, with fewer up atoms and more down atoms until one comes to 180°, where there are only down atoms. It is then clear that, with $\theta = 90°$, half way between the limits of 0° and 180°, half of the atoms will be in the up beam (+m.m.'s) and half will be in the down beam (−m.m.'s).

But what does an individual atom do? It doesn't split in half! *This* atom is deflected up and *that* atom is deflected down. We have no way of predicting or controlling what an individual atom will do; we can only be sure of what will happen, on the average, to very many atoms.

Speaking of averages let's list the average m.m. in μ units, for the transmitted beams of the three arrangements:

			Average
$\theta = 0°$	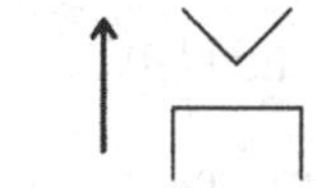	all +m.m.	+1
$\theta = 90°$	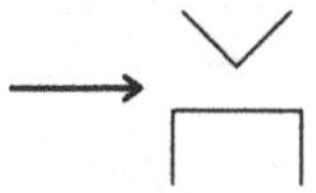	50% +m.m., 50% −m.m.	0
$\theta = 180°$	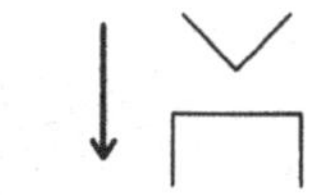	all −m.m.	−1

Can we come up with a reasonable result for any value of θ? The initially measured m.m. along the θ direction can be decomposed into the component along the z direction, and the perpendicular component:

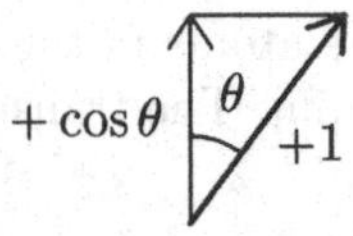

It is natural to assume that the average m.m. measured in the z direction is just the z projection of its known value in the θ direction; that is, $\cos\theta$ which correctly reproduces the values for $\theta = 0°, 90°, 180°$.

The average m.m. is the weighted average of the two possible outcomes, $+1, -1$, the weights being the fraction of a *large* number of atoms in the up, or down, beam. When used in this sense (sufficiently large number of atoms), we speak of the fraction as the probabilities for the respective outcomes. Thus, for the initial selection of a +m.m. in the θ direction,

[L → R]
$$\cos\theta = (+1)p(+,+) + (-1)p(+,-) \tag{1.4.1}$$

and

[L → R]
$$1 = p(+,+) + p(+,-)\,, \tag{1.4.2}$$

so

$$p(+,+) = \frac{1+\cos\theta}{2} = \cos^2\left(\tfrac{1}{2}\theta\right)\,,$$
$$p(+,-) = \frac{1-\cos\theta}{2} = \sin^2\left(\tfrac{1}{2}\theta\right)\,. \tag{1.4.3}$$

The last versions use the trigonometric identities

$$\begin{aligned}\cos\theta = \cos\left(\tfrac{1}{2}\theta + \tfrac{1}{2}\theta\right) &= \cos^2\left(\tfrac{1}{2}\theta\right) - \sin^2\left(\tfrac{1}{2}\theta\right)\\ &= 2\cos^2\left(\tfrac{1}{2}\theta\right) - 1\\ &= 1 - 2\sin^2\left(\tfrac{1}{2}\theta\right)\,.\end{aligned} \tag{1.4.4}$$

Note that the two probabilities in (1.4.3) are really one:

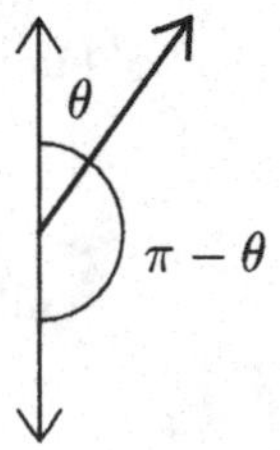

the probabilities of a − outcome is the same as the probability of a + outcome for the angle $\pi - \theta$,

$$\cos^2\left(\tfrac{1}{2}(\pi - \theta)\right) = \sin^2\left(\tfrac{1}{2}\theta\right) ; \tag{1.4.5}$$

the figure

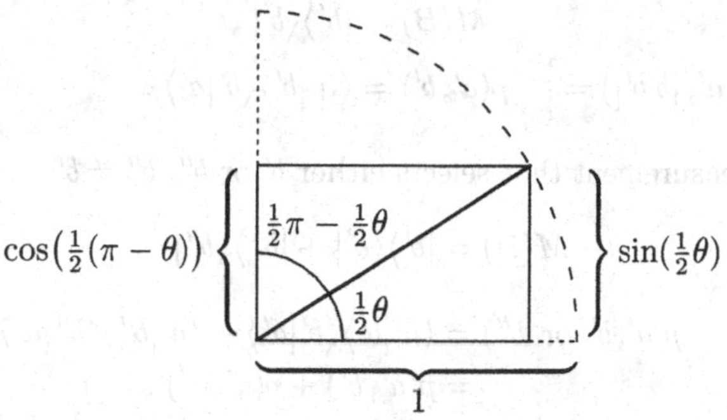

reminds us of its geometrical significance. With this in mind, we can immediately write down the probabilities for an initial choice of −m.m.:

[L → R]

$$p(-,+) = \cos^2\left(\tfrac{1}{2}(\pi - \theta)\right) = \sin^2\left(\tfrac{1}{2}\theta\right) ;$$
$$p(-,-) = \sin^2\left(\tfrac{1}{2}(\pi - \theta)\right) = \cos^2\left(\tfrac{1}{2}\theta\right) . \tag{1.4.6}$$

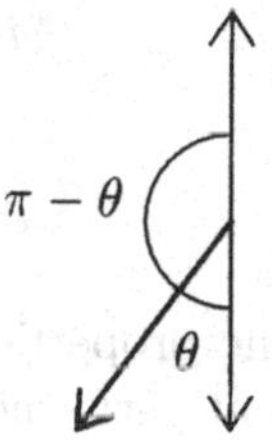

The table

$p(\ ,\ )$	$+$	$-$
$+$	$\cos^2(\frac{1}{2}\theta)$	$\sin^2(\frac{1}{2}\theta)$
$-$	$\sin^2(\frac{1}{2}\theta)$	$\cos^2(\frac{1}{2}\theta)$

displays the four probabilities.

1.5 Probability amplitudes. Interference

Now, more generally, we first measure some property A and select the particular outcome a'; (in L → R reading) we symbolize that by the creation of an a' atom: $\langle a'|$. Then we make an as yet unspecified type of B measurement and symbolize it by $M(B)$, so that we now have $\langle a'|M(B)$. The final step is the annihilation (detection) of the a' atom, which produces the *number*

[L → R]

$$p(a', M(B)) = \langle a'|M(B)|a'\rangle . \tag{1.5.1}$$

We consider three types of $M(B)$:

1. The B measurement that selects only b':

$$M(B) = |b'\rangle\langle b'| \,,$$
$$\left[p(a', |b'b'|) =\right] \quad p(a', b') = \langle a'|b'\rangle\langle b'|a'\rangle \,. \tag{1.5.2}$$

2. The B measurement that selects either b' or b'', $b'' \neq b'$:

$$M(B) = |b'\rangle\langle b'| + |b''\rangle\langle b''| \,,$$
$$\begin{aligned} p(a', b' \text{ or } b'') &= \langle a'|b'\rangle\langle b'|a'\rangle + \langle a'|b''\rangle\langle b''|a'\rangle \\ &= p(a', b') + p(a', b'') \,. \end{aligned} \tag{1.5.3}$$

3. The B measurement that selects all b' without bias:

$$M(B) = \sum_{b'} |b'\rangle\langle b'| = 1 \,,$$
$$\begin{aligned} p(a', 1) &= \sum_{b'} \langle a'|b'\rangle\langle b'|a'\rangle = \langle a'|1|a'\rangle \\ &= \sum_{b'} p(a', b') = 1 \,. \end{aligned} \tag{1.5.4}$$

Here are the properties that qualify $p(a', b')$ to be the probability that a B measurement performed on an a' atom will have the specific outcome b'. For, it should be true that the less specific measurement that selects either b' or b'', that has both b' and b'' atoms in the transmitted beam, has an outcome with the greater probability: $p(a', b') + p(a', b'')$; and that the outcome of the least specific measurement has the greatest probability: $\sum_{b'} p(a', b') = 1$.

We can also verify the probability formula in one simple situation. That is when B is just A, as in the two m.m.'s with $\theta = 0$; or is directly related to A, as in the two m.m.'s with $\theta = 180°$. So suppose that $b' = a''$. Then, surely,

$$p(a', a'') = \left\{\begin{matrix} 1 & \text{if } a' = a'' \\ 0 & \text{if } a' \neq a'' \end{matrix}\right\} = \delta(a', a'') \tag{1.5.5}$$

and indeed

$$\langle a'|a''\rangle\langle a''|a'\rangle = [\delta(a', a'')]^2 = \delta(a', a'') \,. \tag{1.5.6}$$

As we see in the results, and in $\cos^2(\frac{1}{2}\theta)$, a probability is a number in the range between zero and one. What kind of numbers are the $\langle a'|b'\rangle$ so that $\langle a'|b'\rangle\langle b'|a'\rangle$ does lie in this range and is not a negative or a complex number, for example?

There are two possibilities:

1. $\langle a'|b'\rangle$ is a *real* number and $\langle b'|a'\rangle = \langle a'|b'\rangle$. Then

$$p(a',b') = \left[\langle a'|b'\rangle\right]^2 \geq 0 \tag{1.5.7}$$

and automatically $p(a',b') \leq 1$, since the sum of all the non-negative probabilities is equal to one. The objection comes from

$$\langle b'|a'\rangle = \langle a'|b'\rangle \tag{1.5.8}$$

which seems to say that left and right vectors are the same, which they actually are not, physically. Indeed we shall see explicitly that a scheme with only real numbers, in which there is no number with a square equal to -1, does not work.

2. $\langle a'|b'\rangle$ is a complex number and $\langle b'|a'\rangle$ its conjugate, $\langle b'|a'\rangle = \langle a'|b'\rangle^*$. Then

$$p(a',b') = \left|\langle a'|b'\rangle\right|^2 \geq 0\,. \tag{1.5.9}$$

Now left and right vectors are *not* the same, but are interconverted by a process involving complex conjugation (which we shall expand upon later). We accept complex numbers, because the doubling produced by the existence of two kinds of vectors fits perfectly with the doubling provided by the complex numbers.

The construction of probabilities as absolute squares supplies a name for the complex numbers $\langle a'|b'\rangle$: ***probability amplitude***. This is a purely functional description, that real, positive probabilities are derived as absolute squares of complex valued probability amplitudes. However, this relationship is also evocative of analogies, which we proceed to develop.

Consider a sequence of measurements on three (in general different) physical properties : A, B, C. First, perform an A measurement and select the outcome a'; that is, we create an a' atom: $\langle a'|$ (L → R reading). Then perform an as yet unspecified B measurement, symbolized by $M(B)$. Finally we select the outcome c' of the C measurement, symbolized by the annihilation of a c' atom. The number produced by this sequence is $\langle a'|M(B)|c'\rangle$, which is a probability amplitude from which we derive the probability for the success of this compound measurement:

[L → R]
$$p(a',M(B),c') = \left|\langle a'|M(B)|c'\rangle\right|^2 . \tag{1.5.10}$$

We consider three examples.

1. $M(B) = |b'\rangle\langle b'|$, a selective measurement:

$$\Big[p(a', |b'b'|, c') = \Big] \quad p(a', b', c') = |\langle a'|b'\rangle\langle b'|c'\rangle|^2$$
$$= p(a', b')p(b', c') \,. \qquad (1.5.11)$$

   This is as it should be. The beam of a' atoms is subjected to a selective measurement that leaves the fraction of atoms $p(a', b')$. Then these atoms are subjected to another selective measurement, of probability $p(b', c')$, that gives the net fraction, or probability, as the *product*: $p(a', b')p(b', c')$.
2. $M(B) = \sum_{b'} |b'\rangle\langle b'| = 1$, the measurement that accepts everything without discrimination; no measurement. For $\langle a'|1|c'\rangle$ we have, equivalently,

$$\langle a'|c'\rangle = \sum_{b'} \langle a'|b'\rangle\langle b'|c'\rangle \,, \qquad (1.5.12)$$

   and

$$p(a', 1, c') = p(a', c') \,, \qquad (1.5.13)$$

   as it should. Incidentally, (1.5.12) is the counterpart of a familiar three-dimensional vector relation:

$$\boldsymbol{A} \cdot \boldsymbol{C} = \sum_k A_k C_k = \sum_k \boldsymbol{A} \cdot \boldsymbol{e}_k \, \boldsymbol{e}_k \cdot \boldsymbol{C} \,, \qquad (1.5.14)$$

   the evaluation of a scalar product in terms of components relative to some coordinate frame. Note the appearance of the completeness relation (1.3.6).
3. The non-selective B measurement; the B measuring apparatus functions but *no* selection of b' atoms is performed. In the m.m. example, this means that the up and down beams are physically separated, but then the two beams are run along together to the next stage.

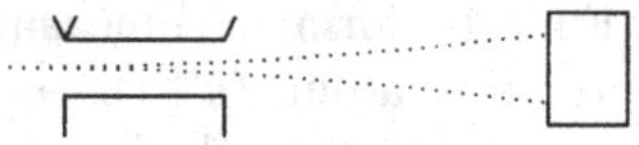

   The result, probability $p(a', b, c')$, is the sum for each independent choice of b', which have been physically distinguished by the experiment:

$$p(a', b, c') = \sum_{b'} p(a', b')p(b', c') \,. \qquad (1.5.15)$$

Neither the B measurement that accepts everything, nor the non-selective B measurement, involves a rejection of atoms at the intermediate stage. So the total fraction, the total probability, for any outcome equals one, for both:

$$\sum_{c'} p(a',1,c') = \sum_{c'} p(a',c') = 1\,,$$
$$\sum_{c'} p(a',b,c') = \underbrace{\sum_{b'} p(a',b')}_{=1} \underbrace{\sum_{c'} p(b',c')}_{=1} = 1\,. \tag{1.5.16}$$

Nevertheless, in general $p(a',1,c') \neq p(a',b,c')$.

As an example consider A, B, C, the m.m. in three directions:

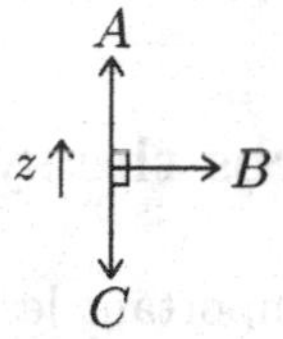

All probabilities for the angle 90° equal $\frac{1}{2}$. So (L → R reading)

$$p(\overset{a}{+},b,\overset{c}{+}) = p(\overset{a}{+},\overset{b}{+})p(\overset{b}{+},\overset{c}{+}) + p(\overset{a}{+},\overset{b}{-})p(\overset{b}{-},\overset{c}{+}) = \tfrac{1}{2}\tfrac{1}{2} + \tfrac{1}{2}\tfrac{1}{2} = \tfrac{1}{2}\,,$$
$$p(+,b,-) = p(+,+)p(+,-) + p(+,-)p(-,-) = \tfrac{1}{2}\tfrac{1}{2} + \tfrac{1}{2}\tfrac{1}{2} = \tfrac{1}{2}\,. \tag{1.5.17}$$

However

$$p(\overset{a}{+},1,\overset{c}{+}) = p(\overset{a}{+},\overset{c}{+}) = 0 \tag{1.5.18}$$

since a m.m. that is + in the z direction is certainly *not* + in the $-z$ direction. And,

$$p(\overset{a}{+},1,\overset{c}{-}) = p(\overset{a}{+},\overset{c}{-}) = 1 \tag{1.5.19}$$

since − in the $-z$ direction is the same as + along $+z$. These probabilities also add up to one.

To see how this difference must come about, let's look in detail at $p(+,1,\pm)$:

$$[\text{L} \to \text{R}] \quad \begin{aligned} p(\overset{a}{+},1,\overset{c}{+}) &= \left|\langle\overset{a}{+}|\overset{b}{+}\rangle\langle\overset{b}{+}|\overset{c}{+}\rangle + \langle\overset{a}{+}|\overset{b}{-}\rangle\langle\overset{b}{-}|\overset{c}{+}\rangle\right|^2\,, \\ p(\overset{a}{+},1,\overset{c}{-}) &= \left|\langle\overset{a}{+}|\overset{b}{+}\rangle\langle\overset{b}{+}|\overset{c}{-}\rangle + \langle\overset{a}{+}|\overset{b}{-}\rangle\langle\overset{b}{-}|\overset{c}{-}\rangle\right|^2. \end{aligned} \tag{1.5.20}$$

Note that the absolute square of each individual term is $\frac{1}{2}\frac{1}{2} = \frac{1}{4}$, and keeping *only* the absolute squares gives the non-selective measurement. Accordingly, each product of probability amplitudes is $\frac{1}{2}$ in *magnitude*. So it must be (we show this explicity later) that in $p(+,1,-)$ the two terms ***add*** up:

$$p(+,1,-) = \left|\tfrac{1}{2} + \tfrac{1}{2}\right|^2 = 1\,, \tag{1.5.21}$$

whereas in $p(+,1,+)$ they *subtract*:

$$p(+,1,+) = \left|\tfrac{1}{2} - \tfrac{1}{2}\right|^2 = 0\,. \tag{1.5.22}$$

In short, we have situations of constructive or destructive *interference*, typical of wave phenomena. On the basis of this *analogy*, one speaks alternatively of a probability amplitude as a wave function.

1.6 "Measurement disturbs the system"

Language aside, there is a very important lesson in the fact that $p(a',1,c')$, in which there is no actual B measurement:

$$p(a',1,c') = \left|\sum_{b'} \langle a'|b'\rangle\langle b'|c'\rangle\right|^2\,, \tag{1.6.1}$$

is generally different from $p(a',b,c')$, in which the B measurement takes place, but all atoms are retained:

$$p(a',b,c') = \sum_{b'} \left|\langle a'|b'\rangle\langle b'|c'\rangle\right|^2\,. \tag{1.6.2}$$

That fact is usually expressed as "measurement disturbs the system". This is very different from the situation in classical physics where it is assumed that in principle, as an idealized limit, measurement does *not* disturb the system. Consider the familiar problem of measuring the electric field at a point, to which one responds by placing a test charge at that point (itself an idealization) and measuring the force on it. To the objection that the presence of the test charge changes the field being measured, one usually says: "yes, but I can make the magnitude of the test charge arbitrarily small." All very well, until one reaches the atom and the electron within it and discovers that (current speculations aside) there are no smaller charges. In short the atomicity of matter – and of physical properties associated with matter – sets a fundamental limit to the basic idealization that is implicit in classical physics. And that's what quantum mechanics is all about.

One gets a quantitative version of "measurement disturbs the system" by asking for a symbol, M_b, of the non-selective B measurement, so that

$$\sum_{b'} \left|\langle a'|b'\rangle\langle b'|c'\rangle\right|^2 = \left|\langle a'|M_b|c'\rangle\right|^2 \tag{1.6.3}$$

in some sense. It helps to go back to $p(a',1,c')$ and write it out as

$$
\begin{aligned}
p(a',1,c') &= \sum_{b'} \langle a'|b'\rangle^* \langle b'|c'\rangle^* \sum_{b''} \langle a'|b''\rangle\langle b''|c'\rangle \\
&= \underbrace{\sum_{b'} \left|\langle a'|b'\rangle\langle b'|c'\rangle\right|^2}_{=p(a',b,c')} + \sum_{b'\neq b''} \langle a'|b'\rangle^*\langle b'|c'\rangle^*\langle a'|b''\rangle\langle b''|c'\rangle\,.
\end{aligned}
\tag{1.6.4}
$$

As we have already seen in an example, $p(a',b,c')$ is derived from $p(a',1,c')$ by keeping only the absolute squares, by omitting the cross products, $b' \neq b''$. We can convey this by choosing

$$
M_b = \sum_{b'} |b'\rangle\, e^{i\varphi(b')} \langle b'|\,, \tag{1.6.5}
$$

where the $\varphi(b')$ are real numbers, phase angles, about which more in a moment. The effect of this in the above is given by the substitutions:

$$
\begin{aligned}
\langle a'|b''\rangle\langle b''|c'\rangle &\longrightarrow \langle a'|b''\rangle\, e^{i\varphi(b'')} \langle b''|c'\rangle\,, \\
\langle a'|b'\rangle^*\langle b'|c'\rangle^* &\longrightarrow \langle a'|b'\rangle^*\, e^{-i\varphi(b')} \langle b'|c'\rangle^*\,.
\end{aligned}
\tag{1.6.6}
$$

So there is no change in the terms with $b'' = b'$,

$$
e^{-i\varphi(b')}\, e^{i\varphi(b')} = \left| e^{i\varphi(b')} \right|^2 = 1\,, \tag{1.6.7}
$$

whereas the terms with $b'' \neq b'$ are multiplied by

$$
e^{-i\varphi(b')}\, e^{i\varphi(b'')}\,. \tag{1.6.8}
$$

We succeed in removing these cross products if each $\varphi(b')$ is not a definite angle, but rather is a randomly distributed quantity. Then a particular value φ, and $\varphi+\pi$, are just as likely to occur and the average is zero ($e^{i\pi} = -1$). It is this randomness of the phases that conveys the *uncontrollable* nature of the disturbance produced by a measurement. That is important because at a higher level of classical measurement, disturbances produced by a measurement can be admitted if they are *known* and therefore could be compensated for.

The combination

$$
\sum_{b''} |b''\rangle\, e^{i\varphi(b'')} \langle b''| \tag{1.6.9}
$$

has a wider range of applicability if one allows some of the $\varphi(b'')$ to assume imaginary values, specifically positive infinite imaginary values, $i\infty$, where

$$
e^{i(i\infty)} = e^{-\infty} = 0\,. \tag{1.6.10}
$$

If one does this for all $b'' \neq b'$, then all those terms disappear and we are left with [$\varphi(b')$ is real]

$$M = \mathrm{e}^{\mathrm{i}\varphi(b')} \left| b'b' \right| \tag{1.6.11}$$

where the remaining phase factor can be ignored in the context of the probability $\left|\langle a' | M | c' \rangle\right|^2$. We have returned to the measurement that selects only b' atoms, all others being discarded, the stopping of those beams now being represented as a kind of complete absorption ($\mathrm{e}^{-\infty}$). The extension of this viewpoint is immediate. If infinite imaginary phases apply for all but b' and $b'' \neq b'$, and these two real phases are locked together: $\varphi(b') = \varphi(b'')$, we effectively produced the measurement $\left|b'b'\right| + \left|b''b''\right|$. And finally, if all $\varphi(b')$ are real, and locked together as a common phase, we arrive at the measurement that accepts everything, $M(B) = 1$.

1.7 Observables

The symbolic representation of measurements and the use of probabilities is a far cry from classical physics where theory is formulated in terms of *the* values of physical properties. But the latter must be latent in the quantum description. Can we find a symbol that is associated with a physical property as a whole?

A natural place to start is the *average* value, or expectation value, of a physical property B, produced by a measurement on a' atoms. This is

[L → R] $$\langle B \rangle_{a'} = \sum_{b'} \underbrace{p(a', b')}_{= \langle a' | b' \rangle \langle b' | a' \rangle} b' = \langle a' | \left(\sum_{b'} | b' \rangle b' \langle b' | \right) | a' \rangle \,, \tag{1.7.1}$$

which exhibits a clear separation between what is measured,

$$B = \sum_{b'} | b' \rangle b' \langle b' | , \tag{1.7.2}$$

and information about the atoms $\langle a' |, | a' \rangle$, namely

$$\langle B \rangle_{a'} = \langle a' | B | a' \rangle \,. \tag{1.7.3}$$

We test the suitability of B as the symbol of property B. First, consider

$$\langle b' | B = \sum_{b''} \underbrace{\langle b' | b'' \rangle}_{= \delta(b', b'')} b'' \langle b'' | = b' \langle b' | \tag{1.7.4}$$

inasmuch as all terms of the b'' summation vanish, except $b'' = b'$. The result can be read (L $\to$ R) as: create a b' atom and measure property B; the outcome is the number b'. An alternative version is

$$B|b''\rangle = \sum_{b'} |b'\rangle b' \underbrace{\langle b'|b''\rangle}_{=\delta(b',b'')} = |b''\rangle b'' . \tag{1.7.5}$$

We might also present these results in another way. Consider $B - b'1$ where any particular value b' multiplies the unit symbol:

$$\begin{aligned} B - b'1 &= \sum_{b''} |b''\rangle b'' \langle b''| - b' \sum_{b''} |b''\rangle\langle b''| \\ &= \sum_{b''} |b''\rangle (b'' - b') \langle b''| . \end{aligned} \tag{1.7.6}$$

Now, we get

$$\langle b'|(B - b') = 0, \qquad (B - b')|b'\rangle = 0 \tag{1.7.7}$$

since $b'' - b' = 0$ for $b'' = b'$. We have also suppressed the unit symbol, its implicit presence being clear in this context.

There is something rather basic we expect of symbol B. If the possible outcomes of measuring B are $b_1, b_2, \ldots, b_n$, then the outcomes of measuring B^2 must be $(b_1)^2, (b_2)^2, \ldots, (b_n)^2$, and similarly for higher powers. Well,

$$B^2 = B \sum_{b'} |b'\rangle b' \langle b'| = \sum_{b'} |b'\rangle (b')^2 \langle b'| \tag{1.7.8}$$

and so on, just as it should. In a related way, consider

$$B - b_1 = \sum_{b'} |b'\rangle (b' - b_1) \langle b'| \tag{1.7.9}$$

and then

$$(B - b_1)(B - b_2) = \sum_{b'} |b'\rangle (b' - b_1)(b' - b_2) \langle b'| \tag{1.7.10}$$

terminating with

$$\begin{aligned} (B - b_1) \cdots (B - b_n) &= \prod_{k=1}^{n} (B - b_k) \\ &= \sum_{b'} |b'\rangle \left[\prod_{k=1}^{n} (b' - b_k) \right] \langle b'| = 0 \end{aligned} \tag{1.7.11}$$

since any b' is one of the b_k's. So B obeys an algebraic equation of degree n. Accordingly, any function of B expressed by a power series is no more than a linear combination, with numerical multiples, of $1, B, B^2, \ldots B^{n-1}$. As such

$$f(B) = \sum_{b'} f(b')|b'b'| = \sum_{b'} |b'\rangle f(b')\langle b'| \,. \tag{1.7.12}$$

This can be taken as the general definition of the function $f(B)$ of the measurement symbol B, or of the observable B; any measurement of the physical property represented by B, measures all functions of B. All that is required of the numerical function $f(b')$ is that it has a well defined value for each potential measurement result b'.

As an illustration of a function of B expressed by a power series of degree $n-1$, consider

$$\delta(B, b') = \prod_{b''(\neq b')} \left(\frac{B - b''}{b' - b''} \right) \tag{1.7.13}$$

a product of $n-1$ linear factors. If B happens to have the value b', this function equals one; if B has a value different from b', the function equals zero, whence the delta symbol designation. And what is this function in terms of measurement symbols? It is

$$\delta(B, b') = \sum_{b''} |b''\rangle \delta(b'', b')\langle b''| = |b'b'| \,, \tag{1.7.14}$$

the symbol of the b' selective measurement. To go along with this, notice that multiplying $\delta(B, b')$ by the factor $B - b'$ produces the polynomial of degree n that is zero. So

$$0 = (B - b')\delta(B, b') = (B - b')|b'\rangle\langle b'| = 0 \,, \tag{1.7.15}$$

as it should.

Incidentally, the unit symbol 1 is also a function of B, one that has the value 1 no matter what the value of B:

$$\langle b'|1 = \langle b'| = 1\langle b'| \,. \tag{1.7.16}$$

We can exhibit the unit symbol as a power series in B by using the power series construction of the $|b'b'|$:

$$1 = \sum_{b'} \prod_{b''(\neq b')} \left(\frac{B - b''}{b' - b''} \right) . \tag{1.7.17}$$

Indeed for *any* value of B, the right side equals $1 + 0 + \cdots + 0 = 1$.

1.8 Algebra of Pauli's operators

In seeking to illustrate this symbolic representation of physical quantities we naturally turn to Stern–Gerlach experiments and μ_z/μ which has the possible values $+1, -1$. Following Pauli* (1927) we designate this physical quantity as σ_z, where $\sigma_z' = +1, -1$. Then

$$(\sigma_z - 1)(\sigma_z + 1) = \sigma_z^2 - 1 = 0 \tag{1.8.1}$$

and (σ_z' and 1 are omitted, the sign suffices)

$$\begin{aligned} |{+}{+}| &= \frac{\sigma_z - (-1)}{1 - (-1)} = \frac{1 + \sigma_z}{2} \,, \\ |{-}{-}| &= \frac{\sigma_z - 1}{-1 - 1} = \frac{1 - \sigma_z}{2} \,. \end{aligned} \tag{1.8.2}$$

Indeed,

$$|{+}{+}| + |{-}{-}| = 1, \qquad |{+}{+}| - |{-}{-}| = \sigma_z \,. \tag{1.8.3}$$

We also note that, from $\sigma_z^2 = 1$, we get

$$\begin{aligned} |{+}{+}||{+}{+}| &= \left(\frac{1 + \sigma_z}{2}\right)^2 = \frac{1}{4}(1 + 1 + 2\sigma_z) = \frac{1 + \sigma_z}{2} = |{+}{+}| \,, \\ |{-}{-}||{-}{-}| &= \left(\frac{1 - \sigma_z}{2}\right)^2 = \frac{1}{4}(1 + 1 - 2\sigma_z) = \frac{1 - \sigma_z}{2} = |{-}{-}| \,, \end{aligned} \tag{1.8.4}$$

and

$$|{+}{+}||{-}{-}| = \frac{1 + \sigma_z}{2}\frac{1 - \sigma_z}{2} = \frac{1 - \sigma_z^2}{4} = 0 \,. \tag{1.8.5}$$

The totality of measurement symbols $|\sigma_z'\sigma_z''|$, $\sigma_z' = \pm 1$, $\sigma_z'' = \pm 1$, is $4 = 2^2$. We have displayed $|{+}{+}|$ and $|{-}{-}|$, or, equivalently, 1 and σ_z. Now, in three-dimensional space there must also be a σ_x and a σ_y. And since all directions in space are equivalent, the outcome of measuring σ_x and σ_y must also be just ± 1, so that we must have

$$\sigma_x^2 = 1, \qquad \sigma_y^2 = 1 \,. \tag{1.8.6}$$

Can we exhibit such physical properties in terms of $|{+}{-}|$ and $|{-}{+}|$?

Let's write out their multiplication properties:

$$\begin{aligned} &|{+}{-}||{+}{-}| = 0 \,, \qquad && |{-}{+}||{-}{+}| = 0 \,, \\ &|{+}{-}||{-}{+}| = |{+}{+}| \,, && |{-}{+}||{+}{-}| = |{-}{-}| \,. \end{aligned} \tag{1.8.7}$$

Since the squares of $|{+}{-}|$ and $|{-}{+}|$ are zero, we must have both present if we are to form an object of unit square. Try

---

*Wolfgang Pauli (1900–1958)

$$\Big[|-+|+|+-|\Big]^2 = \underbrace{|-+||-+|}_{=0} + \underbrace{|+-||-+|}_{=|++|} + \underbrace{|-+||+-|}_{=|--|} + \underbrace{|+-||+-|}_{=0} = 1\,, \qquad (1.8.8)$$

so we define

$$\sigma_x = |-+| + |+-|\,. \qquad (1.8.9)$$

The only combination left is the difference:

$$\Big[|-+|-|+-|\Big]^2 = 0 - |++| - |--| + 0 = -1 \quad ! \qquad (1.8.10)$$

If space where two-dimensional, we could settle for σ_x and σ_z. But in three-dimensional space there is no way to get that last square equal to +1, except by introducing $\mathrm{i}^2 = -1$. Real numbers alone won't work. So σ_z and σ_x are joined by

$$\sigma_y = \mathrm{i}|-+| - \mathrm{i}|+-|\,. \qquad (1.8.11)$$

Note the converse constructions,

$$|+-| = \tfrac{1}{2}\,(\sigma_x + i\sigma_y)\,, \quad |-+| = \tfrac{1}{2}\,(\sigma_x - i\sigma_y) \qquad (1.8.12)$$

for which we get

$$0 = \Big[|+-|\Big]^2 = \tfrac{1}{4}\,(\sigma_x + i\sigma_y)^2 = \tfrac{1}{4}\big[1 - 1 + \mathrm{i}(\sigma_x\sigma_y + \sigma_y\sigma_x)\big]\,, \qquad (1.8.13)$$

or

$$\sigma_x\sigma_y + \sigma_y\sigma_x = 0; \quad \sigma_y\sigma_x = -\sigma_x\sigma_y\,. \qquad (1.8.14)$$

Then we look at $|+-||-+| = |++|$, or

$$\tfrac{1}{4}\,\underbrace{(\sigma_x + \mathrm{i}\sigma_y)(\sigma_x - \mathrm{i}\sigma_y)}_{=1+1-2\mathrm{i}\sigma_x\sigma_y} = \tfrac{1}{2}(1 + \sigma_z)\,, \qquad (1.8.15)$$

which gives

$$\sigma_x\sigma_y = \mathrm{i}\sigma_z\,, \qquad (1.8.16)$$

and, in conjunction with (1.8.14), we get

$$\sigma_x\sigma_z + \sigma_z\sigma_x = 0, \quad \sigma_y\sigma_z + \sigma_z\sigma_y = 0\,. \qquad (1.8.17)$$

Also (by moving the left-hand factor twice):

$$\sigma_x\sigma_y\sigma_z = \mathrm{i} = \sigma_y\sigma_z\sigma_x = \sigma_z\sigma_x\sigma_y \tag{1.8.18}$$

which shows that the indices can be cyclically permuted. So $\sigma_x\sigma_y = \mathrm{i}\sigma_z$ is joined by

$$\sigma_y\sigma_z = \mathrm{i}\sigma_x , \quad \sigma_z\sigma_x = \mathrm{i}\sigma_y . \tag{1.8.19}$$

This direct conversion of the algebra of the four $|\pm\pm|$ into that of the four $1, \sigma_x, \sigma_y, \sigma_z$ is a bit clumsy, but the algebraic properties of Pauli's vector operator $\boldsymbol{\sigma}$ are not. One way to express them compactly considers two numerical vectors, say $\boldsymbol{a}$ and $\boldsymbol{b}$. Then,

$$\begin{aligned}\boldsymbol{\sigma}\cdot\boldsymbol{a}\,\boldsymbol{\sigma}\cdot\boldsymbol{b} &= (\sigma_x a_x + \sigma_y a_y + \sigma_z a_z)(\sigma_x b_x + \sigma_y b_y + \sigma_z b_z) \\ &= a_x b_x + a_y b_y + a_z b_z + \mathrm{i}\sigma_x(a_y b_z - a_z b_y) + \cdots ,\end{aligned} \tag{1.8.20}$$

or

$$\boldsymbol{\sigma}\cdot\boldsymbol{a}\,\boldsymbol{\sigma}\cdot\boldsymbol{b} = \boldsymbol{a}\cdot\boldsymbol{b} + \mathrm{i}\boldsymbol{\sigma}\cdot\boldsymbol{a}\times\boldsymbol{b} . \tag{1.8.21}$$

This unifies all the multiplication properties. In particular, let $\boldsymbol{a} = \boldsymbol{b} = \boldsymbol{n}$, an arbitrarily directed unit vector. Then

$$(\boldsymbol{\sigma}\cdot\boldsymbol{n})^2 = 1 , \tag{1.8.22}$$

which shows that a measurement of the component of $\boldsymbol{\sigma}$ in *any* direction will have the outcomes +1 and −1. The apparent contradiction between *discrete* results of measurements and the possibility of *continuous* change in the direction of measurement is resolved by the non-commutativity of the symbols of different components.

1.9 Adjoint symbols, Hermitian symbols

Now that we have a few more objects to work with, let's develop some more machinery. First, notice that the symbolization of successive acts of measurement by multiplication of corresponding symbols has an element of arbitrariness: we placed them in sequence from left to right (L → R). But that is a cultural bias – we can equally well order them from right to left. As with the freedom of choice of coordinate systems, the validity of a physical statement cannot depend on which conventions we adopt. So a correct relation expressed in one convention must also be a correct relation when we systematically switch to the other convention. We raise this to the status of a mathematical operation, called the adjoint, or Hermitian* conjugation, and symbolize it by the dagger: †.

Incidentally, here is another reason that $\langle a'|b'\rangle$ cannot be identified with a probability, by itself. On reversing the convention, the meaning of $\langle a'|b'\rangle$

*Charles HERMITE (1822–1901)

is taken over by $\langle b'|a'\rangle$. The physical probability must be unchanged, which $\langle a'|b'\rangle$ is not, but $\langle a'|b'\rangle\langle b'|a'\rangle$ *is*.

The measurement symbol $|a'a''|$ has the interpretation of selecting a' atoms and producing a'' atoms, when read from left to right – dextrally. If we read it from right to left – sinistrally – it means selecting a'' atoms and producing a' atoms. We express this by

$$|a'a''|^\dagger = |a''a'| \,, \tag{1.9.1}$$

which gives the new interpretation in the original convention. Read this: the adjoint (or Hermitian conjugate) of $|a'a''|$ is $|a''a'|$. [Compare with $(\mathrm{PAN})^\dagger = \mathrm{NAP}$, for instance.] Note the particular example

$$|a'a'|^\dagger = |a'a'| \,; \tag{1.9.2}$$

any such object is said to be self-adjoint or Hermitian. [Compare with palindromes such as $(\mathrm{MOM})^\dagger = \mathrm{MOM}$.]

The left and right vectors $\langle a'|, |a'\rangle$ are read dextrally as the creation of an a' atom, the destruction of an a' atom, respectively. In the sinistral convention, $\langle a'|$ symbolizes the destruction, and $|a'\rangle$ the creation of the a' atom. So

$$\langle a'|^\dagger = |a'\rangle \,, \quad |a'\rangle^\dagger = \langle a'| \,. \tag{1.9.3}$$

Here, and in general, the repetition of † recovers the original object, as in

$$\left(X^\dagger\right)^\dagger = X \,. \tag{1.9.4}$$

We illustrate another general rule by combining what we have learned:

$$\Big(|a'\rangle\langle a''|\Big)^\dagger = |a''\rangle\langle a'| = \Big(\langle a''|\Big)^\dagger \Big(|a'\rangle\Big)^\dagger \tag{1.9.5}$$

or

$$(XY)^\dagger = Y^\dagger X^\dagger \,; \tag{1.9.6}$$

the adjoint of a product is the product of the adjoints, in the opposite order. Here's a more elaborate example:

$$\Big(|a'b'||c'd'|\Big)^\dagger = |d'c'||b'a'| = |c'd'|^\dagger |a'b'|^\dagger \,. \tag{1.9.7}$$

We learn something new by rewriting the ingredients of the first equality:

$$\begin{aligned} |a'b'||c'd'| &= |a'\rangle\langle b'|c'\rangle\langle d'| = \langle b'|c'\rangle |a'd'| \,, \\ |d'c'||b'a'| &= |d'\rangle\langle c'|b'\rangle\langle a'| = \langle b'|c'\rangle^* |d'a'| \,, \end{aligned} \tag{1.9.8}$$

namely,

$$\left(\langle b'|c'\rangle|a'd'|\right)^{\dagger} = \langle b'|c'\rangle^{*}|a'd'|^{\dagger} ; \tag{1.9.9}$$

in taking adjoints, numbers are replaced by their complex conjugates,

$$(\lambda X)^{\dagger} = \lambda^{*} X^{\dagger} , \tag{1.9.10}$$

where λ is any arbitrary complex number.

We must also remark that † reverses sequence, as expressed by multiplication; it has no effect upon addition:

$$(X + Y)^{\dagger} = X^{\dagger} + Y^{\dagger} . \tag{1.9.11}$$

In consequence the unit symbol is Hermitian,

$$1^{\dagger} = \left(\sum_{a'} |a'a'|\right)^{\dagger} = \sum_{a'} |a'a'|^{\dagger} = \sum_{a'} |a'a'| = 1 . \tag{1.9.12}$$

Indeed, the symbol of any physical property,

$$B = \sum_{b'} |b'\rangle b' \langle b'| \qquad \text{with} \quad b'^{*} = b' , \tag{1.9.13}$$

is Hermitian:

$$B^{\dagger} = \sum_{b'} \langle b'|^{\dagger} b'^{*} |b'\rangle^{\dagger} = \sum_{b'} |b'\rangle b' \langle b'| = B . \tag{1.9.14}$$

Let's try this out on

$$\begin{aligned} 1 &= |{+}{+}| + |{-}{-}| , \\ \sigma_z &= |{+}{+}| - |{-}{-}| , \\ \sigma_x &= |{-}{+}| + |{+}{-}| , \\ \sigma_y &= \mathrm{i}|{-}{+}| - \mathrm{i}|{+}{-}| . \end{aligned} \tag{1.9.15}$$

Now, 1 and σ_z are examples of what we have just shown; they are Hermitian. σ_x and σ_y, however, are not presented in the same form. But surely the adjoint operation is not going to distinguish one direction in space from the other? Fear not:

$$\begin{aligned} \sigma_x^{\dagger} &= |{-}{+}|^{\dagger} + |{+}{-}|^{\dagger} = |{+}{-}| + |{-}{+}| = \sigma_x , \\ \sigma_y^{\dagger} &= -\mathrm{i}|{-}{+}|^{\dagger} + \mathrm{i}|{+}{-}|^{\dagger} = -\mathrm{i}|{+}{-}| + \mathrm{i}|{-}{+}| = \sigma_y . \end{aligned} \tag{1.9.16}$$

Let's also notice how the algebra of the σ's behaves under Hermitian conjugation. The adjoint of ($\boldsymbol{a}$, $\boldsymbol{b}$ are real)

$$\boldsymbol{\sigma} \cdot \boldsymbol{a} \, \boldsymbol{\sigma} \cdot \boldsymbol{b} = \boldsymbol{a} \cdot \boldsymbol{b} + \mathrm{i}\boldsymbol{\sigma} \cdot \boldsymbol{a} \times \boldsymbol{b} \tag{1.9.17}$$

is

$$\begin{aligned} \boldsymbol{\sigma} \cdot \boldsymbol{b} \, \boldsymbol{\sigma} \cdot \boldsymbol{a} &= \boldsymbol{a} \cdot \boldsymbol{b} - \mathrm{i}\boldsymbol{\sigma} \cdot \boldsymbol{a} \times \boldsymbol{b} \\ &= \boldsymbol{b} \cdot \boldsymbol{a} + \mathrm{i}\boldsymbol{\sigma} \cdot \boldsymbol{b} \times \boldsymbol{a} ; \end{aligned} \tag{1.9.18}$$

it is the same statement.

1.10 Matrix representations

The Pauli operators σ_x and σ_y are examples of writing one physical quantity, B, in terms of the measurement symbols of another property, A. We do this generally by exploiting the structure of the unit symbol 1:

$$\begin{aligned} B = 1B1 &= \sum_{a'} |a'\rangle\langle a'|B \sum_{a''} |a''\rangle\langle a''| \\ &= \sum_{a',a''} \langle a'|B|a''\rangle |a'a''| \,. \end{aligned} \tag{1.10.1}$$

This exhibits B as a linear combination of the n^2 measurement symbols $|a'a''|$ as specified by the values of the $n \times n = n^2$ numbers $\langle a'|B|a''\rangle$. These numbers can be displayed in an $n \times n$ square array, or matrix:

$$\begin{matrix} & a'' \to \text{column} \\ \begin{matrix} a' \\ \downarrow \\ \text{row} \end{matrix} & \begin{pmatrix} \langle a_1|B|a_1\rangle & \cdots & \langle a_1|B|a_n\rangle \\ \vdots & & \vdots \\ \langle a_n|B|a_1\rangle & \cdots & \langle a_n|B|a_n\rangle \end{pmatrix} \end{matrix} . \tag{1.10.2}$$

For example

$$\sigma_x : \begin{matrix} & + - \\ \begin{matrix} + \\ - \end{matrix} & \begin{pmatrix} 0 & 1 \\ 1 & 0 \end{pmatrix} \end{matrix} , \quad \sigma_y : \begin{pmatrix} 0 & -\mathrm{i} \\ \mathrm{i} & 0 \end{pmatrix} , \quad \sigma_z : \begin{pmatrix} 1 & 0 \\ 0 & -1 \end{pmatrix} , \quad 1 : \begin{pmatrix} 1 & 0 \\ 0 & 1 \end{pmatrix} , \tag{1.10.3}$$

which are known as the Pauli matrices.

The physical property B is determined by its matrix, the a matrix of B. So all algebraic operations involving B, and other physical properties, can be expressed in terms of their matrices. Here are the basic algebraic operations:

addition of B and C:

$$\langle a'|\,(B + C)\,|a''\rangle = \langle a'|B|a''\rangle + \langle a'|C|a''\rangle \,; \tag{1.10.4}$$

multiplication of B and C:

$$\begin{aligned} \langle a'|BC|a'''\rangle &= \langle a'|B\,1\,C|a'''\rangle \\ &= \sum_{a''} \langle a'|B|a''\rangle\langle a''|C|a'''\rangle \,; \end{aligned} \tag{1.10.5}$$

multiplication of B by a number:

$$\langle a'|\lambda B|a''\rangle = \lambda\langle a'|B|a''\rangle \,. \tag{1.10.6}$$

Notice that the notion of adjoint does not enter here, and B, C can be replaced by not necessarily Hermitian X, Y, where, say, (the change of labeling is for convenience)

$$X = \sum_{a',a''} \langle a''|X|a'\rangle |a''a'| . \tag{1.10.7}$$

Then

$$X^{\dagger} = \sum_{a',a''} \langle a''|X|a'\rangle^{*} |a'a''| , \tag{1.10.8}$$

which is to say

$$\langle a'|X^{\dagger}|a''\rangle = \langle a''|X|a'\rangle^{*} ; \tag{1.10.9}$$

the matrix of an adjoint quantity is the complex conjugate, transposed (row $\leftrightarrow$ column) matrix of the quantity. If X is the Hermitian B, its matrix is restricted by

$$\langle a'|B|a''\rangle = \langle a''|B|a'\rangle^{*} ; \tag{1.10.10}$$

one can check this for the Pauli matrices.

Notice that complex conjugation is the adjoint for numbers; that is, one evaluates it by taking the adjoint of its multiplicative components, as in

$$\langle a'|b'\rangle^{*} = |b'\rangle^{\dagger}\langle a|^{\dagger} = \langle b'|a'\rangle \tag{1.10.11}$$

and

$$\langle a''|X|a'\rangle^{*} = |a'\rangle^{\dagger}X^{\dagger}\langle a''|^{\dagger} = \langle a'|X^{\dagger}|a''\rangle . \tag{1.10.12}$$

1.11 Traces

Here's another useful concept in which one reverses the order of things. The measurement symbol $|a'a''|$ is a product of vectors $|a'\rangle\langle a''|$. If one reverses the multiplication order of the vectors, one gets a number: $\langle a''|a'\rangle = \delta(a', a'')$. This operations is called the *trace* [German: Spur]. So

$$\mathrm{tr}\left\{|a'a''|\right\} = \delta(a', a'') . \tag{1.11.1}$$

Equally well one should have

$$\mathrm{tr}\left\{|b'b''|\right\} = \delta(b', b'') . \tag{1.11.2}$$

But is this consistent with the linear relation between the two types of symbols, namely

$$\begin{aligned} |b'b''| &= \sum_{a',a''} |a'\rangle\langle a'|b'\rangle\langle b''|a''\rangle\langle a''| \\ &= \sum_{a',a''} \langle a'|b'\rangle\langle b''|a''\rangle |a'a''| \,? \end{aligned} \tag{1.11.3}$$

Mindful that the trace has no effect upon addition or on numerical multipliers, we answer this by a symmetric treatment of the first line:

$$\begin{aligned} \operatorname{tr}\{|b'b''|\} &= \sum_{a',a''} \langle b''|a''\rangle\langle a''|a'\rangle\langle a'|b'\rangle \\ &= \langle b''|b'\rangle = \delta(b',b'') \,. \end{aligned} \tag{1.11.4}$$

The trace has a simple meaning in terms of matrix elements:

$$X = \sum_{a',a''} \langle a'|X|a''\rangle |a'a''| \tag{1.11.5}$$

gives

$$\operatorname{tr}\{X\} = \sum_{a',a''} \langle a'|X|a''\rangle \delta(a',a'') = \sum_{a'} \langle a'|X|a'\rangle \,, \tag{1.11.6}$$

the so-called diagonal sum of the matrix. A glance at the Pauli matrices (1.10.3) shows a vanishing trace for each component:

$$\operatorname{tr}\{\boldsymbol{\sigma}\} = 0 \,. \tag{1.11.7}$$

The Pauli matrix for the unit symbol 1 illustrates the general form

$$\langle a'|1|a''\rangle = \delta(a',a'') \tag{1.11.8}$$

from which we get

$$\operatorname{tr}\{1\} = n \,, \tag{1.11.9}$$

the number of different values assumed by property A, or B, or ... .

One can reverse the trace procedure so that any number involving vectors can be presented as a trace. Thus

$$\langle a'|X|a''\rangle = \operatorname{tr}\{X|a''a'|\} = \operatorname{tr}\{|a''a'|X\} \,. \tag{1.11.10}$$

If one multiplies these equivalent forms by $\langle a''|Y|a'\rangle$ and sums over a', a'' it becomes

$$\operatorname{tr}\{XY\} = \operatorname{tr}\{YX\} \,; \tag{1.11.11}$$

the trace of a product of such symbols is independent of the multiplication order. As an example of that symmetry take the trace of $\boldsymbol{\sigma}\cdot\boldsymbol{a}\,\boldsymbol{\sigma}\cdot\boldsymbol{b} = \boldsymbol{a}\cdot\boldsymbol{b} + \mathrm{i}\boldsymbol{\sigma}\cdot\boldsymbol{a}\times\boldsymbol{b}$,

$$\operatorname{tr}\{\boldsymbol{\sigma}\cdot\boldsymbol{a}\,\boldsymbol{\sigma}\cdot\boldsymbol{b}\} = 2\boldsymbol{a}\cdot\boldsymbol{b}\,, \tag{1.11.12}$$

which is indeed symmetrical in $\boldsymbol{a}$ and $\boldsymbol{b}$.

The trace version of a matrix element can be applied to

$$\langle B\rangle_{a'} = \langle a'|B|a'\rangle = \operatorname{tr}\{|a'a'|B\} \tag{1.11.13}$$

in which B can be replaced by any $f(B)$. Consider in particular the function $\delta(B,b') = |b'b'|$. As a quantity that is one, if B is b', and zero otherwise,

$$\langle\delta(B,b')\rangle_{a'} = p(a',b') = \operatorname{tr}\{|a'a'||b'b'|\}\,. \tag{1.11.14}$$

Of course, we check directly that this is $\langle a'|b'\rangle\langle b'|a'\rangle$.

To see the advantages of the trace formula recall that

$$\sigma'_z = +1 : |++| = \frac{1+\sigma_z}{2}\;;\quad \sigma'_z = -1 : |--| = \frac{1-\sigma_z}{2} \tag{1.11.15}$$

or

$$|\sigma'_z\sigma'_z| = \tfrac{1}{2}(1+\sigma'_z\sigma_z)\,. \tag{1.11.16}$$

There is nothing special about the z direction. We apply the same formula to properties A and B, which are $\boldsymbol{\sigma}\cdot\boldsymbol{n}_1$ and $\boldsymbol{\sigma}\cdot\boldsymbol{n}_2$, respectively, where $\boldsymbol{n}_{1,2}$ are unit vectors in two different directions. So

$$\begin{aligned} p(\sigma'_1,\sigma'_2) &= \operatorname{tr}\left\{\frac{1+\sigma'_1\boldsymbol{\sigma}\cdot\boldsymbol{n}_1}{2}\,\frac{1+\sigma'_2\boldsymbol{\sigma}\cdot\boldsymbol{n}_2}{2}\right\} \\ &= \tfrac{1}{4}\operatorname{tr}\{1+\sigma'_1\boldsymbol{\sigma}\cdot\boldsymbol{n}_1+\sigma'_2\boldsymbol{\sigma}\cdot\boldsymbol{n}_2+\sigma'_1\sigma'_2\boldsymbol{\sigma}\cdot\boldsymbol{n}_1\boldsymbol{\sigma}\cdot\boldsymbol{n}_2\} \\ &= \tfrac{1}{2}\left(1+\sigma'_1\sigma'_2\boldsymbol{n}_1\cdot\boldsymbol{n}_2\right) = \tfrac{1}{2}\left(1+\sigma'_1\sigma'_2\cos\theta\right)\,, \end{aligned} \tag{1.11.17}$$

where θ is the angle between the two directions. Hence,

$$\begin{aligned} p(+,+) = p(-,-) &= \frac{1+\cos\theta}{2} = \cos^2(\tfrac{1}{2}\theta)\,, \\ p(+,-) = p(-,+) &= \frac{1-\cos\theta}{2} = \sin^2(\tfrac{1}{2}\theta)\,, \end{aligned} \tag{1.11.18}$$

as we had already surmised.

1.12 Unitary geometry

1.12.1 Column and row vectors, wave functions

It is instructive to look at the σ_z matrix of

$$|\sigma'_n\sigma'_n| = \frac{1+\sigma'_n\boldsymbol{\sigma}\cdot\boldsymbol{n}}{2}\,, \tag{1.12.1}$$

where we use spherical coordinates to parameterize unit vector $\boldsymbol{n}$:

$$\begin{aligned}\boldsymbol{\sigma}\cdot\boldsymbol{n} = \sigma_x &\sin\vartheta\cos\varphi\\ +\sigma_y &\sin\vartheta\sin\varphi\\ +\sigma_z &\cos\vartheta\ .\end{aligned} \tag{1.12.2}$$

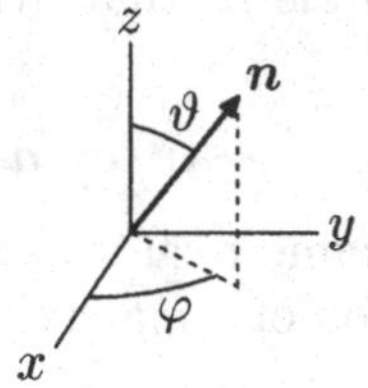

Then $|\sigma'_n\sigma'_n|$ has the matrix

$$\langle\sigma'_z|\boldsymbol{\sigma}\cdot\boldsymbol{n}|\sigma''_z\rangle = \overset{\sigma'_z\backslash\sigma''_z}{\begin{pmatrix}\cos\vartheta & \sin\vartheta\,\mathrm{e}^{-\mathrm{i}\varphi}\\ \sin\vartheta\,\mathrm{e}^{\mathrm{i}\varphi} & -\cos\vartheta\end{pmatrix}}\ . \tag{1.12.3}$$

So

$$\begin{aligned}\langle\sigma'_z|\sigma'_n = +1\rangle\langle\sigma'_n = +1|\sigma''_z\rangle &= \langle\sigma'_z|\frac{1+\boldsymbol{\sigma}\cdot\boldsymbol{n}}{2}|\sigma''_z\rangle\\ &= \begin{pmatrix}\cos^2(\frac{1}{2}\vartheta) & \cos(\frac{1}{2}\vartheta)\sin(\frac{1}{2}\vartheta)\,\mathrm{e}^{-\mathrm{i}\varphi}\\ \cos(\frac{1}{2}\vartheta)\sin(\frac{1}{2}\vartheta)\,\mathrm{e}^{\mathrm{i}\varphi} & \sin^2(\frac{1}{2}\vartheta)\end{pmatrix}\end{aligned} \tag{1.12.4}$$

which must factor, as indicated by $\langle\sigma'_z|\sigma'_n = +1\rangle\langle\sigma'_n = +1|\sigma''_z\rangle$. Indeed, this matrix is the product of a column and a row:

$$\begin{pmatrix}\cos(\frac{1}{2}\vartheta)\\ \sin(\frac{1}{2}\vartheta)\,\mathrm{e}^{\mathrm{i}\varphi}\end{pmatrix}\left(\cos(\tfrac{1}{2}\vartheta)\ ,\ \sin(\tfrac{1}{2}\vartheta)\,\mathrm{e}^{-\mathrm{i}\varphi}\right) \tag{1.12.5}$$

or, alternatively, multiplying the column by $\mathrm{e}^{-\frac{\mathrm{i}}{2}\varphi}$ and the row by $\mathrm{e}^{\frac{\mathrm{i}}{2}\varphi}$, we get

$$\begin{aligned}\langle\sigma'_z|\sigma'_n = +1\rangle &= \begin{pmatrix}\cos(\frac{1}{2}\vartheta)\,\mathrm{e}^{-\frac{\mathrm{i}}{2}\varphi}\\ \sin(\frac{1}{2}\vartheta)\,\mathrm{e}^{\frac{\mathrm{i}}{2}\varphi}\end{pmatrix}\ ;\\ \langle\sigma'_n = +1|\sigma''_z\rangle &= \left(\cos(\tfrac{1}{2}\vartheta)\,\mathrm{e}^{\frac{\mathrm{i}}{2}\varphi}\ ,\ \sin(\tfrac{1}{2}\vartheta)\,\mathrm{e}^{-\frac{\mathrm{i}}{2}\varphi}\right)\ .\end{aligned} \tag{1.12.6}$$

As required by $\langle\sigma'_z|\sigma'_n = +1\rangle^* = \langle\sigma'_n = +1|\sigma'_z\rangle$, the elements of the row are the complex conjugates of the column elements.

The analogous results for $\sigma'_n = -1$ are

$$\begin{aligned}\langle\sigma'_z|\sigma'_n = -1\rangle\langle\sigma'_n = -1|\sigma''_z\rangle &= \langle\sigma'_z|\frac{1-\boldsymbol{\sigma}\cdot\boldsymbol{n}}{2}|\sigma''_z\rangle\\ &= \begin{pmatrix}\sin^2(\frac{1}{2}\vartheta) & -\sin(\frac{1}{2}\vartheta)\cos(\frac{1}{2}\vartheta)\,\mathrm{e}^{-\mathrm{i}\varphi}\\ -\sin(\frac{1}{2}\vartheta)\cos(\frac{1}{2}\vartheta)\,\mathrm{e}^{\mathrm{i}\varphi} & \cos^2(\frac{1}{2}\vartheta)\end{pmatrix}\\ &= \begin{pmatrix}-\sin(\frac{1}{2}\vartheta)\,\mathrm{e}^{-\frac{\mathrm{i}}{2}\varphi}\\ \cos(\frac{1}{2}\vartheta)\,\mathrm{e}^{\frac{\mathrm{i}}{2}\varphi}\end{pmatrix}\left(-\sin(\tfrac{1}{2}\vartheta)\,\mathrm{e}^{\frac{\mathrm{i}}{2}\varphi}\ ,\ \cos(\tfrac{1}{2}\vartheta)\,\mathrm{e}^{-\frac{\mathrm{i}}{2}\varphi}\right)\ .\end{aligned} \tag{1.12.7}$$

The significance of this is expressed generally by writing some vector, say $|b'\rangle$, as a linear combination of the a vectors:

$$|b'\rangle = 1|b'\rangle = \sum_{a'} |a'\rangle\langle a'|b'\rangle \qquad (1.12.8)$$

which exhibits the $\langle a'|b'\rangle$ as components of the vector $|b'\rangle$ relative to the orthonormal system of a vectors – the a coordinate system. An alternative notation that accompanies the notion of 'wave function' is

$$\langle a'|b'\rangle = \psi_{b'}(a') \qquad (1.12.9)$$

so that

$$|b'\rangle = \sum_{a'} |a'\rangle \psi_{b'}(a') \,. \qquad (1.12.10)$$

The adjoint of this is

$$\langle b'| = \sum_{a'} \psi_{b'}(a')^* \langle a'| \,. \qquad (1.12.11)$$

To express the product of the two vectors in terms of their components we can multiply directly:

$$\begin{aligned} \langle b'|b''\rangle &= \sum_{a'a''} \psi_{b'}(a')^* \underbrace{\langle a'|a''\rangle}_{=\delta(a',a'')} \psi_{b''}(a'') \\ &= \sum_{a'} \psi_{b'}(a')^* \psi_{b''}(a') \,, \end{aligned} \qquad (1.12.12)$$

or use $1 = \sum_{a'} |a'\rangle\langle a'|$:

$$\langle b'|b''\rangle = \sum_{a'} \langle b'|a'\rangle\langle a'|b''\rangle = \sum_{a'} \psi_{b'}(a')^* \psi_{b''}(a') \,. \qquad (1.12.13)$$

Notice that the squared length of the b' vector is

$$\langle b'|b'\rangle = \sum_{a'} |\psi_{b'}(a')|^2 > 0 \,. \qquad (1.12.14)$$

Here, for complex-valued components of vectors, is what replaces the sum of the squares of Euclidean geometry: This kind of geometry is called *unitary*. The orthonormality of the b' vectors is conveyed by

$$\begin{aligned} \sum_{a'} |\psi_{b'}(a')|^2 &= 1 \,, \\ b' \neq b'' : \quad \sum_{a'} \psi_{b'}(a')^* \psi_{b''}(a') &= 0 \,. \end{aligned} \qquad (1.12.15)$$

Here, then, are two wave functions:

$$\psi_{\sigma'_n=+1}(\sigma'_z) = \begin{pmatrix} \cos(\frac{1}{2}\vartheta)\, e^{-\frac{i}{2}\varphi} \\ \sin(\frac{1}{2}\vartheta)\, e^{\frac{i}{2}\varphi} \end{pmatrix} ,$$

$$\psi_{\sigma'_n=-1}(\sigma'_z) = \begin{pmatrix} -\sin(\frac{1}{2}\vartheta)\, e^{-\frac{i}{2}\varphi} \\ \cos(\frac{1}{2}\vartheta)\, e^{\frac{i}{2}\varphi} \end{pmatrix} . \tag{1.12.16}$$

Since $\cos^2(\frac{1}{2}\vartheta) + \sin^2(\frac{1}{2}\vartheta) = 1$ it is clear that both of them describe unit vectors of the unitary geometry. And the two vectors are orthogonal:

$$\langle\sigma'_n = +1|\sigma'_n = -1\rangle = \left(\cos(\tfrac{1}{2}\vartheta)\, e^{\frac{i}{2}\varphi}\right)\left(-\sin(\tfrac{1}{2}\vartheta)\, e^{-\frac{i}{2}\varphi}\right) + \left(\sin(\tfrac{1}{2}\vartheta)\, e^{-\frac{i}{2}\varphi}\right)\left(\cos(\tfrac{1}{2}\vartheta)\, e^{\frac{i}{2}\varphi}\right) = 0 . \tag{1.12.17}$$

The two wave functions are really the same wave function for, as before, $\sigma'_n = -1$ is $\sigma'_n = +1$ in the $-\boldsymbol{n}$ direction. According to the side and top views

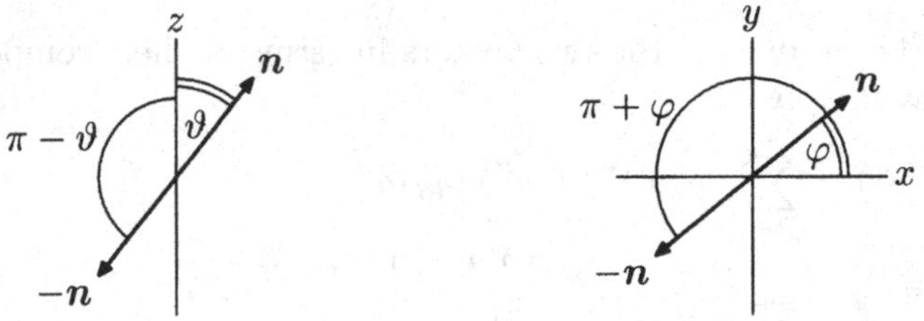

we get $-\boldsymbol{n}$ by the substitution $\vartheta \to \pi - \vartheta$, $\varphi \to \pi + \varphi$, and

$$\psi_{\sigma'_n=+1} \to \begin{pmatrix} \cos\left(\frac{1}{2}(\pi - \vartheta)\right) e^{-\frac{i}{2}(\pi+\varphi)} \\ \sin\left(\frac{1}{2}(\pi - \vartheta)\right) e^{\frac{i}{2}(\pi+\varphi)} \end{pmatrix}$$

$$= i \begin{pmatrix} -\sin(\frac{1}{2}\vartheta)\, e^{-\frac{i}{2}\varphi} \\ \cos(\frac{1}{2}\vartheta)\, e^{\frac{i}{2}\varphi} \end{pmatrix} = i\psi_{\sigma'_n=-1} . \tag{1.12.18}$$

Notice the factor of i. Is that significant? No. The original identification of ψ_{-1} from the product $\psi_{-1}(\sigma'_z)\psi_{-1}(\sigma''_z)^*$ is arbitrary to the extent of a phase factor $e^{i\alpha}$. Any value of α will do. It has no effect upon orthonormality, the statement of the unit symbol, or the physical interpretation in terms of probabilities (see Problem 1-8). The probability amplitude interpretation of a wave function is expressed by

$$p(\sigma'_n, \sigma'_z) = \left|\psi_{\sigma'_n}(\sigma'_z)\right|^2 , \tag{1.12.19}$$

where the results are, of course (ϑ here is the angle between the z direction and $\boldsymbol{n}$):

$$p(+,+) = p(-,-) = \cos^2(\tfrac{1}{2}\vartheta)\ ;$$
$$p(+,-) = p(-,+) = \sin^2(\tfrac{1}{2}\vartheta)\ . \qquad (1.12.20)$$

1.12.2 Two arbitrary components of Pauli's vector operator

It is interesting to use these wave functions to rederive the probabilities associated with two arbitrary directions $\boldsymbol{n}_1$ and $\boldsymbol{n}_2$ (specified by ϑ_1, φ_1 and ϑ_2, φ_2, respectively), according to

$$\langle\sigma_1'|\sigma_2'\rangle = \sum_{\sigma_z'}\langle\sigma_1'|\sigma_z'\rangle\langle\sigma_z'|\sigma_2'\rangle = \sum_{\sigma_z'}\psi_{\sigma_1'}(\sigma_z')^*\psi_{\sigma_2'}(\sigma_z')\ . \qquad (1.12.21)$$

It will suffice to consider

$$\langle +,1|+,2\rangle = \cos(\tfrac{1}{2}\vartheta_1)\cos(\tfrac{1}{2}\vartheta_2)\,\mathrm{e}^{\frac{\mathrm{i}}{2}(\varphi_1-\varphi_2)} + \sin(\tfrac{1}{2}\vartheta_1)\sin(\tfrac{1}{2}\vartheta_2)\,\mathrm{e}^{-\frac{\mathrm{i}}{2}(\varphi_1-\varphi_2)}\ . \qquad (1.12.22)$$

First note two special situations in which $\boldsymbol{n}_1$ and $\boldsymbol{n}_2$ lie in plane.

1. If both are in the x, y plane, we have $\vartheta_1 = \vartheta_2 = \pi/2$ and get

$$\langle +,1|+,2\rangle = \tfrac{1}{2}\mathrm{e}^{\frac{\mathrm{i}}{2}\theta} + \tfrac{1}{2}\mathrm{e}^{-\frac{\mathrm{i}}{2}\theta} = \cos(\tfrac{1}{2}\theta)\ ,$$

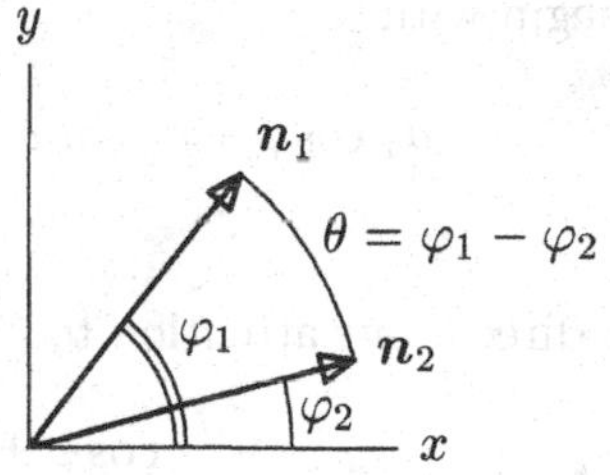

(1.12.23)

   which is a real number and gives the right probability.

2. If both are in the x, z plane, we have $\varphi_1 = \varphi_2 = 0$ and get

$$\langle +,1|+,2\rangle = \cos(\tfrac{1}{2}\vartheta_1)\cos(\tfrac{1}{2}\vartheta_2) + \sin(\tfrac{1}{2}\vartheta_1)\sin(\tfrac{1}{2}\vartheta_2) = \cos(\tfrac{1}{2}\theta)\ ,$$

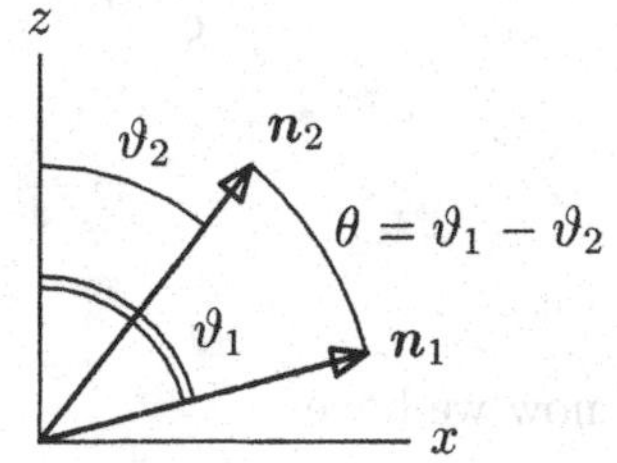

(1.12.24)

   which is a real number and gives the right probability.

We see again that we might get by with real numbers in two dimensions. But, in verifying that we have the right answer in general, we see that complex numbers are essential. Now

$$\begin{aligned}
&\left|\cos(\tfrac{1}{2}\vartheta_1)\cos(\tfrac{1}{2}\vartheta_2)\,\mathrm{e}^{\frac{\mathrm{i}}{2}(\varphi_1-\varphi_2)}+\sin(\tfrac{1}{2}\vartheta_1)\sin(\tfrac{1}{2}\vartheta_2)\,\mathrm{e}^{-\frac{\mathrm{i}}{2}(\varphi_1-\varphi_2)}\right|^2\\
&=\cos^2(\tfrac{1}{2}\vartheta_1)\cos^2(\tfrac{1}{2}\vartheta_2)+\sin^2(\tfrac{1}{2}\vartheta_1)\sin^2(\tfrac{1}{2}\vartheta_2)\\
&\quad+\sin(\tfrac{1}{2}\vartheta_1)\cos(\tfrac{1}{2}\vartheta_1)\sin(\tfrac{1}{2}\vartheta_2)\cos(\tfrac{1}{2}\vartheta_2)\underbrace{\left(\mathrm{e}^{\mathrm{i}(\varphi_1-\varphi_2)}+\mathrm{e}^{-\mathrm{i}(\varphi_1-\varphi_2)}\right)}_{=2\cos(\varphi_1-\varphi_2)}\\
&=\tfrac{1}{2}(1+\cos\vartheta_1)\tfrac{1}{2}(1+\cos\vartheta_2)+\tfrac{1}{2}(1-\cos\vartheta_1)\tfrac{1}{2}(1-\cos\vartheta_2)\\
&\quad+\tfrac{1}{2}\sin\vartheta_1\sin\vartheta_2\cos(\varphi_1-\varphi_2)\\
&=\tfrac{1}{2}\big[1+\underbrace{\cos\vartheta_1\cos\vartheta_2+\sin\vartheta_1\sin\vartheta_2\cos(\varphi_1-\varphi_2)}_{=\cos\theta}\big]=\cos^2(\tfrac{1}{2}\theta)\,. \qquad (1.12.25)
\end{aligned}$$

There's another way of looking at these matters that is suggested by writing out $\boldsymbol{\sigma}\cdot\boldsymbol{n}$ as

$$\boldsymbol{\sigma}\cdot\boldsymbol{n}=(\sigma_x\cos\varphi+\sigma_y\sin\varphi)\sin\vartheta+\sigma_z\cos\vartheta\,. \qquad (1.12.26)$$

We begin with

$$\begin{aligned}
\sigma_x\cos\varphi+\sigma_y\sin\varphi&=\sigma_x\,(\cos\varphi+\mathrm{i}\sigma_z\sin\varphi)\\
&=(\cos\varphi-\mathrm{i}\sigma_z\sin\varphi)\,\sigma_x\,. \qquad (1.12.27)
\end{aligned}$$

That directs our attention to

$$\cos\varphi\pm\mathrm{i}\sigma_z\sin\varphi=\mathrm{e}^{\pm\mathrm{i}\sigma_z\varphi}\,, \qquad (1.12.28)$$

where the use of Euler's* identity involves only the fact that $(\mathrm{i}\sigma_z)^2=-1$, just like $\mathrm{i}^2=-1$. Another approach starts with the exponentiated function of σ_z, and its two possible values:

$$\begin{aligned}
\mathrm{e}^{\pm\mathrm{i}\sigma_z\varphi}&=\underbrace{\frac{1+\sigma_z}{2}}_{=|{+}{+}|}\mathrm{e}^{\mathrm{i}\varphi}+\underbrace{\frac{1-\sigma_z}{2}}_{=|{-}{-}|}\mathrm{e}^{-\mathrm{i}\varphi}\\
&=\cos\varphi\pm\mathrm{i}\sigma_z\sin\varphi\,. \qquad (1.12.29)
\end{aligned}$$

So now we have

$$\sigma_x\cos\varphi+\sigma_y\sin\varphi=\sigma_x\,\mathrm{e}^{\mathrm{i}\varphi\sigma_z}=\mathrm{e}^{-\mathrm{i}\varphi\sigma_z}\sigma_x\,, \qquad (1.12.30)$$

*Leonhard EULER (1707–1783)

where the two forms on the right are connected by the anticommutativity of σ_x and σ_z. We again apply this property in getting a more symmetrical form:

$$\begin{aligned} \sigma_x \, e^{i\varphi\sigma_z} &= \sigma_x \, e^{\frac{i}{2}\varphi\sigma_z} \, e^{\frac{i}{2}\varphi\sigma_z} = e^{-\frac{i}{2}\varphi\sigma_z} \sigma_x \, e^{\frac{i}{2}\varphi\sigma_z} \,, \\ e^{-i\varphi\sigma_z} \sigma_x &= e^{-\frac{i}{2}\varphi\sigma_z} \, e^{-\frac{i}{2}\varphi\sigma_z} \sigma_x = e^{-\frac{i}{2}\varphi\sigma_z} \sigma_x \, e^{\frac{i}{2}\varphi\sigma_z} \,, \end{aligned} \tag{1.12.31}$$

so that

$$\sigma_x \cos\varphi + \sigma_y \sin\varphi = e^{-\frac{i}{2}\varphi\sigma_z} \sigma_x \, e^{\frac{i}{2}\varphi\sigma_z} \tag{1.12.32}$$

and

$$\boldsymbol{\sigma}\cdot\boldsymbol{n} = e^{-\frac{i}{2}\varphi\sigma_z} (\sigma_z \cos\vartheta + \sigma_x \sin\vartheta) \, e^{\frac{i}{2}\varphi\sigma_z} \tag{1.12.33}$$

since

$$e^{-\frac{i}{2}\varphi\sigma_z} \sigma_z \, e^{\frac{i}{2}\varphi\sigma_z} = \sigma_z \, e^{-\frac{i}{2}\varphi\sigma_z} \, e^{\frac{i}{2}\varphi\sigma_z} = \sigma_z \,. \tag{1.12.34}$$

Notice that we have made free use of exponential properties of the form

$$e^{i\lambda_1 A} \, e^{i\lambda_2 A} = e^{i(\lambda_1 + \lambda_2)A} \,, \tag{1.12.35}$$

which we justify by

$$\begin{aligned} e^{i\lambda_1 A} \, e^{i\lambda_2 A} &= \sum_{a'} e^{i\lambda_1 a'} |a'a'| \sum_{a''} e^{i\lambda_2 a''} |a''a''| \\ &= \sum_{a'} e^{i(\lambda_1 + \lambda_2)a'} |a'a'| = e^{i(\lambda_1 + \lambda_2)A} \,. \end{aligned} \tag{1.12.36}$$

In a analogous way: $\varphi \to \vartheta$, $\sigma_x \to \sigma_z$, $\sigma_y \to \sigma_x$, $\sigma_z \to \sigma_y$, we have

$$\sigma_z \cos\vartheta + \sigma_x \sin\vartheta = e^{-\frac{i}{2}\vartheta\sigma_y} \sigma_z \, e^{\frac{i}{2}\vartheta\sigma_y} \tag{1.12.37}$$

and

$$\boldsymbol{\sigma}\cdot\boldsymbol{n} = e^{-\frac{i}{2}\varphi\sigma_z} \, e^{-\frac{i}{2}\vartheta\sigma_y} \sigma_z \, e^{\frac{i}{2}\vartheta\sigma_y} \, e^{\frac{i}{2}\varphi\sigma_z} = U^{-1}\sigma_z U \tag{1.12.38}$$

where

$$U = e^{\frac{i}{2}\vartheta\sigma_y} \, e^{\frac{i}{2}\varphi\sigma_z} \,, \quad U^{-1} = e^{-\frac{i}{2}\varphi\sigma_z} \, e^{-\frac{i}{2}\vartheta\sigma_y} \tag{1.12.39}$$

are indeed inverse since

$$\begin{aligned} UU^{-1} &= e^{\frac{i}{2}\vartheta\sigma_y} \, e^{\frac{i}{2}\varphi\sigma_z} \, e^{-\frac{i}{2}\varphi\sigma_z} \, e^{-\frac{i}{2}\vartheta\sigma_y} = 1 \,, \\ U^{-1}U &= e^{-\frac{i}{2}\varphi\sigma_z} \, e^{-\frac{i}{2}\vartheta\sigma_y} \, e^{\frac{i}{2}\vartheta\sigma_y} \, e^{\frac{i}{2}\varphi\sigma_z} = 1 \,, \end{aligned} \tag{1.12.40}$$

which illustrates the general statement

$$(XY)^{-1} = Y^{-1}X^{-1} . \tag{1.12.41}$$

Before commenting on this relationship between different physical quantities, let's supplement it by a connection between vectors. Write

$$\sigma_z = \sum_{\sigma'} |\sigma'_z = \sigma'\rangle\sigma'\langle\sigma'_z = \sigma'| \tag{1.12.42}$$

and get

$$\begin{aligned}\boldsymbol{\sigma}\cdot\boldsymbol{n} &= \sum_{\sigma'} U^{-1}|\sigma'_z = \sigma'\rangle\sigma'\langle\sigma'_z = \sigma'|U \\ &= \sum_{\sigma'} |\sigma'_n = \sigma'\rangle\sigma'\langle\sigma'_n = \sigma'| ,\end{aligned} \tag{1.12.43}$$

so that

$$\begin{aligned}\langle\sigma'_n = \sigma'| &= \langle\sigma'_z = \sigma'|U , \\ |\sigma'_n = \sigma'\rangle &= U^{-1}|\sigma'_z = \sigma'\rangle .\end{aligned} \tag{1.12.44}$$

The necessary adjoint relation between these left and right vectors tells us that

$$(\langle\sigma'_z|U)^\dagger = U^{-1}|\sigma'_z\rangle , \tag{1.12.45}$$

where the left-hand side can equivalently be presented as

$$U^\dagger\langle\sigma'_z|^\dagger = U^\dagger|\sigma'_z\rangle , \tag{1.12.46}$$

so that

$$U^\dagger = U^{-1} , \quad U^\dagger U = UU^\dagger = 1 . \tag{1.12.47}$$

This one checks directly, first by noting that

$$\begin{aligned}\left(e^{\frac{i}{2}\varphi\sigma_z}\right)^\dagger &= \left(\cos\frac{\varphi}{2} + i\sigma_z\sin\frac{\varphi}{2}\right)^\dagger \\ &= \cos\frac{\varphi}{2} - i\sigma_z\sin\frac{\varphi}{2} \\ &= e^{-\frac{i}{2}\varphi\sigma_z}\end{aligned} \tag{1.12.48}$$

which, once worked through, is seen to be just $i \to -i$. Accordingly,

$$U^\dagger = \left(e^{\frac{i}{2}\vartheta\sigma_y}\, e^{\frac{i}{2}\varphi\sigma_z}\right)^\dagger = e^{-\frac{i}{2}\varphi\sigma_z}\, e^{-\frac{i}{2}\vartheta\sigma_y} = U^{-1} , \tag{1.12.49}$$

as required. The same property of U guarantees that, like σ_z, $\boldsymbol{\sigma}\cdot\boldsymbol{n}$ is Hermitian ($U^{\dagger\dagger} = U$):

$$(U^\dagger \sigma_z U)^\dagger = U^\dagger \sigma_z U\,. \tag{1.12.50}$$

And notice how the requirement $(\boldsymbol{\sigma}\cdot\boldsymbol{n})^2 = 1$ is satisfied

$$(\boldsymbol{\sigma}\cdot\boldsymbol{n})^2 = U^{-1}\sigma_z U U^{-1}\sigma_z U = U^{-1} 1 U = 1\,. \tag{1.12.51}$$

1.13 Unitary operators

We express what we have recognized here more generally. Property A has the possible values $a_1, a_2, \ldots, a_n$. Property B also has the same possible values $b_1 = a_1, \ldots, b_n = a_n$. Now consider

$$\begin{aligned} U_{ab} &= \sum_{k=1}^{n} |a_k\rangle\langle b_k|\,, & U_{ba} &= \sum_k |b_k\rangle\langle a_k|\,, \\ U_{ab}^\dagger &= U_{ba}\,, & U_{ba}^\dagger &= U_{ab}\,, \end{aligned} \tag{1.13.1}$$

for which

$$\begin{aligned} U_{ab}U_{ba} &= \sum_{k,l} |a_k\rangle \underbrace{\langle b_k|b_l\rangle}_{=\delta(k,l)} \langle a_l| = \sum_k |a_k\rangle\langle a_k| = 1\,, \\ U_{ba}U_{ab} &= \sum_{k,l} |b_k\rangle\langle a_k|a_l\rangle\langle b_l| = 1\,, \end{aligned} \tag{1.13.2}$$

so that

$$U_{ab}^\dagger = U_{ba} = U_{ab}^{-1}\,, \quad U_{ba}^\dagger = U_{ab} = U_{ba}^{-1}\,. \tag{1.13.3}$$

Also,

$$\begin{aligned} \langle a_k|U_{ab} &= \langle b_k|\,, & U_{ab}|b_k\rangle &= |a_k\rangle\,, \\ \langle b_k|U_{ba} &= \langle a_k|\,, & U_{ba}|a_k\rangle &= |b_k\rangle\,. \end{aligned} \tag{1.13.4}$$

The U symbols act, or operate, on one orthonormal set of vectors to produce the other orthonormal set. They maintain the *metrical* relations of the unitary geometry:

$$\langle b_k|b_l\rangle = \langle a_k|U_{ab}U_{ba}|a_l\rangle = \langle a_k|a_l\rangle = \delta_{kl} \tag{1.13.5}$$

and so are called unitary operators. Indeed the term operator is applied, mathematically, to every element X of the measurement algebra:

$$X = \sum_{a',a''} \langle a'|X|a''\rangle |a'a''|\,, \tag{1.13.6}$$

which can be thought of as operating on any vector to produce another vector (of the same kind):

$$\langle a'|X = \sum_{a''} \langle a'|X|a''\rangle\langle a''| ,$$
$$X|a''\rangle = \sum_{a'} |a'\rangle\langle a'|X|a''\rangle . \qquad (1.13.7)$$

The word 'operator' is frequently used as synonym for 'measurement symbol' – we have done so above when speaking of Pauli operators, for instance. In particular the Hermitian operator

$$A = \sum |a'\rangle a'\langle a'| \qquad (1.13.8)$$

operates on an a vector to produce a multiple of the same vector (as we know):

$$\langle a'|A = a'\langle a'| , \quad A|a'\rangle = |a'\rangle a' ; \qquad (1.13.9)$$

such vectors are termed eigenvectors of the operator. The numbers a' are correspondingly called eigenvalues.

The Hermitian operator B, the symbol of physical property B, is

$$\begin{aligned} B &= \sum_k |b_k\rangle b_k\langle b_k| \\ &= \sum_k U_{ba}|a_k\rangle a_k\langle a_k|U_{ab} \\ &= U_{ba}AU_{ab} \end{aligned} \qquad (1.13.10)$$

indicating the manner in which U_{ab} operates on A to produce B.

The totality of b vectors is expressed in terms of the a vectors by, for example,

$$|b_l\rangle = \sum_k |a_k\rangle\langle a_k|b_l\rangle \qquad (1.13.11)$$

where the set of n^2 numbers, $\langle a_k|b_l\rangle$, which are generalizations of direction cosines, is called the ab transformation function. That is a mathematical term. Physically these are probability amplitudes or wave functions. They are also the matrix elements

$$\langle a_k|b_l\rangle = \langle a_k|U_{ba}|a_l\rangle . \qquad (1.13.12)$$

We illustrate this for the example of a Stern–Gerlach measurement, where

$$A = \sigma_z , \quad B = \boldsymbol{\sigma}\cdot\boldsymbol{n} , \quad U_{ba} = U_{ab}^{-1} = \mathrm{e}^{-\frac{\mathrm{i}}{2}\varphi\sigma_z}\,\mathrm{e}^{-\frac{\mathrm{i}}{2}\vartheta\sigma_y} , \qquad (1.13.13)$$

so that

$$
\begin{aligned}
\langle\sigma'_z|\sigma''_n\rangle &= \langle\sigma'_z|\,\mathrm{e}^{-\frac{\mathrm{i}}{2}\varphi\sigma_z}\,\mathrm{e}^{-\frac{\mathrm{i}}{2}\vartheta\sigma_y}|\sigma''_z=\sigma''_n\rangle\,, \\
&= \mathrm{e}^{-\frac{\mathrm{i}}{2}\varphi\sigma'_z}\langle\sigma'_z|\left(\cos\frac{\vartheta}{2}-\mathrm{i}\sigma_y\sin\frac{\vartheta}{2}\right)|\sigma''_z\rangle \\
&\quad \sigma'_z\backslash\sigma''_n=\sigma''_z \\
&= \begin{pmatrix} \cos\frac{\vartheta}{2}\,\mathrm{e}^{-\frac{\mathrm{i}}{2}\varphi} & -\sin\frac{\vartheta}{2}\,\mathrm{e}^{-\frac{\mathrm{i}}{2}\varphi} \\ \sin\frac{\vartheta}{2}\,\mathrm{e}^{\frac{\mathrm{i}}{2}\varphi} & \cos\frac{\vartheta}{2}\,\mathrm{e}^{\frac{\mathrm{i}}{2}\varphi} \end{pmatrix}. \qquad (1.13.14)
\end{aligned}
$$

The first column of this matrix is the wave function $\psi_{\sigma_n=+1}(\sigma'_z)$, the second column is $\psi_{\sigma_n=-1}(\sigma'_z)$, consistent with (1.12.6).

1.14 Unitary operator bases. Complementarity

Thus far we have had only one physical example before us – the two-valued magnetic moment of the Stern–Gerlach experiment. The time has come to break out of that beachhead. And we shall do it by examining some apparently innocuous questions.

A unitary operator converts one orthonormal set of vectors into another such set, or, possibly the same set. Certainly unity is a unitary operator:

$$1^\dagger 1 = 11 = 1 \qquad (1.14.1)$$

and indeed

$$\langle a'|1 = \langle a'|\,, \quad 1|a'\rangle = |a'\rangle\,. \qquad (1.14.2)$$

Now suppose we consider a unitary operator that produces the same set of a vectors but in a different order, as produced by numbering them: $a_1, a_2, \ldots, a_n$, and then, for example, cyclically permuting them:

$$
\begin{aligned}
U|a_k\rangle &= |a_{k+1}\rangle\,, \\
U|a_n\rangle &= |a_1\rangle\,.
\end{aligned}
\qquad (1.14.3)
$$

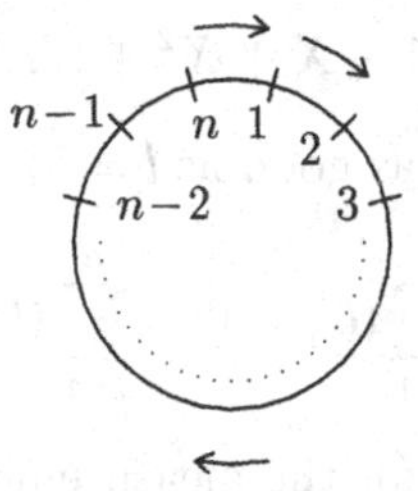

We have the simplest example of this in σ_x:

$$\sigma_x|-\rangle = |+\rangle\,, \quad \sigma_x|+\rangle = |-\rangle\,. \qquad (1.14.4)$$

In this example, repetition gives unity, $\sigma_x^2 = 1$. Now

$$U^2\big|a_k\big\rangle = U\big|a_{k+1}\big\rangle = \big|a_{k+2}\big\rangle \tag{1.14.5}$$

and

$$U^n\big|a_k\big\rangle = \big|a_{k+n}\big\rangle = \big|a_k\big\rangle\,, \tag{1.14.6}$$

where the index k is understood modulo n,

$$\begin{aligned} n+1 &\,\widehat{=}\, n+1 \ (\mathrm{mod}\ n) = 1\,,\\ n+2 &\,\widehat{=}\, n+2 \ (\mathrm{mod}\ n) = 2\,,\\ &\text{et cetera.} \end{aligned} \tag{1.14.7}$$

We have

$$U^n = 1\,; \tag{1.14.8}$$

n is called the period of U. Let's think about the numbers u' that satisfy (1.14.8):

$$(u')^n = 1\,, \quad u' = \mathrm{e}^{2\pi\mathrm{i}\frac{k}{n}}\,; \qquad k = 1, 2, \ldots, n \tag{1.14.9}$$

so that the various u' are the n different n^{th} roots of unity (for $n = 2$, e. g., the two roots are $u' = \mathrm{e}^{2\pi\mathrm{i}\frac{1}{2}} = -1$, $\mathrm{e}^{2\pi\mathrm{i}\frac{2}{2}} = 1$). What we are saying is that U obeys the algebraic equation

$$U^n - 1 = \prod_{k=1}^{n}(U - u_k) = 0 \quad \text{with} \quad u_k = \mathrm{e}^{2\pi\mathrm{i}\frac{k}{n}}\,. \tag{1.14.10}$$

The factorization of $U^n - 1$ into a particular $u - u_k$ and all the rest is done this way:

$$U^n - 1 = (U/u_k)^n - 1 = (U/u_k - 1)\sum_{l=0}^{n-1}(U/u_k)^l \tag{1.14.11}$$

which makes use of the familiar identity

$$(X-1)\left(1 + X + X^2 + \ldots X^{n-1}\right) = X^n - 1\,. \tag{1.14.12}$$

This result ($l = n$ is as good as $l = 0$)

$$(U - u_k)\sum_{l=1}^{n}(U/u_k)^l = \sum_{l=1}^{n}(U/u_k)^l\,(U - u_k) = 0 \tag{1.14.13}$$

suggests the symbol for the measurement of U that yields the value u_k:

$$\big|u_k u_k\big| = \frac{1}{n}\sum_{l=1}^{n}(U/u_k)^l\,; \tag{1.14.14}$$

indeed, $\big|u_k u_k\big| \to 1$ for $U \to u_k$, as it should.

It is important to note that this operator *is* Hermitian:

$$\begin{aligned}|u_k u_k|^\dagger &= \frac{1}{n}\sum_l \left(U^\dagger/u_k^*\right)^l = \frac{1}{n}\sum_l \left(U/u_k\right)^{-l}\\ &= \frac{1}{n}\sum_{l=0}^{n-1}\left(U/u_k\right)^{n-l} = \frac{1}{n}\sum_{l=1}^{n}\left(U/u_k\right)^l\\ &= \left|u_k u_k\right| . \end{aligned} \tag{1.14.15}$$

It is clear that the algebra here is quite the same as with a Hermitian operator and indeed a unitary operator can be considered to be a complex valued function of a Hermitian operator, as in $\mathrm{e}^{\frac{\mathrm{i}}{2}\varphi\sigma_z}$.

Now consider

$$\left|u_k\right\rangle\left\langle u_k\middle|a_n\right\rangle = \frac{1}{n}\sum_{l=1}^{n} u_k^{-l}\underbrace{U^l\left|a_n\right\rangle}_{=\left|a_l\right\rangle} \tag{1.14.16}$$

with the consequence

$$\left\langle a_n\middle|u_k\right\rangle\left\langle u_k\middle|a_n\right\rangle = \left|\left\langle u_k\middle|a_n\right\rangle\right|^2 = \frac{1}{n}u_k^{-n} = \frac{1}{n} . \tag{1.14.17}$$

We choose $\left\langle u_k\middle|a_n\right\rangle = 1/\sqrt{n}$ and get, consistently,

$$\left|u_k\right\rangle = \frac{1}{\sqrt{n}}\sum_{l=1}^{n}\left|a_l\right\rangle \mathrm{e}^{-\frac{2\pi\mathrm{i}}{n}kl} \tag{1.14.18}$$

and

$$\left\langle u_k\right| = \frac{1}{\sqrt{n}}\sum_{l=1}^{n}\mathrm{e}^{\frac{2\pi\mathrm{i}}{n}kl}\left\langle a_l\right| , \tag{1.14.19}$$

for

$$\left\langle u_k\middle|a_l\right\rangle = \frac{1}{\sqrt{n}}\mathrm{e}^{\frac{2\pi\mathrm{i}}{n}kl} ; \quad = \frac{1}{\sqrt{n}} \text{ for } l = n . \tag{1.14.20}$$

Now that we have another set of n orthonormal vectors, let's define a second unitary operator, V, by cyclically permuting them:

$$\begin{aligned}\left\langle u_k\right|V &= \left\langle u_{k+1}\right| ,\\ \left\langle u_k\right|V^2 &= \left\langle u_{k+2}\right| ,\\ &\vdots\\ \left\langle u_k\right|V^m &= \left\langle u_{k+m}\right| ,\\ &\vdots\\ \left\langle u_k\right|V^n &= \left\langle u_k\right| . \end{aligned} \tag{1.14.21}$$

Again

$$V^n = 1 \quad \text{and} \quad v_l = \mathrm{e}^{2\pi \mathrm{i} \frac{l}{n}}\;, \quad l = 1, \ldots n \tag{1.14.22}$$

so that

$$\big|v_l v_l\big| = \frac{1}{n} \sum_{k=1}^{n} (V/v_l)^k\;, \tag{1.14.23}$$

leading to

$$\big\langle u_n \big| v_l \big\rangle \big\langle v_l \big| = \frac{1}{n} \sum_{k=1}^{n} \mathrm{e}^{-\frac{2\pi\mathrm{i}}{n} kl} \big\langle u_k \big|\,, \tag{1.14.24}$$

and then $\big|\big\langle u_n \big| v_l \big\rangle\big|^2 = 1/n$. With $\big\langle u_n \big| v_l \big\rangle = 1/\sqrt{n}$ we get

$$\begin{aligned} \big\langle v_l \big| &= \frac{1}{\sqrt{n}} \sum_{k=1}^{n} \mathrm{e}^{-\frac{2\pi\mathrm{i}}{n} kl} \big\langle u_k \big|\,, \\ \big| v_l \big\rangle &= \frac{1}{\sqrt{n}} \sum_{k=1}^{n} \big| u_k \big\rangle\, \mathrm{e}^{\frac{2\pi\mathrm{i}}{n} kl}\,. \end{aligned} \tag{1.14.25}$$

However,

$$\frac{1}{\sqrt{n}}\, \mathrm{e}^{\frac{2\pi\mathrm{i}}{n} kl} = \big\langle u_k \big| a_l \big\rangle \tag{1.14.26}$$

and therefore

$$\big| v_l \big\rangle = \sum_{k=1}^{n} \big| u_k \big\rangle \big\langle u_k \big| a_l \big\rangle = \big| a_l \big\rangle\;! \tag{1.14.27}$$

We are led back to the initial set of vectors.

What we have, then, is the reciprocal definition of two unitary operators,

$$U \big| v_l \big\rangle = \big| v_{l+1} \big\rangle\,, \quad \big\langle u_k \big| V = \big\langle u_{k+1} \big|\,. \tag{1.14.28}$$

These unitary operators obey

$$U^n = 1\,, \quad V^n = 1\,, \tag{1.14.29}$$

and one more relation derived, for example, by considering

$$\begin{aligned} \big\langle u_k \big| VU &= \big\langle u_{k+1} \big| U = \mathrm{e}^{\frac{2\pi\mathrm{i}}{n}(k+1)} \big\langle u_{k+1} \big|\,, \\ \big\langle u_k \big| UV &= \mathrm{e}^{\frac{2\pi\mathrm{i}k}{n}} \big\langle u_k \big| V = \mathrm{e}^{\frac{2\pi\mathrm{i}}{n}k} \big\langle u_{k+1} \big|\,, \end{aligned} \tag{1.14.30}$$

namely

$$VU = \mathrm{e}^{\frac{2\pi\mathrm{i}}{n}} UV\,. \tag{1.14.31}$$

We also have

$$\langle u_k | v_l \rangle = \frac{1}{\sqrt{n}} \mathrm{e}^{\frac{2\pi\mathrm{i}}{n} kl}\,, \tag{1.14.32}$$

and the same exponential occurs in

$$V^l U^k = \mathrm{e}^{\frac{2\pi\mathrm{i}}{n} kl} U^k V^l\,, \tag{1.14.33}$$

which we get, either by repetition, or from the comparison:

$$\begin{aligned} \langle u_m | V^l U^k &= \langle u_{m+l} | U^k = \mathrm{e}^{\frac{2\pi\mathrm{i}}{n} k(m+l)} \langle u_{m+l} |\,, \\ \langle u_m | U^k V^l &= \mathrm{e}^{\frac{2\pi\mathrm{i}}{n} km} \langle u_m | V^l = \mathrm{e}^{\frac{2\pi\mathrm{i}}{n} km} \langle u_{m+l} |\,. \end{aligned} \tag{1.14.34}$$

For $n = 2$, we have

$$U^2 = 1\,, \quad V^2 = 1\,, \quad UV = -VU \tag{1.14.35}$$

which we recognize as the properties of σ_x and σ_y (or any other pair); these Hermitian operators are also unitary: $\sigma_x^\dagger \sigma_x = (\sigma_x)^2 = 1$. In the progression from $n = 2$ to $n = \infty$, we go from anticommuting U and V to commuting U and V. Commutativity, characteristic of ordinary numbers, is the hall mark of *classical* physics. Indeed, from our viewpoint, in which atomic mechanics, quantum mechanics, is fundamental, we shall *derive* the characteristics of the physical properties that are recognized in the classical limit.

The four basic operators of $n = 2$ are constructed from the fundamental U, V pair as $U^k V^l$, $k, l = 1, 2$:

$$\begin{aligned} U^1 V^1 &= \sigma_x \sigma_y = \mathrm{i}\sigma_z\,, \\ U^1 V^2 &= \sigma_x\,, \\ U^2 V^1 &= \sigma_y\,, \\ U^2 V^2 &= 1\,. \end{aligned} \tag{1.14.36}$$

For general n, $U^k V^l$ with $k, l = 1 \ldots n$, gives n^2 operators, again just the number of measurement symbols. Can we exhibit each of the measurement symbols, say $|u'u''|$ in terms of the $U^k V^l$? We do it by beginning with

$$|u_k\rangle\langle u_k| = \frac{1}{n} \sum_l (U/u_k)^l = \frac{1}{n} \sum_l \mathrm{e}^{-\frac{2\pi\mathrm{i}}{n} kl} U^l\,. \tag{1.14.37}$$

Then multiply on the right by V^{m-k} and get

$$\left|u_k u_m\right| = \frac{1}{n}\sum_l \mathrm{e}^{-\frac{2\pi \mathrm{i}}{n}kl} U^l V^{m-k} , \tag{1.14.38}$$

which also applies for $m < k$ since $V^{-1} = V^{n-1}$, for example. And any operator, say F, which is a linear combination of the n^2 measurement symbols $\left|u'u''\right|$ is also a linear combination of the n^2 unitary operators $U^k V^l$:

$$F = \sum_{k,l=1}^{n} f_{kl} U^k V^l = f(U,V) , \tag{1.14.39}$$

it is a function of the fundamental pair U and V.

Although we can think of U and V as complex valued functions of two physical properties symbolized by Hermitian operators, it is more direct to work with U and V. And so we have the probability $p(u',v')$ that the particular outcome v' will come from a V measurement performed on atoms that have been selected to have the value u' of U. It is $(u' = u_k, v' = v_l)$

$$p(u',v') = \left|\frac{1}{\sqrt{n}}\,\mathrm{e}^{\frac{2\pi \mathrm{i}}{n}kl}\right|^2 = \frac{1}{n} \tag{1.14.40}$$

independent of the particular choice of u' and v'. This statistical aspect of U and V is emphasized by considering a measurement that includes a non-selective measurement:

$$p(u',v,u'') = \sum_{v'} p(u',v')p(v',u'') = \sum_{v'} \frac{1}{n}\frac{1}{n} = \frac{1}{n} , \tag{1.14.41}$$

as compared with

$$p(u',1,u'') = p(u',u'') = \delta(u',u'') ; \tag{1.14.42}$$

the intervening non-selective v measurement has removed all knowledge of the initial u' value – all outcomes are equally probable.

Physical properties whose measurements interfere are called incompatible; U and V display *optimal* incompatibility. That, and the fact that both U and V are needed in the description of the system is what is meant by calling U and V *complementary* properties: Both are required, but the measurement of one wipes out any prior knowledge of the other. Yes, the essence of Bohr's* complementarity (1927) is just that: All quantum objects must possess mutually exclusive properties. That's quantum mechanics.

Here's a nice way to evaluate the trace of an operator F given as a function of U and V:

$$F = f(U,V) = \sum_{k,l} f_{kl} U^k V^l . \tag{1.14.43}$$

*Niels Henrik David BOHR (1885–1962)

Take the matrix element with u' row and v' column:

$$\langle u'|F|v'\rangle = \langle u'|\sum_{k,l} f_{kl}U^k V^l|v'\rangle = f(u',v')\langle u'|v'\rangle \tag{1.14.44}$$

and then

$$\begin{aligned}\operatorname{tr}\{F\} &= \sum_{u'}\langle u'|F|u'\rangle = \sum_{u',v'}\langle u'|F|v'\rangle\langle v'|u'\rangle \\ &= \sum_{u',v'} f(u',v')\left|\langle u'|v'\rangle\right|^2\end{aligned} \tag{1.14.45}$$

or

$$\operatorname{tr}\{f(U,V)\} = \frac{1}{n}\sum_{u',v'} f(u',v')\,. \tag{1.14.46}$$

If F is a physical property, or a function of one, we have

$$F = \sum_{f'} |f'\rangle f'\langle f'| \tag{1.14.47}$$

and, since $\operatorname{tr}\left\{|f'f'|\right\} = 1$,

$$\operatorname{tr}\{F\} = \sum_{f'} f'\,, \tag{1.14.48}$$

that is

$$\sum_{f'} f' = \frac{1}{n}\sum_{u'v'} f(u',v')\,. \tag{1.14.49}$$

Here, then, the sum of all physical values of $f(U,V)$ is given by a summation over all the numbers, $f(u',v')$, that are produced by independent assignment of values to U and V: an *Ergodic Theorem*.

As a preliminary to a trace evaluation, note that

$$\begin{aligned}\sum_l \langle u_k|v_l\rangle\langle v_l|u_m\rangle = \delta_{km} &= \frac{1}{n}\sum_l \mathrm{e}^{\frac{2\pi\mathrm{i}}{n}kl}\,\mathrm{e}^{-\frac{2\pi\mathrm{i}}{n}lm} \\ &= \frac{1}{n}\sum_l \left(\mathrm{e}^{\frac{2\pi\mathrm{i}l}{n}}\right)^{k-m}\end{aligned} \tag{1.14.50}$$

or, equivalently,

$$\frac{1}{n}\sum_{u'} (u')^r = \begin{cases} 1 & \text{for} \quad r = 0\,, \\ 0 & \text{for} \quad 0 < r < n\,, \end{cases} \tag{1.14.51}$$

which is just a property of the nth roots of unity, $u_k = \mathrm{e}^{\frac{2\pi\mathrm{i}}{n}k}$. Now consider

$$\operatorname{tr}\left\{U^k V^l\right\} = \frac{1}{n}\sum_{u',v'} (u')^k (v')^l . \tag{1.14.52}$$

Clearly, for $k = l = n$, we get

$$\operatorname{tr}\{1\} = n , \tag{1.14.53}$$

which is (1.11.9), but for any other choice:

$$\underbrace{k = 1, \dots , n-1 \quad \text{or} \quad l = 1, \dots , n-1}_{n^2 - 1 \text{ in number}} \tag{1.14.54}$$

we have

$$\operatorname{tr}\left\{U^k V^l\right\} = 0 . \tag{1.14.55}$$

This is the generalization of the $n = 2$ property (1.11.7).

1.15 Quantum degrees of freedom

For each choice of $n = 2, 3, 4, \dots$ we get a different pair of complementary quantities, a different physical system. However, for $n = 4 = 2 \times 2$, or $n = 6 = 2 \times 3$, generally $n = n_1 n_2$, it is possible to consider the system to be a composite of two simpler ones characterized by n_1 and n_2. A step in this direction comes from considering

$$\begin{aligned} U_1 &= U^{n_2} , & U_2 &= U^{n_1} , \\ V_1 &= V^{l_1 n_2} , & V_2 &= V^{l_2 n_1} , \end{aligned} \tag{1.15.1}$$

where l_1, l_2 are integers such that

$$\begin{aligned} U_1^{n_1} &= U^n = 1 , & U_2^{n_2} &= U^n = 1 , \\ V_1^{n_1} &= V^{l_1 n} = 1 , & V_2^{n_2} &= V^{l_2 n} = 1 , \end{aligned} \tag{1.15.2}$$

as well as, e. g.,

$$V_1 U_2 = V^{l_1 n_2} U^{n_1} = \mathrm{e}^{\frac{2\pi\mathrm{i}}{n} l_1 n} U_2 V_1 = U_2 V_1 \tag{1.15.3}$$

and

$$V_1 U_1 = V^{l_1 n_2} U^{n_2} = \mathrm{e}^{\frac{2\pi\mathrm{i}}{n} l_1 n_2^2} U_1 V_1 = \mathrm{e}^{\frac{2\pi\mathrm{i}}{n_1}} U_1 V_1 , \tag{1.15.4}$$

provided $l_1 n_2 = 1 \pmod{n_1}$. There is a unique choice of l_1 and l_2, if n_1 and n_2 have no common factors – for $n_1 = 2$, $n_2 = 3$, for instance, one gets $l_1 = 1$,

$l_2 = 2$. However, this does not cover $4 = 2 \times 2$, say. Instead we proceed this way: replace the linear counting $k = 1, \dots, n$ by the rectangular counting,

$$\begin{array}{lcccc} k_1\backslash k_2 & & & & \\ & 1,1 & 1,2 & \dots & 1,n_2 \\ & 2,1 & 2,2 & \dots & 2,n_2 \\ & \vdots & \vdots & & \vdots \\ & n_1,1 & 1,2 & \dots & n_1,n_2 \end{array} \tag{1.15.5}$$

where n_1 rows and n_2 columns account for $n_1 n_2 = n$ pairs k_1, k_2. Then we can introduce operators U_1, V_1 that permute the rows, and independently, operators U_2, V_2 that permute the columns. With

$$u_{1k_1} = \mathrm{e}^{\frac{2\pi \mathrm{i}}{n_1} k_1} = v_{1k_1}\,; \quad u_{2k_2} = \mathrm{e}^{\frac{2\pi \mathrm{i}}{n_2} k_2} = v_{2k_2} \tag{1.15.6}$$

we get the reciprocal definitions:

$$\begin{aligned} \langle u_{1k_1}, u_{2k_2}| V_1 &= \langle u_{1k_1+1}, u_{2k_2}| \,, \\ \langle u_{1k_1}, u_{2k_2}| V_2 &= \langle u_{1k_1}, u_{2k_2+1}| \,, \end{aligned} \tag{1.15.7}$$

and

$$\begin{aligned} U_1 |v_{1k_1}, v_{2k_2}\rangle &= |v_{1k_1+1}, v_{2k_2}\rangle \,, \\ U_2 |v_{1k_1}, v_{2k_2}\rangle &= |v_{1k_1}, v_{2k_2+1}\rangle \,. \end{aligned} \tag{1.15.8}$$

Within each set, U_1, V_1 and U_2, V_2, all is as before,

$$\begin{aligned} U_1^{n_1} = V_1^{n_1} &= 1 \,, & U_2^{n_2} = V_2^{n_2} &= 1 \,, \\ V_1 U_1 &= \mathrm{e}^{\frac{2\pi \mathrm{i}}{n_1}} U_1 V_1 \,, & V_2 U_2 &= \mathrm{e}^{\frac{2\pi \mathrm{i}}{n_2}} U_2 V_2 \,. \end{aligned} \tag{1.15.9}$$

The operators U_1, V_1 act on one index of the vectors; the operators U_2, V_2 act on the other index. It does not matter in which order the two different operator sets act, e. g.,

$$\begin{aligned} \langle u_{1k_1}, u_{2k_2}| U_1 V_2 &= u_{1k_1} \langle u_{1k_1}, u_{2k_2}| V_2 = u_{1k_1} \langle u_{1k_1}, u_{2k_2+1}| \,, \\ \langle u_{1k_1}, u_{2k_2}| V_2 U_1 &= \langle u_{1k_1}, u_{2k_2+1}| U_1 = u_{1k_1} \langle u_{1k_1}, u_{2k_2+1}| \,. \end{aligned} \tag{1.15.10}$$

So

$$\begin{aligned} U_1 U_2 &= U_2 U_1 \,, & V_1 V_2 &= V_2 V_1 \,, \\ U_1 V_2 &= V_2 U_1 \,, & V_1 U_2 &= U_2 V_1 \,. \end{aligned} \tag{1.15.11}$$

Two physical properties such that the measurement of one does not alter prior knowledge of the other, in which the sequence of measurement does not matter – as represented by the commutativity of the associated symbols –

are called compatible. The pair of operators U_1, V_1 is compatible with the operator pair U_2, V_2. It's a useful fiction to think of U_1, V_1 and U_2, V_2 as describing two independent systems, which can be created individually, in either order, or individually destroyed:

$$\begin{aligned} \langle u_1', u_2'| &= \langle u_1'|\langle u_2'| = \langle u_2'|\langle u_1'| \,, \\ |v_1'', v_2''\rangle &= |v_1''\rangle|v_2''\rangle = |v_2''\rangle|v_1''\rangle \,, \end{aligned} \tag{1.15.12}$$

and so

$$\langle u_1', u_2'|v_1'', v_2''\rangle = \langle u_1'|v_1''\rangle\langle u_2'|v_2''\rangle \,, \tag{1.15.13}$$

or explicitly,

$$\begin{aligned} \langle u_{1k_1}, u_{2k_2}|v_{1l_1}, v_{2l_2}\rangle &= \frac{1}{\sqrt{n_1}}\,\mathrm{e}^{\frac{2\pi\mathrm{i}}{n_1}k_1l_1}\,\frac{1}{\sqrt{n_2}}\,\mathrm{e}^{\frac{2\pi\mathrm{i}}{n_2}k_2l_2} \\ &= \frac{1}{\sqrt{n}}\,\mathrm{e}^{2\pi\mathrm{i}\,(k_1l_1/n_1 + k_2l_2/n_2)} \,. \end{aligned} \tag{1.15.14}$$

This also emerges directly by writing the $\langle u_{1k_1}, u_{2k_2}|$ in terms of $\langle v_{1l_1}, u_{2k_2}|$, which involves $\langle u_{1k_1}|v_{1l_1}\rangle$; and then writing the $\langle v_{1l_1}, u_{2k_2}|$ in terms of the $\langle v_{1l_1}, v_{2l_2}|$, which introduces $\langle u_{2k_2}|v_{2l_2}\rangle$.

These references to compatible and incompatible physical properties direct attention to something that has remained implicit so far: We began with the measurement of a single physical property. In general, however, to such a property A_1, we can add another compatible property, A_2. A selective measurement of both is symbolized by

$$|a_1'a_1'||a_2'a_2'| = |a_2'a_2'||a_1'a_1'| \,. \tag{1.15.15}$$

We continue in this way until we reach a *complete* set of compatible physical properties: $A_1, A_2, \ldots, A_{K(A)}$, such that any additional property will be incompatible with at least one member of the complete set. The symbol of such a complete measurement is

$$\underbrace{|a_1'a_1'|\ldots|a_{K(A)}'a_{K(A)}'|}_{\text{in any order}} = |a'a'| \qquad \text{with} \quad a' \equiv \left\{a_1', a_2', \ldots, a_{K(A)}'\right\} \,. \tag{1.15.16}$$

Atoms selected to have the particular set of values a' are said to be in the *state* a'.

From this point on, the entire construction of the measurement algebra and the geometry proceeds as before, the only difference being that the outcome of a measurement is not a single number but a collection of numbers. Thus

$$\sum_{a'} |a'a'| = 1 \,, \quad |a'a'| = |a'\rangle\langle a'| \,, \tag{1.15.17}$$

and

$$\langle a'|a''\rangle = \delta(a', a'') = \prod_{j=1}^{K(A)} \delta(a'_j, a''_j) \,. \tag{1.15.18}$$

Now we reverse things and construct the symbols of measurement of one quantity from that of complete measurements. Thus

$$|a'_1 a'_1| = \sum_{a'_2, \ldots, a'_{K(A)}} |a'a'| \,, \tag{1.15.19}$$

for the measurement that selects atoms with the values a'_1 of A_1 is the less selective complete measurement that selects a'_1 and *any* value of a'_2, a'_3, ..., $a'_{K(A)}$. Of course

$$\sum_{a'_1} |a'_1 a'_1| = \sum_{a'} |a'a'| = 1 \,, \tag{1.15.20}$$

and, e. g.,

$$A_1 = \sum_{a'_1} a'_1 |a'_1 a'_1| = \sum_{a'} |a'\rangle a'_1 \langle a'| \,. \tag{1.15.21}$$

Notice that now $|a'_1 a'_1|$ is *not* a simple product of vectors. That is emphasized by contrasting

$$\mathrm{tr}\,\{|a'a'|\} = \mathrm{tr}\,\{|a'\rangle\langle a'|\} = \langle a'|a'\rangle = 1 \tag{1.15.22}$$

with

$$\mathrm{tr}\,\{|a'_1 a'_1|\} = \mathrm{tr}\left\{ \sum_{a'_2, \ldots, a'_{K(A)}} |a'\rangle\langle a'| \right\} = \sum_{a'_2, \ldots, a'_{K(A)}} 1 = m(a'_1) \,, \tag{1.15.23}$$

the multiplicity of a'_1, the number of a' states that have a'_1 in common. As a result, we have, equivalently,

$$\mathrm{tr}\,\{A_1\} = \sum_{a'_1} a'_1 m(a'_1) = \sum_{a'} a'_1 \,. \tag{1.15.24}$$

Of course, we could start with any other quantity, B_1, in general incompatible with the set $A_1, A_2, \ldots$, and supplement it by compatible properties to form the complete B set, with its sets of values, b', and vectors $\langle b'|, |b'\rangle$. What the incompatible A and B sets have in common is the total number of states, which is the dimensionality of the geometry. That follows from the alternative evaluations:

$$\left.\begin{aligned}\sum_{a'}\sum_{b'}\langle a'|b'\rangle\langle b'|a'\rangle=\sum_{a'}1=n(A)\\ \sum_{b'}\sum_{a'}\langle b'|a'\rangle\langle a'|b'\rangle=\sum_{b'}1=n(B)\end{aligned}\right\}=n\,. \tag{1.15.25}$$

We meet these ideas in a more mathematical way in the context of the eigenvalue problem. Consider an n dimensional space, with a coordinate system $|a'\rangle$. Then given a Hermitian operator B_1, find its eigenvalues and eigenvectors, that is, solve

$$B_1|\ \rangle=|\ \rangle b_1'\,. \tag{1.15.26}$$

Introducing the a coordinate system, we get the system of equations

$$\sum_{a''}\langle a'|B_1|a''\rangle\psi(a'')=\psi(a')b_1' \tag{1.15.27}$$

for the components of the eigenvector,

$$\langle a'|\ \rangle=\psi(a')\,. \tag{1.15.28}$$

When written as

$$\sum_{a''}[\langle a'|B_1|a''\rangle-b_1'\delta(a',a'')]\psi(a'')=0\,, \tag{1.15.29}$$

it is familiar that the equations have no solution other than $\psi(a'')=0$ unless

$$\det[\langle a'|B_1|a''\rangle-b_1'\delta(a',a'')]=0 \tag{1.15.30}$$

or, more explicitly,

$$\det\begin{bmatrix}\langle a_1|B_1|a_1\rangle-b_1' & \cdots & \langle a_1|B_1|a_n\rangle\\ \vdots & & \vdots\\ \langle a_n|B_1|a_1\rangle & \cdots & \langle a_n|B_1|a_n\rangle-b_1'\end{bmatrix}=0\,, \tag{1.15.31}$$

which is an equation of degree n for b_1':

$$(-b_1')^n+(-b_1')^{n-1}\mathrm{tr}\,\{B_1\}+\cdots+\det B_1=0\,. \tag{1.15.32}$$

For a simple example, take $n=2$, $B_1=\sigma_x$, $A=\sigma_z$:

$$\begin{aligned}\det\begin{bmatrix}-b_1' & 1\\ 1 & -b_1'\end{bmatrix}&=(-b_1')^2+(-b_1')\underbrace{\mathrm{tr}\,\{\sigma_x\}}_{=0}+\underbrace{\det\sigma_x}_{=-1}\\ &={b_1'}^2-1=0\,,\end{aligned} \tag{1.15.33}$$

implying $b_1'=\pm 1$.

Given a particular root b_1', one returns to the linear equations for the $\psi(a'')$'s and finds them, to within a constant factor. That is fixed, to within an arbitrary phase, by normalization,

$$\sum_{a'} \left|\psi_{b_1'}(a')\right|^2 = 1 \,. \tag{1.15.34}$$

Two such eigenvectors associated with different roots, or eigenvalues, are orthogonal. That follows from the alternative evaluation

$$\langle b_1'|B_1|b_1''\rangle = b_1'\langle b_1'|b_1''\rangle = \langle b_1'|b_1''\rangle b_1'' \tag{1.15.35}$$

or

$$(b_1' - b_1'')\langle b_1'|b_1''\rangle = 0 \tag{1.15.36}$$

so that

$$b_1' \neq b_1'' : \quad \langle b_1'|b_1''\rangle = \sum_{a'} \psi_{b_1'}(a')^* \psi_{b_1''}(a') = 0 \,. \tag{1.15.37}$$

Now suppose that not all the n roots are distinct, that some of them are multiple. For example, suppose that $n = 4$, and B_1 obeys $B_1^2 = 1$, so that there are only two eigenvalues: $+1$, -1. Then one or both must be multiple, as illustrated by the determinant $(b' - 1)^2(b' + 1)^2$. Let the multiplicity of b_1' be $m(b_1') = 2, 3, \ldots$, so that there are $m(b_1')$ linearly independent solutions of (1.15.27). These vectors, $|b_1', 1\rangle, \ldots, |b_1', m(b_1')\rangle$ are not automatically orthogonal, but they can be made so, as illustrated in two Euclidean dimensions by the projection:

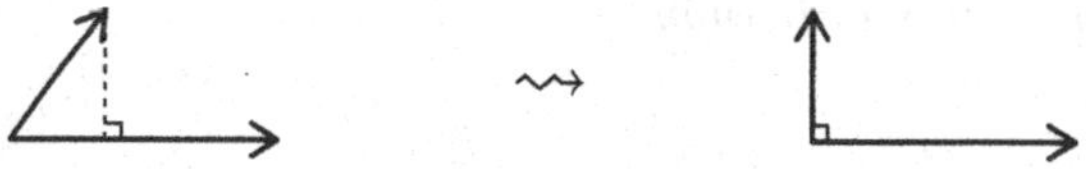

So now we have an orthonormal system: $|b_1', l\rangle$, $l = 1, \ldots, m(b_1')$. Find a second Hermitian operator B_2 that commutes with B_1, $B_1B_2 = B_2B_1$, and

$$B_1B_2|b_1', l\rangle = B_2B_1|b_1', l\rangle = B_2|b_1', l\rangle b_1' \,. \tag{1.15.38}$$

This says that $B_2|b_1', l\rangle$ is a B_1 eigenvector with the eigenvalue b_1', and must therefore be an linear combination of the $m(b_1')$ known vectors:

$$B_2|b_1', l\rangle = \sum_{k=1}^{m(b_1')} |b_1', k\rangle\langle b_1', k|B_2|b_1', l\rangle \tag{1.15.39}$$

where the appropriateness of the matrix element labeling is checked by multiplication with $\langle b_1', k|$.

Within this $m(b_1')$-dimensional space we ask for eigenvectors of B_2

$$B_2|b_1',\ \rangle = |b_1',\ \rangle b_2' \tag{1.15.40}$$

or with $\langle b_1', l|b_1',\ \rangle = \psi_{b_1'}(l)$,

$$\sum_{l=1}^{m(b_1')} \langle b_1', k|B_2|b_1', l\rangle \psi_{b_1'}(l) = \psi_{b_1'}(k) b_2' \tag{1.15.41}$$

leading to

$$\det\left[\langle b_1', k|B_2|b_1', l\rangle - \delta(k,l) b_2'\right] = 0\,. \tag{1.15.42}$$

If all the roots of this equation differ, it provides us with the orthonormal vectors $|b_1', b_2'\rangle$. If not, we find a third commuting operator B_3, and so on.

The U, V operators illustrate this discussion. For a system with n states, the two Hermitian operators implicit in U and V, such that $U^n = V^n = 1$ and $u_k = v_k = \mathrm{e}^{2\pi \mathrm{i} k/n}$ for $k = 1, \dots, n$, each have n distinct eigenvalues, so the multiplicity of any eigenvector is one. In this situation the single operator U, or V, is already complete. If n is a prime integer, that's where it rests. But if $n = n_1 n_2$, we can introduce U_1, V_1 of period n_1, and the commuting U_2, V_2 of period n_2. The eigenvalues of U_1 or V_1 are only $n_1 < n$ in number and must be multiple. We can form various complete sets of compatible properties U_1, U_2; U_1, V_2; V_1, V_2; V_1, U_2 leading to the orthonormal vectors $|u_{1k_1}, u_{2k_2}\rangle$, $|u_{1k_1}, v_{2l_2}\rangle$, etc. If n_1 and/or n_2 is still factorable, this process can be continued, ultimately ending with the various irreducible systems in which the number of states is a prime integer: $\nu = 2, 3, 5, 7, 11, 13, \dots$. These we call *quantum degrees of freedom.*

1.16 The continuum limit

1.16.1 Heisenberg's commutation relation

With the exception of $\nu = 2$, all ν's are odd. Then it's convenient to choose the k in $u_k = v_k = \mathrm{e}^{(2\pi \mathrm{i}/\nu)k}$ as

$$k = 0, \pm 1, \pm 2, \dots, \pm \frac{\nu - 1}{2}\,. \tag{1.16.1}$$

Now let's exhibit the Hermitian operators in U and V. For

$$\frac{2\pi}{\nu} = \epsilon^2 \tag{1.16.2}$$

write

$$U = \mathrm{e}^{\mathrm{i}\epsilon q}\,, \quad V = \mathrm{e}^{\mathrm{i}\epsilon p}\,, \qquad q' = q_k\,, \quad p' = p_k\,,$$

$$q_k = p_k = \epsilon k = 0, \pm\epsilon, \pm 2\epsilon, \ldots, \pm\left(\pi/\epsilon - \tfrac{1}{2}\epsilon\right)\,. \tag{1.16.3}$$

Recall that

$$V^l U^k = \mathrm{e}^{\frac{2\pi\mathrm{i}}{\nu}kl} U^k V^l \tag{1.16.4}$$

or

$$V^{-l} U^k V^l = \mathrm{e}^{-\frac{2\pi\mathrm{i}}{\nu}kl} U^k \tag{1.16.5}$$

and

$$U^k V^l U^{-k} = \mathrm{e}^{-\frac{2\pi\mathrm{i}}{\nu}kl} V^l\,, \tag{1.16.6}$$

which now appear as

$$\mathrm{e}^{-il\epsilon p}\,\mathrm{e}^{ik\epsilon q}\,\mathrm{e}^{il\epsilon p} = \mathrm{e}^{-ik\epsilon l\epsilon}\,\mathrm{e}^{ik\epsilon q} = \mathrm{e}^{ik\epsilon(q - l\epsilon)} \tag{1.16.7}$$

and

$$\mathrm{e}^{ik\epsilon q}\,\mathrm{e}^{il\epsilon p}\,\mathrm{e}^{-ik\epsilon q} = \mathrm{e}^{-ik\epsilon l\epsilon}\,\mathrm{e}^{il\epsilon p} = \mathrm{e}^{il\epsilon(p - k\epsilon)}\,. \tag{1.16.8}$$

With $l\epsilon = q'$, $k\epsilon = p'$ they become

$$\mathrm{e}^{-\mathrm{i}q'p}\,\mathrm{e}^{\mathrm{i}p'q}\,\mathrm{e}^{\mathrm{i}q'p} = \mathrm{e}^{\mathrm{i}p'(q - q')} \tag{1.16.9}$$

and

$$\mathrm{e}^{\mathrm{i}p'q}\,\mathrm{e}^{\mathrm{i}q'p}\,\mathrm{e}^{-\mathrm{i}p'q} = \mathrm{e}^{\mathrm{i}q'(p - p')}\,, \tag{1.16.10}$$

respectively.

We see here unitary transformations on a function of an operator, a function that is defined by a power series. So, with U referring to either unitary operator and A to either q or p, consider

$$\begin{aligned} U^{-1}A^k U &= U^{-1}AAA\cdots U \\ &= U^{-1}AUU^{-1}AU\cdots \\ &= \left(U^{-1}AU\right)^k\,, \end{aligned} \tag{1.16.11}$$

that is, the unitary transform of the function is the function of the transformed operator,

$$U^{-1}f(A)U = f(U^{-1}AU)\,. \tag{1.16.12}$$

This is quite generally true – for any operator function $f(A)$ and any unitary operator U.

In the examples (1.16.9) and (1.16.10), then

$$e^{ip'\left[e^{-iq'p}q\,e^{iq'p}\right]} = e^{ip'(q-q')}\,,$$
$$e^{iq'\left[e^{ip'q}p\,e^{-ip'q}\right]} = e^{iq'(p-p')}\,. \qquad (1.16.13)$$

It is tempting to identify the respective exponents, but it is not generally correct because of the periodicity of the exponentials. However, there is a limit in which this is right: $\nu \to \infty$, $\epsilon \to 0$, with qualifications. It helps to first draw the circle of periodicity ($q_k = p_k = k\epsilon$):

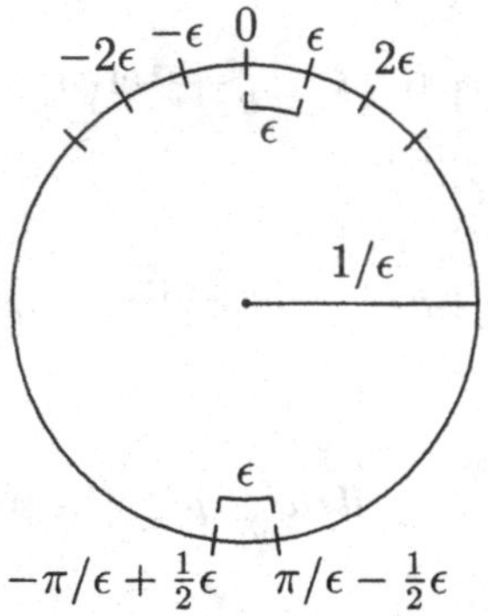

In this limit, as $\epsilon \to 0, 1/\epsilon \to \infty$, any finite portion including 0 is indistinguishable from a continuous straight line:

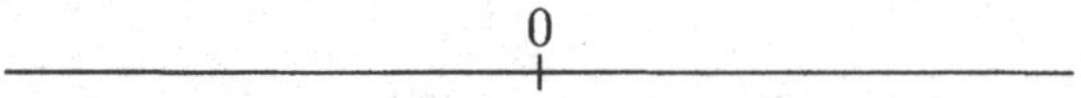

But to avoid the fact that, if the line is continued indefinitely in both directions, the two ends ultimately meet, one must implicitly restrict all applications to *physical* situations where the values of q and p, although possibly very large, remain *finite*. Now the periodicity does not come into play, and the values of q' and p' form a continuum, which permits power by power identification. So

$$e^{-iq'p}\,q\,e^{iq'p} = q - q'\,,$$
$$e^{ip'q}\,p\,e^{-ip'q} = p - p'\,. \qquad (1.16.14)$$

Deferring for a moment the interpretation of these equations we again use the continuous nature of q' and p' permitting a power by power comparison. We want just the first power:

$$\underbrace{(1 - iq'p + \cdots)\,q\,(1 + iq'p + \cdots)}_{= q + iq'(qp - pq) + \cdots} = q - q' \qquad (1.16.15)$$

and

$$\underbrace{(1 + \mathrm{i}p'q + \cdots)\, p\, (1 - \mathrm{i}p'q + \cdots)}_{= p + \mathrm{i}p'(qp - pq) + \cdots} = p - p' \tag{1.16.16}$$

both of which yield the statement of non-commutativity

$$qp - pq = \mathrm{i} \qquad \text{or} \qquad \frac{1}{\mathrm{i}}[q, p] = 1 . \tag{1.16.17}$$

This is the essence of Heisenberg's* discovery of non-commutativity in 1925. To the compact formulation in the form of the commutation relation (1.16.17), Born† contributed substantially.

The action of the unitary operators on vectors:

$$\langle u_k | V^l = \langle u_{k+l} | , \quad U^k | v_l \rangle = | v_{k+l} \rangle \tag{1.16.18}$$

can be relabeled as

$$\langle q' | \, \mathrm{e}^{\mathrm{i}q'p} = \langle q' + q'' | , \quad \mathrm{e}^{\mathrm{i}p''q} | p' \rangle = | p' + p'' \rangle , \tag{1.16.19}$$

after we employ (1.16.2) and (1.16.3) and identify $k\epsilon = q'$, $l\epsilon = q''$ in

$$\begin{aligned} u_k &= \mathrm{e}^{\mathrm{i}\epsilon(k\epsilon)} = \mathrm{e}^{\mathrm{i}\epsilon q'} , \\ V^l &= \mathrm{e}^{\mathrm{i}(l\epsilon)p} = \mathrm{e}^{\mathrm{i}q''p} , \\ u_{k+l} &= \mathrm{e}^{\mathrm{i}\epsilon(k\epsilon + l\epsilon)} = \mathrm{e}^{\mathrm{i}\epsilon(q' + q'')} , \end{aligned} \tag{1.16.20}$$

and $k\epsilon = p''$, $l\epsilon = p'$ in

$$\begin{aligned} U^k &= \mathrm{e}^{\mathrm{i}(k\epsilon)q} = \mathrm{e}^{\mathrm{i}p''q} , \\ v_l &= \mathrm{e}^{\mathrm{i}\epsilon(l\epsilon)} = \mathrm{e}^{\mathrm{i}\epsilon p'} , \\ v_{k+l} &= \mathrm{e}^{\mathrm{i}\epsilon(l\epsilon + k\epsilon)} = \mathrm{e}^{\mathrm{i}\epsilon(p' + p'')} . \end{aligned} \tag{1.16.21}$$

We'll find later, in Section 2.1, that $\langle u_k | \to \sqrt{\epsilon}\langle q' |$, $| v_l \rangle \to | p' \rangle \sqrt{\epsilon}$, but the additional numerical factors simply cancel here.

Here we see, for example, unitary operator $\mathrm{e}^{\mathrm{i}q''p}$ acting on a left eigenvector of q to produce another such eigenvector with a displaced eigenvalue. The statement that $\langle q' | \, \mathrm{e}^{\mathrm{i}q''p}$ is a q eigenvector with the eigenvalue $q' + q''$, or equivalently that it is an eigenvector of $q - q''$ with the eigenvalue q', is

$$\langle q' | \, \mathrm{e}^{\mathrm{i}q''p}(q - q'') = q' \langle q' | \, \mathrm{e}^{\mathrm{i}q''p} = \langle q' | q \, \mathrm{e}^{\mathrm{i}q''p} \tag{1.16.22}$$

*Werner HEISENBERG (1901–1976) †Max BORN (1882–1970)

or

$$e^{iq''p}(q - q'') = q\,e^{iq''p}\,, \tag{1.16.23}$$

which is just the previous operator equation in (1.16.14) (with $q'' \to q'$).

As we see, when there is a statement about q, there is an analogous one for p. That originates in the symmetry (see Problem 1-43a) $U \to V$, $V \to U^{-1}$, or, with $U = e^{i\epsilon q}$, $V = e^{i\epsilon p}$,

$$q \to p\,, \quad p \to -q\,. \tag{1.16.24}$$

Indeed

$$i = qp - pq \quad \longrightarrow \quad p(-q) - (-q)p = qp - pq \tag{1.16.25}$$

and (with $q' \to p'$)

$$\begin{gathered} e^{-iq'p}q\,e^{iq'p} = q - q'\,, \\ \downarrow \\ e^{ip'q}p\,e^{-ip'q} = p - p'\,. \end{gathered} \tag{1.16.26}$$

If we apply this to $\langle q'|\,e^{iq''p} = \langle q' + q''|$, we get $\langle p'|\,e^{-ip''q} = \langle p' + p''|$, the adjoint of $e^{ip''q}|p'\rangle = |p' + p''\rangle$.

1.16.2 Schrödinger's differential-operator representation

Now let's convert the q vector equation to numerical form by multiplying by a right vector $|\;\rangle$:

$$\langle q'|\,e^{iq''p}|\;\rangle = \langle q' + q''|\;\rangle\,. \tag{1.16.27}$$

We assume $|\;\rangle$ to be such that $\langle q' + q''|\;\rangle$ is a continuous function of q'', permitting expansion of both sides in powers of q'':

$$\langle q'|(1 + iq''p + \cdots)|\;\rangle = \langle q'|\;\rangle + q''\frac{\partial}{\partial q'}\langle q'|\;\rangle + \cdots \tag{1.16.28}$$

or

$$\langle q'|p|\;\rangle = \frac{1}{i}\frac{\partial}{\partial q'}\langle q'|\;\rangle\,. \tag{1.16.29}$$

This differential-operator representation of the effect of p is the essence of Schrödinger's discovery of wave mechanics (1926).

To show the equivalence of this with non-commutativity, consider

$$\begin{aligned}\langle q'|(qp-pq)|\ \rangle &= q'\langle q'|p|\ \rangle - \frac{1}{\mathrm{i}}\frac{\partial}{\partial q'}\langle q'|q|\rangle \\ &= \left[q'\frac{1}{\mathrm{i}}\frac{\partial}{\partial q'} - \frac{1}{\mathrm{i}}\frac{\partial}{\partial q'}q'\right]\langle q'|\ \rangle = \langle q'|\mathrm{i}|\ \rangle \end{aligned} \tag{1.16.30}$$

as it should.

The substitution $q \to p$, $p \to -q$ gives us

$$\langle p'|q|\ \rangle = \mathrm{i}\frac{\partial}{\partial p'}\langle p'|\ \rangle \tag{1.16.31}$$

with the adjoint

$$\langle\ |q|p'\rangle = \frac{1}{\mathrm{i}}\frac{\partial}{\partial p'}\langle\ |p'\rangle . \tag{1.16.32}$$

The relation (1.16.29) and

$$\langle q'|f(q)|\ \rangle = f(q')\langle q'|\ \rangle \tag{1.16.33}$$

are combined in

$$\langle q'|f(q,p)|\ \rangle = f\left(q', \frac{1}{\mathrm{i}}\frac{\partial}{\partial q'}\right)\langle q'|\ \rangle \tag{1.16.34}$$

provided $f(q,p)$ can be built up from the basic structures by addition and multiplication. That is, if the differential operator representation is correct for $f_1(q,p)$ and $f_2(q,p)$, as it is for $f(q)$ and p, then it is also correct for $f_1(q,p) + f_2(q,p)$ and $f_1(q,p)f_2(q,p)$. Indeed

$$\begin{aligned}\langle q'|\left[f_1(q,p) + f_2(q,p)\right]|\ \rangle &= \langle q'|f_1(q,p)|\ \rangle + \langle q'|f_2(q,p)|\ \rangle \\ &= \left[f_1\left(q', \frac{1}{\mathrm{i}}\frac{\partial}{\partial q'}\right) + f_2\left(q', \frac{1}{\mathrm{i}}\frac{\partial}{\partial q'}\right)\right]\langle q'|\ \rangle \end{aligned} \tag{1.16.35}$$

and

$$\begin{aligned}\langle q'|f_1(q,p)f_2(q,p)|\ \rangle &= f_1\left(q', \frac{1}{\mathrm{i}}\frac{\partial}{\partial q'}\right)\langle q'|f_2(q,p)|\ \rangle \\ &= f_1\left(q', \frac{1}{\mathrm{i}}\frac{\partial}{\partial q'}\right) f_2\left(q', \frac{1}{\mathrm{i}}\frac{\partial}{\partial q'}\right)\langle q'|\ \rangle . \end{aligned} \tag{1.16.36}$$

The corresponding statements about p eigenvectors,

$$\begin{aligned}\langle p'|f(q,p)|\ \rangle &= f\left(\mathrm{i}\frac{\partial}{\partial p'}, p'\right)\langle p'|\ \rangle , \\ \langle\ |f(q,p)|p'\rangle &= f\left(\frac{1}{\mathrm{i}}\frac{\partial}{\partial p'}, p'\right)\langle\ |p'\rangle , \end{aligned} \tag{1.16.37}$$

are obvious analogs.

Problems

1-1 What is the root mean square speed of silver atoms passing through the slits of the Stern–Gerlach apparatus after evaporating in the oven at 2 227° C?

1-2 The torque $\boldsymbol{\mu} \times \boldsymbol{B}$ that results from the action of the magnetic field $\boldsymbol{B}$ on the magnetic moment $\boldsymbol{\mu}$ changes the atomic angular momentum vector $\boldsymbol{J}$ in accordance with

$$\frac{\mathrm{d}}{\mathrm{d}t}\boldsymbol{J} = \boldsymbol{\mu} \times \boldsymbol{B} .$$

Suppose that $\boldsymbol{\mu} = \gamma \boldsymbol{J}$ and let the homogeneous field $\boldsymbol{B}$ point in the z direction. How do the components of $\boldsymbol{J}$ change in time? Assuming that $J_z = 0$ initially, after what lapse of time will $\boldsymbol{J}$ reverse direction? Express this as a distance d traveled in the field at the mean speed v. Take $\gamma = 1.76 \times 10^7$ and $|\boldsymbol{B}| = B_z = 0.01$ [cgs units] and find d.

1-3 On page 30, speaking of the classical anticipation of the outcome of the Stern–Gerlach experiment, the text says: "the Ag atoms certainly have their magnetic moments (m.m.) distributed uniformly in all directions, which is to say that the distribution of μ_z should be uniform between the limits μ and $-\mu$." Justify the latter conclusion.

1-4 The Stern–Gerlach experiment shows that the possible values of μ_z are $\pm\mu$. What then would be measured for μ_z^2, μ_x^2, μ_y^2, and

$$\boldsymbol{\mu}^2 = \mu_x^2 + \mu_y^2 + \mu_z^2 \ ?$$

How does this compare with the classical picture in which $\boldsymbol{\mu}$ should have some specific direction in space?

1-5 A certain textbook, seeking to define the addition of measurement symbols, $|a'a'| + |a''a''|$, $a' \neq a''$, says: "This is the case of the Stern–Gerlach experiment followed by a screen with two holes." Comment on this assertion.

1-6 Prove that $(a' \neq a'')$

$$\Big[|a'a'| + |a''a''|\Big]^2 = |a'a'| + |a''a''| .$$

1-7 Evaluate the commutator

$$\Big[|a'a''|, |a'''a^{\text{iv}}|\Big] .$$

Check that your answer has zero trace. Why should that be so?

1-8 Show that the vectors

$$\langle\overline{a'}| = \lambda(a')\langle a'| , \qquad |\overline{a''}\rangle = |a''\rangle\,[\lambda(a'')]^{-1}$$

also satisfy

$$\langle\overline{a'}|\overline{a''}\rangle = \delta(a', a'')$$

for arbitrary numerical values of the $\lambda(a')$, and verify that

$$\sum_{a'} |\overline{a'}\rangle\langle\overline{a'}| = 1 .$$

Similar statements apply to the vector sets $\langle\overline{b'}|$ and $|\overline{b''}\rangle$, with numbers $\lambda(b')$. Find what restrictions on $\lambda(a')$ and $\lambda(b')$ are required so that

$$\langle\overline{b'}|\overline{a'}\rangle = \langle\overline{a'}|\overline{b'}\rangle^* .$$

Use this arbitrariness – this freedom of definition – to argue that $\langle a'|b'\rangle$ cannot itself be a probability, whereas $\langle a'|b'\rangle\langle b'|a'\rangle$ can be, and indeed is.

1-9 Concerning page 43: Show that the apparently more general possibility for real numbers,

$$\langle b'|a'\rangle = \gamma\langle a'|b'\rangle , \qquad p(a', b') = \gamma\left[\langle a'|b'\rangle\right]^2$$

with γ real and positive, can be reduced to what was considered by an acceptable redefinition of the a vectors, or the b vectors.

1-10 An initial Stern–Gerlach measurement selects the + beam. A second + selection is made on this beam, in a direction differing by angle ε. A third + selection is made on that beam in a direction again differing by angle ε. This change in direction is repeated n times in all to produce the final + selection at an angle $\theta = n\varepsilon$ relative to the initial direction. In the limit as $\varepsilon \to 0$ and $n = \theta/\varepsilon \to \infty$, what fraction of the initially chosen beam emerges from the final selection?

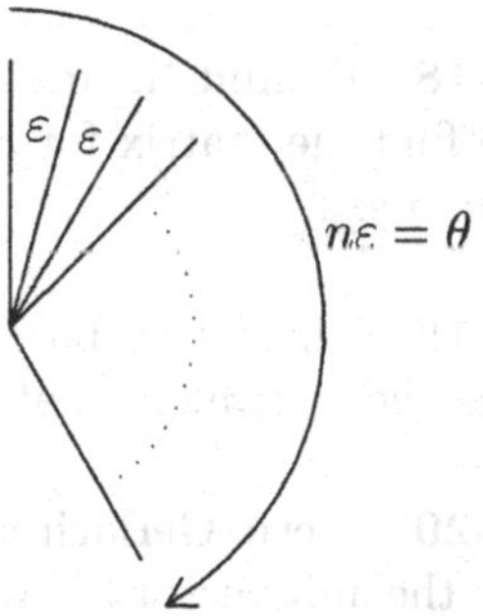

1-11 Write M_b, the symbol of a non-selective B measurement, as a function of operator B. What kind of operator is this M_b?

1-12 Physical property s has the possible values $+1, 0, -1$. What algebraic equation does the symbol s obey? Exhibit the measurement symbols $|++|$, $|00|$, $|--|$ as functions of s.

1-13 Given that

$$(\sigma_x \cos\varphi + \sigma_y \sin\varphi)^2 = 1$$

for all angles φ, and that

$$\sigma_z = -\mathrm{i}\sigma_x\sigma_y ,$$

deduce all the algebraic properties of σ_x, σ_y, σ_z.

1-14 Evaluate

$$\boldsymbol{\sigma}\cdot\boldsymbol{a}\,\boldsymbol{\sigma}\cdot\boldsymbol{b}\,\boldsymbol{\sigma}\cdot\boldsymbol{c}$$

in two ways indicated by $(\boldsymbol{\sigma}\cdot\boldsymbol{a}\,\boldsymbol{\sigma}\cdot\boldsymbol{b})\,\boldsymbol{\sigma}\cdot\boldsymbol{c}$ and $\boldsymbol{\sigma}\cdot\boldsymbol{a}\,(\boldsymbol{\sigma}\cdot\boldsymbol{b}\,\boldsymbol{\sigma}\cdot\boldsymbol{c})$.

1-15 Evaluate the commutator of $\boldsymbol{\sigma}\cdot\boldsymbol{a}$ and $\boldsymbol{\sigma}\cdot\boldsymbol{b}$, and also their anticommutator.

1-16 The magnetic moment in the Stern–Gerlach experiment is $\boldsymbol{\mu} = \mu\boldsymbol{\sigma}$. What can you say about the possible outcomes of a measurement of $\boldsymbol{\mu}^2$? Compare with Problem 1-4.

1-17 Use the fact that

$$\sigma_x = \mathrm{e}^{-\mathrm{i}\frac{\pi}{4}\sigma_y}\,\sigma_z\,\mathrm{e}^{\mathrm{i}\frac{\pi}{4}\sigma_y}$$

to find the eigenvectors of σ_x in terms of the σ_z eigenvectors. What is the probability interpretation of these σ_x eigenvectors?

1-18 Exhibit the matrices for $\boldsymbol{\sigma}\cdot\boldsymbol{a}$ and $\boldsymbol{\sigma}\cdot\boldsymbol{b}$. Then use matrix multiplication to find the matrix for $\boldsymbol{\sigma}\cdot\boldsymbol{a}\,\boldsymbol{\sigma}\cdot\boldsymbol{b}$, thereby checking the multiplication law for the σ's.

1-19 The Pauli matrices in (1.10.3) refer to the eigenvectors of σ_z. What are the equivalent matrices referring to eigenvectors of σ_x, or of σ_y.

1-20 Stern–Gerlach measurements are made on atoms that are symbolized by the unit vector $|\ \rangle$, with σ'_z components $\psi(\sigma'_z = \pm 1)$. Write the expectation values of σ_x, σ_y, σ_z (comprising the three-dimensional vector $\langle\boldsymbol{\sigma}\rangle$) in terms of the wave function $\psi(\sigma'_z)$, which is arbitrary apart from the statement of normalization. Prove that $\langle\boldsymbol{\sigma}\rangle$ is a unit three-dimensional vector.

1-21a Verify the commutator identity

$$[X, YZ] = Y[X, Z] + [X, Y]Z .$$

What is the analogous version of $[XY, Z]$? What does the phrase "the commutator $[X, Y]$ is linear in X and in Y" imply about

$$[X_1 + X_2, Y_1 + Y_2] \ ?$$

1-21b Check that

$$[X,[Y,Z]] + [Y,[Z,X]] + [Z,[X,Y]] = 0 \, .$$

What statement about three-dimensional vectors emerges from the choice

$$X = \boldsymbol{\sigma} \cdot \boldsymbol{x} \, , \qquad Y = \boldsymbol{\sigma} \cdot \boldsymbol{y} \, , \qquad Z = \boldsymbol{\sigma} \cdot \boldsymbol{z} \ ?$$

Verify it.

1-22a State the adjoints of the measurement symbols

$$X^\dagger X \, , \quad X^\dagger Y + Y^\dagger X \, , \quad \mathrm{i}(X^\dagger Y - Y^\dagger X) \, ,$$

and of the eigenvector equation $\langle a' | A = a' \langle a' |$.

1-22b X is a non-Hermitian operator. What kind of operators are $X^\dagger X$ and $XX^\dagger$? Verify that

$$A \equiv \tfrac{1}{2} \left(X + X^\dagger \right) \, , \qquad B \equiv \tfrac{1}{2\mathrm{i}} \left(X - X^\dagger \right)$$

are Hermitian, which gives the construction

$$X = A + \mathrm{i}B \, .$$

Apply your results concerning $X^\dagger X$ and $XX^\dagger$ to learn the adjoint nature of $(1/\mathrm{i})[A, B]$.

1-23 Hermitian operator A has eigenvalues a' and eigenvectors $|a'\rangle$; Hermitian operator B is arbitrary. Show that the matrix elements $\langle a' | B | a'' \rangle$ obey the sum rule

$$\sum_{a''} (a'' - a') \left| \langle a' | B | a'' \rangle \right|^2 = \tfrac{1}{2} \langle a' | \left[B, [A, B] \right] | a' \rangle \, .$$

1-24 Use the matrix significance of the trace to prove that

$$\mathrm{tr}\, \{XY\} = \mathrm{tr}\, \{YX\} \, .$$

What is $\mathrm{tr}\, \{[X, Y]\}$? Evaluate $\mathrm{tr}\, \{f(A)\}$ in terms of the numbers $f(a')$.

1-25a Use the construction of the $|b'b'|$ by means of the $|a'a'|$ to show that

$$\sum_{b'} |b'b'| = \sum_{a'} |a'a'| \qquad [= 1] .$$

1-25b By transforming from b vectors to a vectors prove that

$$\sum_{b'} \langle b'|X|b'\rangle = \sum_{a'} \langle a'|X|a'\rangle \qquad [= \operatorname{tr}\{X\}] .$$

1-26 The trace has been defined as a numerical property of an operator, not by any particular matrix realization. As a check of this show that there is a unitary operator U such that the a matrix of $U^{-1}XU$ is the b matrix of X. Conclude from $\operatorname{tr}\{U^{-1}XU\} = \operatorname{tr}\{X\}$ (why?) that the trace is independent of the matrix realization used to compute it.

1-27 Show that

$$\operatorname{tr}\{X^\dagger X\} = \sum_{a',a''} |\langle a'|X|a''\rangle|^2 \geq 0 \qquad [= 0 \text{ only if } X = 0]$$

which is the squared length of operator X expressed in terms of components relative to the orthonormal operator set $|a'a''|$. Verify the inequality

$$\operatorname{tr}\{X^\dagger X\} \operatorname{tr}\{Y^\dagger Y\} \geq |\operatorname{tr}\{X^\dagger Y\}|^2 .$$

When does the equal sign hold?

1-28a The unit symbol and the three Pauli operators are used to define another set of four Hermitian operators,

$$\begin{aligned}
B_{00} &= \tfrac{1}{2}(1 + \sigma_x + \sigma_y + \sigma_z) , \\
B_{10} &= \sigma_x B_{00} \sigma_x , \\
B_{01} &= \sigma_z B_{00} \sigma_z , \\
B_{11} &= \sigma_z \sigma_x B_{00} \sigma_x \sigma_z .
\end{aligned}$$

Verify that they are indeed Hermitian, and find the eigenvalues of B_{jk} $(j, k = 0, 1)$. Show that

$$\operatorname{tr}\{B_{jk}\} = 1 \qquad \text{and} \qquad \operatorname{tr}\{B_{jk} B_{j'k'}\} = 2\delta_{jj'}\delta_{kk'} .$$

1-28b The B_{jk}'s are complete, which is to say that any function X of Pauli's vector operator $\boldsymbol{\sigma}$ can be written in the form

$$X = \sum_{j,k} x_{jk} B_{jk} \, .$$

Justify this statement, and find what is needed in $x_{jk} = \operatorname{tr}\{(?)X\}$. Are the coefficients x_{jk} unique for a given X? Write 1, σ_x, σ_y, σ_z as sums of the B_{jk}'s.

1-28c Operators X and Y are specified by their respective coefficients x_{jk} and y_{jk}. Express the traces $\operatorname{tr}\{X\}$, $\operatorname{tr}\{X^\dagger X\}$, and $\operatorname{tr}\{Y^\dagger X\}$ in terms of these coefficients.

1-29 Work out the value of

$$\operatorname{tr}\left\{\frac{1+(\sigma_x+\sigma_y)/\sqrt{2}}{2}\,\frac{1+(-\sigma_x+\sigma_y)/\sqrt{2}}{2}\right\} .$$

What physical meaning can you give to this number?

1-30a Expectation value is a linear concept:

$$\langle (X+Y) \rangle = \langle X \rangle + \langle Y \rangle \, , \qquad \langle \lambda X \rangle = \lambda \langle X \rangle \, ,$$

and the expectation value of the Hermitian operator representing a physical quantity is real:

$$\langle A \rangle^* = \langle A \rangle \, , \; \langle B \rangle^* = \langle B \rangle \, , \; \ldots \, ,$$

or, more generally,

$$\langle (A+\mathrm{i}B) \rangle^* = \langle (A-\mathrm{i}B) \rangle \, , \text{ that is: } \langle X \rangle^* = \left\langle X^\dagger \right\rangle .$$

Also, $\langle 1 \rangle = 1$. Define the ***probability operator*** P (frequently called density matrix) by

$$\left\langle |a'a''| \right\rangle = \langle a''|P|a' \rangle$$

(does $\left\langle |b'b''| \right\rangle$ have the same form?) so that the expectation value of

$$X = \sum_{a',a''} \langle a'|X|a'' \rangle \, |a'a''|$$

is

$$\langle X \rangle = \sum_{a',a''} \langle a'|X|a'' \rangle \langle a''|P|a' \rangle = \operatorname{tr}\{XP\} \, .$$

Prove that P is Hermitian and that $\operatorname{tr}\{P\} = 1$.

1-30b Each $|a'a'|$ symbolizes a physical quantity with the non-negative values 0 and 1. So

$$\left\langle |a'a'| \right\rangle = \langle a'|P|a'\rangle \geq 0 .$$

If the Hermitian operator P is regarded as symbolizing some physical property, conclude that one can write

$$P = \sum_{p'} p' |p'p'| \qquad \text{with} \quad p' \geq 0 \,, \; \sum_{p'} p' = 1 \,.$$

Recognize that

$$\langle X \rangle_P = \sum_{p'} p' \, \langle X \rangle_{p'}$$

is a weighted average of the more familiar

$$\langle X \rangle_{p'} = \operatorname{tr} \left\{ X |p'p'| \right\} = \langle p'|X|p'\rangle \,.$$

What happens if $P^2 = P$?

1-31 Property B is measured and atoms selected with $B = b'$. What is the corresponding probability operator P? What is it for a non-selective B measurement?

1-32 Stern–Gerlach measurements are made to determine the expectation values of σ_x, σ_y, σ_z for atoms emerging from a certain source. Show that

$$P = \frac{1}{2}(1 + \langle \boldsymbol{\sigma} \rangle \cdot \boldsymbol{\sigma})$$

is the probability operator characterizing these atoms. Compare with Problem 1-20.

1-33a The determinant $\det\{X\}$ of operator X can be defined by the differential statement

$$\delta \log \det \{X\} = \operatorname{tr} \left\{ X^{-1} \delta X \right\}$$

together with $\det\{1\} = 1$. If this is to produce a unique $\det\{X\}$ the differential expression must be integrable, which means that

$$\delta_1 \delta_2 \log \det \{X\} = \delta_2 \delta_1 \log \det \{X\} \;,$$

where $\delta_1 X$ and $\delta_2 X$ are independent infinitesimal variations of X. (You will need an expression for δX^{-1}, the variation of the inverse of operator X).

1-33b What is $\delta(XY)$ in terms of δX and δY? Use this to prove that

$$\det\{XY\} = \det\{X\}\det\{Y\} \; .$$

What is the relation between $\det\{\lambda X\}$ and $\det\{X\}$, where λ is a number?

1-34 In Section 1.5 we anticipated that the probabilities

$$p(\overset{a}{+},\overset{b}{1},\overset{c}{-}) \qquad \text{and} \qquad p(\overset{a}{+},\overset{b}{1},\overset{c}{+})$$

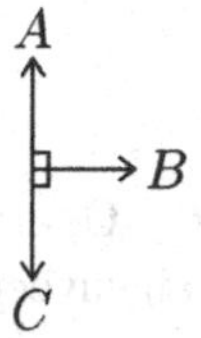

show constructive and destructive interference, respectively. In Section 1.12 we found the required probability amplitudes. Demonstrate that this works out as expected. (For simplicity take the three directions to be in the x, z plane, with azimuthal angles $\varphi = 0$.)

1-35 U and V are unitary operators. Show that UV and VU are also unitary.

1-36a Any member of the 2×2 measurement algebra – any such operator – is a linear combination of 1 and the three σ's. Consider

$$U = u_0 + \mathrm{i}\boldsymbol{u} \cdot \boldsymbol{\sigma} \; ,$$

where the number u_0 is real. Under what restrictions on u_0 and the numerical vector $\boldsymbol{u}$ is U unitary?

1-36b Unitary operators U and V are given by the numerical parameters $u_0, \boldsymbol{u}$ and $v_0, \boldsymbol{v}$, respectively. What are the parameters for UV and VU? Verify that they obey the conditions for a unitary operator.

1-36c Satisfy the unitarity conditions on $u_0, \boldsymbol{u}$ in such a way the U is seen to be an exponential function of i times a Hermitian operator.

1-36d By what simple modification of $u_0, \boldsymbol{u}$ does one remove the restriction that u_0 is real?

1-37 With what restriction on the number λ is

$$\frac{1 + \mathrm{i}\lambda\boldsymbol{\sigma} \cdot \boldsymbol{n}}{1 - \mathrm{i}\lambda\boldsymbol{\sigma} \cdot \boldsymbol{n}} \qquad (\text{where } \boldsymbol{n}^* = \boldsymbol{n},\ \boldsymbol{n} \cdot \boldsymbol{n} = 1)$$

a unitary operator? Write it in the form of Problem 1-36a; in the form of Problem 1-36c.

1-38 Evaluate $\mathrm{e}^{\frac{\pi \mathrm{i}}{2}\sigma_x}\, \mathrm{e}^{\frac{\pi \mathrm{i}}{2}\sigma_y}\, \mathrm{e}^{\frac{\pi \mathrm{i}}{2}\sigma_z}$.

1-39 Show that the adjoint relation between eigenvectors of unitary operator U,

$$\langle u'|^\dagger = |u'\rangle ,$$

is consistent with the eigenvector equation

$$\langle u'|U = u'\langle u'| .$$

1-40 Operator U permutes vectors $|a_k\rangle$ cyclically [cf. (1.14.3)]. Construct the eigenvectors

$$|u_l\rangle = \sum_k |a_k\rangle\langle a_k|u_l\rangle$$

by solving the eigenvector equation, thereby determining the eigenvalues u_l and, with the normalization requirement, the $\langle a_k|u_l\rangle$.

1-41 Use

$$U^l|a_k\rangle = |a_{k+l}\rangle$$

to arrive at

$$U^l = \sum_{m,n} |a_m\rangle\delta_{m-n,l}\langle a_n| .$$

Prove (1.14.14) in this way.

1-42 Non-Hermitian A commutes with its adjoint,

$$AA^\dagger = A^\dagger A .$$

(Operators with this property are called *normal.*) Show that A can be understood as a (complex) function of a Hermitian operator B,

$$A = f(B) = f_1(B) + \mathrm{i}f_2(B) , \qquad A^\dagger = f^*(B) = f_1(B) - \mathrm{i}f_2(B) ,$$

with real functions $f_{1,2}$. What is the relevance thereof in the context of (1.14.15)?

1-43a Unitary operators U and V reciprocally defined as in Section 1.14: Check that the substitution $U \to V$, $V \to U^{-1}$ leaves the algebraic properties of U and V intact. Then find what happens to the reciprocal definitions of U and V by permutation of eigenvectors.

1-43b Unitary operator W has the defining property

$$W\left|v_k\right\rangle = \left|u_k\right\rangle \qquad \text{for} \quad k = 1, 2, \ldots, n\,.$$

Show that

$$W\left|u_k\right\rangle = \left|v_{n-k}\right\rangle ,$$

and then

$$W^{-1}UW = V\,, \qquad W^{-1}VW = U^{-1}\,.$$

1-43c Proceed from

$$W = \sum_{k=1}^{n} \left|u_k v_k\right| = \sum_{k=1}^{n} \left|u_k u_k\right| \frac{1}{\left\langle u_k\middle|v_k\right\rangle} \left|v_k v_k\right|$$

and find the coefficients w_{kl} in

$$W = \sum_{k,l=1}^{n} w_{kl} U^k V^l\,.$$

1-44 Show the consistency of the uv transformation function (1.14.32) with the algebraic properties of U and V by evaluating and comparing the matrix elements

$$\left\langle u_k\middle|UV\middle|v_l\right\rangle , \quad \left\langle u_k\middle|VU\middle|v_l\right\rangle .$$

1-45 Recall [cf. (1.14.10)]

$$u^n - 1 = \prod_{u'}(u - u')$$

with $u' = u_k = \mathrm{e}^{(2\pi\mathrm{i}/n)k}$ for $k = 1, \ldots, n$. By taking the logarithm of this equation and expanding in powers of $1/u$, prove that

$$\sum_{u'}(u')^r = 0 \qquad \text{for } 0 < r < n\,.$$

1-46 Use the cyclic property of the trace,

$$\operatorname{tr}\{XYZ\} = \operatorname{tr}\{YZX\} = \operatorname{tr}\{ZXY\}\,,$$

for an alternative demonstration of (1.14.55), that is

$$\operatorname{tr}\left\{U^k V^l\right\} = 0 \qquad \text{for} \quad 0 < k < n \quad \text{or} \quad 0 < l < n\,.$$

1-47 Evaluate

$$p(u',v') = \text{tr}\left\{|u'u'||v'v'|\right\}$$

by using the construction

$$|u'u'| = \frac{1}{n}\sum_{l}(U/u')^{l}$$

and the similar one for $|v'v'|$.

1-48 Apply the result of Problem 1-45 to prove the completeness of the $|u'u'|$ measurement symbols,

$$\sum_{u'}|u'u'| = 1\,.$$

1-49a Consider

$$F = f(U,V) = \sum_{k,l} f_{kl}U^{k}V^{l} = \sum_{k,l}\overline{f}_{kl}V^{l}U^{k}\,.$$

Express the coefficients $\overline{f}_{kl}$ in terms of the f_{kl}'s.

1-49b Show that an F which commutes with U, or $U^{-1}FU = F$, can be only a function of U, and similarly that an F which commutes with V can only be a function of V. Conclude that if F commutes with U and V, it must be just a numerical multiple of the unit symbol 1.

1-50 Prove that

$$\text{tr}\left\{|a'a''|^{\dagger}|a'''a^{\text{iv}}|\right\} = \delta(a',a''')\,\delta(a'',a^{\text{iv}})\,,$$

an orthonormality statement for the n^2 measurement symbols, regarded as vectors in operator space. Then show the analogous property for the n^2 powers of U and V, that is

$$\frac{1}{n}\text{tr}\left\{\left(U^{k}V^{l}\right)^{\dagger}U^{k'}V^{l'}\right\} = \delta_{kk'}\delta_{ll'}\,.$$

1-51 Demonstrate that

$$\sum_{a',a''}|a'a''|^{\dagger}\,X\,|a'a''| = 1\,\text{tr}\,\{X\}$$

and then prove the analog

$$\frac{1}{n}\sum_{k,l}(U^k V^l)^\dagger\, X\, U^k V^l = 1\,\mathrm{tr}\,\{X\}$$

by, for example, applying Problem 1-49b, or by considering $X = U^r V^s$.

1-52 Use both versions of Problem 1-51, for $n = 2$, to show that

$$\frac{1}{2}(X + \boldsymbol{\sigma}\cdot X\boldsymbol{\sigma}) = 1\,\mathrm{tr}\,\{X\}\,.$$

Check this for $X = 1, \sigma_k$.

1-53 Construct $\langle q'|p'\rangle$, apart from the numerical factor, by using

$$\frac{1}{\mathrm{i}}\frac{\partial}{\partial q'}\langle q'| = \langle q'|p\,, \qquad \frac{1}{\mathrm{i}}\frac{\partial}{\partial p'}|p'\rangle = q|p'\rangle\,.$$

1-54 Convert

$$\mathrm{e}^{\mathrm{i}p'q}\,p\,\mathrm{e}^{-\mathrm{i}p'q} = p - p'$$

into

$$\left[p, \mathrm{e}^{-\mathrm{i}p'q}\right] = \frac{1}{\mathrm{i}}\frac{\partial}{\partial q}\,\mathrm{e}^{-\mathrm{i}p'q}\,.$$

Check the generalization to

$$[p, f(q)] = \frac{1}{\mathrm{i}}\frac{\mathrm{d}f(q)}{\mathrm{d}q}$$

by introducing the q Schrödinger description. Write the analog with q and p interchanged.

1-55 Arrange

$$\mathrm{e}^{-\mathrm{i}f(q)}\,\mathrm{e}^{\mathrm{i}q'p}\,\mathrm{e}^{\mathrm{i}f(q)}$$

in different ways to show that

$$\begin{aligned}\mathrm{e}^{\mathrm{i}q'[p+f'(q)]} &= \mathrm{e}^{\mathrm{i}q'p}\,\mathrm{e}^{\mathrm{i}[f(q)-f(q-q')]}\\ &= \mathrm{e}^{\mathrm{i}[f(q+q')-f(q)]}\,\mathrm{e}^{\mathrm{i}q'p}\,,\end{aligned}$$

where

$$f'(q) = \frac{\mathrm{d}f(q)}{\mathrm{d}q} = \mathrm{i}\left[p, f(q)\right]\,.$$

Apply this to $f'(q) = -(p'/q')q$ to recover familiar statements, and find new ones for $f'(q) = \kappa q^2/2$.

1-56 Generalize the results of Problem 1-54 to arrive at

$$\frac{\partial}{\partial q} f(q,p) = \mathrm{i}\left[p, f(q,p)\right] , \qquad \frac{\partial}{\partial p} f(q,p) = \mathrm{i}\left[f(q,p), q\right] .$$

Then show that the order of differentiation doesn't matter,

$$\frac{\partial}{\partial p}\frac{\partial}{\partial q} f(q,p) = \frac{\partial}{\partial q}\frac{\partial}{\partial p} f(q,p) .$$

2. Continuous q, p Degree of Freedom

2.1 Wave functions

The completeness of the U or V vectors is expressed, for example, by the summation

$$1 = \sum_k |u_k\rangle\langle u_k| \,. \tag{2.1.1}$$

But, in introducing $q_k = k\epsilon$ and proceeding to the limit $\epsilon \to 0$, where $q'(= q_k)$ becomes a continuous variable, the summation must be replaced by an integration. Accordingly we write

$$\langle u_k| = \sqrt{\epsilon}\langle q'| \,, \quad |u_k\rangle = |q'\rangle\sqrt{\epsilon} \,, \tag{2.1.2}$$

and

$$1 = \sum |q'\rangle\epsilon\langle q'| \,, \qquad \text{with} \quad \epsilon = q_{k+1} - q_k = \Delta q' \,, \tag{2.1.3}$$

and, in the limit, arrive at

$$1 = \int_{-\infty}^{\infty} |q'\rangle \, \mathrm{d}q' \, \langle q'| \,, \tag{2.1.4}$$

which states the completeness of the q vectors. This symbolic statement implies the numerical one

$$\langle 1|2\rangle = \int_{-\infty}^{\infty} \underbrace{\langle 1|q'\rangle}_{=\psi_1(q')^*} \, \mathrm{d}q' \, \underbrace{\langle q'|2\rangle}_{=\psi_2(q')} \,. \tag{2.1.5}$$

In particular, for a single vector,

$$\langle\;|\;\rangle = \int_{-\infty}^{\infty} \psi(q')^* \, \mathrm{d}q' \, \psi(q') = \int_{-\infty}^{\infty} \mathrm{d}q' \, |\psi(q')|^2 = 1 \,, \tag{2.1.6}$$

if it is a unit vector.

We get to the same point from the probability interpretation of

$$\left|\langle u_k|\ \rangle\right|^2 = \epsilon\left|\langle q'|\ \rangle\right|^2 \to \mathrm{d}q'\left|\psi(q')\right|^2 , \tag{2.1.7}$$

being the probability $\mathrm{d}P$ that a q measurement carried out on the system represented by the vector $|\ \rangle$, or the wave function $\psi(q')$, shall yield a value in the range $\mathrm{d}q'$ about q'. [We owe this probability interpretation of Schrödinger wave functions to Born.] The total probability of any outcome must be unity:

$$\int \mathrm{d}P = \int_{-\infty}^{\infty} \mathrm{d}q'\ \left|\psi(q')\right|^2 = 1 . \tag{2.1.8}$$

The necessary existence of this integral requires that $\psi(q') \to 0$ as $|q'| \to \infty$. For a physical system, truly infinite values of q' do not occur.

The qp transformation function is produced by relabeling the uv transformation function (1.14.32),

$$\langle u_k|v_l\rangle = \sqrt{\epsilon}\langle q'|p'\rangle\sqrt{\epsilon} = \frac{1}{\sqrt{2\pi/\epsilon^2}}\,\mathrm{e}^{\mathrm{i}(k\epsilon)(l\epsilon)} , \tag{2.1.9}$$

or

$$\langle q'|p'\rangle = \frac{1}{\sqrt{2\pi}}\,\mathrm{e}^{\mathrm{i}q'p'} . \tag{2.1.10}$$

What about $\langle u_k|u_l\rangle = \delta_{kl}$? Dividing by $\epsilon = \sqrt{\epsilon}\sqrt{\epsilon}$ we get

$$\langle q'|q''\rangle = \frac{\delta(q', q'')}{\epsilon} \to \delta(q' - q'') , \tag{2.1.11}$$

which introduces Dirac's delta function, where ($\epsilon \to 0$)

$$\begin{aligned} q' \neq q'' ,\ q' - q'' \neq 0 :\ \delta(q' - q'') = 0 , \\ q' = q'' ,\ q' - q'' = 0 :\ \delta(q' - q'') = \infty . \end{aligned} \tag{2.1.12}$$

One more property of the delta function comes from

$$\sum_l \underbrace{\langle u_k|u_l\rangle}_{\to\epsilon\delta(q' - q'')} \psi(u_l) = \psi(u_k) ,$$

$$\longrightarrow \int_{-\infty}^{\infty} \delta(q' - q'')\,\mathrm{d}q''\ \psi(q'') = \psi(q') \tag{2.1.13}$$

for arbitrary $\psi(u_k)$. Inasmuch as $\delta(q' - q'') = 0$ if $q'' \neq q'$, only the value of $\psi(q'')$ for $q'' = q'$ occurs in the integral, and we learn that

$$\int_{-\infty}^{\infty} \delta(q' - q'')\,\mathrm{d}q'' = \int_{-\infty}^{\infty} \mathrm{d}q'\,\delta(q' - q'') = 1 . \tag{2.1.14}$$

Let's see how all this fits together,

$$\sum_l \underbrace{\langle u_k | v_l \rangle \langle v_l | u_m \rangle}_{\rightarrow (\sqrt{\epsilon}\langle q'|p'\rangle\sqrt{\epsilon})(\sqrt{\epsilon}\langle p'|q''\rangle\sqrt{\epsilon})} = \langle u_k | u_m \rangle = \delta_{km} \rightarrow \epsilon\delta(q' - q'') ,$$

$$\longrightarrow \int_{-\infty}^{\infty} \langle q'|p'\rangle \, \mathrm{d}p' \, \langle p'|q''\rangle = \delta(q' - q'') , \tag{2.1.15}$$

which is formally correct as a combination of

$$\int |p'\rangle \mathrm{d}p' \langle p'| = 1 \quad \text{and} \quad \langle q'|q''\rangle = \delta(q' - q'') , \tag{2.1.16}$$

which state the completeness of the p vectors and the orthonormality of the q vectors. But how does it work numerically? We have

$$\langle q'|p'\rangle = \frac{1}{\sqrt{2\pi}} \mathrm{e}^{\mathrm{i}q'p'} , \quad \langle p'|q''\rangle = \frac{1}{\sqrt{2\pi}} \mathrm{e}^{-\mathrm{i}p'q''} \tag{2.1.17}$$

so that

$$\delta(q' - q'') = \int_{-\infty}^{\infty} \frac{\mathrm{d}p'}{2\pi} \mathrm{e}^{\mathrm{i}p'(q' - q'')} . \tag{2.1.18}$$

But the integral from $-\infty$ to ∞ does not converge!

$$\begin{aligned} \int_{-\infty}^{\infty} \frac{\mathrm{d}p'}{2\pi} \mathrm{e}^{\mathrm{i}p'(q' - q'')} &= \int_0^{\infty} \frac{\mathrm{d}p'}{\pi} \cos\bigl(p'(q' - q'')\bigr) \\ &= \int_0^{\infty} \frac{\mathrm{d}p'}{\pi} \frac{\partial}{\partial p'} \frac{\sin\bigl(p'(q' - q'')\bigr)}{q' - q''} \\ &= \lim_{P \to \infty} \frac{1}{\pi} \frac{\sin\bigl(P(q' - q'')\bigr)}{q' - q''} = ? . \end{aligned} \tag{2.1.19}$$

Evidently the difficulty is that we have not made explicit the restriction that $P = \infty$ is not allowed.

To see that this is only an apparent problem, let's restate what has been done in terms of the individual $\langle q'|p'\rangle$ and $\langle p'|q'\rangle$ transformation functions:

$$\begin{aligned} \langle q'| \;\rangle &= \int \langle q'|p'\rangle \mathrm{d}p' \langle p'| \;\rangle , \\ \langle p'| \;\rangle &= \int \langle p'|q'\rangle \mathrm{d}q' \langle q'| \;\rangle , \end{aligned} \tag{2.1.20}$$

or

$$\begin{aligned} \psi(q') &= \int_{-\infty}^{\infty} \mathrm{e}^{\mathrm{i}q'p'} \frac{\mathrm{d}p'}{\sqrt{2\pi}} \psi(p') , \\ \psi(p') &= \int_{-\infty}^{\infty} \mathrm{e}^{-\mathrm{i}p'q'} \frac{\mathrm{d}q'}{\sqrt{2\pi}} \psi(q') . \end{aligned} \tag{2.1.21}$$

If we eliminate $\psi(p')$, for example, we get

$$\psi(q') = \int_{-\infty}^{\infty} \mathrm{e}^{\mathrm{i}q'p'}\frac{\mathrm{d}p'}{2\pi}\int_{-\infty}^{\infty} \mathrm{e}^{-\mathrm{i}p'q''}\,\mathrm{d}q''\psi(q'') \tag{2.1.22}$$

which is Fourier's* integral theorem: 'any' function $\psi(q')$ can be written as a linear combination of the functions $\mathrm{e}^{\mathrm{i}q'p'}$, $-\infty < p' < \infty$, with the indicated coefficients. Should we interchange the integration and integrate first over p', we come to the integral for $\delta(q'-q'')$. But what happens if we use it as is?

For example,

$$\psi(q') = \frac{1}{\pi^{1/4}}\,\mathrm{e}^{-\frac{1}{2}q'^2} \tag{2.1.23}$$

is a normalized wave function since

$$\int_{-\infty}^{\infty} \mathrm{d}q'\,\mathrm{e}^{-q'^2} = \sqrt{\pi}\,, \tag{2.1.24}$$

the basic Gaussian† integral. Its Fourier transform

$$\begin{aligned}\psi(p') &= \int_{-\infty}^{\infty} \mathrm{e}^{-\mathrm{i}p'q'}\frac{\mathrm{d}q'}{\sqrt{2\pi}}\psi(q')\\ &= \frac{1}{\sqrt{2\pi}}\frac{1}{\pi^{1/4}}\int_{-\infty}^{\infty} \mathrm{e}^{-\mathrm{i}p'q'}\,\mathrm{d}q'\,\mathrm{e}^{-\frac{1}{2}q'^2}\\ &= \frac{1}{\sqrt{2\pi}}\frac{1}{\pi^{1/4}}\underbrace{\int_{-\infty}^{\infty} \mathrm{d}q'\,\mathrm{e}^{-\frac{1}{2}(q'+\mathrm{i}p')^2}}_{=\sqrt{2\pi}}\,\mathrm{e}^{-\frac{1}{2}p'^2}\\ &= \frac{1}{\pi^{1/4}}\,\mathrm{e}^{-\frac{1}{2}p'^2}\end{aligned} \tag{2.1.25}$$

is obviously also normalized. And then

$$\begin{aligned}\psi(q') &= \frac{1}{\pi^{1/4}}\frac{1}{\sqrt{2\pi}}\int_{-\infty}^{\infty} \underbrace{\mathrm{e}^{\mathrm{i}q'p'}\,\mathrm{d}p'\,\mathrm{e}^{-\frac{1}{2}p'^2}}_{=\mathrm{d}p'\,\mathrm{e}^{-\frac{1}{2}(p'-\mathrm{i}q')^2}\,\mathrm{e}^{-\frac{1}{2}q'^2}}\\ &= \frac{1}{\pi^{1/4}}\,\mathrm{e}^{-\frac{1}{2}q'^2}\,,\end{aligned} \tag{2.1.26}$$

which is back where we started, without the slightest difficulty. [More about such Gaussian Fourier integrals later, in particular in Sections 5.4 and 6.5.]

It is clear that, for a physical $\psi(q')$, the actual range of p' is quite finite, so no harm should be done in leaving P large but finite in (2.1.19). Accordingly, let's check that, in this sense,

---

*Jean Baptiste Joseph FOURIER (1768–1830) †Karl Friedrich GAUSS (1777–1855)

$$\psi(q') = \int_{-\infty}^{\infty} \frac{1}{\pi} \frac{\sin\bigl(P(q'-q'')\bigr)}{q'-q''} \, \mathrm{d}q'' \, \psi(q'') \,. \tag{2.1.27}$$

Put $q'' = q' + \chi/P$, $\mathrm{d}q'' = \mathrm{d}\chi/P$:

$$\psi(q') \overset{?}{=} \int_{-\infty}^{\infty} \frac{\mathrm{d}\chi}{\pi} \frac{\sin\chi}{\chi} \psi(q' + \chi/P) \,. \tag{2.1.28}$$

We are content here to argue that, because of destructive interference ($\sin\chi$) for large χ, the relevant χ values are limited by $|\chi| \sim 1$, and for sufficiently large P the replacement $\psi\,(q' + \sim 1/P) \to \psi(q')$ is permissible, a statement of continuity. Then

$$\psi(q') = \psi(q') \underbrace{\int_{-\infty}^{\infty} \frac{\mathrm{d}\chi}{\pi} \frac{\sin\chi}{\chi}}_{=1} = \psi(q') \tag{2.1.29}$$

as it should.

More generally, the restriction to possibly large but finite values of p' is expressed by the more precise definition

$$\delta(q'-q'') = \int_{-\infty}^{\infty} \frac{\mathrm{d}p'}{2\pi} K(\epsilon p') \, \mathrm{e}^{\mathrm{i}p'(q'-q'')} \Big|_{\epsilon \to 0} \tag{2.1.30}$$

where

$$\begin{aligned} &K(\epsilon p') \to 0 \text{ as } |p'| \to \infty \text{ for given } \epsilon \ \ (\text{i.e., } |\epsilon p'| \to \infty) \,, \\ &K(\epsilon p') \to 1 \text{ as } \ \ \epsilon \to 0 \ \ \text{ for given } p' \ (\text{i.e., } \epsilon p' \to 0) \,, \end{aligned} \tag{2.1.31}$$

and the eventual limit $\epsilon \to 0$ is reserved until all integrations are performed. The example just discussed is

$$K(\epsilon p') = \begin{Bmatrix} 0 & \text{for} & \epsilon|p'| > 1 \\ 1 & \text{for} & \epsilon|p'| < 1 \end{Bmatrix} \qquad \text{with } \epsilon = 1/P \,. \tag{2.1.32}$$

Another example is

$$K(\epsilon p') = \mathrm{e}^{-\epsilon|p'|} \tag{2.1.33}$$

so that ($\epsilon > 0$, eventual limit $\epsilon \to 0$ understood)

$$\begin{aligned} \delta(q'-q'') &= \int_0^{\infty} \frac{\mathrm{d}p'}{2\pi} \mathrm{e}^{-\epsilon p'} \left[\mathrm{e}^{\mathrm{i}p'(q'-q'')} + \mathrm{e}^{-\mathrm{i}p'(q'-q'')} \right] \\ &= \mathrm{Re} \frac{1}{\pi} \int_0^{\infty} \mathrm{d}p' \, \mathrm{e}^{-[\epsilon - \mathrm{i}(q'-q'')]p'} \\ &= \mathrm{Re} \frac{1}{\pi} \frac{1}{\epsilon - \mathrm{i}(q'-q'')} \\ &= \frac{1}{\pi} \frac{\epsilon}{(q'-q'')^2 + \epsilon^2} \,. \end{aligned} \tag{2.1.34}$$

Indeed

$$\left\{\begin{array}{l} q'-q''\neq 0\colon\ \delta(q'-q'')\to 0 \\ q'-q''=0\colon\ \delta(q'-q'')\to\infty \end{array}\right\} \quad \text{as } \epsilon\to 0\,, \tag{2.1.35}$$

and (substitute $q'-q''=\epsilon x$ and $x=\tan\vartheta$)

$$\begin{aligned}\int_{-\infty}^{\infty}\mathrm{d}q'\,\delta\left(q'-q''\right) &= \int_{-\infty}^{\infty}\mathrm{d}q'\frac{1}{\pi}\frac{\epsilon}{\left(q'-q''\right)^2+\epsilon^2} = \frac{1}{\pi}\int_{-\infty}^{\infty}\mathrm{d}x\,\frac{1}{x^2+1} \\ &= \frac{1}{\pi}\int_{-\pi/2}^{\pi/2}\mathrm{d}\vartheta = 1\,. \end{aligned} \tag{2.1.36}$$

Yet another,

$$K(\epsilon p') = \frac{\sin(\frac{1}{2}\epsilon p')}{\frac{1}{2}\epsilon p'} \to \left\{\begin{array}{l} 0 \text{ as } |\epsilon p'|\to\infty\,, \\ 1 \text{ as } \epsilon p'\to 0\,. \end{array}\right. \tag{2.1.37}$$

Now,

$$\begin{aligned}\delta\left(q'-q''\right) &= \int_{-\infty}^{\infty}\frac{\mathrm{d}p'}{2\pi}\frac{\sin(\frac{1}{2}\epsilon p')}{\frac{1}{2}\epsilon p'}\,\mathrm{e}^{\mathrm{i}p'\left(q'-q''\right)} \\ &= \int_0^{\infty}\frac{\mathrm{d}p'}{\pi}\frac{\sin(\frac{1}{2}\epsilon p')}{\frac{1}{2}\epsilon p'}\cos\bigl(p'(q'-q'')\bigr) \\ &= \int_0^{\infty}\frac{\mathrm{d}p'}{\pi}\frac{1}{\epsilon p'}\left[\sin\bigl(p'(q'-q''+\tfrac{1}{2}\epsilon)\bigr)-\sin\bigl(p'(q'-q''-\tfrac{1}{2}\epsilon)\bigr)\right] \\ &= \frac{1}{2\epsilon}\left[\left\{\begin{array}{r} 1 \text{ for } q'-q''>-\frac{1}{2}\epsilon \\ -1 \text{ for } q'-q''<-\frac{1}{2}\epsilon \end{array}\right\}-\left\{\begin{array}{r} 1 \text{ for } q'-q''>\frac{1}{2}\epsilon \\ -1 \text{ for } q'-q''<\frac{1}{2}\epsilon \end{array}\right\}\right] \\ &= \frac{1}{\epsilon}\left\{\begin{array}{lll} 0 & \text{for} & q'-q''>\frac{1}{2}\epsilon\,, \\ 1 & \text{for} & -\frac{1}{2}\epsilon<q'-q''<\frac{1}{2}\epsilon\,, \\ 0 & \text{for} & q'-q''<-\frac{1}{2}\epsilon\,, \end{array}\right. \end{aligned} \tag{2.1.38}$$

which is graphically represented by

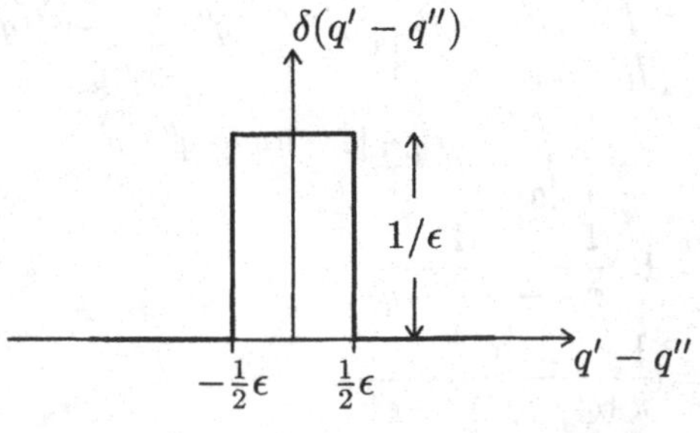

Evidently all conditions are satisfied as $\epsilon \to 0$. What we have here is the continuum version of the initial introduction of the delta function as the discrete $(1/\epsilon)\delta(q', q'')$, which is $1/\epsilon$ for $q' - q'' = 0$, and zero for $|q' - q''| = \epsilon, 2\epsilon, \ldots$.

Now notice this:

$$\begin{aligned} & \langle q'|q''\rangle = \delta(q' - q'') , \\ q' \neq q'' : \quad & \langle q'|q''\rangle = 0, \end{aligned} \tag{2.1.39}$$

states the orthogonality of the q eigenstates, BUT

$$q' = q'' : \quad \langle q'|q'\rangle = \infty \, ! \tag{2.1.40}$$

We have lost the normalization to unity. What we have here is a reminder that an *exact* value of q' is an overidealization; any measurement of q, with its continuous spectrum, will locate it within a certain range, which can be very small, but never zero. The simplest example of such states is

$$|q', \Delta q'\rangle = \int_{q' - \frac{1}{2}\Delta q'}^{q' + \frac{1}{2}\Delta q'} dq'' \frac{1}{\sqrt{\Delta q'}} |q''\rangle \tag{2.1.41}$$

or

$$\psi_{q',\Delta q'}(q'') = \frac{1}{\sqrt{\Delta q'}} \begin{cases} 1 & \text{for} \quad |q'' - q'| < \frac{1}{2}\Delta q' , \\ 0 & \text{for} \quad |q'' - q'| > \frac{1}{2}\Delta q' , \end{cases} \tag{2.1.42}$$

where

$$\int_{-\infty}^{\infty} dq'' |\psi_{q',\Delta q'}(q'')|^2 = 1 , \tag{2.1.43}$$

with a graphical representation of the integrand given by

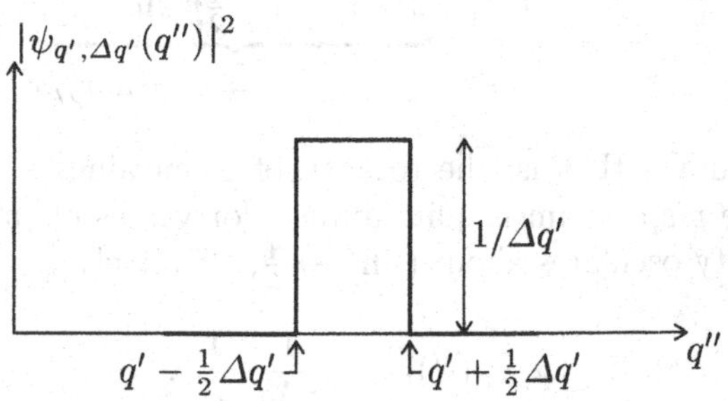

What is the p description of this state? We evaluate [this is, of course, essentially the inverse of the Fourier transform in (2.1.38)]

$$\begin{aligned}\psi_{q',\Delta q'}(p') &= \int_{-\infty}^{\infty} \mathrm{d}q'' \frac{1}{\sqrt{2\pi}} \mathrm{e}^{-\mathrm{i}p'q''} \psi_{q',\Delta q'}(q'') \\ &= \frac{1}{\sqrt{2\pi\Delta q'}} \int_{q'-\frac{1}{2}\Delta q'}^{q'+\frac{1}{2}\Delta q'} \mathrm{d}q''\, \mathrm{e}^{-\mathrm{i}p'q''} \\ &= \frac{1}{\sqrt{2\pi\Delta q'}} \mathrm{e}^{-\mathrm{i}p'q'} \frac{\sin(\frac{1}{2}p'\Delta q')}{\frac{1}{2}p'} \end{aligned} \tag{2.1.44}$$

and get

$$|\psi_{q',\Delta q'}(p')|^2 = \frac{\Delta q'}{2\pi} \left[\frac{\sin(\frac{1}{2}p'\Delta q')}{\frac{1}{2}p'\Delta q'}\right]^2 , \tag{2.1.45}$$

which is graphically represented by

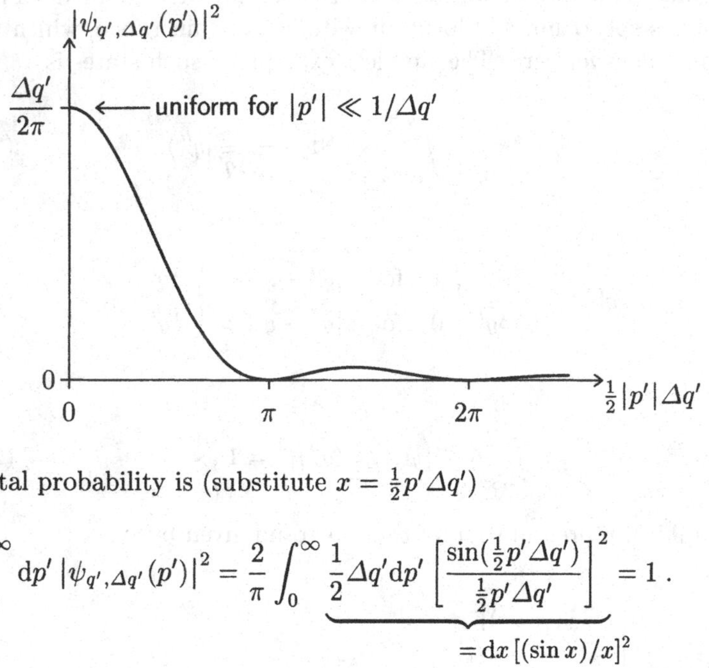

The total probability is (substitute $x = \frac{1}{2}p'\Delta q'$)

$$\int_{\infty}^{\infty} \mathrm{d}p'\, |\psi_{q',\Delta q'}(p')|^2 = \frac{2}{\pi} \int_0^{\infty} \underbrace{\frac{1}{2}\Delta q' \mathrm{d}p' \left[\frac{\sin(\frac{1}{2}p'\Delta q')}{\frac{1}{2}p'\Delta q'}\right]^2}_{= \mathrm{d}x\, [(\sin x)/x]^2} = 1\, . \tag{2.1.46}$$

The q spectrum – that is: the totality of eigenvalues – is sharply limited, whereas the p spectrum is quite diffuse: for values of $|p'| \gg 1/\Delta q'$ the probability density oscillates about ($\sin^2 \to \frac{1}{2}$, effectively)

$$|\psi_{q',\Delta q'}(p')|^2 \sim \frac{1}{\pi\Delta q'} \frac{1}{p'^2}\, . \tag{2.1.47}$$

The decrease to zero as $|p'| \to \infty$ is sufficient that the total probability exists (and equals one), but the average of p'^2 does not,

$$\left\langle p'^2 \right\rangle_{q',\Delta q'} = \int_{-\infty}^{\infty} \mathrm{d}p'\, p'^2 \left|\psi_{q',\Delta q'}(p')\right|^2 = \infty\,. \tag{2.1.48}$$

This unphysical behavior reflects the unnatural nature of the sharply limited q spectrum. It invites our attention to physical states for which both the q and p spectra are effectively bounded.

2.2 Expectation values and their spreads

For this purpose we examine the expectation or average values of q and p in the physical state $|\ \rangle$:

$$\langle q\rangle = \langle\ |q|\ \rangle, \quad \langle p\rangle = \langle\ |p|\ \rangle \tag{2.2.1}$$

and the mean square deviation from these averages:

$$\begin{aligned}(\delta q)^2 &= \langle\ |(q-\langle q\rangle)^2|\ \rangle \equiv \langle 1|1\rangle\,,\\ (\delta p)^2 &= \langle\ |(p-\langle p\rangle)^2|\ \rangle \equiv \langle 2|2\rangle\,,\end{aligned} \tag{2.2.2}$$

where

$$\langle 1| = \langle\ |(q-\langle q\rangle)\,, \qquad |2\rangle = (p-\langle p\rangle)|\ \rangle\,. \tag{2.2.3}$$

Now we consider

$$(\delta q)^2(\delta p)^2 = \langle 1|1\rangle\langle 2|2\rangle \geq |\langle 1|2\rangle|^2\,. \tag{2.2.4}$$

The latter inequality is the unitary counterpart of the 3-dimensional Euclidean statement $\boldsymbol{a}\cdot\boldsymbol{a}\,\boldsymbol{b}\cdot\boldsymbol{b} \geq (\boldsymbol{a}\cdot\boldsymbol{b})^2$, or $\cos^2\theta \leq 1$. To prove it, think of the vector

$$|1/2\rangle = |1\rangle - |2\rangle\frac{\langle 2|1\rangle}{\langle 2|2\rangle}$$

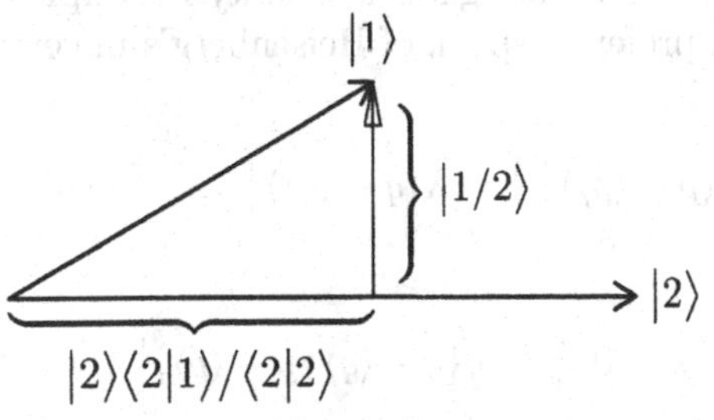

(2.2.5)

such that

$$\langle 2|1/2\rangle = \langle 2|1\rangle - \langle 2|2\rangle\frac{\langle 2|1\rangle}{\langle 2|2\rangle} = 0\,. \tag{2.2.6}$$

We know that $\langle 1/2|1/2\rangle \geq 0$ where the equality holds only if $|1/2\rangle = 0$, that is, when $|2\rangle$ is a numerical multiple of $|1\rangle$; when they are parallel. Now, as a consequence of (2.2.6),

$$\langle 1/2|1/2\rangle = \langle 1|1/2\rangle = \langle 1|1\rangle - \frac{|\langle 1|2\rangle|^2}{\langle 2|2\rangle} \geq 0\,, \tag{2.2.7}$$

which is the inequality stated in (2.2.4). So

$$(\delta q)^2 (\delta p)^2 \geq |\langle\ |(q - \langle q\rangle)(p - \langle p\rangle)|\ \rangle|^2 \tag{2.2.8}$$

where the equal sign applies only if $(p - \langle p\rangle)|\ \rangle$ is parallel to $(q - \langle q\rangle)|\ \rangle$,

$$(p - \langle p\rangle)|\ \rangle = \lambda(q - \langle q\rangle)|\ \rangle\,. \tag{2.2.9}$$

Now for the product of Hermitian operators A and B, one can use identity (1.2.25),

$$AB = \tfrac{1}{2}\{A, B\} + \tfrac{\mathrm{i}}{2}\tfrac{1}{\mathrm{i}}[A, B]\,, \tag{2.2.10}$$

which is a non-Hermitian combination of the two Hermitian operators $\{A, B\}$ and $\frac{1}{\mathrm{i}}[A, B]$. The diagonal matrix elements of a Hermitian operator – expectation values – are real. So, with $A = q - \langle q\rangle$ and $B = p - \langle p\rangle$, the absolute square occurring in (2.2.8) is the sum of the squared real and imaginary parts:

$$(\delta q\, \delta p)^2 \geq \underbrace{\left[\langle\ |\tfrac{1}{2}\{q - \langle q\rangle, p - \langle p\rangle\}|\ \rangle\right]^2}_{\geq 0} + \Big[\tfrac{1}{2}\underbrace{\langle\ |\tfrac{1}{\mathrm{i}}[q, p]|\ \rangle}_{=1}\Big]^2 \tag{2.2.11}$$

or

$$\delta q\, \delta p \geq \frac{1}{2}\,. \tag{2.2.12}$$

This is Heisenberg's Uncertainty Principle of 1927 (in Weyl's* form of 1928); we'll prefer to speak of Heisenberg's uncertainty *relation*. The equality holds only if

$$(p - \langle p\rangle)|\ \rangle = \lambda(q - \langle q\rangle)|\ \rangle, \quad \langle\ |(p - \langle p\rangle) = \lambda^*\langle\ |(q - \langle q\rangle) \tag{2.2.13}$$

and

$$0 = \langle\ |\tfrac{1}{2}\{q - \langle q\rangle, p - \langle p\rangle\}|\ \rangle = \tfrac{1}{2}(\lambda + \lambda^*)(\delta q)^2\,, \tag{2.2.14}$$

so that $\lambda = \mathrm{i}\gamma$ with γ real. Notice that

$$\underbrace{\langle\ |(p - \langle p\rangle)^2|\ \rangle}_{=(\delta p)^2} = \gamma^2 \underbrace{\langle\ |(q - \langle q\rangle)^2|\ \rangle}_{=(\delta q)^2}\,,$$

*Claus Hugo Hermann Weyl (1885–1955)

and therefore

$$\gamma = \frac{\delta p}{\delta q} = \frac{1}{2}\frac{1}{(\delta q)^2} = 2\,(\delta p)^2\,, \tag{2.2.15}$$

which anticipates that γ must be positive.

2.3 States of minimal uncertainty

To study the optimal situation, $\delta q \delta p = \frac{1}{2}$, let's use $\langle q'|$ components:

$$\langle q'|\,(p - \langle p\rangle)\,|\;\rangle = \mathrm{i}\gamma\langle q'|\,(q - \langle q\rangle)\,|\;\rangle \tag{2.3.1}$$

or

$$\left(\frac{1}{\mathrm{i}}\frac{\partial}{\partial q'} - \langle p\rangle\right)\psi(q') = \mathrm{i}\gamma\,(q' - \langle q\rangle)\,\psi(q')\,. \tag{2.3.2}$$

Therefore

$$\mathrm{d}\log\psi = \frac{\mathrm{d}\psi}{\psi} = [\mathrm{i}\,\langle p\rangle - \gamma\,(q' - \langle q\rangle)]\,\mathrm{d}\,(q' - \langle q\rangle) \tag{2.3.3}$$

and

$$\psi = C\,\mathrm{e}^{\mathrm{i}\,\langle p\rangle\,(q' - \langle q\rangle)}\,\mathrm{e}^{-\frac{1}{2}\gamma\,(q' - \langle q\rangle)^2}\,. \tag{2.3.4}$$

After determining the value of the integration constant C from the normalization of ψ,

$$\begin{aligned} 1 &= \int_{-\infty}^{\infty} \mathrm{d}q'\,|\psi|^2 = |C|^2 \int_{-\infty}^{\infty} \mathrm{d}q'\,\mathrm{e}^{-\gamma\,(q' - \langle q\rangle)^2} \\ &= |C|^2\sqrt{\frac{\pi}{\gamma}} = |C|^2\sqrt{2\pi}\,\delta q\,, \end{aligned} \tag{2.3.5}$$

(Note that $\gamma > 0$ is essential here.) we arrive at

$$\psi(q') = (2\pi)^{-\frac{1}{4}}\frac{1}{\sqrt{\delta q}}\,\mathrm{e}^{\mathrm{i}\,\langle p\rangle\,(q' - \langle q\rangle)}\,\mathrm{e}^{-\frac{(q'-\langle q\rangle)^2}{4(\delta q)^2}}\,\mathrm{e}^{\frac{\mathrm{i}}{2}\,\langle p\rangle\,\langle q\rangle}\,, \tag{2.3.6}$$

where the phase constant is introduced in anticipation of a symmetry property. The substitution $q \to p, p \to -q$ gives us directly

$$\psi(p') = (2\pi)^{-\frac{1}{4}}\frac{1}{\sqrt{\delta p}}\,\mathrm{e}^{\mathrm{i}\,\langle q\rangle\,(p' - \langle p\rangle)}\,\mathrm{e}^{-\frac{(p'-\langle p\rangle)^2}{4(\delta p)^2}}\,\mathrm{e}^{-\frac{\mathrm{i}}{2}\,\langle p\rangle\,\langle q\rangle}\,, \tag{2.3.7}$$

which should also emerge as

$$\psi(p') = \int \mathrm{e}^{-\mathrm{i}p'q'} \frac{\mathrm{d}q'}{\sqrt{2\pi}} \psi(q')$$
$$= \mathrm{e}^{-\mathrm{i}p' \langle q \rangle} \int \frac{\mathrm{d}q'}{\sqrt{2\pi}} \mathrm{e}^{-\mathrm{i}\,(p' - \langle p \rangle)\,(q' - \langle q' \rangle)}$$
$$\times (2\pi)^{-\frac{1}{4}} \frac{1}{\sqrt{\delta q}} \mathrm{e}^{-\frac{(q' - \langle q \rangle)^2}{4(\delta q)^2}} \mathrm{e}^{\frac{\mathrm{i}}{2}\langle p \rangle \langle q \rangle} \qquad (2.3.8)$$

or, after completing a square,

$$\psi(p') = \mathrm{e}^{-\mathrm{i}\,\langle q \rangle\,(p' - \langle p \rangle)}\, \mathrm{e}^{\frac{\mathrm{i}}{2}\langle p \rangle \langle q \rangle}$$
$$\times \frac{1}{\sqrt{\delta q}} \int \frac{\mathrm{d}q'}{\sqrt{2\pi}} \mathrm{e}^{-\frac{1}{4(\delta q)^2}\left[q' - \langle q \rangle + 2\mathrm{i}\,(p' - \langle p \rangle)\,(\delta q)^2\right]^2}$$
$$\times (2\pi)^{-\frac{1}{4}} \mathrm{e}^{-\,(\delta q)^2\,(p - \langle p \rangle)^2}$$
$$= (2\pi)^{-\frac{1}{4}} \underbrace{\sqrt{2\delta q}}_{=1/\sqrt{\delta p}}\, \mathrm{e}^{-\mathrm{i}\,\langle q \rangle\,(p' - \langle p \rangle)}\, \underbrace{\mathrm{e}^{-\,(\delta q)^2\,(p' - \langle p \rangle)^2}}_{=\,\mathrm{e}^{-\,(p' - \langle p \rangle)^2/(2\delta p)^2}}\, \mathrm{e}^{-\frac{\mathrm{i}}{2}\langle p \rangle \langle q \rangle}\,. \qquad (2.3.9)$$

Now we can see the precise sense in which wave functions such as $\langle q' | p'' \rangle = (2\pi)^{-\frac{1}{2}} \mathrm{e}^{\mathrm{i}q'p''}$ and $\langle p' | p'' \rangle = \delta(p' - p'')$ are idealizations. First, let's write

$$\psi(q') = \left[\frac{(2\pi)^{\frac{1}{4}}}{\sqrt{\delta q}} \mathrm{e}^{-\frac{\mathrm{i}}{2}\langle p \rangle \langle q \rangle} \mathrm{e}^{-\frac{(q' - \langle q \rangle)^2}{4(\delta q)^2}}\right] \frac{1}{\sqrt{2\pi}} \mathrm{e}^{\mathrm{i}q'p''} \quad \text{with } p'' = \langle p \rangle\,,$$
$$\psi(p') = \left[(2\pi)^{\frac{1}{4}} \sqrt{\delta p}\, \mathrm{e}^{-\frac{\mathrm{i}}{2}\langle p \rangle \langle q \rangle} \mathrm{e}^{-\mathrm{i}\,\langle q \rangle\,(p' - p'')}\right] \delta\,(p' - p'', \delta q)\,, \qquad (2.3.10)$$

where

$$\delta\,(p' - p'', \delta p) = \frac{\pi^{-\frac{1}{2}}}{2\delta p} \mathrm{e}^{-\frac{(p' - p'')^2}{4(\delta p)^2}} \qquad (2.3.11)$$

obeys (substitute $p' - p'' = 2x\delta p$)

$$\int_{-\infty}^{\infty} \mathrm{d}p'\, \delta\,(p' - p'', \delta p) = \frac{1}{\sqrt{\pi}} \int_{-\infty}^{\infty} \mathrm{d}x\, \mathrm{e}^{-x^2} = 1\,. \qquad (2.3.12)$$

For $\delta p \to 0$, one has $\delta\,(p' - p'', \delta p) \to \delta\,(p' - p'')$ and $\mathrm{e}^{-\mathrm{i}\,\langle q \rangle\,(p' - p'')} \to 1$ follows. Also in that limit, $\delta q = 1/\delta p \to \infty$, and then $\mathrm{e}^{-(q' - \langle q \rangle)^2/(2\delta q)^2} \to 1$. This leaves, to within a common (arbitrarily small) factor, just the wave functions $\langle q' | p'' \rangle$ and $\langle p' | p'' \rangle$, respectively.

More practically, suppose δp is sufficiently small that $\langle q \rangle\, \delta p \ll 1$, or [since $1/(2\delta p) = \delta q$]

$$\langle q \rangle \ll \delta q \, , \tag{2.3.13}$$

and that the excursions of q' about $\langle q \rangle$ are limited by

$$|q' - \langle q \rangle \,| \ll \delta q \, . \tag{2.3.14}$$

This situation is essentially indiscernible from $\delta p \to 0$, $\delta q \to \infty$. Yet the value of $\delta\,(p' - p'', \delta p)$ for $p' - p'' = 0$ is not infinite ($\sim 1/\delta p$) and the integral of $|\psi(p')|^2$ is one, in contrast to that of $[\delta\,(p' - p'')]^2$. And the Gaussian function of q' drops below unity for $|q' - \langle q \rangle \,| \sim \delta q$, resulting in unit value for the integral of $|\psi(q')|^2$, in contrast to that of $|\langle q'|p''\rangle|^2$, which is infinite.

Let's return to the characterization of the state of minimal uncertainty, $\delta q \delta p = \frac{1}{2}$, namely

$$(p - \langle p \rangle) \,|\,\rangle = \mathrm{i}\frac{\delta p}{\delta q}(q - \langle q \rangle) \,|\,\rangle \tag{2.3.15}$$

or, equivalently,

$$\left(\frac{q - \langle q \rangle}{\delta q} + \mathrm{i}\frac{p - \langle p \rangle}{\delta p} \right) |\,\rangle = 0 \, . \tag{2.3.16}$$

Notice that

$$\left\langle \left(\frac{q - \langle q \rangle}{\sqrt{2}\,\delta q} \right)^2 \right\rangle = \frac{1}{2} \, , \qquad \left\langle \left(\frac{p - \langle p \rangle}{\sqrt{2}\,\delta p} \right)^2 \right\rangle = \frac{1}{2} \, , \tag{2.3.17}$$

and

$$\frac{1}{\mathrm{i}}\left[\frac{q - \langle q \rangle}{\sqrt{2}\,\delta q}, \frac{p - \langle p \rangle}{\sqrt{2}\,\delta p} \right] = \frac{1}{2\delta q \delta p} \times \frac{1}{\mathrm{i}}[q, p] = 1 \, . \tag{2.3.18}$$

So there is no loss in generality on redefining q and p,

$$\frac{q - \langle q \rangle}{\sqrt{2}\,\delta q} \to q \, , \qquad \frac{p - \langle p \rangle}{\sqrt{2}\,\delta p} \to p \, , \tag{2.3.19}$$

so that

$$\begin{aligned} \langle q \rangle = 0 \, , \qquad (\delta q)^2 &= \left\langle q^2 \right\rangle = \frac{1}{2} \, , \\ \langle p \rangle = 0 \, , \qquad (\delta p)^2 &= \left\langle p^2 \right\rangle = \frac{1}{2} \, , \end{aligned} \tag{2.3.20}$$

and also $\frac{1}{\mathrm{i}}[q, p] = 1$ hold for the redefined variables, and $|\,\rangle$ is characterized by

$$\underbrace{\frac{1}{\sqrt{2}}(q + \mathrm{i}p)}_{\equiv y}|\,\rangle = 0 \, . \tag{2.3.21}$$

So $|\ \rangle$ appears as an eigenvector of the *non*-Hermitian operator y, with eigenvalue 0. Henceforth we write $|0\rangle$ for this state. But we can also regard $|0\rangle$ as an eigenvector of the Hermitian operator $y^\dagger y$, with the eigenvalue zero. Certainly multiplication on the left by the operator

$$y^\dagger = \frac{1}{\sqrt{2}}(q - \mathrm{i}p) \tag{2.3.22}$$

gives

$$y^\dagger y|0\rangle = 0\,. \tag{2.3.23}$$

And this is equivalent to the original statement, since the consequence

$$0 = \langle 0|y^\dagger y|0\rangle = \left(y|0\rangle\right)^\dagger \left(y|0\rangle\right) \tag{2.3.24}$$

says that the vector $y|0\rangle$ is of *zero length*: $y|0\rangle = 0$. The Hermitian operator introduced here is

$$\begin{aligned} y^\dagger y &= \tfrac{1}{2}(q - \mathrm{i}p)(q + \mathrm{i}p) = \tfrac{1}{2}\left(q^2 + p^2 - \frac{1}{\mathrm{i}}[q,p]\right) \\ &= \tfrac{1}{2}\left(q^2 + p^2 - 1\right)\,. \end{aligned} \tag{2.3.25}$$

Notice that

$$\begin{aligned} yy^\dagger &= \tfrac{1}{2}(q + \mathrm{i}p)(q - \mathrm{i}p) = \tfrac{1}{2}\left(q^2 + p^2 + \frac{1}{\mathrm{i}}[q,p]\right) \\ &= \tfrac{1}{2}\left(q^2 + p^2 + 1\right)\,, \end{aligned} \tag{2.3.26}$$

so

$$yy^\dagger - y^\dagger y = \left[y, y^\dagger\right] = 1\,. \tag{2.3.27}$$

As a reminder, for this state

$$\psi_0(q') = \pi^{-\frac{1}{4}}\,\mathrm{e}^{-\frac{1}{2}q'^2}\,, \qquad \psi_0(p') = \pi^{-\frac{1}{4}}\,\mathrm{e}^{-\frac{1}{2}p'^2}\,, \tag{2.3.28}$$

which are (2.3.6) and (2.3.7) for $\langle q\rangle = \langle p\rangle = 0$ and $\delta q = \delta p = 1/\sqrt{2}$.

2.4 States of stationary uncertainty

The minimum value of $2\delta q\delta p = 1$ is attained for the unique state $|0\rangle$. Are there other states for which $2\delta q\delta p$ is neither a maximum nor a minimum, but is *stationary* for small changes of the state?

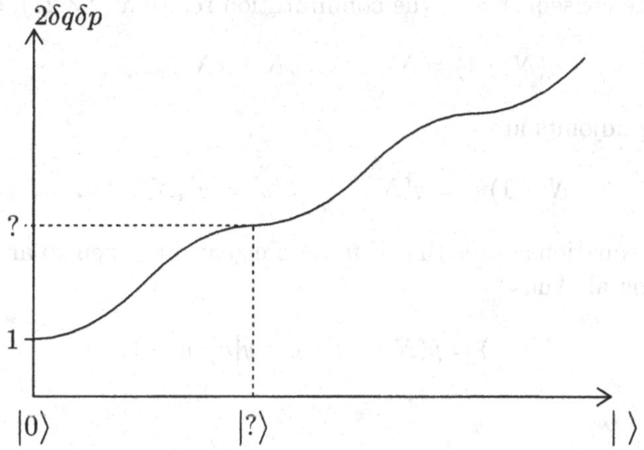

If we again use the permissible simplification $\langle q\rangle = \langle p\rangle = 0$, we have

$$(\delta q\,\delta p)^2 = \langle\ |q^2|\ \rangle\langle\ |p^2|\ \rangle\,; \qquad \langle\ |\ \rangle = 1\,. \tag{2.4.1}$$

Now consider a small change, $\Delta\langle\ |$ and $\Delta|\ \rangle$, leading to

$$\begin{aligned}(\text{a})\ \Delta\,(\delta q\,\delta p)^2 = \left(\Delta\langle\ |\right)\,[q^2(\delta p)^2 + p^2(\delta q)^2]\,|\ \rangle \\ +\langle\ |\,[q^2(\delta p)^2 + p^2(\delta q)^2]\left(\Delta|\ \rangle\right) \stackrel{?}{=} 0\end{aligned}$$

$$\text{where (b)}\qquad \Delta\langle\ |\ \rangle = \left(\Delta\langle\ |\right)|\ \rangle + \langle\ |\left(\Delta|\ \rangle\right) = 0\,. \tag{2.4.2}$$

At this point we adopt $(\delta q)^2 = (\delta p)^2$, another permissible simplification. Then (a) is satisfied in consequence of (b) if

$$\frac{q^2+p^2}{2}|\ \rangle = \lambda|\ \rangle\,, \qquad \langle\ |\frac{q^2+p^2}{2} = \langle\ |\lambda\,. \tag{2.4.3}$$

We verify later that $\langle q^2\rangle = \langle p^2\rangle$ follows from this.

Let $N = y^\dagger y$ so that [cf. (2.3.25)]

$$\frac{q^2+p^2}{2} = N + \frac{1}{2}\,. \tag{2.4.4}$$

Then, with $\lambda = n + \frac{1}{2}$, the eigenvector characterization of $|\ \rangle$ reads

$$N|\ \rangle = |\ \rangle n\,, \qquad \langle\ |N = n\langle\ |\,. \tag{2.4.5}$$

We know that $n = 0$ is an eigenvalue. What are the others? First observe that

$$yN - Ny = \left(yy^\dagger - y^\dagger y\right)y = y\,, \tag{2.4.6}$$

an immediate consequence of the commutation relation (2.3.27), or

$$y(N-1) = Ny\ , \qquad yN = (N+1)y\ , \tag{2.4.7}$$

of which the adjoints are

$$(N-1)y^\dagger = y^\dagger N\ , \qquad Ny^\dagger = y^\dagger(N+1)\ . \tag{2.4.8}$$

What these equations say is that if n is an eigenvalue, then so are $n+1$ and $n-1$, in general. Thus

$$Ny\big|n\big\rangle = y(N-1)\big|n\big\rangle = y\big|n\big\rangle(n-1) \tag{2.4.9}$$

and

$$\big\langle n\big|yN = \big\langle n\big|(N+1)y = (n+1)\big\langle n\big|y\ . \tag{2.4.10}$$

So

$$y\big|n\big\rangle = \big|n-1\big\rangle C_n\ , \qquad \big\langle n\big|y = C_{n+1}\big\langle n+1\big|\ , \tag{2.4.11}$$

where the relationship of the coefficients follows from

$$\big\langle n-1\big|y\big|n\big\rangle = C_n\ , \qquad \big\langle n\big|y\big|n+1\big\rangle = C_{n+1}\ . \tag{2.4.12}$$

Then we have

$$|C_n|^2 = \big\langle n\big|\underbrace{y^\dagger y}_{=N}\big|n\big\rangle = n \quad \text{so that} \quad C_n = \sqrt{n}\ , \tag{2.4.13}$$

if we agree on $C_n \geq 0$, and therefore

$$y\big|n\big\rangle = \big|n-1\big\rangle\sqrt{n}\ , \quad \big\langle n-1\big|y = \sqrt{n}\big\langle n\big| \tag{2.4.14}$$

and

$$\big\langle n\big|y^\dagger = \sqrt{n}\big\langle n-1\big|, \quad y^\dagger\big|n-1\big\rangle = \big|n\big\rangle\sqrt{n}\ . \tag{2.4.15}$$

From the fact that

$$n = \big\langle n\big|y^\dagger y\big|n\big\rangle = \Big(y\big|n\big\rangle\Big)^\dagger\Big(y\big|n\big\rangle\Big) \geq 0 \tag{2.4.16}$$

we conclude there must be a non-negative least value n_0 of n, which must be such that $y\big|n_0\big\rangle = 0$. Of course, from the above equations, it follows that $n_0 = 0$; the eigenvalue spectrum of N is $n = 0, 1, 2, \ldots$.

The observations that

$$\langle n|y|n\rangle = \sqrt{n+1}\langle n+1|n\rangle = 0\,,$$
$$\langle n|y^2|n\rangle = \langle n|yy|n\rangle = \sqrt{n+1}\langle n+1|n-1\rangle\sqrt{n} = 0\,, \qquad (2.4.17)$$

where

$$y = \frac{1}{\sqrt{2}}(q+\mathrm{i}p)\,,$$
$$y^2 = \frac{1}{2}(q+\mathrm{i}p)^2 = \frac{q^2-p^2}{2} + \mathrm{i}\frac{qp+pq}{2}\,, \qquad (2.4.18)$$

tell us that

$$\langle q\rangle_n = 0\,,\quad \langle p\rangle_n = 0\,,\qquad \langle(qp+pq)\rangle_n = 0\,,\quad \Big\langle(q^2-p^2)\Big\rangle_n = 0\,, \qquad (2.4.19)$$

which imply

$$(\delta q)^2_n = (\delta p)^2_n\,, \qquad (2.4.20)$$

as promised at (2.4.3). From the eigenvalue equation

$$\tfrac{1}{2}(q^2+p^2)|n\rangle = |n\rangle(n+\tfrac{1}{2}) \qquad (2.4.21)$$

we get

$$\Big\langle\tfrac{1}{2}(q^2+p^2)\Big\rangle_n = \Big\langle q^2\Big\rangle_n = (\delta q)^2_n = (\delta q\,\delta p)_n = n+\tfrac{1}{2}\,, \qquad (2.4.22)$$

which gives all the stationary values of the uncertainty product $\delta q\delta p$.

The eigenvectors for eigenvalue n are constructed as

$$\langle n| = \langle n-1|\frac{y}{\sqrt{n}} = \langle n-2|\frac{y}{\sqrt{n-1}}\frac{y}{\sqrt{n}} = \cdots = \langle 0|\frac{y^n}{\sqrt{n!}} \qquad (2.4.23)$$

and

$$|n\rangle = \frac{(y^\dagger)^n}{\sqrt{n!}}|0\rangle\,. \qquad (2.4.24)$$

These symbolic constructions are realized numerically by the wave functions

$$\psi_n(q') = \langle q'|n\rangle = \langle q'|\frac{(q-\mathrm{i}p)^n}{\sqrt{2^n n!}}|0\rangle$$
$$= (2^n\,n!)^{-\frac{1}{2}}\left(q' - \frac{\mathrm{d}}{\mathrm{d}q'}\right)^n\langle q'|0\rangle$$
$$= (2^n\,n!)^{-\frac{1}{2}}\left(q' - \frac{\mathrm{d}}{\mathrm{d}q'}\right)^n \pi^{-\frac{1}{4}}\,\mathrm{e}^{-\frac{1}{2}q'^2} \qquad (2.4.25)$$

and

$$\begin{aligned}\psi_n(p') = \langle p'|n\rangle &= \langle p'|\frac{(q-\mathrm{i}p)^n}{\sqrt{2^n n!}}|0\rangle \\ &= (-\mathrm{i})^n\,(2^n\,n!)^{-\frac{1}{2}}\left(p' - \frac{\mathrm{d}}{\mathrm{d}p'}\right)^n \langle p'|0\rangle \\ &= (-\mathrm{i})^n\,(2^n\,n!)^{-\frac{1}{2}}\left(p' - \frac{\mathrm{d}}{\mathrm{d}p'}\right)^n \pi^{-\frac{1}{4}}\,\mathrm{e}^{-\frac{1}{2}p'^2}\,. \end{aligned} \tag{2.4.26}$$

These wave functions should be connected by

$$\psi_n(p') = \int \mathrm{d}q'\,\frac{\mathrm{e}^{-\mathrm{i}p'q'}}{\sqrt{2\pi}}\,\psi_n(q')\,. \tag{2.4.27}$$

Note that

$$\int \mathrm{d}q'\,\mathrm{e}^{-\mathrm{i}p'q'}\left(q' - \frac{\partial}{\partial q'}\right)\cdots = -\mathrm{i}\left(p' - \frac{\partial}{\partial p'}\right)\int \mathrm{d}q'\,\mathrm{e}^{-\mathrm{i}p'q'}\ldots\,; \tag{2.4.28}$$

the n-fold repetition of the transformation, combined with the truth of this relation for $n = 0$ produces the general verification of (2.4.27).

2.5 Hermite polynomials

Now observe that

$$q' - \frac{\mathrm{d}}{\mathrm{d}q'} = \mathrm{e}^{\frac{1}{2}q'^2}\left(-\frac{\mathrm{d}}{\mathrm{d}q'}\right)\mathrm{e}^{-\frac{1}{2}q'^2} \tag{2.5.1}$$

from which follows

$$\left(q' - \frac{\mathrm{d}}{\mathrm{d}q'}\right)^n = \mathrm{e}^{\frac{1}{2}q'^2}\left(-\frac{\mathrm{d}}{\mathrm{d}q'}\right)^n \mathrm{e}^{-\frac{1}{2}q'^2} \tag{2.5.2}$$

and

$$\psi_n(q') = \frac{\pi^{-\frac{1}{4}}}{\sqrt{2^n n!}}\,\mathrm{e}^{\frac{1}{2}q'^2}\left(-\frac{\mathrm{d}}{\mathrm{d}q'}\right)^n \mathrm{e}^{-q'^2}\,. \tag{2.5.3}$$

Define the nth Hermite polynomial $\mathrm{H}_n(q')$ by

$$\left(-\frac{\mathrm{d}}{\mathrm{d}q'}\right)^n \mathrm{e}^{-q'^2} = \mathrm{H}_n(q')\,\mathrm{e}^{-q'^2}\,, \tag{2.5.4}$$

giving

$$\psi_n(q') = \frac{\pi^{-\frac{1}{4}}}{\sqrt{2^n n!}} \mathrm{H}_n(q')\, \mathrm{e}^{-\frac{1}{2}q'^2} \tag{2.5.5}$$

and

$$\psi_n(p') = (-\mathrm{i})^n \frac{\pi^{-\frac{1}{4}}}{\sqrt{2^n n!}} \mathrm{H}_n(p')\, \mathrm{e}^{-\frac{1}{2}p'^2} . \tag{2.5.6}$$

The first few polynomials are

$$\begin{aligned} &\mathrm{H}_0(q') = 1\,, && \mathrm{H}_3(q') = (2q')^3 - 6(2q')\,, \\ &\mathrm{H}_1(q') = 2q'\,, && \mathrm{H}_4(q') = (2q')^4 - 12(2q')^2 + 12\,, \\ &\mathrm{H}_2(q') = (2q')^2 - 2\,, && \mathrm{H}_5(q') = (2q')^5 - 20(2q')^3 + 60(2q')\,, \end{aligned} \tag{2.5.7}$$

the highest power being $(2q')^n$, generally, and

$$\mathrm{H}_n(-q') = (-1)^n \mathrm{H}_n(q') . \tag{2.5.8}$$

Note the differential recurrence relation

$$-\frac{\mathrm{d}}{\mathrm{d}q'}\left(\mathrm{H}_n(q')\,\mathrm{e}^{-q'^2}\right) = \mathrm{H}_{n+1}(q')\,\mathrm{e}^{-q'^2} \tag{2.5.9}$$

or

$$\mathrm{H}_{n+1}(q') = \left(-\frac{\mathrm{d}}{\mathrm{d}q'} + 2q'\right) \mathrm{H}_n(q') \tag{2.5.10}$$

from which all polynomials can be constructed successively. Another differential recurrence relation is found by differentiation:

$$\begin{aligned} \frac{\mathrm{d}}{\mathrm{d}q'}\left(\mathrm{H}_n(q')\,\mathrm{e}^{-q'^2}\right) &= (-1)^n \left(\frac{\mathrm{d}}{\mathrm{d}q'}\right)^n (-2q')\,\mathrm{e}^{-q'^2} \\ &= -2q'\mathrm{H}_n(q')\,\mathrm{e}^{-q'^2} + 2n\mathrm{H}_{n-1}(q')\,\mathrm{e}^{-q'^2} \end{aligned} \tag{2.5.11}$$

or

$$\frac{\mathrm{d}}{\mathrm{d}q'}\mathrm{H}_n(q') = 2n\mathrm{H}_{n-1}(q') . \tag{2.5.12}$$

The two relations are combined in

$$\frac{\mathrm{d}}{\mathrm{d}q'}\left(-\frac{\mathrm{d}}{\mathrm{d}q'} + 2q'\right) \mathrm{H}_n(q') = 2(n+1)\mathrm{H}_n(q') \tag{2.5.13}$$

which is the differential equation

$$\left(\frac{\mathrm{d}^2}{\mathrm{d}q'^2} - 2q'\frac{\mathrm{d}}{\mathrm{d}q'} + 2n\right) \mathrm{H}_n(q') = 0 . \tag{2.5.14}$$

Of course, this must be a consequence of the eigenvector equation (2.4.21) as represented by

$$\frac{1}{2}\left(q'^2 - \frac{d^2}{dq'^2}\right)\psi_n(q') = (n + \tfrac{1}{2})\psi_n(q') \,. \tag{2.5.15}$$

Indeed the replacement of $\psi_n(q')$ by $e^{-\frac{1}{2}q'^2}H_n(q')$ gives, according to

$$\left(\frac{d^2}{dq'^2} - q'^2\right) e^{-\frac{1}{2}q'^2} H_n(q') = e^{-\frac{1}{2}q'^2}\left(\frac{d^2}{dq'^2} - 2q'\frac{d}{dq'} + 2n\right) H_n(q') \,, \tag{2.5.16}$$

just the previous equation (2.5.14).

Here's another way of getting at the structure of the polynomials

$$H_n(q') = 2^n \prod_{k=1}^{n} (q' - q_k) \,, \tag{2.5.17}$$

where the q_k are the n roots of the polynomial of degree n. Write

$$\frac{d}{dq'} H_n(q') = \phi_n(q') H_n(q') \,, \tag{2.5.18}$$

so that

$$\phi_n(q') = \frac{d}{dq'} \log H_n(q') = \sum_k \frac{1}{q' - q_k} \,. \tag{2.5.19}$$

Then, since

$$\frac{d^2}{dq'^2} H_n(q') = \left(\frac{d}{dq'}\phi_n(q') + [\phi_n(q')]^2\right) H_n(q') \tag{2.5.20}$$

we get the first-order non-linear differential equation

$$\frac{d}{dq'}\phi_n(q') + [\phi_n(q')]^2 - 2q'\phi_n(q') + 2n = 0 \,. \tag{2.5.21}$$

The insertion of the general form of ϕ_n gives

$$-\sum_k \frac{1}{(q' - q_k)^2} + \sum_{k,l} \frac{1}{q' - q_k}\frac{1}{q' - q_l} - 2\sum_k \frac{q'}{q' - q_k} + 2n = 0 \,. \tag{2.5.22}$$

The identities

$$\sum_k \frac{q'}{q' - q_k} - n = \sum_k \frac{q' - (q' - q_k)}{q' - q_k} = \sum_k \frac{q_k}{q' - q_k} \tag{2.5.23}$$

and

$$\sum_{k,l}\frac{1}{q'-q_k}\frac{1}{q'-q_l}-\sum_k\frac{1}{(q'-q_k)^2}=\sum_{k\neq l}\frac{1}{q'-q_k}\frac{1}{q'-q_l}$$
$$=\sum_{k\neq l}\left(\frac{1}{q'-q_k}-\frac{1}{q'-q_l}\right)\frac{1}{q_k-q_l}$$
$$=2\sum_k\frac{1}{q'-q_k}\sum_{l(\neq k)}\frac{1}{q_k-q_l} \tag{2.5.24}$$

then lead to

$$q_k=\sum_{l(\neq k)}\frac{1}{q_k-q_l}\,. \tag{2.5.25}$$

Two general results that follow immediately are

$$\sum_k q_k=\sum_{k\neq l}\frac{1}{q_k-q_l}=0 \tag{2.5.26}$$

and

$$\sum_k q_k^2=\sum_{k\neq l}\frac{\frac{1}{2}(q_k+q_l)+\frac{1}{2}(q_k-q_l)}{q_k-q_l}=\frac{1}{2}(n-1)n\,; \tag{2.5.27}$$

the first is a consequence of the symmetry property (2.5.8): if q_k is a root, so is $-q_k$.

Let n be even, $n=2r$, so there are r distinct roots in magnitude, both signs occurring. Then

$$q_k=\sum_{l(\neq k)=1}^{r}\frac{1}{q_k-q_l}+\sum_{l(\neq k)=1}^{r}\frac{1}{q_k+q_l}+\frac{1}{2q_k} \tag{2.5.28}$$

or

$$1=2\sum_{l(\neq k)=1}\frac{1}{q_k^2-q_l^2}+\frac{1}{2q_k^2}\quad\text{for}\quad k=1,2,\ldots,r \tag{2.5.29}$$

from which follows

$$r=\frac{1}{2}\sum_{k=1}^{r}\frac{1}{q_k^2}\,. \tag{2.5.30}$$

The first non-trivial examples are: $n=2$, $r=1$, where $q_1^2=\frac{1}{2}$ is immediate and gives $\mathrm{H}_2=2^2(q'^2-q_1^2)=(2q')^2-2$; $n=4$, $r=2$, where

$$1 = 2\frac{1}{q_1^2 - q_2^2} + \frac{1}{2q_1^2}, \qquad 1 = -2\frac{1}{q_1^2 - q_2^2} + \frac{1}{2q_2^2}, \tag{2.5.31}$$

so that

$$2 = \frac{1}{2q_1^2} + \frac{1}{2q_2^2} \quad \text{or} \quad q_1^2 + q_2^2 = 4q_1^2 q_2^2 \tag{2.5.32}$$

and

$$\frac{4}{q_1^2 - q_2^2} = \frac{1}{2q_2^2} - \frac{1}{2q_1^2} \quad \text{or} \quad \left(q_1^2 - q_2^2\right)^2 = 8q_1^2 q_2^2 . \tag{2.5.33}$$

This gives q_1^2 and q_2^2 as $\frac{1}{4}(1+x)$ where $(x-1)^2 = 8x$, that is

$$q_1^2 = \tfrac{3}{2} - \sqrt{\tfrac{3}{2}}, \qquad q_2^2 = \tfrac{3}{2} + \sqrt{\tfrac{3}{2}}. \tag{2.5.34}$$

So

$$\begin{aligned} H_4 &= 2^4 \left(q'^2 - \tfrac{3}{2} + \sqrt{\tfrac{3}{2}}\right)\left(q'^2 - \tfrac{3}{2} - \sqrt{\tfrac{3}{2}}\right) \\ &= 2^4 \left(q'^4 - 3q'^2 + \tfrac{3}{4}\right) = (2q')^4 - 12(2q')^2 + 12 . \end{aligned} \tag{2.5.35}$$

If n is odd, $n = 2r + 1$, there are r pairs of opposite signed roots, and zero. Now

$$q_k = \sum_{l \neq k = 1}^{r} \frac{1}{q_k - q_l} + \frac{1}{q_k} + \sum_{l \neq k = 1}^{r} \frac{1}{q_k + q_l} + \frac{1}{2q_k} \tag{2.5.36}$$

and

$$1 = 2\sum_{l \neq q} \frac{1}{q_k^2 - q_l^2} + \frac{3}{2q_k^2}, \qquad r = \frac{3}{2}\sum_{k=1}^{r} \frac{1}{q_k^2}. \tag{2.5.37}$$

With $n = 3$, $r = 1$ we have $q_1^2 = \frac{3}{2}$ and $\mathrm{H}_3 = 2^3 q' \left(q' - \frac{3}{2}\right) = (2q')^3 - 6(2q')$. For $n = 5$, $r = 2$,

$$1 = 2\frac{1}{q_1^2 - q_2^2} + \frac{3}{2}\frac{1}{q_1^2}, \qquad 1 = -2\frac{1}{q_1^2 - q_2^2} + \frac{3}{2}\frac{1}{q_2^2}, \tag{2.5.38}$$

so that here

$$3\left(q_1^2 + q_2^2\right) = 4q_1^2 q_2^2 , \qquad 3\left(q_1^2 - q_2^2\right)^2 = 8q_1^2 q_2^2 , \tag{2.5.39}$$

which gives q_1^2 and q_2^2 as $\frac{3}{4}(1+x)$ where $(x-1)^2 = \frac{8}{3}x$,

$$q_1^2 = \tfrac{5}{2} - \sqrt{\tfrac{5}{2}}, \qquad q_2^2 = \tfrac{5}{2} + \sqrt{\tfrac{5}{2}}, \tag{2.5.40}$$

and finally

$$\begin{aligned}\mathrm{H}_5 &= 2^5 q' \left(q'^2 - \tfrac{5}{2} + \sqrt{\tfrac{5}{2}}\right)\left(q'^2 - \tfrac{5}{2} - \sqrt{\tfrac{5}{2}}\right)\\ &= (2q')^5 - 20(2q')^3 + 60(2q') \,. \end{aligned} \tag{2.5.41}$$

All polynomials in (2.5.7) are thus written in the form (2.5.17).

2.6 Completeness of stationary-uncertainty states

How do we know that the vectors $|n\rangle$, $\langle n|$, are complete, that there are no other eigenvalues of N than $n = 0, 1, 2, \ldots$? That is asking whether

$$\sum_{n=0}^{\infty} \big|nn\big| = 1 \,. \tag{2.6.1}$$

To answer, let's begin by exhibiting the $\big|nn\big|$ as algebraic functions of N. The symbol $\big|00\big|$ should have the value 1 if N is 0, and zero otherwise. Notice that a function of the number $n = 0, 1, 2, \ldots$ that has this behavior is

$$(1-1)^n = \begin{cases} 1 \text{ for } n = 0\,, \\ 0 \text{ for } n = 1, 2, \ldots \,. \end{cases} \tag{2.6.2}$$

Accordingly, replacing n by N and using the binomial expansion:

$$\big|00\big| = \sum_{k=0}^{\infty} (-1)^k \frac{N!}{k!(N-k)!} = 1 - N + \frac{N(N-1)}{2} + \cdots \,. \tag{2.6.3}$$

Similarly,

$$n(1-1)^{n-1} = \begin{cases} 1 \text{ for } n = 1\,, \\ 0 \text{ for } n = 0, 2, 3 \ldots\,, \end{cases} \tag{2.6.4}$$

and

$$\tfrac{1}{2} n(n-1)(1-1)^{n-2} = \begin{cases} 1 \text{ for } n = 2\,, \\ 0 \text{ for } n = 0, 1, 3 \ldots \,. \end{cases} \tag{2.6.5}$$

So

$$\begin{aligned} \big|11\big| &= \sum_{k=0}^{\infty} (-1)^k \frac{N!}{k!(N-1-k)!} = N - N(N-1) + \cdots \,, \\ \big|22\big| &= \sum_{k=0}^{\infty} (-1)^k \frac{1}{2} \frac{N!}{k!(N-2-k)!} = \tfrac{1}{2} N(N-1) + \cdots \,, \\ \big|nn\big| &= \sum_{k=0}^{\infty} (-1)^k \frac{1}{n!} \frac{N!}{k!(N-n-k)!} \,. \end{aligned} \tag{2.6.6}$$

Now try

$$\sum_{n=0}^{\infty}\big|nn\big| = \sum_{k=0}^{\infty}(-1)^k\frac{1}{k!}\frac{N!}{(N-k)!}\underbrace{\sum_{n=0}^{\infty}\frac{(N-k)!}{n!(N-k-n)!}}_{=(1+1)^{N-k}}$$

$$= \sum_{k=0}^{\infty}\frac{N!}{k!(N-k)!}(-1)^k 2^{N-k} = (2-1)^N = 1^N = 1\,, \qquad (2.6.7)$$

all right.

How is this related to the operator constructions of (2.4.23) and (2.4.24), where

$$\big|nn\big| = \frac{(y^\dagger)^n}{\sqrt{n!}}\big|0\big\rangle\big\langle 0\big|\frac{y^n}{\sqrt{n!}} = \frac{1}{n!}(y^\dagger)^n\big|00\big|y^n \qquad (2.6.8)$$

and

$$\sum_{n=0}^{\infty}\big|nn\big| = \sum_{n=0}^{\infty}\frac{1}{n!}(y^\dagger)^n\big|00\big|y^n \equiv \exp\left(y^\dagger;\big|00\big|;y\right)\ ? \qquad (2.6.9)$$

The notation is intended to indicate the insertion of $\big|00\big|$ between the powers of $y^\dagger$, on the left, and y, on the right. It helps to notice that $y^\dagger y = N$, and

$$(y^\dagger)^2y^2 = y^\dagger N y = y^\dagger y(N-1) = N(N-1) \qquad (2.6.10)$$

followed by

$$\begin{aligned}(y^\dagger)^3y^3 &= y^\dagger N(N-1)y = y^\dagger N y(N-2) = y^\dagger y(N-1)(N-2)\\ &= N(N-1)(N-2)\,; \end{aligned} \qquad (2.6.11)$$

in general

$$(y^\dagger)^n y^n = \frac{N!}{(N-k)!}\,. \qquad (2.6.12)$$

Then we can write

$$\begin{aligned}\big|00\big| &= \sum_{k=0}^{\infty}(-1)^k\frac{N!}{k!(N-k)!} = \sum_{k=0}^{\infty}(-1)^k\frac{(y^\dagger)^k y^k}{k!}\\ &\equiv \exp\left(-y^\dagger;y\right)\,, \end{aligned} \qquad (2.6.13)$$

a basic ordered exponential function. And now

$$\sum_{n=0}^{\infty}\big|nn\big| = \exp\Big(y^\dagger;\exp\left(-y^\dagger;y\right);y\Big) = 1 \qquad (2.6.14)$$

because the two ordered exponentials are inverses of each other. That would be obvious if we did not have to worry about the non-commutativity of $y^\dagger$ and y, but in fact, since all $y^\dagger$'s stand on the left, and all y's on the right, that non-commutativity never enters. This is illustrated by the first terms

$$\begin{aligned}\exp\Bigl(y^\dagger ; \exp\left(-y^\dagger ; y\right) ; y\Bigr) &= \mathrm{e}^{-y^\dagger ; y} + y^\dagger \,\mathrm{e}^{-y^\dagger ; y} y + \frac{1}{2} y^{\dagger 2}\, \mathrm{e}^{-y^\dagger ; y} y^2 + \cdots \\ &= (1 - y^\dagger y + \frac{1}{2} y^{\dagger 2} y^2 + \cdots) \\ &\quad + y^\dagger (1 - y^\dagger y + \cdots) y + \frac{1}{2} y^{\dagger 2} (1 + \cdots) y^2 + \cdots \\ &= 1 \,. \end{aligned} \tag{2.6.15}$$

2.7 Eigenvectors of non-Hermitian operators

Nevertheless, things would be more immediately apparent if we could replace the operators by numbers, by introducing eigenvectors and eigenvalues of y and $y^\dagger$:

$$y|y'\rangle = |y'\rangle y' , \qquad \langle y^{\dagger\prime}| y^\dagger = y^{\dagger\prime} \langle y^{\dagger\prime}| \tag{2.7.1}$$

for then

$$\begin{aligned}\langle y^{\dagger\prime}| \Bigl(\sum_n |nn|\Bigr) |y'\rangle &= \langle y^{\dagger\prime}| \,\mathrm{e}^{y^{\dagger\prime} y'}\, \mathrm{e}^{-y^{\dagger\prime} y'} |y'\rangle \\ &= \langle y^{\dagger\prime}| 1 |y'\rangle \,. \end{aligned} \tag{2.7.2}$$

We know that there are zero eigenvalues,

$$y|0\rangle = 0 , \qquad \langle 0| y^\dagger = 0 \,. \tag{2.7.3}$$

So we have only to *change* the eigenvalues: $0 \to y'$, $0 \to y^{\dagger\prime}$. That sounds familiar [we use the adjoints of the relations stated in (1.16.19)]:

$$|q' + q''\rangle = \mathrm{e}^{-\mathrm{i} q'' p} |q'\rangle , \quad \langle p' + p''| = \langle p'| \,\mathrm{e}^{-\mathrm{i} p'' q} \,, \tag{2.7.4}$$

illustrating operators that change the eigenvalues of q and p. These relations are consequences of $[q,p] = \mathrm{i}$. But $[y, \mathrm{i}y^\dagger] = \mathrm{i}$, suggesting an analogy: $q \to y$, $p \to \mathrm{i}y^\dagger$. Accordingly we expect

$$\begin{aligned} q' \to 0 \,,\ q'' \to y' ,\ p \to \mathrm{i}y^\dagger : &\quad |y'\rangle = \mathrm{e}^{y^\dagger y'} |0\rangle \,, \\ p' \to 0 \,,\ p'' \to \mathrm{i}y^{\dagger\prime} ,\ q \to y : &\quad \langle y^{\dagger\prime}| = \langle 0| \,\mathrm{e}^{y^{\dagger\prime} y} \,. \end{aligned} \tag{2.7.5}$$

Unlike $|q'\rangle$, $\langle p'|$, these vectors are in adjoint relation.

First, notice that the adjoint of

$$y\big|y'\big\rangle = \big|y'\big\rangle y' \tag{2.7.6}$$

is

$$\big|y'\big\rangle^\dagger y^\dagger = y'^* \big|y'\big\rangle^\dagger , \tag{2.7.7}$$

so

$$\big|y'\big\rangle^\dagger = \big\langle y^{\dagger\prime}\big| \quad \text{if} \quad y^{\dagger\prime} = y'^* . \tag{2.7.8}$$

Clearly the eigenvalues of non-Hermitian $y = 2^{-\frac{1}{2}}(q + \mathrm{i}p)$ are complex numbers. Even if we thought they were real, we would see that $\overline{y} = \mathrm{e}^{\mathrm{i}\alpha} y$, $\overline{y}^\dagger = \mathrm{e}^{-\mathrm{i}\alpha} y^\dagger$, $[\overline{y}, \overline{y}^\dagger] = 1$, which are also a perfectly good pair of non-Hermitian operators, have complex eigenvalues: $\overline{y}' = \mathrm{e}^{\mathrm{i}\alpha} y'$. So it follows that

$$\left(\mathrm{e}^{y^\dagger y'}\big|0\big\rangle\right)^\dagger = \big\langle 0\big|\,\mathrm{e}^{y^{\dagger\prime} y} \quad \text{if} \quad y^{\dagger\prime} = y'^* . \tag{2.7.9}$$

Now let's verify what the analogy suggests:

$$y\big|y'\big\rangle = y\,\mathrm{e}^{y^\dagger y'}\big|0\big\rangle = \mathrm{e}^{y^\dagger y'}\left(\mathrm{e}^{-y^\dagger y'} y\, \mathrm{e}^{y^\dagger y'}\right)\big|0\big\rangle . \tag{2.7.10}$$

Then consider

$$\begin{aligned} \frac{\mathrm{d}}{\mathrm{d}y'}\left(\mathrm{e}^{-y^\dagger y'} y\, \mathrm{e}^{y^\dagger y'}\right) &= \mathrm{e}^{-y^\dagger y'} y \left(y^\dagger \mathrm{e}^{y^\dagger y'}\right) + \left(\mathrm{e}^{-y^\dagger y'}(-y^\dagger)\right) y\, \mathrm{e}^{y^\dagger y'} \\ &= \mathrm{e}^{-y^\dagger y'} \underbrace{[y, y^\dagger]}_{=1} \mathrm{e}^{y^\dagger y'} = 1 , \end{aligned} \tag{2.7.11}$$

so

$$\mathrm{e}^{-y^\dagger y'} y\, \mathrm{e}^{y^\dagger y'} = y' + y \tag{2.7.12}$$

which incorporates the obvious integration constant. Then, since

$$\mathrm{e}^{y^\dagger y'}(y' + \underset{\substack{\downarrow \\ 0}}{y})\big|0\big\rangle = \mathrm{e}^{y^\dagger y'}\big|0\big\rangle y' = \big|y'\big\rangle y' , \tag{2.7.13}$$

we have checked that $y\big|y'\big\rangle = \big|y'\big\rangle y'$. The eigenvector is

$$\big|y'\big\rangle = \sum_{n=0}^{\infty} \frac{(y^\dagger)^n}{\sqrt{n!}} \frac{(y')^n}{\sqrt{n!}} \big|0\big\rangle = \sum_{n=0}^{\infty} \big|n\big\rangle \frac{(y')^n}{\sqrt{n!}} , \tag{2.7.14}$$

which employs (2.4.24), and taking the adjoint gives ($y'^* \to y^{\dagger\prime}$)

$$\langle y^{\dagger\prime}| = \sum_{n=0}^{\infty} \frac{(y^{\dagger\prime})^n}{\sqrt{n!}} \langle n| \,, \tag{2.7.15}$$

from which we get

$$\langle y^{\dagger\prime}|y''\rangle = \sum_{n=0}^{\infty} \frac{(y^{\dagger\prime})^n}{\sqrt{n!}} \frac{(y'')^n}{\sqrt{n!}} = \mathrm{e}^{y^{\dagger\prime} y''} \,. \tag{2.7.16}$$

Apart from the numerical factor, this is the analog of

$$\langle p'|q'\rangle = \frac{1}{\sqrt{2\pi}} \mathrm{e}^{-\mathrm{i}p'q'} \,. \tag{2.7.17}$$

Notice the special examples

$$\langle 0|y'\rangle = 1, \qquad \langle y^{\dagger\prime}|0\rangle = 1 \,. \tag{2.7.18}$$

The construction of the y, $y^\dagger$ eigenvectors in terms of n states is conveyed by the components

$$\langle n|y'\rangle = \frac{(y')^n}{\sqrt{n!}} \,, \qquad \langle y^{\dagger\prime}|n\rangle = \frac{(y^{\dagger\prime})^n}{\sqrt{n!}} \,. \tag{2.7.19}$$

Evidently we have arrived at y', and $y^{\dagger\prime}$, wave functions that represent the abstract n vectors of (2.4.23) and (2.4.24). Indeed, from the latter we get

$$\langle y^{\dagger\prime}|n\rangle = \langle y^{\dagger\prime}| \frac{(y^\dagger)^n}{\sqrt{n!}} |0\rangle = \frac{(y^{\dagger\prime})^n}{\sqrt{n!}} \underbrace{\langle y^{\dagger\prime}|0\rangle}_{=1} \,. \tag{2.7.20}$$

And we can understand the statement $\langle y^{\dagger\prime}|0\rangle = 1$ as the numerical realization of a vector property, namely

$$y|0\rangle = 0 \,. \tag{2.7.21}$$

First, notice that

$$\begin{aligned} \frac{\partial}{\partial y^{\dagger\prime}} \langle y^{\dagger\prime}| &= \frac{\partial}{\partial y^{\dagger\prime}} \langle 0|\, \mathrm{e}^{y^{\dagger\prime} y} = \langle 0|\, \mathrm{e}^{y^{\dagger\prime} y} y \\ &= \langle y^{\dagger\prime}| y \,, \end{aligned} \tag{2.7.22}$$

the analog of

$$\mathrm{i}\frac{\partial}{\partial p'} \langle p'| = \langle p'| q \,. \tag{2.7.23}$$

Then

$$\frac{\partial}{\partial y^{\dagger\prime}}\langle y^{\dagger\prime}|0\rangle = \langle y^{\dagger\prime}|y|0\rangle = 0 \tag{2.7.24}$$

and

$$\langle y^{\dagger\prime}|0\rangle = \langle 0|0\rangle = 1 . \tag{2.7.25}$$

Having recognized these wave functions, note that

$$\langle y^{\dagger\prime}|y''\rangle = \sum_{n=0}^{\infty} \frac{(y^{\dagger\prime})^n}{\sqrt{n!}} \frac{(y'')^n}{\sqrt{n!}} = \sum_{n=0}^{\infty} \langle y^{\dagger\prime}|n\rangle\langle n|y''\rangle \tag{2.7.26}$$

repeats the properties of completeness for the n states.

What is the q' description of the $|y'\rangle$ vectors, that is: what is $\langle q'|y'\rangle$? We construct differential operators according to

$$\frac{\partial}{\partial q'}\langle q'| = \langle q'|\mathrm{i}p, \qquad \frac{\partial}{\partial y'}|y'\rangle = y^\dagger|y'\rangle \tag{2.7.27}$$

the latter being the adjoint of the earlier statement about $\langle y^{\dagger\prime}|$ in (2.7.22) or the analog of $(\partial/\partial q')|q'\rangle = -\mathrm{i}p|q'\rangle$. So

$$\begin{aligned} \frac{\partial}{\partial q'}\langle q'|y'\rangle &= \langle q'|\mathrm{i}p|y'\rangle = \langle q'|\left(\sqrt{2}\,y - q\right)|y'\rangle \\ &= \left(\sqrt{2}\,y' - q'\right)\langle q'|y'\rangle \end{aligned} \tag{2.7.28}$$

and

$$\begin{aligned} \frac{\partial}{\partial y'}\langle q'|y'\rangle &= \langle q'|y^\dagger|y'\rangle = \langle q'|\left(\sqrt{2}\,q - y\right)|y'\rangle \\ &= \left(\sqrt{2}\,q' - y'\right)\langle q'|y'\rangle , \end{aligned} \tag{2.7.29}$$

leading to

$$\langle q'|y'\rangle = C\,\mathrm{e}^{-\frac{1}{2}q'^2 + \sqrt{2}\,q'y' - \frac{1}{2}y'^2} . \tag{2.7.30}$$

The complex conjugate of this is

$$\langle y^{\dagger\prime}|q'\rangle = C^*\,\mathrm{e}^{-\frac{1}{2}q'^2 + \sqrt{2}\,q'y^{\dagger\prime} - \frac{1}{2}(y^{\dagger\prime})^2} . \tag{2.7.31}$$

We can determine the integration constant C in several ways. First, put $y' = 0$, which must give us the known wave function

$$\psi_0(q') = \pi^{-\frac{1}{4}}\,\mathrm{e}^{-\frac{1}{2}q'^2} , \quad \text{so that } C = \pi^{-\frac{1}{4}} . \tag{2.7.32}$$

Or, we can use the completeness of the q' states:

$$\begin{aligned}\langle y^\dagger|y''\rangle &= \int \langle y^{\dagger\prime}|q'\rangle \mathrm{d}q' \langle q'|y''\rangle \\ &= |C|^2 \int \mathrm{d}q'\, \mathrm{e}^{-q'^2 + \sqrt{2}\,q'(y^{\dagger\prime} + y'') - \frac{1}{2}(y^{\dagger\prime 2} + y''^2)} \\ &= |C|^2 \underbrace{\int \mathrm{d}q'\, \mathrm{e}^{-[q' - (y^{\dagger\prime} + y'')/\sqrt{2}]^2}}_{=\sqrt{\pi}} \underbrace{\mathrm{e}^{y^{\dagger\prime} y''}}_{=\langle y^\dagger|'y''\rangle}\,, \end{aligned} \tag{2.7.33}$$

so that $C = \pi^{-\frac{1}{4}}$ again (to within a here undetermined phase).

As an application of the transformation function $\langle q'|y'\rangle$ that touches a number of bases, introduce the n states via

$$\langle q'|y'\rangle = \sum_{n=0}^{\infty} \langle q'|n\rangle\langle n|y'\rangle\,, \tag{2.7.34}$$

so that

$$\pi^{-\frac{1}{4}}\,\mathrm{e}^{-\frac{1}{2}q'^2 + \sqrt{2}\,q'y' - \frac{1}{2}y'^2} = \sum_{n=0}^{\infty} \psi_n(q') \frac{(y')^n}{\sqrt{n!}}\,. \tag{2.7.35}$$

This appears as a generating function for the ψ_n, which are selected by the power of y'. Think of this as a Taylor* series expansion and identify

$$\begin{aligned}\psi_n(q') &= \frac{1}{\sqrt{n!}}\pi^{-\frac{1}{4}} \left(\frac{\mathrm{d}}{\mathrm{d}y'}\right)^n \mathrm{e}^{-\frac{1}{2}\left(y' - \sqrt{2}q'\right)^2}\, \mathrm{e}^{\frac{1}{2}q'^2}\bigg|_{y'=0} \\ &= \frac{1}{\sqrt{n!}}\pi^{-\frac{1}{4}}\,\mathrm{e}^{\frac{1}{2}q'^2} \left(\frac{\mathrm{d}}{\mathrm{d}(-\sqrt{2}q')}\right)^n \mathrm{e}^{-q'^2} \\ &= \frac{\pi^{-\frac{1}{4}}}{\sqrt{2^n n!}}\,\mathrm{e}^{\frac{1}{2}q'^2} \left(-\frac{\mathrm{d}}{\mathrm{d}q'}\right)^n \mathrm{e}^{-q'^2} \\ &= \frac{\pi^{-\frac{1}{4}}}{\sqrt{2^n n!}} \mathrm{H}_n(q')\,\mathrm{e}^{-\frac{1}{2}q'^2}\,, \end{aligned} \tag{2.7.36}$$

which agrees with the previous result (2.5.5), indeed.

To say more about the physical interpretation of the states $|y'\rangle$, $\langle y^{\dagger\prime}|$, we must notice that, with the exception of $|0\rangle$, these vectors are of finite, but *not* unit length,

$$y^{\dagger\prime} = y'^* : \qquad \langle y^{\dagger\prime}|y'\rangle = \mathrm{e}^{|y'|^2} \geq 1\,. \tag{2.7.37}$$

Thus a normalized vector is

*Brook TAYLOR (1685–1731)

$$|q',p'\rangle = |y'\rangle\, \mathrm{e}^{-\frac{1}{2}|y'|^2} \quad \text{with} \quad y' = \frac{1}{\sqrt{2}}(q' + \mathrm{i}p')\,, \tag{2.7.38}$$

(frequently called a *coherent state*, a term coined by Glauber*) and an expectation value in this state is computed as

$$\langle X\rangle_{q',p'} = \mathrm{e}^{-|y'|^2} \langle y^{\dagger\prime}|X|y'\rangle\,. \tag{2.7.39}$$

Since $(y - y')|y'\rangle = 0$, we have

$$\langle y\rangle_{q',p'} = y' \qquad \text{or} \quad \langle q\rangle_{q',p'} = q' \quad \text{and} \quad \langle p\rangle_{q',p'} = p'\,, \tag{2.7.40}$$

and from

$$\begin{gathered} \langle (y-y')^2\rangle_{q',p'} = 0\,, \qquad \left\langle (y^\dagger - y^{\dagger\prime})^2 \right\rangle_{q',p'} = 0\,, \\ \left\langle (y^\dagger - y^{\dagger\prime})(y-y') \right\rangle_{q',p'} = 0\,, \end{gathered} \tag{2.7.41}$$

we get

$$(\delta q)^2 = (\delta p)^2 = \tfrac{1}{2} \qquad \text{and} \qquad \delta q\, \delta p = \tfrac{1}{2}\,. \tag{2.7.42}$$

The y eigenvectors are all minimal-uncertainty states, which we have already discussed in Section 2.3. To see the connection we return to our earlier results, with just the simplifications $\delta q = \delta p\ (= 1/\sqrt{2})$:

$$\gamma = \frac{\delta p}{\delta q} = 1: \qquad (p - \langle p\rangle)|\ \rangle = \mathrm{i}(q - \langle q\rangle)|\ \rangle \tag{2.7.43}$$

or

$$\left[\frac{q + \mathrm{i}p}{\sqrt{2}} - \frac{\langle q\rangle + \mathrm{i}\,\langle p\rangle}{\sqrt{2}}\right]|\ \rangle = 0\,, \tag{2.7.44}$$

which is $(y - y')|\ \rangle = 0$, with $q' = \langle q\rangle$, $p' = \langle p\rangle$. Also [cf. (2.3.6)]

$$\psi(q'') = \pi^{-\frac{1}{4}}\, \mathrm{e}^{\mathrm{i}p'(q''-q')}\, \mathrm{e}^{-\frac{1}{2}(q''-q')^2}\, \mathrm{e}^{\frac{\mathrm{i}}{2}q'p'}\,, \tag{2.7.45}$$

as compared with

$$\begin{aligned} \langle q''|q',p'\rangle &= \langle q''|y'\rangle\, \mathrm{e}^{-\frac{1}{2}|y'|^2} \\ &= \pi^{-\frac{1}{4}}\, \mathrm{e}^{-\frac{1}{2}q''^2 + \sqrt{2}q''y' - \frac{1}{2}y'^2 - \frac{1}{2}|y'|^2} \\ &= \pi^{-\frac{1}{4}}\, \mathrm{e}^{-\frac{1}{2}q''^2 + q''(q'+\mathrm{i}p') - \frac{1}{4}(q'+\mathrm{i}p')^2 - \frac{1}{4}(q'^2+p'^2)} \\ &= \pi^{-\frac{1}{4}}\, \mathrm{e}^{\mathrm{i}p'(q''-q')}\, \mathrm{e}^{\frac{\mathrm{i}}{2}q'p'}\, \mathrm{e}^{-\frac{1}{2}(q''-q')^2}\,; \end{aligned} \tag{2.7.46}$$

they are the same.

---

*Roy Jay GLAUBER (b. 1925)

Given vectors like the $|q',p'\rangle$ and $\langle q',p'|$ we naturally ask about completeness and orthogonality. Let's look first at

$$\langle q''|q',p'\rangle\langle q',p'|q'''\rangle = \pi^{-\frac{1}{2}}\,\mathrm{e}^{\mathrm{i}p'(q''-q''')}\,\mathrm{e}^{-\frac{1}{2}(q''-q')^2}\,\mathrm{e}^{-\frac{1}{2}(q'''-q')^2} \tag{2.7.47}$$

where

$$\int \langle q''|q',p'\rangle \frac{\mathrm{d}p'}{2\pi}\langle q',p'|q'''\rangle = \pi^{-\frac{1}{2}}\delta(q''-q''')\,\mathrm{e}^{-(q''-q')^2} \tag{2.7.48}$$

and then

$$\int \langle q''|q',p'\rangle \frac{\mathrm{d}q'\mathrm{d}p'}{2\pi}\langle q',p'|q'''\rangle = \delta(q''-q''') = \langle q''|1|q'''\rangle \tag{2.7.49}$$

which is a completeness statement in appearance:

$$1 = \int |q',p'\rangle \frac{\mathrm{d}q'\mathrm{d}p'}{2\pi}\langle q',p'|\,. \tag{2.7.50}$$

Indeed, as learned in Problem 2-29, there are many different completeness statements for the $|q',p'\rangle$ vectors: they are over-complete.

But now consider

$$\begin{aligned}\langle q',p'|q'',p''\rangle &= \mathrm{e}^{-\frac{1}{2}|y'|^2}\underbrace{\langle y^{\dagger\prime}|y''\rangle}_{=\exp(y^{\dagger\prime}y'')}\mathrm{e}^{-\frac{1}{2}|y''|^2} \\ &= \mathrm{e}^{\frac{\mathrm{i}}{2}(q'p''-p'q'')}\,\mathrm{e}^{-\frac{1}{4}(q'-q'')^2}\,\mathrm{e}^{-\frac{1}{4}(p'-p'')^2}\,.\end{aligned} \tag{2.7.51}$$

This equals one for $q',p' = q'',p''$, but it is *not* zero for $q',p' \neq q'',p''$. It only becomes small for appreciable values of $|q'-q''|$ or $|p'-p''|$. But that is as it should be; the state $\langle q',p'|$ describes a situation in which the average values of q and p measurements are q' and p', but the individual results have spreads of $\delta q = \delta p = 1/\sqrt{2}$. A change of q' or p' that is small on this scale is not a significantly different state. A change that is large, resulting in very small values of one or both Gaussian functions, produces a different state. The completeness expression suggests that one can approximately associate a state with each range $\Delta q'$, $\Delta p'$ such that

$$\frac{\Delta q'\Delta p'}{2\pi} = 1\,. \tag{2.7.52}$$

It is not necessary to identify individual states if we ask for probabilities in a range $\Delta q'\Delta p'/2\pi \gg 1$. It is in this sense that we assert, from $[\psi(q',p') = \langle q',p'|\ \rangle$ for unit vector $|\ \rangle]$

$$1 = \int \frac{\mathrm{d}q'\mathrm{d}p'}{2\pi}|\psi(q',p')|^2\,, \tag{2.7.53}$$

that

$$\frac{\mathrm{d}q'\mathrm{d}p'}{2\pi}|\psi(q',p')|^2 \tag{2.7.54}$$

is the probability that q and p measurements of minimal uncertainty will yield values q', p' in the interval $\mathrm{d}q'$, $\mathrm{d}p'$.

2.8 Classical limit

As an example, consider

$$\psi_n(q',p') = \langle q',p'|n\rangle = \mathrm{e}^{-\frac{1}{2}|y'|^2} \underbrace{\langle y^{\dagger\prime}|n\rangle}_{=(y^{\dagger\prime})^n/\sqrt{n!}} \tag{2.8.1}$$

and

$$|\psi_n(q',p')|^2 = \frac{1}{n!}\left(\frac{q'^2+p'^2}{2}\right)^n \mathrm{e}^{-\frac{1}{2}(q'^2+p'^2)} = \frac{\eta^n}{n!}\,\mathrm{e}^{-\eta}$$

$$\text{with} \quad \eta = \frac{q'^2+p'^2}{2} \,. \tag{2.8.2}$$

We also have (polar coordinates)

$$\begin{aligned} q' &= \sqrt{2\eta}\,\cos\phi \,, \\ p' &= \sqrt{2\eta}\,\sin\phi \,, \\ \frac{\mathrm{d}q'\mathrm{d}p'}{2\pi} &= \frac{\mathrm{d}\phi}{2\pi}\mathrm{d}\eta \to \mathrm{d}\eta \,, \end{aligned} \tag{2.8.3}$$

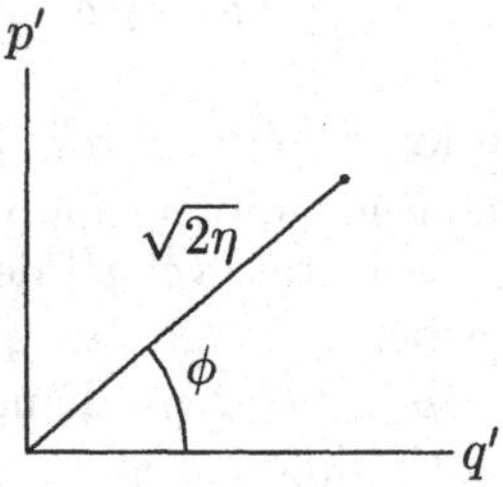

so one quantum state corresponds approximately to $\Delta\eta = 1$ (more about this in a moment). As it should,

$$\int \frac{\mathrm{d}q'\mathrm{d}p'}{2\pi}|\psi_n(q',p')|^2 = \int_0^\infty \mathrm{d}\eta\,\frac{\eta^n}{n!}\,\mathrm{e}^{-\eta} = 1 \,. \tag{2.8.4}$$

Also notice that

$$\sum_{n=0}^{\infty}|\psi_n(q',p')|^2 = \sum_{n=0}^{\infty}\frac{\eta^n}{n!}\,\mathrm{e}^{-\eta} = \mathrm{e}^{\eta}\,\mathrm{e}^{-\eta} = 1 \,, \tag{2.8.5}$$

which could have been anticipated:

$$\sum_{n=0}^{\infty} |\psi_n(q',p')|^2 = \sum_{n=0}^{\infty} \langle q',p'|n\rangle\langle n|q',p'\rangle$$
$$= \langle q',p'|q',p'\rangle = 1 , \qquad (2.8.6)$$

where the completeness of the $|nn|$ measurement symbols and the normalization of the $|q',p'\rangle$ vectors enter.

For a given n the maximum value of $|\psi_n|^2$ occurs at η such that

$$\frac{\mathrm{d}}{\mathrm{d}\eta} \log |\psi_n|^2 = \frac{\mathrm{d}}{\mathrm{d}\eta}[n \log \eta - \eta] = \frac{n}{\eta} - 1 = 0 , \qquad (2.8.7)$$

or, at $\eta = n$. The value at the maximum is

$$|\psi_n|^2_{\eta=n} = \frac{n^n}{n!} \mathrm{e}^{-n} . \qquad (2.8.8)$$

According to Stirling's* asymptotic expansion for large n,

$$n! \sim \sqrt{2\pi n}\, n^n \mathrm{e}^{-n} \left(1 + \frac{1}{12n} + \cdots\right) , \qquad (2.8.9)$$

[the displayed terms give $1! \cong 0.9990$, $2! \cong 2 \times 0.9995$, $3! \cong 6 \times 0.9997$] and

$$|\psi_n|^2 \cong \frac{1}{\sqrt{2\pi n}} \frac{1}{1 + \frac{1}{12n}} \qquad (2.8.10)$$

at the maximum. The average values of η and η^2 are

$$\langle\eta\rangle = \int_0^\infty \mathrm{d}\eta \frac{\eta^{n+1}}{n!} \mathrm{e}^{-\eta} = \frac{(n+1)!}{n!} = n + 1 ,$$
$$\left\langle\eta^2\right\rangle = \int_0^\infty \mathrm{d}\eta \frac{\eta^{n+2}}{n!} \mathrm{e}^{-\eta} = \frac{(n+2)!}{n!} = (n+2)(n+1)$$
$$= \langle\eta\rangle^2 + \langle\eta\rangle , \qquad (2.8.11)$$

giving the dispersion

$$(\delta\eta)^2 = \left\langle(\eta - \langle\eta\rangle)^2\right\rangle = \left\langle\eta^2\right\rangle - \langle\eta\rangle^2 = \langle\eta\rangle \qquad (2.8.12)$$

or

$$\delta\eta = \sqrt{\langle\eta\rangle} . \qquad (2.8.13)$$

So, for very large n, the *relative* dispersion of η approaches zero,

$$\frac{\delta\eta}{\langle\eta\rangle} = \frac{1}{\sqrt{\langle\eta\rangle}} = \frac{1}{\sqrt{n+1}} \begin{cases} \cong 1/\sqrt{n} & \text{for } n \gg 1 , \\ \to 0 & \text{for } n \to \infty . \end{cases} \qquad (2.8.14)$$

*James Stirling (1692–1770)

To see what the probability density does, we put $\eta = n(1+\nu)$, and use the leading approximation of (2.8.9):

$$\begin{aligned}|\psi_n|^2 &\cong \frac{\mathrm{e}^{n\log[n(1+\nu)] - n - n\nu}}{\sqrt{2\pi n}\, n^n\, \mathrm{e}^{-n}} \\ &\cong \frac{1}{\sqrt{2\pi n}}\, \mathrm{e}^{n\log(1+\nu) - n\nu} \\ &\cong \frac{\mathrm{e}^{-\frac{1}{2}n\nu^2}}{\sqrt{2\pi n}}\,, \end{aligned} \tag{2.8.15}$$

because $\log(1+\nu) \cong \nu - \frac{1}{2}\nu^2$ for $|\nu| \ll 1$, which is a consistent approximation. Now,

$$\int \mathrm{d}\eta |\psi_n|^2 \cong \int_{-\infty}^{\infty} \mathrm{d}\nu \sqrt{\frac{n}{2\pi}}\, \mathrm{e}^{-\frac{1}{2}n\nu^2} = 1 \tag{2.8.16}$$

and, since

$$\lim_{n\to\infty} \sqrt{\frac{n}{2\pi}}\, \mathrm{e}^{-\frac{1}{2}n\nu^2} = \begin{cases} \infty \text{ for } \nu = 0\,, \\ 0 \;\text{ for } \nu \neq 0\,, \end{cases} \tag{2.8.17}$$

we see that

$$\lim_{n\to\infty} \mathrm{d}\eta |\psi_n(q',p')|^2 = \mathrm{d}\nu\, \delta(\nu)\,. \tag{2.8.18}$$

The exact asymptotic correspondence of $\eta = \frac{1}{2}(q'^2 + p'^2)$ with n identifies this as the classical limit, in which the non-commutativity of q and p, and the equivalent uncertainty $\delta q \delta p = \frac{1}{2}$, is unobservable.

Now let's return to (2.8.6) and proceed to integrate over η from 0 to $\overline{\eta}$:

$$\sum_{n=0}^{\infty} \int_0^{\overline{\eta}} \mathrm{d}\eta |\psi_n(q',p')|^2 = \overline{\eta}\,. \tag{2.8.19}$$

Let $\overline{\eta} \gg 1$. For any $n \ll \overline{\eta}$, the integral covers the entire $|\psi_n|^2$ distribution and produces a unit contribution. For large values n, comparable with $\overline{\eta}$, we can write, approximately,

$$\mathrm{d}\eta |\psi_n|^2 \cong \mathrm{d}\nu\, \delta(\nu) = \mathrm{d}\eta\, \delta(\eta - n)\,, \tag{2.8.20}$$

so those values of $n < \overline{\eta}$ make a unit contribution, whereas those greater than $\overline{\eta}$ do not contribute. The result is to identify asymptotically the count of the number of states having $n < \overline{\eta}$ with $\overline{\eta}$. It is in this sense that increasing $\overline{\eta}$ by one, $\Delta\eta = 1$, adds one additional state.

2.9 More about stationary-uncertainty states

We first found the spectrum of $N = y^\dagger y$, $n = 0, 1, 2\ldots$ and then we constructed the eigenvectors. Here is a method that gives the eigenvalues and eigenvectors (expressed by wave functions) together. Consider

$$\langle y^{\dagger\prime}|\,\mathrm{e}^{-\beta N}|y''\rangle \qquad \text{with } \beta > 0\,, \tag{2.9.1}$$

which we proceed to differentiate with respect to β:

$$-\frac{\partial}{\partial\beta}\langle y^{\dagger\prime}|\,\mathrm{e}^{-\beta N}|y''\rangle = \langle y^{\dagger\prime}|y^\dagger y\,\mathrm{e}^{-\beta N}|y''\rangle\,. \tag{2.9.2}$$

Now (see Problem 2-32)

$$yf(N) = f(N+1)y\,, \tag{2.9.3}$$

so

$$y\,\mathrm{e}^{-\beta N} = \mathrm{e}^{-\beta N}\,\mathrm{e}^{-\beta}y \tag{2.9.4}$$

and

$$\begin{aligned}\langle y^{\dagger\prime}|y^\dagger y\,\mathrm{e}^{-\beta N}|y''\rangle &= \langle y^{\dagger\prime}|y^\dagger\,\mathrm{e}^{-\beta N}\,\mathrm{e}^{-\beta}y|y''\rangle\\ &= y^{\dagger\prime}\,\mathrm{e}^{-\beta}y''\langle y^{\dagger\prime}|\,\mathrm{e}^{-\beta N}|y''\rangle\,.\end{aligned} \tag{2.9.5}$$

Now the differentiated equation is ($\partial\beta \to \mathrm{d}\beta$)

$$\frac{\mathrm{d}\langle y^{\dagger\prime}|\,\mathrm{e}^{-\beta N}|y''\rangle}{\langle y^{\dagger\prime}|\,\mathrm{e}^{-\beta N}|y''\rangle} = -\mathrm{d}\beta\, y^{\dagger\prime}\,\mathrm{e}^{-\beta}y'' = \mathrm{d}[y^{\dagger\prime}\,\mathrm{e}^{-\beta}y''] \tag{2.9.6}$$

or

$$\langle y^{\dagger\prime}|\,\mathrm{e}^{-\beta N}|y''\rangle = \mathrm{e}^{y^{\dagger\prime}\,\mathrm{e}^{-\beta}y''}\,; \tag{2.9.7}$$

the constant of integration has already been picked to satisfy

$$\langle y^{\dagger\prime}|\,\mathrm{e}^{-\beta N}|y''\rangle \to \langle y^{\dagger\prime}|y''\rangle = \mathrm{e}^{y^{\dagger\prime}y''} \qquad \text{for } \beta \to 0. \tag{2.9.8}$$

Now if we introduce the eigenvalues of N (temporarily called N') and the eigenvectors, we get

$$\sum_{N'}\langle y^{\dagger\prime}|N'\rangle\,\mathrm{e}^{-\beta N'}\langle N'|y''\rangle = \sum_{n=0}^{\infty}\frac{(y^{\dagger\prime})^n}{\sqrt{n!}}\,\mathrm{e}^{-\beta n}\frac{(y'')^n}{\sqrt{n!}}\,, \tag{2.9.9}$$

from which we learn, once more, that

$$N' = n = 0, 1, \ldots\,, \tag{2.9.10}$$

and

$$\langle y^{\dagger\prime}|n\rangle = \frac{(y^{\dagger\prime})^n}{\sqrt{n!}}\,, \quad \langle n|y''\rangle = \frac{(y'')^n}{\sqrt{n!}}\,. \tag{2.9.11}$$

Problems

2-1 It has been stated, at (2.1.28) and (2.1.29), that

$$\lim_{P\to\infty}\int_{-\infty}^{\infty}\frac{d\chi}{\pi}\frac{\sin\chi}{\chi}\psi(q'+\chi/P)=\psi(q')\,.$$

Check this by explicitly evaluating the integral for the example

$$\psi(q')=e^{ikq'}\,,$$

where k is a constant, say positive.

2-2 Show that, for any ϵ,

$$\int_{-\infty}^{\infty}dq'\,\delta(q'-q'')=1$$

for $\delta(q'-q'')$ of the form (2.1.30), provided that $K(\epsilon p')$ obeys (2.1.31). Check this explicitly for the example $K(x)=1/(1+x^2)$.

2-3 Evaluate

$$\int_{-\infty}^{\infty}\frac{dp'}{2\pi}\,e^{ip'(q'-q'')}\,e^{-\frac{1}{2}(\epsilon p')^2}$$

and verify that it has the delta function properties in the limit $\epsilon\to 0$.

2-4 Differentiate

$$x\delta(x)=0$$

to establish

$$x\frac{d}{dx}\delta(x)=-\delta(x)\,.$$

Find an analogous expression for $x^2(d/dx)^2\delta(x)$.

2-5 In a similar way, proceed from

$$f(x)\delta(x-a)=f(a)\delta(x-a)$$

to show

$$f(x)\frac{d}{dx}\delta(x-a)=f(a)\frac{d}{dx}\delta(x-a)-\frac{df}{dx}(a)\delta(x-a)$$

and the corresponding expression for $f(x)(d/dx)^2\delta(x-a)$.

2-6 Begin with (1.14.46), that is

$$\operatorname{tr}\{f(U,V)\} = \frac{1}{\nu}\sum_{u',v'} f(u';v') ,$$

with the semicolon indicating the UV ordering of (1.14.43), and arrive at

$$\operatorname{tr}\{f(q,p)\} = \int \frac{\mathrm{d}q'\,\mathrm{d}p'}{2\pi}\, f(q';p') .$$

2-7a Consider the normalized matrix elements

$$f_1(q',p') = \frac{\langle q'|f(q,p)|p'\rangle}{\langle q'|p'\rangle} , \qquad f_2(p',q') = \frac{\langle p'|f(q,p)|q'\rangle}{\langle p'|q'\rangle} ,$$

and show that

$$f(q,p) = f_1(q;p) = f_2(p;q) .$$

Given $f_1(q',p')$, what is $f_2(p',q')$?

2-7b Find the qp ordered form of $f(q,p) = pq^2p$ by using $[q,p] = \mathrm{i}$, and also by evaluating $f_1(q',p')$. Repeat for the pq ordered form.

2-8a Evaluate

$$U(q',p') = \mathrm{e}^{\frac{\mathrm{i}}{2}\frac{p'}{q'}q^2}\,\mathrm{e}^{\mathrm{i}q'p}\,\mathrm{e}^{-\frac{\mathrm{i}}{2}\frac{p'}{q'}q^2}$$

in various ways to arrive at

$$U(q',p') = \mathrm{e}^{\mathrm{i}(q'p-p'q)} = \mathrm{e}^{\mathrm{i}q'p}\,\mathrm{e}^{-\mathrm{i}p'q}\,\mathrm{e}^{\frac{\mathrm{i}}{2}q'p'} = \mathrm{e}^{-\mathrm{i}p'q}\,\mathrm{e}^{\mathrm{i}q'p}\,\mathrm{e}^{-\frac{\mathrm{i}}{2}q'p'} .$$

2-8b Proceed from (cf. Problem 1-51)

$$\frac{1}{\nu}\operatorname{tr}\left\{\left(U^kV^l\right)^\dagger\, U^{k'}V^{l'}\right\} = \delta_{kk'}\delta_{ll'}$$

$$\text{with } k,l,k',l' = 0,\pm1,\pm2,\ldots,\pm\tfrac{1}{2}(\nu-1)$$

and derive

$$\operatorname{tr}\left\{U(q',p')^\dagger U(q'',p'')\right\} = 2\pi\delta(q'-q'')\delta(p'-p'') .$$

2-8c Use Problem 2-6 to prove that

$$\operatorname{tr}\{U(q',p')\} = 2\pi\delta(q')\delta(p') .$$

Can you produce the result of Problem 2-8b from this?

2-9a Verify

$$qU(q',p') + U(q',p')q = 2\mathrm{i}\frac{\partial}{\partial p'}U(q',p') ,$$

and find the analogous statement for $pU + Up$. Use them to show that the *reflection operator*

$$R = \int \frac{\mathrm{d}q'\,\mathrm{d}p'}{4\pi} U(q',p')$$

is such that

$$Rq = -qR , \qquad Rp = -pR .$$

2-9b Show that R is both Hermitian and unitary. What, then, are the eigenvalues of R?

2-9c Use Problems 2-7 to find the qp ordered form of R. Can you then get the pq ordered version immediately?

2-10a Use the Taylor series expansion of $\mathrm{e}^{-\lambda B} A\,\mathrm{e}^{\lambda B}$ to prove that

$$\mathrm{e}^{-B} A\,\mathrm{e}^{B} = A + [A, B] + \frac{1}{2!}\Big[[A, B], B\Big] + \cdots .$$

Note how this simplifies if $[A, B]$ commutes with B.

2-10b Recognize in the results of Problem 2-8a a realization of the following statements, which apply when $[A, B]$ commutes with A and B:

$$\mathrm{e}^{A+B} = \mathrm{e}^{A}\,\mathrm{e}^{B}\,\mathrm{e}^{-\frac{1}{2}[A, B]} = \mathrm{e}^{B}\,\mathrm{e}^{A}\,\mathrm{e}^{\frac{1}{2}[A, B]} .$$

2-10c Provide a direct proof of this statement by starting with

$$\frac{\mathrm{d}}{\mathrm{d}\lambda}\mathrm{e}^{\lambda(A+B)} = (A + B)\,\mathrm{e}^{\lambda(A+B)}$$

and then writing, for example, $\mathrm{e}^{\lambda(A+B)} = \mathrm{e}^{\lambda B} X(\lambda)$.

2-11 Extend Problems 2-10b and 2-10c to the situation where $[A, B]$ commutes with B and $[[A, B], A]$ commutes with A.

2-12a As an alternative to the symmetrical $\nu \to \infty$ limit of Sections 1.16 and 2.1, reconsider

$$\sum_m \langle u_k | v_m \rangle \langle v_m | u_l \rangle = \delta_{kl}$$

and define

$$\phi = \frac{2\pi k}{\nu}\,, \qquad \phi' = \frac{2\pi l}{\nu}\,,$$
$$\text{with} \quad k, l = 0, \pm 1, \pm 2, \ldots, \pm\tfrac{1}{2}(\nu - 1)\,.$$

Show that in the limit $\nu \to \infty$ one then gets

$$\sum_{m=-\infty}^{\infty} \frac{1}{2\pi}\, \mathrm{e}^{\mathrm{i}m(\phi - \phi')} = \delta(\phi - \phi') \quad \text{with} \quad -\pi < \phi, \phi' < \pi\,.$$

2-12b Identify thus vectors $\langle \phi |$ and $| m \rangle$ such that

$$\langle \phi | m \rangle = \frac{1}{\sqrt{2\pi}}\, \mathrm{e}^{\mathrm{i}\phi m}$$

with

$$-\pi < \phi < \pi \qquad \text{and} \qquad m = 0, \pm 1, \pm 2, \ldots\,,$$

and verify the relations

$$\int_{-\pi}^{\pi} \mathrm{d}\phi\, \langle m | \phi \rangle \langle \phi | m' \rangle = \delta_{m,m'}\,,$$
$$\sum_{m=-\infty}^{\infty} \langle \phi | m \rangle \langle m | \phi' \rangle = \delta(\phi - \phi')\,.$$

Explain why these statements express the orthogonality and completeness of the vector sets.

2-12c Define Hermitian operators Φ and M in accordance with

$$\Phi = \int_{-\pi}^{\pi} \mathrm{d}\phi\, | \phi \rangle \phi \langle \phi |\,, \qquad M = \sum_{m=-\infty}^{\infty} | m \rangle m \langle m |\,.$$

State the analogs of (1.14.28) and (1.14.33) in terms of the unitary operators $\mathrm{e}^{\mathrm{i}m'\Phi}$ and $\mathrm{e}^{\mathrm{i}\phi' M}$. [You will need to introduce the convention that the vectors $\langle \phi |$ and $\langle \phi' |$ are identical if $\phi - \phi' = 2\pi$. Why?]

2-13 Begin with the inequality

$$\frac{1}{2} \sum_{a',a''} \left| \psi_1(a')\psi_2(a'') - \psi_2(a')\psi_1(a'') \right|^2 \geq 0$$

and arrive at

$$\langle 1|1\rangle\langle 2|2\rangle \geq |\langle 1|2\rangle|^2$$

along with the condition for the equal sign to hold.

2-14 Hermitian operators A and B have expectation values $\langle A\rangle$, $\langle B\rangle$ and spreads δA, δB, respectively. Define

$$X = \delta B\,(A - \langle A\rangle) + \mathrm{i}\,\delta A\,(B - \langle B\rangle)$$

and exploit (cf. Problem 1-27)

$$\left\langle X^\dagger X\right\rangle \geq 0\,, \qquad \left\langle X X^\dagger\right\rangle \geq 0$$

to establish

$$\delta A\,\delta B \geq \frac{1}{2}\Bigl|\mathrm{i}\bigl[A, B\bigr]\Bigr|$$

along with the condition for the equal sign to hold. This is Robertson's* general form (1929) of Heisenberg's uncertainty relation. Apply it to the pairs $(A, B) = (q, p)$ and $(A, B) = (\sigma_x, \sigma_y)$.

2-15 The result $A\bigl|1\bigr\rangle$ of applying a Hermitian operator A to a normalized vector $\bigl|1\bigr\rangle$ is, quite generally, of the form

$$A\bigl|1\bigr\rangle = \bigl|1\bigr\rangle a + \bigl|2\bigr\rangle b$$

where $\bigl|2\bigr\rangle$ is a normalized vector orthogonal to $\bigl|1\bigr\rangle$. Show that a and $|b|$ are the expectation value and the spread of A, respectively, if $\bigl|1\bigr\rangle$ specifies the state of the quantum system. What changes if A is non-Hermitian?

2-16 By exponentiating a small multiple of the Hermitian operator A, one gets a unitary operator $\exp(\mathrm{i}\varepsilon A)$, $|\varepsilon| \ll 1$, that differs little from the unit symbol 1. Evaluate the expectation value $\bigl\langle \exp(\mathrm{i}\varepsilon A)\bigr\rangle$ to second order in ε, and demonstrate that

$$\Bigl|\Bigl\langle \mathrm{e}^{\mathrm{i}\varepsilon A}\Bigr\rangle\Bigr|^2 = 1 - \varepsilon^2(\delta A)^2 + O(\varepsilon^4)\,.$$

Apply this to the situation of Problem 1-10.

2-17a In Section 2.4, the "permissible simplifications" $\langle q\rangle = \langle p\rangle = 0$ and $\delta q = \delta p$ are made. As a justification, consider the general situation, in which the expectation values $\langle q\rangle$, $\langle p\rangle$, as well as

---

*Howard Percy ROBERTSON (1903–1961)

$$\left\langle q^2 \right\rangle = \langle q \rangle^2 + (\delta q)^2 ,$$
$$\left\langle p^2 \right\rangle = \langle p \rangle^2 + (\delta p)^2 ,$$
$$\left\langle \tfrac{1}{2}(qp + pq) \right\rangle = \langle q \rangle \langle p \rangle + \Delta$$

are arbitrary except for the restriction (why?)

$$(\delta q \, \delta p)^2 \geq \tfrac{1}{4} + \Delta^2 .$$

Three consecutive linear transformations,

$$\begin{pmatrix} q \\ p \end{pmatrix} \to \begin{pmatrix} \overline{q} \\ \overline{p} \end{pmatrix} = U_1^{-1} \begin{pmatrix} q \\ p \end{pmatrix} U_1 = \begin{pmatrix} q - \langle q \rangle \\ p - \langle p \rangle \end{pmatrix} ,$$
$$\begin{pmatrix} \overline{q} \\ \overline{p} \end{pmatrix} \to \begin{pmatrix} \overline{\overline{q}} \\ \overline{\overline{p}} \end{pmatrix} = U_2^{-1} \begin{pmatrix} \overline{q} \\ \overline{p} \end{pmatrix} U_2 = \begin{pmatrix} \cos\phi & \sin\phi \\ -\sin\phi & \cos\phi \end{pmatrix} \begin{pmatrix} \overline{q} \\ \overline{p} \end{pmatrix} ,$$
$$\begin{pmatrix} \overline{\overline{q}} \\ \overline{\overline{p}} \end{pmatrix} \to \begin{pmatrix} Q \\ P \end{pmatrix} = U_3^{-1} \begin{pmatrix} \overline{\overline{q}} \\ \overline{\overline{p}} \end{pmatrix} U_3 = \begin{pmatrix} \mathrm{e}^{-\kappa} \overline{\overline{q}} \\ \mathrm{e}^{\kappa} \overline{\overline{p}} \end{pmatrix} ,$$

are then chosen such that

$$\langle Q \rangle = 0 , \quad \langle P \rangle = 0 , \qquad \langle Q^2 \rangle = \langle P^2 \rangle ,$$
$$\left\langle \tfrac{1}{2}(QP + PQ) \right\rangle = 0 .$$

Verify first that each tranformation preserves the fundamental commutation relation, that is

$$[\overline{q}, \overline{p}] = [\overline{\overline{q}}, \overline{\overline{p}}] = [Q, P] = [q, p] = \mathrm{i} .$$

Then note that

$$\langle \overline{q} \rangle = 0 , \quad \langle \overline{p} \rangle = 0$$

by construction, so that

$$\left\langle \overline{\overline{q}} \right\rangle = 0 , \quad \left\langle \overline{\overline{p}} \right\rangle = 0 , \qquad \langle Q \rangle = 0 , \quad \langle P \rangle = 0 .$$

Next, determine the angle ϕ such that

$$\left\langle \tfrac{1}{2}(\overline{\overline{q}}\,\overline{\overline{p}} + \overline{\overline{p}}\,\overline{\overline{q}}) \right\rangle = 0 .$$

Finally, find the value of κ that ensures $\left\langle Q^2 \right\rangle = \left\langle P^2 \right\rangle$. Can you thus justify the simplifications $\langle q \rangle = 0$, $\langle p \rangle = 0$, $\left\langle q^2 \right\rangle = \left\langle p^2 \right\rangle$, and $\left\langle (qp + pq) \right\rangle = 0$?

2-17b Express $(\delta q)^2$, $(\delta p)^2$, and Δ in terms of ϕ, κ, and $\epsilon \equiv 2\left\langle Q^2 \right\rangle = 2\left\langle P^2 \right\rangle \geq 1$. What do you get for $(\delta q \, \delta p)^2 - \Delta^2$? Conclusion?

2-17c Verify that the unitary operators U_1, U_2, U_3 are given by

$$U_1(q,p) = e^{i(p\langle q\rangle - q\langle p\rangle)} ,$$
$$U_2(\overline{q},\overline{p}) = e^{-i\phi(\overline{q}^2 + \overline{p}^2)/2} ,$$
$$U_3(\overline{\overline{q}},\overline{\overline{p}}) = e^{i\kappa(\overline{\overline{q}}\,\overline{\overline{p}} + \overline{\overline{p}}\,\overline{\overline{q}})/2} .$$

Show that

$$U_1(q,p) = U_1(\overline{q},\overline{p}) , \quad U_2(\overline{q},\overline{p}) = U_2(\overline{\overline{q}},\overline{\overline{p}}) , \quad U_3(\overline{\overline{q}},\overline{\overline{p}}) = U_3(Q,P) ,$$

both by an explicit check and by a general argument.

2-17d The total transformation

$$\begin{pmatrix} q \\ p \end{pmatrix} \to \begin{pmatrix} Q \\ P \end{pmatrix} = U^{-1}\begin{pmatrix} q \\ p \end{pmatrix} U$$

involves

$$U = U_1(q,p)U_2(\overline{q},\overline{p})U_3(\overline{\overline{q}},\overline{\overline{p}}) .$$

Without invoking the explicit forms of the U_k's, demonstrate that

$$U = U_3(q,p)U_2(q,p)U_1(q,p) .$$

This expresses U in terms of the original variables q and p. Can you write U as a function of the final variables Q and P?

2-17e If the second transformation is left out ($\phi = 0$), which value of κ ensures $\langle Q^2\rangle = \langle P^2\rangle$? Argue that this $\phi = 0$ case is sufficient to justify the simplifications $\langle q\rangle = 0$, $\langle p\rangle = 0$, $\delta q = \delta p$.

2-18 Verify the sum rule of Problem 1-23 for $A = y^\dagger y$ and $B = \alpha y^\dagger + \alpha^* y$ with arbitrary complex α.

2-19a Use the definition (2.5.4) to derive

$$e^{2tq - t^2} = \sum_{n=0}^{\infty} \frac{t^n}{n!} \mathrm{H}(q) ,$$

a generating function for the Hermite polynomials, and recognize it in (2.7.35).

2-19b Rederive (2.5.8), (2.5.10), and (2.5.12) directly from this generating function.

2-19c Employ the generating function to prove the orthogonality relation

$$\int_{-\infty}^{\infty} dq\, e^{-q^2} H_n(q) H_m(q) = \delta_{nm} \sqrt{\pi}\, 2^n n!\,.$$

2-20 Rearrange the binomial expansion (2.6.3) to show that

$$N\big|00\big| = 0\,,$$

and, similarly, that

$$(N - n)\big|nn\big| = 0\,.$$

2-21 What are the analogs of Problem 1-54 for the operators y and $y^\dagger$? Combine them with $\big|00\big| = \exp(-y^\dagger; y)$ to prove that

$$y\big|00\big| = 0\,, \qquad \big|00\big|y^\dagger = 0\,.$$

Now check that

$$\big|00\big|\,\big|00\big| = \big|00\big|\,.$$

2-22a Prove that

$$e^{-\lambda y^\dagger; y} = (1 - \lambda)^N\,, \qquad N = y^\dagger y\,.$$

For $\lambda \to 1$ these are expressions for $\big|00\big|$. Put

$$\lambda = 1 - e^{-\beta}$$

and get

$$e^{-\beta N} = e^{-\left(1 - e^{-\beta}\right) y^\dagger; y}\,;$$

use this to identify the eigenvalues and eigenvectors of N.

2-22b Discuss the interpretation of this relation for $\beta \ll 1$ and for $\beta \to \infty$.

2-23 Here is yet another method for finding the eigenvalues and eigenvectors of $N = y^\dagger y$. First show that, for any vector $|\ \rangle$, the wave function $\psi(y^{\dagger\prime}) = \langle y^{\dagger\prime}|\ \rangle$ is an entire function of $y^{\dagger\prime}$, that is: $\psi(y^{\dagger\prime})$ is analytical everywhere in the complex $y^{\dagger\prime}$ plane. Then use this global analyticity to get the eigenvalues and eigenvectors of N. [Dirac calls $\psi(y^{\dagger\prime})$ the Fock* representation of the vector $|\ \rangle$. In today's terminology, however, the set of probability amplitudes $\psi_n = \langle n|\ \rangle$ constitutes the Fock representation, and the stationary-uncertainty states $|n\rangle$ are commonly termed Fock states.]

2-24 Show that

$$y\,\mathrm{e}^{-\lambda y^\dagger;y} = \mathrm{e}^{-\lambda y^\dagger;y}(1-\lambda)y\,,$$

and state the adjoint. Then verify that

$$R = \mathrm{e}^{-2y^\dagger;y} = (-1)^{y^\dagger y}$$

is the reflection operator of Problems 2-9.

2-25 Exhibit $y^k y^{\dagger k}$ for $k = 1, 2, \ldots$ as polynomials in N. Use them to prove that

$$\mathrm{e}^{\lambda y;y^\dagger} = (1-\lambda)^{-N-1} \qquad \text{for } |\lambda| < 1\,.$$

Replace the λ in Problem 2-22a by $-\lambda/(1-\lambda)$ and conclude that

$$\mathrm{e}^{\lambda y;y^\dagger} = \frac{1}{1-\lambda}\,\mathrm{e}^{\frac{\lambda}{1-\lambda}y^\dagger;y}\,.$$

Check that both sides give the same result when applied to $|0\rangle$.

2-26 By differentiating with respect to $y^\dagger$, and y, show that Y, the equivalent ordered operators of Problem 2-25, is such that

$$yY = Y\frac{y}{1-\lambda}\,, \qquad Yy^\dagger = \frac{y^\dagger}{1-\lambda}Y\,.$$

Accordingly, generalize Problem 2-25 to

$$f(y)\,\mathrm{e}^{\lambda y;y^\dagger} = \frac{1}{1-\lambda}\,\mathrm{e}^{\frac{\lambda}{1-\lambda}y^\dagger;y} f\left(\frac{y}{1-\lambda}\right)\,,$$

$$\mathrm{e}^{\lambda y;y^\dagger} f(y^\dagger) = \frac{1}{1-\lambda} f\left(\frac{y^\dagger}{1-\lambda}\right)\mathrm{e}^{\frac{\lambda}{1-\lambda}y^\dagger;y}\,.$$

2-27a In the continuum limit, the analog of operator W in Problems 1-43 is

$$W = \int \mathrm{d}x\,|q' = x\rangle\langle p' = x|\,.$$

*Vladimir Alexandrovich Fock (1898–1974)

Find $W|p'\rangle$, $W|q'\rangle$, and state the analogs of the various forms for W in Problem 1-43c.

2-27b Show that

$$Wy = \mathrm{i}yW\,, \qquad y^\dagger W = W\mathrm{i}y^\dagger\,.$$

Use them to express W as a function of $y^\dagger$ and y.

2-28a The operator

$$\gamma = \tfrac{1}{2}(\sigma_x + \mathrm{i}\sigma_y)y + y^\dagger\tfrac{1}{2}(\sigma_x - \mathrm{i}\sigma_y)$$

is a Hermitian combination (show this) of $y, y^\dagger$ with the non-Hermitian Pauli operators $\frac{1}{2}(\sigma_x \pm \mathrm{i}\sigma_y)$. Find the eigenvalues of γ. [Hint: What is γ^2?] Then express the eigenvectors of γ in terms of the common eigenvectors $|n,\pm\rangle$ of $y^\dagger y$ and σ_z.

2-28b Repeat for $\gamma = \frac{1}{2}(\sigma_x + \mathrm{i}\sigma_y)y + y^\dagger\frac{1}{2}(\sigma_x - \mathrm{i}\sigma_y) + \epsilon\sigma_z$.

2-29 Verify that

$$\int \frac{\mathrm{d}q'\,\mathrm{d}p'}{4\pi}\,|q',0\rangle\,\mathrm{e}^{\frac{1}{4}(q'^2 + 2\mathrm{i}q'p' + p'^2)}\langle 0,p'| = 1$$

holds for the coherent states of (2.7.38). How does this tell us that the $|q', p' = 0\rangle$ vectors are complete by themselves, and also the $|q' = 0, p'\rangle$ vectors? Can you find more completeness statements for the coherent states?

2-30 Use (2.9.7) and the completeness statement (2.7.50) to get

$$\begin{aligned}\mathrm{tr}\left\{\mathrm{e}^{-\beta N}\right\} &= \int_{-\infty}^{\infty}\frac{\mathrm{d}q'\,\mathrm{d}p'}{2\pi}\,\mathrm{e}^{-|y'|^2}\langle y^{\dagger\prime}|\,\mathrm{e}^{-\beta N}|y'\rangle\Big|_{y^{\dagger\prime} = y'^*}\\ &= \frac{1}{1-\mathrm{e}^{-\beta}}\,.\end{aligned}$$

Is this result expected from the spectrum of N? Employ the completeness relation of Problem 2-29 for an alternative derivation.

2-31 Use the result of Problem 2-30 to evaluate

$$\langle N\rangle_\beta = \mathrm{tr}\,\{NP\} \qquad \text{with} \quad P = \frac{\mathrm{e}^{-\beta N}}{\mathrm{tr}\left\{\mathrm{e}^{-\beta N}\right\}}\,.$$

Does this P obey the general conditions imposed on the probability operator (cf. Problems 1-30)?

2-32 Exploit the completeness of the $|nn|$ measurement symbols to prove that $yf(N) = f(N+1)y$ and $f(N)y^\dagger = y^\dagger f(N+1)$.

2-33 Evaluate $\mathrm{tr}\,\{\mathrm{e}^{-\beta N}\}$ by differentiating with respect to β, then applying the result of Problem 2-32, followed by the use of the cyclic property of the trace. You will need a boundary condition for $\beta \to \infty$. Does your final result agree with that of Problem 2-30?

2-34 Use the result of Problem 2-30 to find

$$\mathrm{tr}\left\{\mathrm{e}^{-\beta N}\right\} \qquad \text{with} \quad N = y_1^\dagger y_1 + y_2^\dagger y_2 \,,$$

where 1 and 2 refer to different degrees of freedom. From your result derive the eigenvalues of this N along with their multiplicities. Check the latter by an elementary argument. Can you supply the corresponding answers for ν degrees of freedom,

$$N = \sum_{k=1}^{\nu} y_k^\dagger y_k \quad ?$$

2-35 In view of the cyclic property of the trace (cf. Problems 1-24 and 1-46), is it true that

$$\mathrm{tr}\,\{yy^\dagger\} - \mathrm{tr}\,\{y^\dagger y\} = 0 \quad ?$$

In particular, give the explicit numerical forms of the two sides. Then, compare with the explicit numerical forms of ($\lambda > 0$)

$$\mathrm{tr}\left\{y\,\mathrm{e}^{-\lambda y^\dagger y} y^\dagger\right\} - \mathrm{tr}\left\{y^\dagger y\,\mathrm{e}^{-\lambda y^\dagger y}\right\} .$$

Is this zero? What is the moral?

2-36 Why haven't we talked about a vector obeying

$$y^\dagger |\ \rangle = 0\,?$$

Answer by multiplying on the left by y, and also by using the q' description to construct the wave function representing the vector. Nevertheless, there is a sense in which $|y^{\dagger\prime}\rangle$ and $\langle y'|$ can be defined. Consider

$$|y'| > 0 : \qquad \langle y'| = \int_0^\infty \mathrm{d}y^{\dagger\prime\prime}\ \mathrm{e}^{-y^{\dagger\prime\prime} y'} \langle y^{\dagger\prime\prime}| \,,$$

with the integration path extended to ∞ so as to produce convergence. Then (check this)

$$\langle y'|(y - y') = \int_0^\infty \mathrm{d}y^{\dagger\prime\prime} \frac{\partial}{\partial y^{\dagger\prime\prime}} \left[\mathrm{e}^{-y^{\dagger\prime\prime} y'} \langle y^{\dagger\prime\prime}| \right]$$
$$= -\langle y^{\dagger\prime\prime} = 0| \;;$$

not zero, but a constant vector. So

$$\langle y'| = \langle 0| \frac{1}{y' - y} \,, \qquad \langle y'|y''\rangle = \frac{1}{y' - y''} \,.$$

As a check of consistency, compare

$$\langle y'|y|y''\rangle = \langle y'|y''\rangle y'' = \frac{y''}{y' - y''}$$

with

$$\langle y'|y|y''\rangle = y'\langle y'|y''\rangle - \langle 0|y''\rangle$$
$$= \frac{y'}{y' - y''} - 1 = \frac{y''}{y' - y''} \,.$$

Also, from (prove this)

$$-\frac{\partial}{\partial y'}\langle y'| = \langle y'|y^\dagger$$

we get

$$\frac{(-\partial/\partial y')^n}{\sqrt{n!}} \underbrace{\langle y'|0\rangle}_{=1/y'} = \langle y'| \frac{(y^\dagger)^n}{\sqrt{n!}} |0\rangle = \langle y'|n\rangle$$

or

$$\langle y'|n\rangle = \frac{\sqrt{n!}}{(y')^{n+1}} \;; \qquad \langle n|y''\rangle = \frac{(y'')^n}{\sqrt{n!}} \,.$$

Then

$$\langle y'|y''\rangle = \sum_{n=0}^\infty \langle y'|n\rangle\langle n|y''\rangle = \sum_{n=0}^\infty \frac{(y'')^n}{(y')^{n+1}}$$
$$= \frac{1}{y' - y''} \,,$$

for $|y'| > |y''|$.

3. Angular Momentum

3.1 Infinitesimal unitary transformations

Physical properties or combinations of them are symbolized by operators $X, Y, \ldots$ obeying algebraic relations, $X+Y = Z, XY = Z$; states are symbolized by vectors $\langle\ |, |\ \rangle$, with algebraic relations, $X|1\rangle = |2\rangle$, $|1\rangle + |2\rangle = |3\rangle$, all this subject to the adjoint relations, such as $A^\dagger = A$ for a Hermitian operator and $\langle 1|X^\dagger = \langle 2|$. There are numbers formed by the vectors and operators: $\langle 1|2\rangle, \langle 1|X|2\rangle$, or equivalent traces, e.g., tr $\{X|2\rangle\langle 1|\}$. Suppose one systematically redefines all vectors and operators:

$$\overline{\langle\ |} = \langle\ |U\,, \quad \overline{|\ \rangle} = U^{-1}|\ \rangle\,, \quad \overline{X} = U^{-1}XU\,, \tag{3.1.1}$$

where $U^\dagger = U^{-1}$ is a unitary operator. Then all algebraic adjoint and numerical relations are maintained:

$$\left.\begin{aligned} X+Y &= Z \\ XY &= Z \\ |1\rangle + |2\rangle &= |3\rangle \\ X|1\rangle &= |2\rangle \\ A^\dagger &= A \end{aligned}\right\} \to \left\{\begin{aligned} &\overline{X} + \overline{Y} = \overline{Z} \\ &\overline{X}\,\overline{Y} = \overline{Z} \\ &\overline{|1\rangle} + \overline{|2\rangle} = \overline{|3\rangle} \\ &\overline{X}\,\overline{|1\rangle} = \overline{|2\rangle} \\ &\overline{A}^\dagger = \overline{A} \end{aligned}\right. \tag{3.1.2}$$

and

$$\overline{\langle\ |}^\dagger = \overline{|\ \rangle}\,, \quad \langle 1|2\rangle = \overline{\langle 1|2\rangle} = \overline{\langle 1|} \times \overline{|2\rangle}\,, \quad \langle 1|X|2\rangle = \overline{\langle 1|}\,\overline{X}\,\overline{|2\rangle}\,. \tag{3.1.3}$$

Thus the symbolic and numerical representations of the atomic phenomena permit the freedom of unitary transforms, in much the same way as there is a freedom of coordinate systems. Indeed the two are linked: the change of description produced by introducing a new coordinate system is represented by a unitary transformation.

$U = 1$ is the identity transformation. We are interested in transformations that differ infinitesimally from the identity:

$$U = 1 + \mathrm{i}G\,, \tag{3.1.4}$$

where G is an infinitesimal operator of order ϵ; in

$$U^\dagger = 1 - \mathrm{i}G^\dagger \,,$$
$$U^\dagger U = (1 - \mathrm{i}G^\dagger)(1 + \mathrm{i}G) = 1 + \mathrm{i}(G - G^\dagger) + G^\dagger G \qquad (3.1.5)$$

the product $G^\dagger G$ is of second order, $\propto \epsilon^2$, and is neglected. Therefore

$$G^\dagger = G\,; \qquad (3.1.6)$$

the *generator* G is a Hermitian operator.

In what follows we use a notation based on analogy with this familiar change in coordinate system:

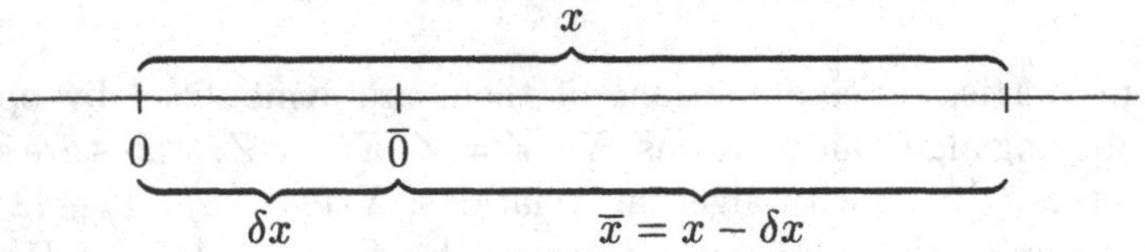

The displacement of the origin by δx changes coordinate x into $x - \delta x$. So, we write

$$|\;\rangle - \delta|\;\rangle = \overline{|\;\rangle} = U^{-1}|\;\rangle = (1 - \mathrm{i}G)|\;\rangle\,,$$
$$\langle\;| - \delta\langle\;| = \overline{\langle\;|} = \langle\;|U = \langle\;|(1 + \mathrm{i}G)\,,$$
$$X - \delta X = \overline{X} = U^{-1}XU = (1 - \mathrm{i}G)X(1 + \mathrm{i}G) = X - \frac{1}{\mathrm{i}}(XG - GX)\,, \qquad (3.1.7)$$

and get

$$\delta|\;\rangle = \mathrm{i}G|\;\rangle\,, \qquad \delta\langle\;| = -\langle\;|\mathrm{i}G\,, \qquad \delta X = \frac{1}{\mathrm{i}}[X, G]\,. \qquad (3.1.8)$$

3.2 Infinitesimal rotations

Consider an infinitesimal rotation of the coordinate system, about the z axis through the angle $\delta\varphi$,

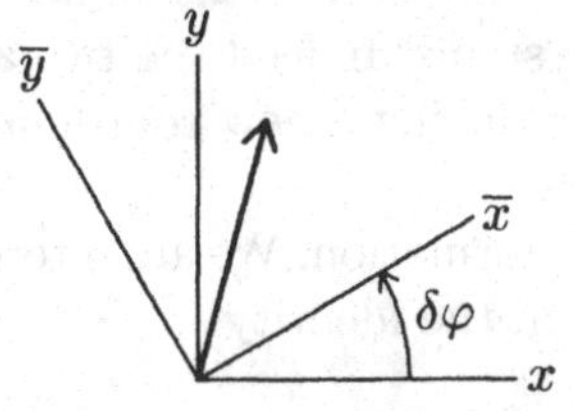

The components of any vector $\boldsymbol{V}$ are changed to

$$\begin{aligned}\overline{V}_x &= V_x + \delta\varphi V_y\ ,\\ \overline{V}_y &= V_y - \delta\varphi V_x\ ,\\ \overline{V}_z &= V_z\ ,\end{aligned} \tag{3.2.1}$$

together:

$$\overline{\boldsymbol{V}} = \boldsymbol{V} - \delta\boldsymbol{\omega} \times \boldsymbol{V}\ , \tag{3.2.2}$$

where $\delta\boldsymbol{\omega}$ here is the product of $\delta\varphi$ with the unit z vector. Here then

$$\delta\boldsymbol{V} = \delta\boldsymbol{\omega} \times \boldsymbol{V}\ . \tag{3.2.3}$$

The infinitesimal nature of G is clearly set by $\delta\varphi$; we write

$$G = \delta\varphi J_z = \delta\boldsymbol{\omega} \cdot \boldsymbol{J} \tag{3.2.4}$$

for this generator of infinitesimal rotations. What name do we give to the vector $\boldsymbol{J}$? How about angular momentum? The appropriateness for this classical mechanics *terminology* (which is *all* that we borrow) will become clear later.

We have learned that

$$\delta\boldsymbol{\omega} \times \boldsymbol{V} = \frac{1}{\mathrm{i}}[\boldsymbol{V}, \delta\boldsymbol{\omega} \cdot \boldsymbol{J}] \tag{3.2.5}$$

which supplies the commutators relating the components of any vector, with the components of $\boldsymbol{J}$. As one way of expressing this, we multiply by numerical vector $\boldsymbol{a}$ and get ($\delta\boldsymbol{\omega} \to \boldsymbol{b}$)

$$\frac{1}{\mathrm{i}}[\boldsymbol{a} \cdot \boldsymbol{V}, \boldsymbol{b} \cdot \boldsymbol{J}] = \boldsymbol{a} \times \boldsymbol{b} \cdot \boldsymbol{V}\ . \tag{3.2.6}$$

Thus, parallel components of $\boldsymbol{V}$ and $\boldsymbol{J}$ commute,

$$[V_k, J_k] = 0\ , \tag{3.2.7}$$

perpendicular components have commutators proportional to the component of $\boldsymbol{V}$ in the third direction:

$$\frac{1}{\mathrm{i}}[V_j, J_k] = \begin{cases} +V_l & \text{if } jkl \text{ is a cyclic permutation of } xyz,\\ -V_l & \text{if it's an anticyclic permutation,}\end{cases} \tag{3.2.8}$$

as illustrated by

$$\frac{1}{\mathrm{i}}[V_x, J_y] = V_z\ , \qquad \frac{1}{\mathrm{i}}[V_y, J_x] = -V_z\ . \tag{3.2.9}$$

These relations must also apply to vector $\boldsymbol{J}$:

$$\frac{1}{\mathrm{i}}[\boldsymbol{a}\cdot\boldsymbol{J},\boldsymbol{b}\cdot\boldsymbol{J}] = \boldsymbol{a}\times\boldsymbol{b}\cdot\boldsymbol{J} \tag{3.2.10}$$

thus

$$J_x J_y - J_y J_x = \mathrm{i}J_z\ , \tag{3.2.11}$$

for example, or

$$\boldsymbol{J}\times\boldsymbol{J} = \mathrm{i}\boldsymbol{J}\ . \tag{3.2.12}$$

The answer to Problem 1-36a,

$$[\boldsymbol{a}\cdot\boldsymbol{\sigma},\boldsymbol{b}\cdot\boldsymbol{\sigma}] = 2\mathrm{i}\boldsymbol{a}\times\boldsymbol{b}\cdot\boldsymbol{\sigma}\ , \tag{3.2.13}$$

gives us an example of an angular momentum vector $\boldsymbol{J}$:

$$\boldsymbol{J} = \tfrac{1}{2}\boldsymbol{\sigma}. \tag{3.2.14}$$

We began with vectors, but even simpler is a scalar: S. It is unchanged by the rotation, $\delta S = 0$, so

$$0 = \frac{1}{\mathrm{i}}[S, \delta\boldsymbol{\omega}\cdot\boldsymbol{J}] \qquad \text{or} \quad [S,\boldsymbol{J}] = 0\ . \tag{3.2.15}$$

The scalar product of two vectors is a scalar. Are these characterizations consistent? We are asking whether it is true that

$$\delta(\boldsymbol{V_1}\cdot\boldsymbol{V_2}) = \delta\boldsymbol{V_1}\cdot\boldsymbol{V_2} + \boldsymbol{V_1}\cdot\delta\boldsymbol{V_2}\ ? \tag{3.2.16}$$

which amounts to

$$\delta\boldsymbol{\omega}\times\boldsymbol{V_1}\cdot\boldsymbol{V_2} + \underbrace{\boldsymbol{V_1}\cdot\delta\boldsymbol{\omega}\times\boldsymbol{V_2}}_{=\boldsymbol{V_1}\times\delta\boldsymbol{\omega}\cdot\boldsymbol{V_2}} = 0\ . \tag{3.2.17}$$

Put more generally: Does

$$[XY,Z] = X[Y,Z] + [X,Z]Y \tag{3.2.18}$$

hold? Yes, see Problem 1-21a.

3.3 Common eigenvectors of $\boldsymbol{J}^2$ and J_z

The square of the angular momentum vector, $\boldsymbol{J}^2$, is a scalar, so

$$[\boldsymbol{J}^2,\boldsymbol{J}] = 0\ . \tag{3.3.1}$$

In particular $\boldsymbol{J}^2$ and J_z commute; they are compatible properties. But we cannot add J_x, or J_y to this list; they commute with $\boldsymbol{J}^2$, but not J_z. We now ask for the eigenvectors of $\boldsymbol{J}^2$ and J_z:

$$\boldsymbol{J}^2\big|\boldsymbol{J}^{2\prime}, J_z'\big\rangle = \big|\boldsymbol{J}^{2\prime}, J_z'\big\rangle \boldsymbol{J}^{2\prime} , \quad J_z\big|\boldsymbol{J}^{2\prime}, J_z'\big\rangle = \big|\boldsymbol{J}^{2\prime}, J_z'\big\rangle J_z' . \tag{3.3.2}$$

It's convenient to introduce the non-Hermitian operators (frequently called *ladder operators*)

$$J_+ = J_x + \mathrm{i}J_y , \quad J_- = J_x - \mathrm{i}J_y = J_+^\dagger \tag{3.3.3}$$

such that

$$\begin{aligned}[J_z, J_+] &= \mathrm{i}J_y + J_x = J_+ , \\ [J_z, J_-] &= \mathrm{i}J_y - J_x = -J_- .\end{aligned} \tag{3.3.4}$$

We also have

$$\begin{aligned}J_+J_- &= (J_x + \mathrm{i}J_y)(J_x - \mathrm{i}J_y) = J_x^2 + J_y^2 - \mathrm{i}[J_x, J_y] \\ &= \boldsymbol{J}^2 - J_z^2 + J_z , \\ J_-J_+ &= (J_x - \mathrm{i}J_y)(J_x + \mathrm{i}J_y) = J_x^2 + J_y^2 + \mathrm{i}[J_x, J_y] \\ &= \boldsymbol{J}^2 - J_z^2 - J_z ,\end{aligned} \tag{3.3.5}$$

so that

$$[J_+, J_-] = 2J_z . \tag{3.3.6}$$

For simplicity, write $J_z' = m$, and, for reasons we see in a moment we shall write $\boldsymbol{J}^{2\prime} = j(j+1)$ with $j \geq 0$ by convention. Then from

$$J_zJ_+ - J_+J_z = J_+ \qquad \text{or} \qquad J_zJ_+ = J_+(J_z + 1) \tag{3.3.7}$$

we get

$$J_zJ_+\big|j, m\big\rangle = J_+(J_z + 1)\big|j, m\big\rangle = J_+\big|j, m\big\rangle(m + 1) . \tag{3.3.8}$$

Now $\boldsymbol{J}^2$, commuting with $\boldsymbol{J}$, commutes with J_+, so that

$$\boldsymbol{J}^2J_+\big|j, m\big\rangle = J_+\boldsymbol{J}^2\big|j, m\big\rangle = J_+\big|j, m\big\rangle j(j + 1) , \tag{3.3.9}$$

which says that

$$J_+\big|j, m\big\rangle = \big|j, m + 1\big\rangle A_{jm} \tag{3.3.10}$$

with a numerical factor A_{jm} to be found; if j, m are eigenvalues, so, in general, are $j, m + 1$. But this increase in m cannot go on forever. Note that

$$\langle j,m|J_-J_+|j,m\rangle = \begin{cases} \left(J_+|j,m\rangle\right)^\dagger\left(J_+|j,m\rangle\right) \geq 0\,, \\ j(j+1)-m(m+1)\,, \end{cases} \tag{3.3.11}$$

so $m(m+1)$ cannot exceed $j(j+1)$. In fact since the largest value $\overline{m}$ of m is such that $J_+|j,\overline{m}\rangle = 0$, we have just

$$\overline{m} = j \tag{3.3.12}$$

($-j-1$ is *not* $\overline{m}$). Incidentally, the above relation also tells us that

$$j(j+1)-m(m+1) = (j-m)(j+m+1) = |A_{jm}|^2 \tag{3.3.13}$$

where

$$A_{jm} = \langle j,m+1|J_+|j,m\rangle = \langle j,m|J_-|j,m+1\rangle^*\,. \tag{3.3.14}$$

Now, starting with

$$J_zJ_- = J_-(J_z-1) \tag{3.3.15}$$

we have

$$J_zJ_-|j,m\rangle = J_-|j,m\rangle(m-1) \tag{3.3.16}$$

and

$$\boldsymbol{J}^2J_-|j,m\rangle = J_-|j,m\rangle j(j+1) \tag{3.3.17}$$

implying

$$J_-|j,m\rangle = |j,m-1\rangle B_{jm} \tag{3.3.18}$$

where

$$B_{jm} = \langle j,m-1|J_-|j,m\rangle = A^*_{jm-1}\,. \tag{3.3.19}$$

This tells us that

$$|B_{jm}|^2 = (j+m)(j+1-m)\,. \tag{3.3.20}$$

So, in general, if j,m are eigenvalues, so are $j,m-1$. But the decrease of m cannot go on forever, or positive $|B_{jm}|^2$ would turn negative. Indeed, the lowest value $\underline{m}$ of m is such that $J_-|j,\underline{m}\rangle = 0$, or

$$\underline{m} = -j\,. \tag{3.3.21}$$

The descent from $\overline{m} = j$ to $\underline{m} = -j$ occurs in $n = 0,1,2,\ldots$ steps. Therefore $2j = n$,

$$j = \tfrac{1}{2}n = 0, \tfrac{1}{2}, 1, \tfrac{3}{2}, 2, \dots \; . \tag{3.3.22}$$

For each n, or j, the possible values of m are

$$m = j, j-1, \dots, -(j-1), -j \; , \tag{3.3.23}$$

which are $n + 1 = 2j + 1$ in number,

$$\begin{array}{c|cccccc} j & 0 & \frac{1}{2} & 1 & \frac{3}{2} & 2 & \dots \\ \hline 2j+1 & 1 & 2 & 3 & 4 & 5 & \dots \end{array} \tag{3.3.24}$$

In particular $j = \frac{1}{2}$, $2j+1 = 2$, $m = \pm\frac{1}{2}$, is clearly the Stern–Gerlach system: $\boldsymbol{J} = \frac{1}{2}\boldsymbol{\sigma}$, $\boldsymbol{J}^2 = \frac{3}{4} = \frac{1}{2}(\frac{1}{2}+1)$.

To summarize:

$$\boldsymbol{J}^{2\prime} = j(j+1) \, , \quad j = \tfrac{1}{2}n \, , \quad J'_z = m = j, \dots, -j \, ,$$

$$\begin{aligned} J_+ \big|j, m\big\rangle &= \big|j, m+1\big\rangle \sqrt{(j-m)(j+m+1)} \, , \\ J_- \big|j, m\big\rangle &= \big|j, m-1\big\rangle \sqrt{(j+m)(j-m+1)} \, , \end{aligned} \tag{3.3.25}$$

where the usual phase convention is adopted, namely that $A_{jm} \geq 0$ and $B_{jm} \geq 0$.

3.4 Decomposition into spins

The combinations

$$\begin{aligned} j - m &\equiv n_- = 0, 1, \dots \, , \\ j + m &\equiv n_+ = 0, 1, \dots \, , \end{aligned} \tag{3.4.1}$$

direct our attention to another formulation. First note that

$$j = \tfrac{1}{2}(n_+ + n_-) = n \, , \quad m = \tfrac{1}{2}(n_+ - n_-) \, . \tag{3.4.2}$$

Then, relabeling the states, we get

$$\begin{aligned} J_+ \big|n_+, n_-\big\rangle &= \big|n_+ + 1, n_- - 1\big\rangle \sqrt{n_-(n_+ + 1)} \, , \\ J_- \big|n_+, n_-\big\rangle &= \big|n_+ - 1, n_+ + 1\big\rangle \sqrt{n_+(n_- + 1)} \, . \end{aligned} \tag{3.4.3}$$

Doesn't this set the bells ringing? Recall (2.4.14) and (2.4.15):

$$y\big|n\big\rangle = \big|n-1\big\rangle\sqrt{n} \, , \quad y^\dagger\big|n\big\rangle = \big|n+1\big\rangle\sqrt{n+1} \, ! \tag{3.4.4}$$

It is as though there are two independent systems, $y_+, y_+^\dagger$, and $y_-, y_-^\dagger$, with composite states $\big|n_+, n_-\big\rangle = \big|n_+\big\rangle\big|n_-\big\rangle$, and

$$J_+ = y_+^\dagger y_- \,, \quad J_- = y_-^\dagger y_+ \,. \tag{3.4.5}$$

In addition,

$$J_z = \tfrac{1}{2}(y_+^\dagger y_+ - y_-^\dagger y_-) \,, \qquad J_z' = m = \tfrac{1}{2}(n_+ - n_-) \,, \tag{3.4.6}$$

and $[j = \sqrt{j(j+1)+\frac{1}{4}} - \frac{1}{2}]$

$$\sqrt{\boldsymbol{J}^2 + \tfrac{1}{4}} - \tfrac{1}{2} = \tfrac{1}{2}(y_+^\dagger y_+ + y_-^\dagger y_-) \,. \tag{3.4.7}$$

Notice also that

$$\begin{aligned} J_x &= \tfrac{1}{2}(J_+ + J_-) = \tfrac{1}{2}(y_+^\dagger y_- + y_-^\dagger y_+) \,, \\ J_y &= \tfrac{1}{2\mathrm{i}}(J_+ - J_-) = \tfrac{1}{2}(\mathrm{i}y_-^\dagger y_+ - \mathrm{i}y_+^\dagger y_-) \,. \end{aligned} \tag{3.4.8}$$

Then, if we introduce the 2-component objects

$$y = \begin{pmatrix} y_+ \\ y_- \end{pmatrix} \,, \qquad y^\dagger = (y_+^\dagger, y_-^\dagger) \,, \tag{3.4.9}$$

we find that

$$\boldsymbol{J} = y^\dagger \tfrac{1}{2}\boldsymbol{\sigma} y \,, \quad \sqrt{\boldsymbol{J}^2 + \tfrac{1}{4}} - \tfrac{1}{2} = \tfrac{1}{2} y^\dagger y \,, \tag{3.4.10}$$

where the cartesian* components of vector $\boldsymbol{\sigma}$ are the Pauli matrices of (1.10.3).

The explicit construction of the eigenvectors that follows from

$$\big|n\big\rangle = \frac{(y^\dagger)^n}{\sqrt{n!}}\big|0\big\rangle \qquad \text{with} \quad y\big|0\big\rangle = 0 \tag{3.4.11}$$

is $(j \pm m = n_\pm)$

$$\big|j,m\big\rangle = \frac{(y_+^\dagger)^{j+m}(y_-^\dagger)^{j-m}}{\sqrt{(j+m)!(j-m)!}}\big|0\big\rangle \qquad \text{with} \quad \boldsymbol{J}\big|0\big\rangle = 0 \,, \tag{3.4.12}$$

where $\big|0\big\rangle \equiv \big|j=0, m=0\big\rangle = \big|n_+ = 0, n_- = 0\big\rangle$. This state $\big|0\big\rangle$, with angular momentum equal to zero (notice there is no conflict with the commutation relations if *all* components of $\boldsymbol{J}$ have the value 0) is unchanged by rotations:

$$\delta\big|0\big\rangle = \mathrm{i}\delta\boldsymbol{\omega}\cdot\boldsymbol{J}\big|0\big\rangle = 0 \,; \tag{3.4.13}$$

it is rotationally invariant.

For $j = \frac{1}{2}$, $m = \pm\frac{1}{2}$, we get

$$\big|\tfrac{1}{2},\tfrac{1}{2}\big\rangle = y_+^\dagger\big|0\big\rangle \,, \quad \big|\tfrac{1}{2},-\tfrac{1}{2}\big\rangle = y_-^\dagger\big|0\big\rangle \,. \tag{3.4.14}$$

*René DESCARTES (1596–1650)

We can say that $y_\sigma^\dagger, \sigma = \pm 1$, acts on the $|0\rangle$ state to *create* a $j = \frac{1}{2}$ state of angular momentum, the one with $m = \frac{1}{2}\sigma$. That tells us why the σ matrices appear in the construction of $\boldsymbol{J}$. For

$$\langle \tfrac{1}{2}, m|\boldsymbol{J}|\tfrac{1}{2}, m'\rangle\big|_{m=\frac{1}{2}\sigma, m'=\frac{1}{2}\sigma'} = \langle 0|y_\sigma \underbrace{y^\dagger \tfrac{1}{2}\boldsymbol{\sigma} y}\, y_{\sigma'}^\dagger|0\rangle . \tag{3.4.15}$$
$$= \sum_{\sigma'',\sigma'''} y_{\sigma''}^\dagger \langle \sigma''|\tfrac{1}{2}\boldsymbol{\sigma}|\sigma'''\rangle y_{\sigma'''}$$

Now the commutator properties of y and $y^\dagger$ are summarized by

$$[y_\sigma, y_{\sigma'}^\dagger] = \delta_{\sigma\sigma'} ; \quad [y_\sigma, y_{\sigma'}] = 0 ; \quad [y_\sigma^\dagger, y_{\sigma'}^\dagger] = 0 . \tag{3.4.16}$$

So

$$y_{\sigma'''} y_{\sigma'}^\dagger |0\rangle = (\delta_{\sigma'''\sigma'} + y_{\sigma'}^\dagger y_{\sigma'''})|0\rangle = |0\rangle \delta_{\sigma'''\sigma'} , \tag{3.4.17}$$

the adjoint of which is ($\sigma' \to \sigma, \sigma''' \to \sigma''$)

$$\langle 0|y_\sigma y_{\sigma''}^\dagger = \delta_{\sigma\sigma''}\langle 0| . \tag{3.4.18}$$

Therefore

$$\langle \tfrac{1}{2}, \tfrac{1}{2}\sigma|\boldsymbol{J}|\tfrac{1}{2}, \tfrac{1}{2}\sigma'\rangle = \langle \sigma|\tfrac{1}{2}\boldsymbol{\sigma}|\sigma'\rangle \tag{3.4.19}$$

which of course says that, for $j = \frac{1}{2}, \boldsymbol{J} = \frac{1}{2}\boldsymbol{\sigma}$.

The basic angular momentum, $j = \frac{1}{2}$, is called a *spin*. The construction for $|j, m\rangle$ in (3.4.12) can now be described by saying that angular momentum j, m is produced by creating $n = (j + m) + (j - m) = 2j$ spins, where the combination of $j + m$ $+\frac{1}{2}$-spins and $j - m$ $-\frac{1}{2}$-spins gives for J_z':

$$+\tfrac{1}{2}(j + m) - \tfrac{1}{2}(j - m) = m . \tag{3.4.20}$$

For thoroughness we ought to verify directly that

$$\boldsymbol{J} = \sum_{\sigma,\sigma'} y_\sigma^\dagger \langle \sigma|\tfrac{1}{2}\boldsymbol{\sigma}|\sigma'\rangle y_{\sigma'} \tag{3.4.21}$$

obeys the $\boldsymbol{J} \times \boldsymbol{J} = \mathrm{i}\boldsymbol{J}$ commutation relations. Consider then

$$\left[y_\sigma^\dagger y_{\sigma'}, y_{\sigma''}^\dagger y_{\sigma'''}\right] = y_\sigma^\dagger (y_{\sigma''}^\dagger y_{\sigma'} + \delta_{\sigma'\sigma''}) y_{\sigma'''} - y_{\sigma''}^\dagger (\delta_{\sigma'''\sigma} + y_\sigma^\dagger y_{\sigma'''}) y_{\sigma'}$$
$$= \delta_{\sigma'\sigma''} y_\sigma^\dagger y_{\sigma'''} - \delta_{\sigma'''\sigma} y_{\sigma''}^\dagger y_{\sigma'} , \tag{3.4.22}$$

where the cancellation occurs because the order of multiplication of $y_+^\dagger$ and $y_-^\dagger$, or y_+ and y_-, does not matter. Notice that this commutation relation has the same structure as that of the spin measurement symbols (Stern–Gerlach)

$$\left[|\sigma\sigma'|, |\sigma''\sigma'''|\right] = \delta_{\sigma'\sigma''}|\sigma\sigma'''| - \delta_{\sigma'''\sigma'}|\sigma''\sigma'| \,. \tag{3.4.23}$$

Of course, the latter also have the multiplication property

$$|\sigma\sigma'||\sigma''\sigma'''| = \delta_{\sigma'\sigma''}|\sigma\sigma'''| \tag{3.4.24}$$

whereas [see (3.4.16) above]

$$y^\dagger_\sigma y_{\sigma'} y^\dagger_{\sigma''} y_{\sigma'''} = \delta_{\sigma'\sigma''} y^\dagger_\sigma y_{\sigma'''} + y^\dagger_\sigma y^\dagger_{\sigma''} y_{\sigma'} y_{\sigma'''} \,. \tag{3.4.25}$$

However, as applied to *one* spin,

$$y_{\sigma'} y_{\sigma'''}|n=1\rangle = 0 \qquad \text{and} \qquad \langle n=1|y^\dagger_\sigma y^\dagger_{\sigma''} = 0 \,, \tag{3.4.26}$$

and they are the same. So $\langle\sigma'|$, which, acting to the right, annihilates the spin of type σ' is replaced by $y_{\sigma'}$ which, acting to the right, annihilates *a* spin of type σ'. Similarly, for $|\sigma\rangle$ and $y^\dagger_\sigma$. This kind of extension from one system to any number of similar systems is usually (somewhat erroneously) called second quantization. Now, we know that the product of two operators (here, for one spin) is represented by the product of their matrices, the commutator of the operators by the commutator of the matrices. So, since the algebra of the commutators is the same, $\boldsymbol{J}$ commutators are given by

$$\boldsymbol{J} \times \boldsymbol{J} = \sum_{\sigma\sigma'} y^\dagger_\sigma \langle\sigma| \underbrace{\tfrac{1}{2}\boldsymbol{\sigma} \times \tfrac{1}{2}\boldsymbol{\sigma}}_{=\mathrm{i}\frac{1}{2}\boldsymbol{\sigma}} |\sigma'\rangle y_{\sigma'} = \mathrm{i}\boldsymbol{J} \,, \tag{3.4.27}$$

indeed.

3.5 Angular momentum of a composite system

In recognizing the significance of $\boldsymbol{J}$ as a collection of n spins we took it as intuitively obvious that J'_z is the sum of the individual $\pm\frac{1}{2}$'s. Now let's look more generally at the angular momentum of a composite system. Consider two parts, with angular momenta $\boldsymbol{J_1}$ and $\boldsymbol{J_2}$, which of course commute. An infinitesimal reference frame rotation acts similarly and independently on the two parts

$$\begin{aligned} U = U_1 U_2 = U_2 U_1 &= (1 + \mathrm{i}\delta\boldsymbol{\omega}\cdot\boldsymbol{J_1})(1 + \mathrm{i}\delta\boldsymbol{\omega}\cdot\boldsymbol{J_2}) \\ &= 1 + \mathrm{i}\delta\boldsymbol{\omega}\cdot(\boldsymbol{J_1} + \boldsymbol{J_2}) \,. \end{aligned} \tag{3.5.1}$$

Thus, the total angular momentum is

$$\boldsymbol{J} = \boldsymbol{J_1} + \boldsymbol{J_2} \,. \tag{3.5.2}$$

We verify that

$$\begin{aligned} \boldsymbol{J} \times \boldsymbol{J} &= (\boldsymbol{J}_1 + \boldsymbol{J}_2) \times (\boldsymbol{J}_1 + \boldsymbol{J}_2) \\ &= \underbrace{\boldsymbol{J}_1 \times \boldsymbol{J}_1}_{=\mathrm{i}\boldsymbol{J}_1} + \underbrace{\boldsymbol{J}_1 \times \boldsymbol{J}_2 + \boldsymbol{J}_2 \times \boldsymbol{J}_1}_{=0} + \underbrace{\boldsymbol{J}_2 \times \boldsymbol{J}_2}_{=\mathrm{i}\boldsymbol{J}_2} \\ &= \mathrm{i}\boldsymbol{J} \,. \end{aligned} \tag{3.5.3}$$

How are the possible values of j [in $\boldsymbol{J}^{2'} = j(j+1)$] determined by j_1 and j_2? Here's a simple counting argument. The largest value of $m = m_1 + m_2$ is $m = j_1 + j_2$, which is unique. That is, only $(m_1, m_2) = (j_1, j_2)$ will produce it. Hence the largest value of j is $j_{\max} = j_1 + j_2$; it has $m = j_1 + j_2$ and $j_1+j_2-1, \ldots, -(j_1+j_2-1), -(j_1+j_2)$. Now consider $m = j_1+j_2-1$; it can be produced in two ways: $(m_1, m_2) = (j_1-1, j_2)$, and $(m_1, m_2) = (j_1, j_2-1)$. One such state is already contained in $j = j_1 + j_2$, $m = j_1 + j_2 - 1$. Hence there must be another state for which $m = j_1 + j_2 - 1$ is the maximum value, thus it has $j = j_1 + j_2 - 1$. For $m = j_1 + j_2 - 2$ there are (in general) three ways: $(m_1, m_2) = (j_1 - 2, j_2), (j_1 - 1, j_2 - 1), (j_1, j_2 - 2)$. Two of these accompany $j = j_1 + j_2$ and $j = j_1 + j_2 - 1$, so there must be a third state for which $m = j_1 + j_2 - 2$ is maximum; it is $j = j_1 + j_2 - 2$. There are only a finite number of states; where does this stop? Consider $m = j_1 + j_2 - \Delta j$: $(m_1, m_2) = (j_1 - \Delta j, j_2), \ldots, (j_1, j_2 - \Delta j)$; $\Delta j + 1$ in number. Suppose that $j_2 \leq j_1$. Then if $\Delta j = 2j_2$, the values of m_2 range over the full gamut, $m_2 = j_2, \ldots, -j_2$. And this means that if we now increase Δj by unity, to $2j_2 + 1$, we will *not* get an additional state; the lowest value of j is $j_{\min} = j_1+j_2-2j_2 = |j_1 - j_2|$, the latter holding generally. We can verify this result,

$$\begin{aligned} j_1 + j_2 \geq j \geq |j_1 - j_2| \,, \\ j \geq m \geq -j \,, \end{aligned} \tag{3.5.4}$$

by counting the total number of states. It is $(j_1 \geq j_2)$

$$\sum_{j=j_1-j_2}^{j_1+j_2} (2j+1) = \underbrace{(2j_1+1)}_{\text{aver. value}} \times \underbrace{(2j_2+1)}_{\text{no. of terms}} \,, \tag{3.5.5}$$

as it should.

To illustrate this consider two spins, $j_1 = j_2 = \frac{1}{2}$. Then we start with

$$m = j = 1: \quad |1,1\rangle \equiv |{+}{+}\rangle \,, \text{ i.e., } m_1 = m_2 = +\tfrac{1}{2} \tag{3.5.6}$$

which, one should notice, is perfectly symmetrical between the two spins: $1 \leftrightarrow 2$ leaves this vector intact. Now we use

$$J_-|j,m\rangle = |j,m\rangle\sqrt{(j+m)(j-m+1)}$$

[cf. (3.3.25)], here for $j = 1$,

$$J_-|1,1\rangle = |1,0\rangle\sqrt{2} \,; \qquad J_-|1,0\rangle = |1,-1\rangle\sqrt{2} \tag{3.5.7}$$

to construct the other states of $j = 1$. For this we use the $j = \frac{1}{2}$ properties

$$J_- \underbrace{\left|\tfrac{1}{2}, \tfrac{1}{2}\right\rangle}_{\equiv |+\rangle} = \underbrace{\left|\tfrac{1}{2}, -\tfrac{1}{2}\right\rangle}_{\equiv |-\rangle}, \quad J_- \underbrace{\left|\tfrac{1}{2}, -\tfrac{1}{2}\right\rangle}_{=|-\rangle} = 0\,. \tag{3.5.8}$$

So, from $|1,1\rangle = |{+}{+}\rangle$, we get

$$\begin{aligned} |1,0\rangle &= \frac{1}{\sqrt{2}}\left(J_{1-} + J_{2-}\right)|{+}{+}\rangle \\ &= \frac{1}{\sqrt{2}}\left(|{-}{+}\rangle + |{+}{-}\rangle\right)\,, \end{aligned} \tag{3.5.9}$$

which is a unit vector, and is symmetrical in 1 and 2. Then

$$\begin{aligned} |1,-1\rangle &= \frac{1}{\sqrt{2}}\left(J_{1-} + J_{2-}\right)\frac{1}{\sqrt{2}}\left(|{-}{+}\rangle + |{+}{-}\rangle\right) \\ &= \frac{1}{2}\left(|{-}{-}\rangle + |{-}{-}\rangle\right) = |{-}{-}\rangle\,, \end{aligned} \tag{3.5.10}$$

which we could have written directly (there is only one way to get $m = -1$), although we did have to see how the phase came out. Here are the $2j+1 = 3$ states of $j = \frac{1}{2} + \frac{1}{2} = 1$. The only other possibility among the $2 \times 2 = 4$ states is the $2j+1 = 1$ state of $j = \frac{1}{2} - \frac{1}{2} = 0$. This $m = 0$ state is the combination orthogonal to $|1,0\rangle$, which is (to within a phase factor)

$$|0,0\rangle = \frac{1}{\sqrt{2}}\left(|{-}{+}\rangle - |{+}{-}\rangle\right)\,. \tag{3.5.11}$$

Under the interchange $1 \leftrightarrow 2$, this vector turns into its negative – it is antisymmetrical.

The symmetry properties of the 3 + 1 states can be expressed by an operator. First we note that

$$\boldsymbol{J}^{2\prime} = j(j+1) = \begin{cases} 2 \text{ for } j = 1\,, \\ 0 \text{ for } j = 0\,, \end{cases} \tag{3.5.12}$$

or

$$(\boldsymbol{J}^2 - 1)' = \begin{cases} +1 \text{ for } j = 1\,, \\ -1 \text{ for } j = 0\,. \end{cases} \tag{3.5.13}$$

so the permutation operator P_{12}, which tests symmetry or antisymmetry under the exchange $1 \leftrightarrow 2$ is

$$\begin{aligned} P_{12} &= \left(\tfrac{1}{2}\boldsymbol{\sigma}_1 + \tfrac{1}{2}\boldsymbol{\sigma}_2\right)^2 - 1 = \tfrac{3}{4} + \tfrac{1}{2}\boldsymbol{\sigma}_1 \cdot \boldsymbol{\sigma}_2 + \tfrac{3}{4} - 1 \\ &= \tfrac{1}{2}\left(1 + \boldsymbol{\sigma}_1 \cdot \boldsymbol{\sigma}_2\right)\,. \end{aligned} \tag{3.5.14}$$

This operator is such that

$$P_{12}\left|\sigma_1',\sigma_2''\right\rangle = \left|\sigma_1'',\sigma_2'\right\rangle . \tag{3.5.15}$$

For n spins, the state with $m = j = \frac{1}{2}n$ is

$$\left|\underbrace{+ + \cdots +}_{n}\right\rangle \tag{3.5.16}$$

which is totally symmetrical in all n spins. All the other m's for $j = \frac{1}{2}n$, produced by successive application of J_-, which is symmetrical in all the spins, are also symmetrical. It is that total symmetry that picks out the maximum value of j. And that is what is displayed by the construction [cf. (3.4.12)]

$$\left|j,m\right\rangle = \frac{1}{\sqrt{(j+m)!(j-m)!}}\underbrace{y_+^\dagger y_+^\dagger \cdots}_{j+m}\underbrace{y_-^\dagger y_-^\dagger \cdots}_{j-m}\left|0\right\rangle , \tag{3.5.17}$$

for all these operators are commutative, and have no association with any individual spin.

The repetition of an infinitesimal rotation, say about the z axis, produces a finite rotation:

$$\begin{aligned} U(\varphi) &= \underbrace{(1+\mathrm{i}\delta\varphi J_z)(1+\mathrm{i}\delta\varphi J_z)\cdots(1+\mathrm{i}\delta\varphi J_z)}_{\varphi/\delta\varphi \gg 1 \text{ factors}} \\ &= \lim_{\delta\varphi\to 0}(1+\mathrm{i}\delta\varphi J_z)^{\varphi/\delta\varphi} = \mathrm{e}^{\mathrm{i}\varphi J_z} . \end{aligned} \tag{3.5.18}$$

The first use we put this to is a more elegant derivation of the addition of two angular momenta. Begin with the trace

$$\begin{aligned} \operatorname{tr}_{(j)}\left\{\mathrm{e}^{\mathrm{i}\varphi J_z}\right\} &= \sum_{m=-j}^{j} \mathrm{e}^{\mathrm{i}m\varphi} = \frac{\mathrm{e}^{\mathrm{i}(j+1)\varphi} - \mathrm{e}^{-\mathrm{i}j\varphi}}{\mathrm{e}^{\mathrm{i}\varphi}-1} \\ &= \frac{\sin\left((j+\frac{1}{2})\varphi\right)}{\sin(\frac{1}{2}\varphi)} \end{aligned} \tag{3.5.19}$$

and then consider $\boldsymbol{J} = \boldsymbol{J_1} + \boldsymbol{J_2}$, so

$$\mathrm{e}^{\mathrm{i}\varphi J_z} = \mathrm{e}^{\mathrm{i}\varphi J_{1z}}\,\mathrm{e}^{\mathrm{i}\varphi J_{2z}} . \tag{3.5.20}$$

This operator has the matrix (diagonal elements)

$$\begin{aligned} \left\langle j_1,m_1;j_2,m_2\right|\mathrm{e}^{\mathrm{i}\varphi J_z}\left|j_1,m_1;j_2,m_2\right\rangle =& \left\langle j_1,m_1\right|\mathrm{e}^{\mathrm{i}\varphi J_{1z}}\left|j_1,m_1\right\rangle \\ &\times \left\langle j_2,m_2\right|\mathrm{e}^{\mathrm{i}\varphi J_{2z}}\left|j_2,m_2\right\rangle , \end{aligned} \tag{3.5.21}$$

leading to the trace for given j_1, j_2, the sum over m_1 and m_2:

$$\mathrm{tr}_{(j_1,j_2)}\left\{\mathrm{e}^{\mathrm{i}\varphi J_z}\right\} = \mathrm{tr}_{(j_1)}\left\{\mathrm{e}^{\mathrm{i}\varphi J_{1z}}\right\}\mathrm{tr}_{(j_2)}\left\{\mathrm{e}^{\mathrm{i}\varphi J_{2z}}\right\} = \sum_{m_1,m_2} \mathrm{e}^{\mathrm{i}(m_1+m_2)\varphi} . \tag{3.5.22}$$

Now

$$\begin{aligned}\mathrm{tr}_{(j_1)}\left\{\mathrm{e}^{\mathrm{i}\varphi J_{1z}}\right\}\mathrm{tr}_{(j_2)}\left\{\mathrm{e}^{\mathrm{i}\varphi J_{2z}}\right\} &= \frac{\sin\left((j_1+\frac{1}{2})\varphi\right)}{\sin(\frac{1}{2}\varphi)}\frac{\sin\left((j_2+\frac{1}{2})\varphi\right)}{\sin(\frac{1}{2}\varphi)} \\ &= \frac{-\cos\left((j_1+j_2+1)\varphi\right)+\cos\left(|j_1-j_2|\varphi\right)}{2\sin^2(\frac{1}{2}\varphi)} \\ &= \frac{-\sum\limits_{j=|j_1-j_2|}^{j_1+j_2}\left[\cos\left((j+1)\varphi\right)-\cos(j\varphi)\right]}{2\sin^2(\frac{1}{2}\varphi)}\end{aligned} \tag{3.5.23}$$

or $\left[j+1=(j+\frac{1}{2})+\frac{1}{2}, j=(j+\frac{1}{2})-\frac{1}{2}\right]$

$$\begin{aligned}\mathrm{tr}_{(j_1)}\left\{\mathrm{e}^{\mathrm{i}\varphi J_{1z}}\right\}\mathrm{tr}_{(j_2)}\left\{\mathrm{e}^{\mathrm{i}\varphi J_{2z}}\right\} &= \sum_{j=|j_1-j_2|}^{j_1+j_2}\frac{\sin\left((j+\frac{1}{2})\varphi\right)}{\sin(\frac{1}{2}\varphi)} \\ &= \sum_{j=|j_1-j_2|}^{j_1+j_2}\sum_{m=-j}^{j}\mathrm{e}^{\mathrm{i}m\varphi}\end{aligned} \tag{3.5.24}$$

showing that the totality of m_1, m_2 values can be classified by total angular momentum eigenvalues:

$$j \geq m \geq -j, \quad j_1+j_2 \geq j \geq |j_1-j_2| ; \tag{3.5.25}$$

see (3.5.4). Putting $\varphi = 0$ repeats the alternative, equivalent, count of states

$$(2j_1+1)(2j_2+1) = \sum_{j=|j_1-j_2|}^{j_1+j_2}(2j+1) ; \tag{3.5.26}$$

see (3.5.5).

3.6 Finite rotations. Eulerian angles

By changing the coordinate system, the rotation around the z axis becomes the rotation about the direction given by unit vector $\boldsymbol{n}$. Then if we call the rotation angle γ we get the unitary operator representing an arbitrary rotation

$$U = e^{i\gamma \boldsymbol{n} \cdot \boldsymbol{J}} . \tag{3.6.1}$$

Notice that three angles are required to specify this rotation, two to define $\boldsymbol{n}$, and γ. Other ways of choosing the three angles can be useful. The one introduced by Euler uses three successive rotations of the coordinate system.

1. Rotate through angle ϕ about original z axis:

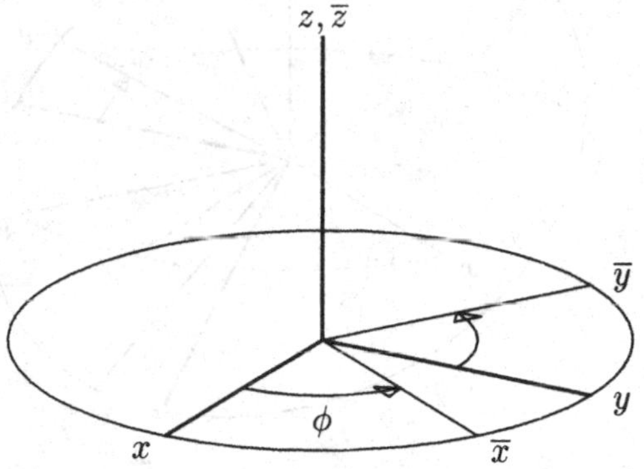

$$U_1 = e^{i\phi J_z} , \qquad \overline{\boldsymbol{J}} = U_1^{-1} \boldsymbol{J} U_1 . \tag{3.6.2}$$

2. Rotate through angle θ about new y axis:

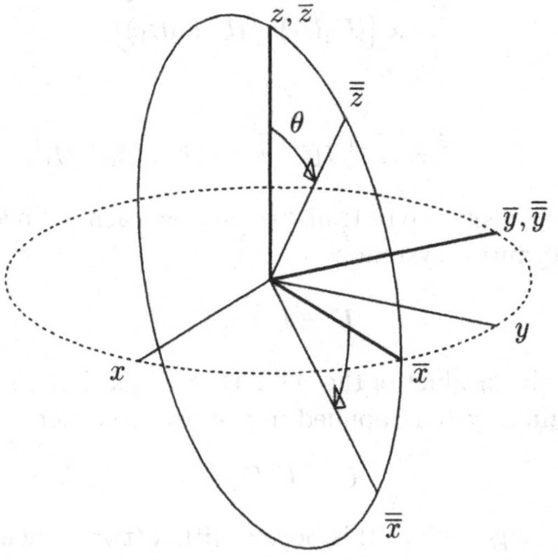

$$\overline{U}_2 = \mathrm{e}^{\mathrm{i}\theta \overline{J}_y} = U_1^{-1} \underbrace{\mathrm{e}^{\mathrm{i}\theta J_y}}_{=U_2} U_1 ,$$

$$\overline{\overline{\boldsymbol{J}}} = \overline{U}_2^{-1} \overline{\boldsymbol{J}} \overline{U}_2 = \overline{U}_2^{-1} U_1^{-1} \boldsymbol{J} U_1 \overline{U}_2 = U_1^{-1} U_2^{-1} \boldsymbol{J} U_2 U_1 \,. \quad (3.6.3)$$

3. Rotate through angle ψ about new z axis:

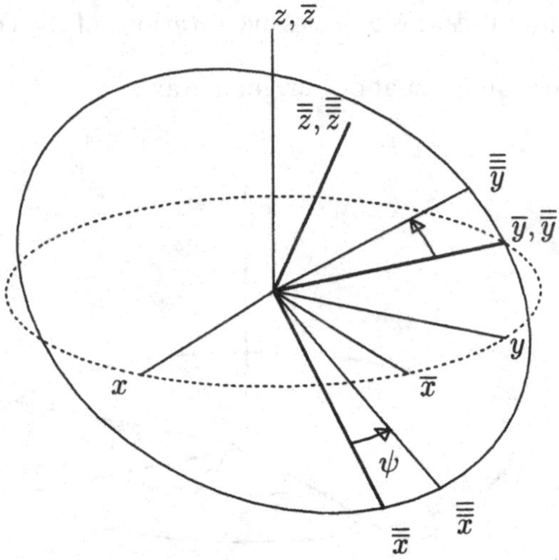

$$\overline{\overline{U}}_3 = \mathrm{e}^{\mathrm{i}\psi \overline{\overline{J}}_z} = U_1^{-1} U_2^{-1} \underbrace{\mathrm{e}^{\mathrm{i}\psi J_z}}_{=U_3} U_2 U_1 = (U_2U_1)^{-1}(U_3U_2U_1) ,$$

$$\begin{aligned}\overline{\overline{\overline{\boldsymbol{J}}}} = \overline{\overline{U}}_3^{-1} \overline{\overline{\boldsymbol{J}}} \overline{\overline{U}}_3 = &\left[(U_3U_2U_1)^{-1}(U_2U_1)\right] \left[(U_2U_1)^{-1}\boldsymbol{J}(U_2U_1)\right] \\ &\times \left[(U_2U_1)^{-1}(U_3U_2U_1)\right] \end{aligned} \quad (3.6.4)$$

or

$$\overline{\overline{\overline{\boldsymbol{J}}}} = U^{-1}\boldsymbol{J}U \quad \text{with } U = U_3U_2U_1 \,. \quad (3.6.5)$$

We have applied a sequence of transformations, each defined by its action on the previous coordinate system:

$$U = U_1 \overline{U}_2 \overline{\overline{U}}_3 \quad (3.6.6)$$

and ended with the product of the three transformations defined on the same system, the original system, applied in the reverse order

$$U = U_3U_2U_1 \,. \quad (3.6.7)$$

Another way of getting at this begins with a (two-dimensional) Euclidean analogy:

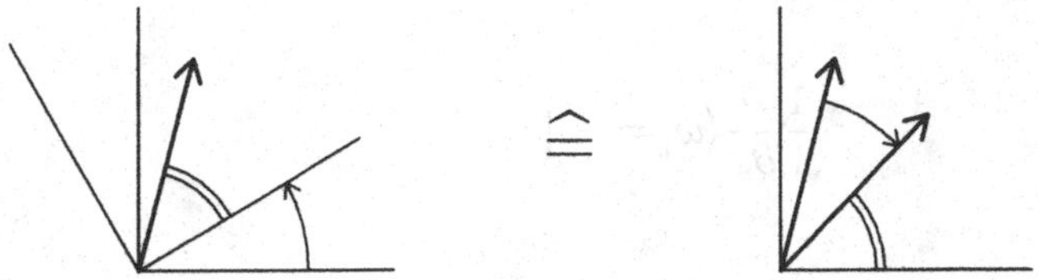

The final relation between vector and coordinate system is the same whether the vector is fixed and the coordinate system is rotated (counterclockwise) or the coordinate system is fixed and the vector is rotated (clockwise). Now consider, say, $\langle a'|\ \rangle$ where $\langle a'|$ is a coordinate system for vectors such as $|\ \rangle$, and change the coordinate system (as will be induced by a change of reference frame)

$$\langle a'| \to \overline{\langle a'|} = \langle a'|U_1 \,. \tag{3.6.8}$$

Then

$$\langle a'|\ \rangle \to \langle a'|U_1|\ \rangle \tag{3.6.9}$$

and we have in effect retained the coordinate system but changed the vector:

$$|\ \rangle \to U_1|\ \rangle \,. \tag{3.6.10}$$

Now make a second change of coordinate system, $\langle a| \to \langle a'|U_2$, so that

$$\langle a'|\ \rangle \underset{U_1}{\longrightarrow} \langle a'|U_1|\ \rangle \underset{U_2}{\longrightarrow} \langle a'|U_2U_1|\ \rangle \tag{3.6.11}$$

which shows quite clearly that for any number of successive changes, the net operator is

$$U = \cdots U_3U_2U_1 \,. \tag{3.6.12}$$

So, for Eulerian angles,

$$U(\underbrace{\psi,\theta,\phi}_{\equiv\omega}) = \mathrm{e}^{\mathrm{i}\psi J_z}\,\mathrm{e}^{\mathrm{i}\theta J_y}\,\mathrm{e}^{\mathrm{i}\phi J_z} \,. \tag{3.6.13}$$

See Problem 2-17d for another example illustrating these matters.

The dependence on the three angles lets us produce differential-operator realizations for the three components of $\boldsymbol{J}$. Consider $\langle\omega| = \langle\ |U(\omega)$. Then, since

$$\frac{1}{\mathrm{i}}\frac{\partial}{\partial\phi}U(\omega) = U(\omega)J_z \,, \tag{3.6.14}$$

we get

$$\frac{1}{\mathrm{i}}\frac{\partial}{\partial\phi}\langle\omega| = \langle\omega|J_z \ . \tag{3.6.15}$$

Next

$$\frac{1}{\mathrm{i}}\frac{\partial}{\partial\theta}U(\omega) = U(\omega)\,\mathrm{e}^{-\mathrm{i}\phi J_z}\,J_y\,\mathrm{e}^{\mathrm{i}\phi J_z} \tag{3.6.16}$$

which introduces one of the two statements about a z rotation:

$$\begin{aligned} J_x(\phi) &= \mathrm{e}^{-\mathrm{i}\phi J_z}\,J_x\,\mathrm{e}^{\mathrm{i}\phi J_z} = J_x\cos\phi + J_y\sin\phi \ , \\ J_y(\phi) &= \mathrm{e}^{-\mathrm{i}\phi J_z}\,J_y\,\mathrm{e}^{\mathrm{i}\phi J_z} = J_y\cos\phi - J_x\sin\phi \ . \end{aligned} \tag{3.6.17}$$

Geometry aside, one can check this by producing differential equations:

$$\begin{aligned} \frac{\partial}{\partial\phi}J_x(\phi) &= \mathrm{e}^{-\mathrm{i}\phi J_z}\mathrm{i}\,[J_x, J_z]\,\mathrm{e}^{\mathrm{i}\phi J_z} = J_y(\phi) \ , \\ \frac{\partial}{\partial\phi}J_y(\phi) &= \mathrm{e}^{-\mathrm{i}\phi J_z}\mathrm{i}\,[J_y, J_z]\,\mathrm{e}^{\mathrm{i}\phi J_z} = -J_x(\phi) \ . \end{aligned} \tag{3.6.18}$$

Define $J_\pm(\phi) = J_x(\phi) \pm \mathrm{i}J_y(\phi)$, which obey

$$\frac{\partial}{\partial\phi}J_+(\phi) = -\mathrm{i}J_+(\phi) \ , \quad \frac{\partial}{\partial\phi}J_-(\phi) = \mathrm{i}J_-(\phi) \ , \tag{3.6.19}$$

so

$$J_+(\phi) = \mathrm{e}^{-\mathrm{i}\phi}J_+ \ , \quad J_-(\phi) = \mathrm{e}^{\mathrm{i}\phi}J_- \ , \tag{3.6.20}$$

and $J_x(\phi) = \frac{1}{2}[J_+(\phi) + J_-(\phi)]$, $J_y(\phi) = \frac{1}{2\mathrm{i}}[J_+(\phi) - J_-(\phi)]$ give (3.6.17). So we have learned that

$$\frac{1}{\mathrm{i}}\frac{\partial}{\partial\theta}\langle\omega| = \langle\omega|\left(-J_x\sin\phi + J_y\cos\phi\right) \ . \tag{3.6.21}$$

Finally,

$$\frac{1}{\mathrm{i}}\frac{\partial}{\partial\psi}U(\omega) = \mathrm{e}^{\mathrm{i}\psi J_z}\,J_z\,\mathrm{e}^{\mathrm{i}\theta J_y}\,\mathrm{e}^{\mathrm{i}\phi J_z} = U(\omega)\,\mathrm{e}^{-\mathrm{i}\phi J_z}\,\mathrm{e}^{-\mathrm{i}\theta J_y}\,J_z\,\mathrm{e}^{\mathrm{i}\theta J_y}\,\mathrm{e}^{\mathrm{i}\phi J_z} \tag{3.6.22}$$

or with $x \to z$, $z \to y$, $y \to x$, $\phi \to \theta$ in (3.6.17),

$$\begin{aligned} \frac{1}{\mathrm{i}}\frac{\partial}{\partial\psi}U(\omega) &= U(\omega)\,\mathrm{e}^{-\mathrm{i}\phi J_z}\,(J_z\cos\theta + J_x\sin\theta)\,\mathrm{e}^{\mathrm{i}\phi J_z} \\ &= U(\omega)\,(J_x\sin\theta\cos\phi + J_y\sin\theta\sin\phi + J_z\cos\theta) \ , \end{aligned} \tag{3.6.23}$$

leading to

$$\frac{1}{\mathrm{i}}\frac{\partial}{\partial\psi}\langle\omega| = \langle\omega|\,(J_x \sin\theta\cos\phi + J_y \sin\theta\sin\phi + J_z\cos\theta) \tag{3.6.24}$$

and then to

$$\frac{1}{\sin\theta}\left(\frac{1}{\mathrm{i}}\frac{\partial}{\partial\psi} - \cos\theta\frac{1}{\mathrm{i}}\frac{\partial}{\partial\phi}\right)\langle\omega| = \langle\omega|\,(J_x\cos\phi + J_y\sin\phi)\;. \tag{3.6.25}$$

Equation (3.6.24) could have been anticipated. Angle ψ refers to rotations around the new z axis, which points in the direction given by spherical coordinates θ, ϕ. This is just the component of $\boldsymbol{J}$ that appears here.

Now multiply the θ equation (3.6.21) by $\pm\mathrm{i}$ and add to the above; the outcome is

$$\mathrm{e}^{\pm\mathrm{i}\phi}\left[\pm\frac{\partial}{\partial\theta} + \frac{1}{\sin\theta}\left(\frac{1}{\mathrm{i}}\frac{\partial}{\partial\psi} - \cos\theta\frac{1}{\mathrm{i}}\frac{\partial}{\partial\phi}\right)\right]\langle\omega| = \langle\omega|J_\pm\;. \tag{3.6.26}$$

We can also construct a differential operator representing $\boldsymbol{J}^2$:

$$\boldsymbol{J}^2 = J_-J_+ + J_z^2 + J_z\;, \tag{3.6.27}$$

so

$$\begin{aligned}\langle\omega|\boldsymbol{J}^2 = \Bigg\{&\mathrm{e}^{-\mathrm{i}\phi}\left[-\frac{\partial}{\partial\theta} + \frac{1}{\sin\theta}\left(\frac{1}{\mathrm{i}}\frac{\partial}{\partial\psi} - \cos\theta\frac{1}{\mathrm{i}}\frac{\partial}{\partial\phi}\right)\right]\\ &\times \mathrm{e}^{\mathrm{i}\phi}\left[\frac{\partial}{\partial\theta} + \frac{1}{\sin\theta}\left(\frac{1}{\mathrm{i}}\frac{\partial}{\partial\psi} - \cos\theta\frac{1}{\mathrm{i}}\frac{\partial}{\partial\phi}\right)\right]\\ &+\left(\frac{1}{\mathrm{i}}\frac{\partial}{\partial\phi}\right)^2 + \frac{1}{\mathrm{i}}\frac{\partial}{\partial\phi}\Bigg\}\langle\omega|\end{aligned} \tag{3.6.28}$$

or

$$\langle\omega|\boldsymbol{J}^2 = \left[-\frac{\partial^2}{\partial\theta^2} - \cot\theta\frac{\partial}{\partial\theta} - \frac{1}{\sin^2\theta}\left(\frac{\partial^2}{\partial\phi^2} - 2\cos\theta\frac{\partial}{\partial\phi}\frac{\partial}{\partial\psi} + \frac{\partial^2}{\partial\psi^2}\right)\right]\langle\omega|\;. \tag{3.6.29}$$

We make just one application of this differential equation, for which it is helpful to write

$$\frac{\partial^2}{\partial\theta^2} + \cot\theta\frac{\partial}{\partial\theta} = \frac{1}{\sin\theta}\frac{\partial}{\partial\theta}\sin\theta\frac{\partial}{\partial\theta}\;. \tag{3.6.30}$$

We are going to integrate over all angles, combining the familiar solid angle $\sin\theta\mathrm{d}\theta\mathrm{d}\phi$ with $\mathrm{d}\psi$. What is the range of integration? It is $0 \to \pi$ for θ, set by the vanishing of $\sin\theta$. But for ϕ and ψ, appearing in $\mathrm{e}^{\mathrm{i}\phi J_z}$, $\mathrm{e}^{\mathrm{i}\psi J_z}$, the fact that J'_z can be $\pm\frac{1}{2}, \pm\frac{3}{2}, \ldots$, in addition to integers, leads us to integrate over the range $\phi : 0 \to 4\pi$; $\psi : 0 \to 4\pi$ in order to attain periodicity. So if we define the normalized "volume"

$$d\omega = \frac{1}{2}\sin\theta d\theta \frac{1}{4\pi} d\phi \frac{1}{4\pi} d\psi \,, \quad \int d\omega = 1 \,, \tag{3.6.31}$$

the result of integrating the differential equation obeyed by $U(\omega)$ is to remove all $\partial/\partial\phi$, $\partial/\partial\psi$ terms which have no net change, and, as for the θ derivative we get

$$\int_0^\pi d\theta \frac{\partial}{\partial\theta}\left(\sin\theta \frac{\partial}{\partial\theta} U(\omega)\right) = \sin\theta \frac{\partial}{\partial\theta} U \bigg|_0^\pi = 0 \,. \tag{3.6.32}$$

In short, we learn that

$$\int d\omega \, U(\omega) \boldsymbol{J}^2 = 0 \tag{3.6.33}$$

which tells us that

$$\int d\omega \, U(\omega) = |00| \,, \tag{3.6.34}$$

the symbol of the selection of the $j = 0$ state, the one with $\boldsymbol{J}^{2\prime} = 0$. As for whether the numerical factor is right, note that

$$|00||00| = |00| \int d\omega \, U(\omega) \,. \tag{3.6.35}$$

But the selection of $\boldsymbol{J}^{2\prime} = 0$, for which $\boldsymbol{J} = 0$, means: $U(\omega) = 1$, and $\int d\omega = 1$, which shows that the right side is just $|00|$. More simply put: $|00|$ has the value one for a state with $\boldsymbol{J}^2 = 0$.

3.7 Rotated angular-momentum eigenvectors

We are given vectors

$$\langle j, m| = \langle 0| \frac{y_+^{j+m} y_-^{j-m}}{\sqrt{(j+m)!(j-m)!}} \,. \tag{3.7.1}$$

What are these vectors in the rotated frame,

$$\begin{aligned}\langle j, m; \omega| &= \langle j, m|U(\omega) = \langle 0| \frac{y_+^{j+m} y_-^{j-m}}{\sqrt{(j+m)!(j-m)!}} U(\omega) \\ &= \langle 0| \frac{\overline{y}_+^{j+m} \overline{y}_-^{j-m}}{\sqrt{(j+m)!(j-m)!}} \,?\end{aligned} \tag{3.7.2}$$

In the last version we have used $\langle 0|U(\omega) = \langle 0|$ and defined

$$\overline{y} = U^{-1} y U \,. \tag{3.7.3}$$

[In the notation of (3.6.2)–(3.6.5) we'd write $\overline{\overline{\overline{y}}}$, rather than $\overline{y}$.] To evaluate the latter, let's differentiate the form (3.6.1) for $U(\omega)$,

$$\frac{1}{\mathrm{i}}\frac{\partial}{\partial\gamma}\overline{y} = \frac{1}{\mathrm{i}}\frac{\partial}{\partial\gamma}\mathrm{e}^{-\mathrm{i}\gamma\boldsymbol{n}\cdot\boldsymbol{J}}\, y\, \mathrm{e}^{\mathrm{i}\gamma\boldsymbol{n}\cdot\boldsymbol{J}} = U^{-1}\left[y, \boldsymbol{n}\cdot\boldsymbol{J}\right]U \,. \tag{3.7.4}$$

Now

$$\begin{aligned}
[y_\sigma, \boldsymbol{n}\cdot\boldsymbol{J}] &= \Big[y_\sigma, \sum_{\sigma',\sigma''} y^\dagger_{\sigma'}\langle\sigma'|\boldsymbol{n}\cdot\tfrac{1}{2}\boldsymbol{\sigma}|\sigma''\rangle y_{\sigma''}\Big] \\
&= \sum_{\sigma',\sigma''}\underbrace{[y_\sigma, y^\dagger_{\sigma'}]}_{=\delta(\sigma,\sigma')}\langle\sigma'|\boldsymbol{n}\cdot\tfrac{1}{2}\boldsymbol{\sigma}|\sigma''\rangle y_{\sigma''} \\
&= \sum_{\sigma''}\langle\sigma|\boldsymbol{n}\cdot\tfrac{1}{2}\boldsymbol{\sigma}|\sigma''\rangle y_{\sigma''} = \left(\boldsymbol{n}\cdot\tfrac{1}{2}\boldsymbol{\sigma}y\right)_\sigma \,,
\end{aligned} \tag{3.7.5}$$

so

$$\frac{1}{\mathrm{i}}\frac{\partial}{\partial\gamma}\overline{y} = \boldsymbol{n}\cdot\tfrac{1}{2}\boldsymbol{\sigma}\overline{y} \tag{3.7.6}$$

and

$$\overline{y} = \mathrm{e}^{\mathrm{i}\gamma\boldsymbol{n}\cdot\frac{1}{2}\boldsymbol{\sigma}}y = \mathcal{U}y \,, \tag{3.7.7}$$

where $\mathcal{U}$ is the 2×2 matrix to the $j=\frac{1}{2}$ version of U.

When we use Eulerian angles, $\mathcal{U}$ is

$$\mathcal{U} = \mathrm{e}^{\mathrm{i}\psi\frac{1}{2}\sigma_z}\,\mathrm{e}^{\mathrm{i}\theta\frac{1}{2}\sigma_y}\,\mathrm{e}^{\mathrm{i}\phi\frac{1}{2}\sigma_z} \,, \tag{3.7.8}$$

which is the matrix

$$\langle\sigma|\mathcal{U}|\sigma'\rangle = \overset{\sigma\backslash\sigma'}{\begin{pmatrix} \mathrm{e}^{\mathrm{i}\frac{\psi}{2}}\cos\frac{\theta}{2}\,\mathrm{e}^{\mathrm{i}\frac{\phi}{2}} & \mathrm{e}^{\mathrm{i}\frac{\psi}{2}}\sin\frac{\theta}{2}\,\mathrm{e}^{-\mathrm{i}\frac{\phi}{2}} \\ -\mathrm{e}^{-\mathrm{i}\frac{\psi}{2}}\sin\frac{\theta}{2}\,\mathrm{e}^{\mathrm{i}\frac{\phi}{2}} & \mathrm{e}^{-\mathrm{i}\frac{\psi}{2}}\cos\frac{\theta}{2}\,\mathrm{e}^{-\mathrm{i}\frac{\phi}{2}} \end{pmatrix}} \,. \tag{3.7.9}$$

We now see that the $\langle j,m;\omega|$ are given as a linear combination of the $\langle j,m'|$,

$$\langle j,m;\omega| = \sum_{m'}\underbrace{\langle j,m;\omega|j,m'\rangle}_{=\langle j,m|U(\omega)|j,m'\rangle}\langle j,m'| \,, \tag{3.7.10}$$

by expanding

$$\frac{\overline{y}_+^{j+m}\,\overline{y}_-^{j-m}}{\sqrt{(j+m)!(j-m)!}} = \sum_{m'} \langle j,m|U(\omega)|j,m'\rangle \frac{y_+^{j+m'}\,y_-^{j-m'}}{\sqrt{(j+m')!(j-m')!}} \tag{3.7.11}$$

in which

$$\begin{aligned} \overline{y}_+ &= \mathrm{e}^{\mathrm{i}\frac{\psi}{2}}\left(\cos\frac{\theta}{2}\,\mathrm{e}^{\mathrm{i}\frac{\phi}{2}}y_+ + \sin\frac{\theta}{2}\,\mathrm{e}^{-\mathrm{i}\frac{\phi}{2}}y_-\right), \\ \overline{y}_- &= \mathrm{e}^{-\mathrm{i}\frac{\psi}{2}}\left(-\sin\frac{\theta}{2}\,\mathrm{e}^{\mathrm{i}\frac{\phi}{2}}y_+ + \cos\frac{\theta}{2}\,\mathrm{e}^{-\mathrm{i}\frac{\phi}{2}}y_-\right). \end{aligned} \tag{3.7.12}$$

We should be able to simplify the ψ and ϕ dependence, because

$$\begin{aligned} \langle j,m|U(\omega)|j,m'\rangle &= \langle j,m|\,\mathrm{e}^{\mathrm{i}\psi J_z}\,\mathrm{e}^{\mathrm{i}\theta J_y}\,\mathrm{e}^{\mathrm{i}\phi J_z}\,|j,m'\rangle \\ &= \mathrm{e}^{\mathrm{i}m\psi}\langle j,m|\,\mathrm{e}^{\mathrm{i}\theta J_y}\,|j,m'\rangle\,\mathrm{e}^{\mathrm{i}m'\phi}. \end{aligned} \tag{3.7.13}$$

Indeed, we see on the left the factor

$$\left(\mathrm{e}^{\mathrm{i}\frac{\psi}{2}}\right)^{j+m}\left(\mathrm{e}^{-\mathrm{i}\frac{\psi}{2}}\right)^{j-m} = \mathrm{e}^{\mathrm{i}m\psi} \tag{3.7.14}$$

and also just the combinations $\mathrm{e}^{\mathrm{i}\frac{\phi}{2}}y_+$, $\mathrm{e}^{-\mathrm{i}\frac{\phi}{2}}y_-$ occur, as required for

$$\mathrm{e}^{\mathrm{i}m'\phi}y_+^{j+m'}y_-^{j-m'} = \left(\mathrm{e}^{\mathrm{i}\frac{\phi}{2}}y_+\right)^{j+m'}\left(\mathrm{e}^{-\mathrm{i}\frac{\phi}{2}}y_-\right)^{j-m'}. \tag{3.7.15}$$

So it suffices to put $\psi = 0$, $\phi = 0$, and get

$$\begin{aligned} &\frac{\left(\cos\frac{\theta}{2}y_+ + \sin\frac{\theta}{2}y_-\right)^{j+m}\left(-\sin\frac{\theta}{2}y_+ + \cos\frac{\theta}{2}y_-\right)^{j-m}}{\sqrt{(j+m)!(j-m)!}} \\ &\qquad = \sum_{m'} \underbrace{\langle j,m|\,\mathrm{e}^{\mathrm{i}\theta J_y}\,|j,m'\rangle}_{=U^{(j)}_{m,m'}(\theta)} \frac{y_+^{j+m'}\,y_-^{j-m'}}{\sqrt{(j+m')!(j-m')!}}. \end{aligned} \tag{3.7.16}$$

As an example consider $j = 1$. Then

$\underline{m = 1:}$

$$\begin{aligned} &\frac{1}{\sqrt{2}}\left(\cos\frac{\theta}{2}\,y_+ + \sin\frac{\theta}{2}\,y_-\right)^2 \\ &\qquad = \cos^2\frac{\theta}{2}\frac{y_+^2}{\sqrt{2}} + \sqrt{2}\cos\frac{\theta}{2}\sin\frac{\theta}{2}\,y_+y_- + \sin^2\frac{\theta}{2}\frac{y_-^2}{\sqrt{2}} \end{aligned} \tag{3.7.17}$$

or

$$U^{(1)}_{1,1}(\theta) = \cos^2\frac{\theta}{2}, \quad U^{(1)}_{1,0}(\theta) = \sqrt{2}\cos\frac{\theta}{2}\sin\frac{\theta}{2}, \quad U^{(1)}_{1,-1}(\theta) = \sin^2\frac{\theta}{2}, \tag{3.7.18}$$

and

$$\underline{m=0:}\qquad \left(\cos\frac{\theta}{2}\,y_+ + \sin\frac{\theta}{2}\,y_-\right)\left(-\sin\frac{\theta}{2}\,y_+ + \cos\frac{\theta}{2}\,y_-\right)$$
$$= -\frac{1}{\sqrt{2}}\sin\theta\,\frac{y_+^2}{\sqrt{2}} + \cos\theta\,y_+y_- + \frac{1}{\sqrt{2}}\sin\theta\,\frac{y_-^2}{\sqrt{2}} \qquad (3.7.19)$$

or

$$U^{(1)}_{0,1}(\theta) = -\frac{1}{\sqrt{2}}\sin\theta\,, \quad U^{(1)}_{0,0}(\theta) = \cos\theta\,, \quad U^{(1)}_{0,-1}(\theta) = \frac{1}{\sqrt{2}}\sin\theta\,, \qquad (3.7.20)$$

as well as

$$\underline{m=-1:}\qquad \frac{1}{\sqrt{2}}\left(-\sin\frac{\theta}{2}\,y_+ + \cos\frac{\theta}{2}\,y_-\right)^2$$
$$= \sin^2\frac{\theta}{2}\frac{y_+^2}{\sqrt{2}} - \sqrt{2}\sin\frac{\theta}{2}\cos\frac{\theta}{2}\,y_+y_- + \cos^2\frac{\theta}{2}\frac{y_-^2}{\sqrt{2}} \qquad (3.7.21)$$

or

$$U^{(1)}_{-1,1}(\theta) = \sin^2\theta\,, \quad U^{(1)}_{-1,0}(\theta) = -\sqrt{2}\sin\frac{\theta}{2}\cos\frac{\theta}{2}\,, \quad U^{(1)}_{-1,-1}(\theta) = \cos^2\frac{\theta}{2}\,. \qquad (3.7.22)$$

What we have here is an example of the probabilities relating measurements of angular momentum, or magnetic moment, in two different directions related by angle θ:

$$p(m,m';\theta) = \left|\langle j,m;\omega|j,m'\rangle\right|^2 = \left[U^{(j)}_{m,m'}(\theta)\right]^2\,. \qquad (3.7.23)$$

The table

$j=1:$ (3.7.24)

m \ m'	1	0	−1
1	$\cos^4\frac{\theta}{2}$	$2\cos^2\frac{\theta}{2}\sin^2\frac{\theta}{2}$	$\sin^4\frac{\theta}{2}$
0	$\frac{1}{2}\sin^2\theta$	$\cos^2\theta$	$\frac{1}{2}\sin^2\theta$
−1	$\sin^4\frac{\theta}{2}$	$2\sin^2\frac{\theta}{2}\cos^2\frac{\theta}{2}$	$\cos^4\frac{\theta}{2}$

summarizes the $j=1$ probabilities. Evidently,

$$p(m,m') = p(m',m)\,, \quad \sum_m p(m,m') = 1\,, \quad \sum_{m'} p(m,m') = 1\,. \qquad (3.7.25)$$

For another specialization put $m = 0$, which requires that j be an integer: $l = 0, 1, 2, \ldots$. Then $(m' \to m)$

$$\frac{1}{l!}\left[\left(\cos\frac{\theta}{2}y_+ + \sin\frac{\theta}{2}y_-\right)\left(-\sin\frac{\theta}{2}y_+ + \cos\frac{\theta}{2}y_-\right)\right]^l$$
$$= \sum_{m=-l}^{l} U_{0,m}^{(l)}(\theta)\frac{y_+^{l+m}y_-^{l-m}}{\sqrt{(l+m)!(l-m)!}} . \tag{3.7.26}$$

Notice that the left side is multiplied by $(-1)^l$ when $y_+ \to y_-$, $y_- \to -y_+$. So the right side is also equal to

$$(-1)^l \sum_{m=-l}^{l} U_{0,m}^{(l)}(\theta)\frac{y_-^{l+m}(-y_+)^{l-m}}{\sqrt{(l+m)!(l-m)!}}$$
$$\underset{\underbrace{\uparrow}_{m\to -m}}{=} \sum_{m=-l}^{l} (-1)^m U_{0,-m}^{(l)}(\theta)\frac{y_+^{l+m}y_-^{l-m}}{\sqrt{(l+m)!(l-m)!}} \tag{3.7.27}$$

and

$$U_{0,m}^{(l)}(\theta) = (-1)^m U_{0,-m}^{(l)}(\theta) \tag{3.7.28}$$

follows, for which $U_{0,1}^{(1)} = -U_{0,-1}^{(1)}$ in (3.7.20) is an example. To get an expression for $U_{0,m}^{(l)}(\theta)$, put

$$y_+ = \sin\theta\ , \qquad y_- = t - \cos\theta \tag{3.7.29}$$

so that

$$\begin{aligned}\cos\frac{\theta}{2}y_+ + \sin\frac{\theta}{2}y_- &= (t+1)\sin\frac{\theta}{2}\ ,\\ -\sin\frac{\theta}{2}y_+ + \cos\frac{\theta}{2}y_- &= (t-1)\cos\frac{\theta}{2}\ ,\end{aligned} \tag{3.7.30}$$

and

$$(\sin\theta)^l\frac{(t^2-1)^l}{2^l l!} = \sum_{m=-l}^{l} U_{0,m}^{(l)}(\theta)\frac{(\sin\theta)^{l+m}(t-\cos\theta)^{l-m}}{\sqrt{(l+m)!(l-m)!}}\ , \tag{3.7.31}$$

which has the form of a Taylor series expansion. So

$$(\sin\theta)^l\left(\frac{\mathrm{d}}{\mathrm{d}t}\right)^{l-m}\frac{(t^2-1)^l}{2^l l!}\Bigg|_{t=\cos\theta} = U_{0,m}^{(l)}(\theta)(\sin\theta)^{l+m}\sqrt{\frac{(l-m)!}{(l+m)!}} \tag{3.7.32}$$

or

$$U^{(l)}_{0,m}(\theta) = \sqrt{\frac{(l+m)!}{(l-m)!}}(\sin\theta)^{-m}\left(\frac{\mathrm{d}}{\mathrm{d}\cos\theta}\right)^{l-m}\frac{(\cos^2\theta-1)^l}{2^l l!} . \tag{3.7.33}$$

In particular, $U^{(l)}_{0,0}$ is given by Legendre's* polynomial,

$$U^{(l)}_{0,0}(\theta) = \left(\frac{\mathrm{d}}{\mathrm{d}\cos\theta}\right)^{l}\frac{(\cos^2\theta-1)^l}{2^l l!} = \mathrm{P}_l(\cos\theta) , \tag{3.7.34}$$

as illustrated by

$$U^{(1)}_{0,0}(\theta) = \cos\theta = \mathrm{P}_1(\cos\theta) . \tag{3.7.35}$$

Is it accidental that we run into Legendre's polynomial here? Hardly. To see why this happened, let's restore the angle ϕ:

$$y_+ \to y_+\,\mathrm{e}^{\mathrm{i}\frac{\phi}{2}} , \qquad y_- \to y_-\,\mathrm{e}^{-\mathrm{i}\frac{\phi}{2}} , \qquad y_+^{l+m}y_-^{l-m} \to \mathrm{e}^{\mathrm{i}m\phi}y_+^{l+m}y_-^{l-m} . \tag{3.7.36}$$

We now meet the lth power of

$$\begin{aligned}
&\left(\cos\frac{\theta}{2}\,\mathrm{e}^{\mathrm{i}\frac{\phi}{2}}y_+ + \sin\frac{\theta}{2}\,\mathrm{e}^{-\mathrm{i}\frac{\phi}{2}}y_-\right)\left(-\sin\frac{\theta}{2}\,\mathrm{e}^{\mathrm{i}\frac{\phi}{2}}y_+ + \cos\frac{\theta}{2}\,\mathrm{e}^{-\mathrm{i}\frac{\phi}{2}}y_-\right)\\
&\quad = -\frac{1}{2}\sin\theta\,\mathrm{e}^{\mathrm{i}\phi}y_+^2 + \frac{1}{2}\sin\theta\,\mathrm{e}^{-\mathrm{i}\phi}y_-^2 + \cos\theta\,y_+y_-\\
&\quad = \frac{1}{2}\Big[\underbrace{\sin\theta\cos\phi}_{=x/r}\,\underbrace{(y_-^2-y_+^2)}_{=\nu_x} + \underbrace{\sin\theta\sin\phi}_{=y/r}\,\underbrace{(-\mathrm{i}y_-^2-\mathrm{i}y_+^2)}_{=\nu_y} + \underbrace{\cos\theta}_{=z/r}\,\underbrace{2y_+y_-}_{=\nu_z}\Big]\\
&\quad = \frac{1}{2}\frac{\boldsymbol{r}}{r}\cdot\boldsymbol{\nu}
\end{aligned} \tag{3.7.37}$$

where r,θ,ϕ are the spherical coordinates of point $\boldsymbol{r}$ [cf. the illustration in (1.12.2) for unit vector $\boldsymbol{n}=\boldsymbol{r}/r$], and the cartesian components of vector $\boldsymbol{\nu}$ are as indicated,

$$\boldsymbol{\nu} = (y_-^2 - y_+^2, -\mathrm{i}y_-^2 - \mathrm{i}y_+^2, 2y_+y_-) . \tag{3.7.38}$$

Notice that

$$\boldsymbol{\nu}^2 = (y_-^2 - y_+^2)^2 + (\mathrm{i}y_-^2 + \mathrm{i}y_+^2)^2 + (2y_+y_-)^2 = 0 , \tag{3.7.39}$$

$\boldsymbol{\nu}$ is a complex vector of zero length, a null vector. Its components are presented, in terms of $y = \begin{pmatrix} y_+ \\ y_- \end{pmatrix}$ and its transpose $y^{\mathrm{T}} = (y_+, y_-)$, in matrix form:

*Adrien Marie LEGENDRE (1752–1833)

$$\nu_x = y^{\mathrm{T}}(-\sigma_z)y\,, \qquad \nu_y = y^{\mathrm{T}}(-\mathrm{i})y\,, \qquad \nu_z = y^{\mathrm{T}}\sigma_x y \tag{3.7.40}$$

or

$$\boldsymbol{\nu} = y^{\mathrm{T}}\frac{1}{\mathrm{i}}\sigma_y \boldsymbol{\sigma} y\,. \tag{3.7.41}$$

Now we have

$$\frac{(\boldsymbol{r}\cdot\boldsymbol{\nu})^l}{2^l l!} = \sum_{m=-l}^{l} r^l Y_{lm}(\theta,\phi)\sqrt{\frac{4\pi}{2l+1}}\frac{y_+^{l+m}y_-^{l-m}}{\sqrt{(l+m)!(l-m)!}} \tag{3.7.42}$$

where we have written

$$U_{0,m}^{(l)}(\theta)\,\mathrm{e}^{\mathrm{i}m\phi} = \sqrt{\frac{4\pi}{2l+1}}Y_{lm}(\theta,\phi)\,. \tag{3.7.43}$$

Consider the Laplacian* of the left side:

$$\boldsymbol{\nabla}^2(\boldsymbol{r}\cdot\boldsymbol{\nu})^l = \boldsymbol{\nabla}\cdot l(\boldsymbol{r}\cdot\boldsymbol{\nu})^{l-1}\boldsymbol{\nu} = l(l-1)(\boldsymbol{r}\cdot\boldsymbol{\nu})^{l-2}\,\nu^2 = 0\,, \tag{3.7.44}$$

it is a solution of Laplace's equation. Solutions that are homogeneous in $\boldsymbol{r}$, of degree l are called solid harmonics; the $r^l Y_{lm}(\theta,\phi)$ are solid harmonics [the term spherical harmonics is applied to $Y_{lm}(\theta,\phi)$]. It is familiar that Legendre's polynomial gives the spherical harmonics that are independent of ϕ; $m = 0$. Incidentally, the factor introduced in defining $Y_{lm}(\theta,\phi)$ by the generating function (3.7.42) is such that the orthonormality statement

$$\int \mathrm{d}\Omega\, Y_{lm}(\theta,\phi)^* Y_{l'm'}(\theta,\phi) = \delta_{ll'}\delta_{mm'} \quad \text{with } \mathrm{d}\Omega = \sin\theta\mathrm{d}\theta\mathrm{d}\phi \tag{3.7.45}$$

has a simple appearance. For a proof based on (3.7.42), see Problems 3-14.

The quantity $\boldsymbol{\nu}$ has been identified as a vector, but is it really? Suppose one replaces y by $\overline{y} = \mathcal{U}y$ then

$$\overline{\boldsymbol{\nu}} = (\mathcal{U}y)^{\mathrm{T}}\frac{1}{\mathrm{i}}\sigma_y\boldsymbol{\sigma}(\mathcal{U}y) = y^{\mathrm{T}}\mathcal{U}^{\mathrm{T}}\frac{1}{\mathrm{i}}\sigma_y\boldsymbol{\sigma}\mathcal{U}y\,. \tag{3.7.46}$$

Next, the effect of transposition on the Pauli matrices (1.10.3) is given by

$$\sigma_x^{\mathrm{T}} = \sigma_x\,, \qquad \sigma_y^{\mathrm{T}} = -\sigma_y\,, \qquad \sigma_z^{\mathrm{T}} = \sigma_z \tag{3.7.47}$$

or

$$\boldsymbol{\sigma}^{\mathrm{T}} = -\sigma_y\boldsymbol{\sigma}\sigma_y\,. \tag{3.7.48}$$

*Marquis de Pierre Simon LAPLACE (1749–1827)

Then, with

$$\mathcal{U} = e^{i\gamma \boldsymbol{n} \cdot \frac{1}{2}\boldsymbol{\sigma}} = \cos\frac{\gamma}{2} + i\boldsymbol{n}\cdot\boldsymbol{\sigma}\sin\frac{\gamma}{2} \tag{3.7.49}$$

we have

$$\mathcal{U}^{\mathrm{T}} = \cos\frac{\gamma}{2} - i\sigma_y \boldsymbol{n}\cdot\boldsymbol{\sigma}\sigma_y \sin\frac{\gamma}{2} = \sigma_y \mathcal{U}^{-1}\sigma_y \tag{3.7.50}$$

or

$$\mathcal{U}^{\mathrm{T}}\sigma_y = \sigma_y \mathcal{U}^{-1}, \tag{3.7.51}$$

giving

$$\overline{\boldsymbol{\nu}} = y^{\mathrm{T}}\frac{1}{i}\sigma_y \mathcal{U}^{-1}\boldsymbol{\sigma}\mathcal{U}y . \tag{3.7.52}$$

We recognize in $\mathcal{U}^{-1}\boldsymbol{\sigma}\mathcal{U}$ the transformation that produces $\overline{\boldsymbol{\sigma}}$. Specifically, for infinitesimal transformations, we have

$$\mathcal{U} = 1 + i\delta\boldsymbol{\omega}\cdot\frac{1}{2}\boldsymbol{\sigma} , \tag{3.7.53}$$

and

$$\begin{aligned}\overline{\boldsymbol{\sigma}} &= (1 - i\delta\boldsymbol{\omega}\cdot\frac{1}{2}\boldsymbol{\sigma})\boldsymbol{\sigma}(1 + i\delta\boldsymbol{\omega}\cdot\frac{1}{2}\boldsymbol{\sigma}) \\ &= \boldsymbol{\sigma} - \frac{1}{2i}[\boldsymbol{\sigma}, \delta\boldsymbol{\omega}\cdot\boldsymbol{\sigma}] = \boldsymbol{\sigma} - \delta\boldsymbol{\omega}\times\boldsymbol{\sigma} ,\end{aligned} \tag{3.7.54}$$

yielding

$$\overline{\boldsymbol{\nu}} = \boldsymbol{\nu} - \delta\boldsymbol{\omega}\times\boldsymbol{\nu} ; \tag{3.7.55}$$

$\boldsymbol{\nu}$ is a vector.

Having seen how things work for $m = 0$, let's return to the general situation, using the same substitutions. After using (3.7.29) and (3.7.30) in (3.7.16),

$$\begin{aligned}&\frac{[(t+1)\sin\frac{\theta}{2}]^{j+m}[(t-1)\cos\frac{\theta}{2}]^{j-m}}{\sqrt{(j+m)!(j-m)!}} \\ &\qquad = \sum_{m'}(\sin\theta)^{j+m'}\frac{(t-\cos\theta)^{j-m'}}{\sqrt{(j+m')!(j-m')!}}U^{(j)}_{m,m'}(\theta) ,\end{aligned} \tag{3.7.56}$$

we get

$$U^{(j)}_{m,m'}(\theta) = \sqrt{\frac{(j+m')!}{(j-m')!}}\,\frac{\left(\sin\frac{\theta}{2}\right)^{m-m'}\left(\cos\frac{\theta}{2}\right)^{-m-m'}}{2^{j+m'}\sqrt{(j+m)!(j-m)!}}$$
$$\times\left(\frac{\mathrm{d}}{\mathrm{d}\cos\theta}\right)^{j-m'}\left[(\cos\theta+1)^{j+m}(\cos\theta-1)^{j-m}\right]\,. \qquad (3.7.57)$$

As we see by writing

$$\frac{1}{2^{m'}}\left(\sin\frac{\theta}{2}\right)^{m-m'}\left(\cos\frac{\theta}{2}\right)^{-m-m'} = \left(\tan\frac{\theta}{2}\right)^{m}(\sin\theta)^{-m'} \qquad (3.7.58)$$

we recover the known form for $m = 0$. Note the simple example

$$U^{(j)}_{m,j}(\theta) = \sqrt{\frac{(2j)!}{(j+m)!(j-m)!}}\left(\cos\frac{\theta}{2}\right)^{j+m}\left(-\sin\frac{\theta}{2}\right)^{j-m}\,, \qquad (3.7.59)$$

which can be immediately read off from the original expansion.

Given two angular momenta, $\boldsymbol{J}_1$ and $\boldsymbol{J}_2$, what is the state of zero total angular momentum? As we see from the restriction $|j_1 - j_2| \le j$, $j = 0$ is only attainable for $j_1 = j_2$ (we have already met $j = 0$ for $j_1 = j_2 = \frac{1}{2}$). So (omitting $j_1 = j_2$)

$$\langle 0| = \sum_{m_1,m_2}\langle 0|m_1,m_2\rangle\langle m_1,m_2|\,, \qquad (3.7.60)$$

which must obey

$$\langle 0|(\boldsymbol{J}_1 + \boldsymbol{J}_2) = 0\,. \qquad (3.7.61)$$

The z component of this relation

$$\sum_{m_1,m_2}\langle 0|m_1,m_2\rangle(m_1+m_2)\langle m_1,m_2| = 0 \qquad (3.7.62)$$

tells us that only values obeying

$$m = m_1 + m_2 = 0 \qquad (3.7.63)$$

occur. Therefore, in a simplified notation,

$$\langle 0| = \sum_m c_m\langle m,-m|\,. \qquad (3.7.64)$$

Now consider

$$\begin{aligned} 0 = \langle 0|(\boldsymbol{J}_{1-}+\boldsymbol{J}_{2-}) &= \sum_m c_m\langle m,-m|(\boldsymbol{J}_{1-}+\boldsymbol{J}_{2-}) \\ &= \sum_m c_m\Big\{\sqrt{(j_1-m)(j_1+m+1)}\langle m+1,-m| \qquad (3.7.65)\\ &\qquad +\sqrt{(j_2+m)(j_2-m+1)}\langle m,-m+1|\Big\}\,, \end{aligned}$$

so that ($j_2 = j_1$)

$$0 = \sum_m \sqrt{(j_1 - m)(j_1 + m + 1)}(c_m + c_{m+1})\langle m+1, -m| \,, \qquad (3.7.66)$$

implying $c_{m+1} = -c_m$. Therefore c_m alternates in sign as m changes by unity, or

$$\langle 0| = \sum_{m=-j_1}^{j_1} \frac{1}{\sqrt{2j_1+1}}(-1)^{j_1 - m}\langle m, -m| \qquad (3.7.67)$$

which already incorporates the normalization factor for a sum of $2j_1 + 1$ orthogonal unit vectors. For the example $j_1 = \frac{1}{2}$, we get

$$\langle 0| = \frac{1}{\sqrt{2}}\left[\langle +, -| - \langle -, +|\right] \,, \qquad (3.7.68)$$

which is just the previous result in (3.5.11), apart from a minus sign.

Problems

3-1 Use the $y_+^{\dagger\,\prime}$, $y_-^{\dagger\,\prime}$ description to find differential operators that represent the components of $\boldsymbol{J}$. Apply the latter to the appropriate angular momentum wave functions to arrive at the known results for $\boldsymbol{J}|j, m\rangle$.

3-2 There are two angular momenta, $j_1 = 1$, $j_2 = \frac{1}{2}$. Begin with the total angular momentum state $m = j = \frac{3}{2}$ in terms of the individual momenta, and then construct the other states of $j = \frac{3}{2}$. Next find the $m = j = \frac{1}{2}$ state as the one orthogonal to the $m = \frac{1}{2}$, $j = \frac{3}{2}$ state. Finally, construct the other $j = \frac{1}{2}$ state and verify that it is orthogonal to the $m = -\frac{1}{2}$, $j = \frac{3}{2}$ state.

3-3 What is the $j = m = 2$ state for the two angular momenta $j_1 = j_2 = 1$? Find the four other states with $j = 2$. How would you get the states with $j = 1$ and $j = 0$? Construct all of them.

3-4a Iso(topic) spin T: The nucleon is a particle of isospin $T = \frac{1}{2}$; the state with $T_3 = \frac{1}{2}$ is the proton (p), the state with $T_3 = -\frac{1}{2}$ is the neutron (n). Electric charge is given by $Q = \frac{1}{2} + T_3$. The π meson, or pion, has isospin $T = 1$, and electric charge $Q = T_3$, so there are three kinds of pions with different electric charge: $T_3 = 1$ (π^+), $T_3 = 0$ (π^0), $T_3 = -1$ (π^-).

Consider the system of a nucleon and a pion. The electric charge of this system is $Q = \frac{1}{2} + T_3$. Check that a system of charge 2, $T_3 = \frac{3}{2}$, is $\mathrm{p} + \pi^+$, according to the isospin assignments. Now, if the system is in the state $T = \frac{3}{2}$, $T_3 = \frac{1}{2}$, what is the probability of finding a proton? What is the accompanying π-meson? Answer the same questions for $T = \frac{1}{2}$, $T_3 = -\frac{1}{2}$.

3-4b The unstable particle Δ, with $T = \frac{3}{2}$, decays into $T = \frac{3}{2}$ states of one nucleon and one pion. What fraction of such events for Δ^+ produces a π^+, a π^0, a π^-?

3-4c Suppose nucleon 1 and nucleon 2 (charge $Q = 1 + T_3$ for this system) are in the $T = 0$ state. What is the probability that nucleon 1 is a proton?

3-5 The states of angular momentum $j = 1$ can be constructed from two $j = \frac{1}{2}$ spins, in a symmetrical state. Use these two-spin states to evaluate the probabilities $p(m, m', \theta)$ connecting measurements of components of $\boldsymbol{J}$ in two directions related by angle θ. Do this for $m = m' = \pm 1$ and $m = -m' = \pm 1$. Then apply probability normalization to find all other probabilities. Independently evaluate $p(0, 0, \theta)$. [The question refers specifically to two spins, not an application of a general result.]

3-6a For $j = 1$ evaluate

$$\begin{aligned} &\operatorname{tr}\{J_k\} \quad \text{for } k = x, y, z\,, \\ &\operatorname{tr}\{J_k J_l\} \quad \text{for } k, l = x, y, z\,, \\ &\operatorname{tr}\left\{J_k^2 J_l^2\right\} \quad \text{for } k, l = x, y, z\,. \end{aligned}$$

3-6b Use measurement symbols and the trace formula to evaluate the probability for finding $(J_z \cos\theta + J_x \sin\theta)' = 0$ if $J_z' = 0$ is prepared.

3-7a For $j = \frac{1}{2}$, two orthogonal cartesian components of $\boldsymbol{J}$, such as J_x and J_z, are complementary. Show that they are not for $j = 1$.

3-7b How about two non-orthogonal components, such as J_z and $\boldsymbol{e} \cdot \boldsymbol{J}$ with some unit vector $\boldsymbol{e}$?

3-8a Suppose $\boldsymbol{J} = \frac{1}{2}\boldsymbol{\sigma}_1 + \frac{1}{2}\boldsymbol{\sigma}_2$, then what are the associated eigenvalues of $(\boldsymbol{\sigma}_1 \cdot \boldsymbol{\sigma}_2)$? Use them to show directly that

$$(\boldsymbol{\sigma}_1 \cdot \boldsymbol{\sigma}_2)^2 = 3 - \boldsymbol{\sigma}_1 \cdot \boldsymbol{\sigma}_2$$

and that the permutation operator P_{12} of (3.5.14) obeys $P_{12}^2 = 1$.

3-8b Use the algebraic properties of $\boldsymbol{\sigma}_1$ and $\boldsymbol{\sigma}_2$ to verify that

$$\begin{aligned} \boldsymbol{\sigma}_1 P_{12} &= P_{12} \boldsymbol{\sigma}_2 \quad \text{or} \quad \boldsymbol{\sigma}_2 = P_{12}^{-1} \boldsymbol{\sigma}_1 P_{12}\,, \\ \text{and} \quad \boldsymbol{\sigma}_2 P_{12} &= P_{12} \boldsymbol{\sigma}_1 \quad \text{or} \quad \boldsymbol{\sigma}_1 = P_{12}^{-1} \boldsymbol{\sigma}_2 P_{12}\,. \end{aligned}$$

How does (3.5.15) follow?

3-9 For $\boldsymbol{J} = \frac{1}{2}\boldsymbol{\sigma}_1 + \frac{1}{2}\boldsymbol{\sigma}_2$, show that J_x^2, J_y^2, J_z^2 commute with each other. Find their common eigenvectors (sometimes call Bell* states), expressed in terms of the common eigenvectors of σ_{1z} and σ_{2z}. How are they related to the common eigenvectors of $\boldsymbol{J}^2$ and J_z?

3-10a System 1 has arbitrary angular momentum $\boldsymbol{J}_1$, system 2 has $\boldsymbol{J}_2 = \frac{1}{2}\boldsymbol{\sigma}$, so that $j = j_1 \pm \frac{1}{2}$ for total angular momentum $\boldsymbol{J} = \boldsymbol{J}_1 + \boldsymbol{J}_2$. Show that the eigenvalues of $\boldsymbol{\sigma} \cdot \boldsymbol{J}_1$ are given by

$$(\boldsymbol{\sigma} \cdot \boldsymbol{J}_1)' = \begin{cases} j_1 & \text{for } j = j_1 + \frac{1}{2}\,, \\ -(j_1 + 1) & \text{for } j = j_1 - \frac{1}{2}\,, \end{cases}$$

and verify algebraically that

$$(\boldsymbol{\sigma} \cdot \boldsymbol{J}_1)^2 + \boldsymbol{\sigma} \cdot \boldsymbol{J}_1 - j_1(j_1 + 1) = (\boldsymbol{\sigma} \cdot \boldsymbol{J}_1 - j_1)(\boldsymbol{\sigma} \cdot \boldsymbol{J}_1 + j_1 + 1) = 0\,.$$

Then show that

$$\frac{\boldsymbol{\sigma} \cdot \boldsymbol{J}_1 + j_1 + 1}{2j_1 + 1} \quad \text{and} \quad \frac{j_1 - \boldsymbol{\sigma} \cdot \boldsymbol{J}_1}{2j_1 + 1}$$

are measurement symbols for $j = j_1 + \frac{1}{2}$ and $j = j_1 - \frac{1}{2}$, respectively.

3-10b Employ

$$\boldsymbol{\sigma} \cdot \boldsymbol{J}_1 = \sigma_z J_{1z} + \tfrac{1}{2}(\sigma_x + \mathrm{i}\sigma_y)(J_{1x} - \mathrm{i}J_{1y}) + \tfrac{1}{2}(\sigma_x - \mathrm{i}\sigma_y)(J_{1x} + \mathrm{i}J_{1y})$$

to show that

$$\frac{\boldsymbol{\sigma} \cdot \boldsymbol{J}_1 + j_1 + 1}{2j_1 + 1}\left|m_1 = m - \tfrac{1}{2}, m_2 = \tfrac{1}{2}\right\rangle = \left|j = j_1 + \tfrac{1}{2}, m\right\rangle\sqrt{\frac{j_1 + m + \frac{1}{2}}{2j_1 + 1}}\,,$$

$$\frac{j_1 - \boldsymbol{\sigma} \cdot \boldsymbol{J}_1}{2j_1 + 1}\left|m_1 = m - \tfrac{1}{2}, m_2 = \tfrac{1}{2}\right\rangle = \left|j = j_1 - \tfrac{1}{2}, m\right\rangle\sqrt{\frac{j_1 - m + \frac{1}{2}}{2j_1 + 1}}\,.$$

Use them to check the orthogonality of $\left|j = j_1 + \frac{1}{2}, m\right\rangle$ and $\left|j = j_1 - \frac{1}{2}, m\right\rangle$.

3-10c Evaluate the expectation values of $\langle\sigma_z\rangle_{j=j_1\pm\frac{1}{2},m}$ and show that the same result is obtained for the z component of the projected operator

$$\frac{\boldsymbol{\sigma} \cdot \boldsymbol{J}}{\boldsymbol{J}^2}\boldsymbol{J}\,.$$

3-11 Express $\gamma\boldsymbol{n}$ of (3.6.1) in terms of the Eulerian angles ϕ, θ, and ψ.

*John Stewart BELL (1928–1990)

3-12 Give alternative evaluations for

$$\mathrm{e}^{\mathrm{i}\pi J_y}\,\mathrm{e}^{-\mathrm{i}\pi J_z}\,\mathrm{e}^{-\mathrm{i}\pi J_y}\,\mathrm{e}^{\mathrm{i}\pi J_z}$$

by applying the two different unitary transformations describing finite rotations that appear here. Repeat with

$$\mathrm{e}^{\mathrm{i}\pi J_y}\,\mathrm{e}^{\mathrm{i}\pi J_z}\,\mathrm{e}^{-\mathrm{i}\pi J_y}\,\mathrm{e}^{-\mathrm{i}\pi J_z}\,.$$

Conclusion?

3-13 Justify the appearance of $\left[U^{(j)}_{m,m'}\right]^2$ rather than $|U^{(j)}_{m,m'}|^2$ in (3.7.23) by showing that these matrix elements are real.

3-14a Use arguments that exploit the vector nature of null vector $\boldsymbol{\nu}$ to argue that

$$\int \mathrm{d}\Omega \left(\frac{\mathbf{r}}{r}\cdot\boldsymbol{\nu}^\dagger\right)^l \left(\frac{\mathbf{r}}{r}\cdot\boldsymbol{\nu}\right)^{l'} = \delta_{ll'}\, c_l \left(\boldsymbol{\nu}^\dagger\cdot\boldsymbol{\nu}\right)^l$$

where the eventual application to $\langle y_+^{\dagger\,\prime}, y_-^{\dagger\,\prime}|\ldots|y'_+, y'_-\rangle$ is understood so that $\boldsymbol{\nu}$ and $\boldsymbol{\nu}^\dagger$ can be regarded as numerical complex zero-length vectors. Then find the value of c_l by a special choice for $\boldsymbol{\nu}$, such as $\nu_x = 1$, $\nu_y = \mathrm{i}$, $\nu_z = 0$.

3-14b Now show that

$$\left(\boldsymbol{\nu}^\dagger\cdot\boldsymbol{\nu}\right)^l = 2^l(2l)!\sum_{m=-l}^{l} \frac{\left(y_+^\dagger y_+\right)^{l+m}}{(l+m)!}\,\frac{\left(y_-^\dagger y_-\right)^{l-m}}{(l-m)!}$$

and compare this with what (3.7.42) gives for the integral in Problem 3-14a. You should arrive at (3.7.45).

3-15 Recognize the orthogonality statement for Legendre polynomials,

$$\frac{1}{2}\int_{-1}^{1} \mathrm{d}x\, \mathrm{P}_l(x)\mathrm{P}_{l'}(x) = \frac{\delta_{ll'}}{2l+1}$$

as a special case of (3.7.45). Use

$$\mathrm{P}_0(x) = 1\,, \quad \mathrm{P}_1(x) = x\,, \quad \mathrm{P}_2(x) = \tfrac{3}{2}x^2 - \tfrac{1}{2}$$

to check this explicitly for $l, l' = 0, 1, 2$.

3-16 Show that

$$\int_0^{2\pi} \frac{\mathrm{d}\phi}{2\pi}\left(\frac{\mathbf{r}}{r}\cdot\boldsymbol{\nu}\right)^l = \nu_z^l\,\mathrm{P}_l(\cos\theta)\,.$$

Which choice for ν does then imply the integral representation

$$P_l(\cos\theta) = \int_0^{2\pi} \frac{d\phi}{2\pi} \left(\cos\theta + i\sin\theta\cos\phi\right)^l$$

of Legendre's polynomial? Use it to derive the familiar generating function

$$\frac{1}{\sqrt{1-2xt+t^2}} = \sum_{l=0}^{\infty} t^l P_l(x) ,$$

which can serve as a convenient alternative definition of the Legendre polynomials.

4. Galilean Invariance

4.1 Generators of infinitesimal transformations

The freedom of choice for reference frames includes more than rotations: one can displace the origin, translate it by a constant vector; or one can let that translation grow proportionally with *time*; the two frames are in relative motion at constant velocity. We'll consider only relative speeds that are small on the scale of the speed of light; see Problems 4-3 and 4-4 for other circumstances. Then time has an absolute significance (Galilean*–Newtonian relativity) apart from the freedom of displacing its origin. The infinitesimal transformations of these types are displayed by the space-time changes

$$\begin{aligned} \bar{t} &= t - \delta t\,, \\ \bar{\boldsymbol{r}} &= \boldsymbol{r} - \delta\boldsymbol{r}\,, \\ \text{with} \quad \delta\boldsymbol{r} &= \delta\boldsymbol{\epsilon} + \delta\boldsymbol{\omega}\times\boldsymbol{r} + \delta\boldsymbol{v}\,t\,, \end{aligned} \tag{4.1.1}$$

where δt is a constant, as are the vectors $\delta\boldsymbol{\epsilon}$, $\delta\boldsymbol{\omega}$, $\delta\boldsymbol{v}$. The accompanying unitary operator is

$$U = 1 + \mathrm{i}G \tag{4.1.2}$$

where, now

$$G = \delta\boldsymbol{\epsilon}\cdot\boldsymbol{P} + \delta\boldsymbol{\omega}\cdot\boldsymbol{J} + \delta\boldsymbol{v}\cdot\boldsymbol{N} - \delta t\,H + \delta\varphi\,, \tag{4.1.3}$$

and we want to recognize that we always have the freedom of a phase transformation. The names for the generators are derived from classical mechanics:

- $\boldsymbol{P}$: linear momentum vector,
- $\boldsymbol{J}$: (already familiar) angular momentum vector,
- H: energy; Hamiltonian (or Hamilton† operator),
- $\boldsymbol{N}$: no classical name, perhaps *booster*?

But now we have to notice something. If we write $U = 1 + \mathrm{i}G$, it is clear that G is dimensionless – it is given by pure numbers. But $\delta\boldsymbol{\epsilon}\cdot\boldsymbol{P}$, the product

*Galileo GALILEI (1564–1642) †Sir William Rowan HAMILTON (1805–1865)

of length $[L]$ by momentum $[ML/T]$, or $[ML^2/T]$ – or equally well $-\delta t H$: time $[T]$ times energy $[ML^2/T^2]$, not to mention $\delta\boldsymbol{\omega}\cdot\boldsymbol{J}$: angle (dimensionless) times angular momentum $[ML^2/T]$ – has dimensions, those of *action*. It is clear that up to now we have been employing natural atomic units, not the arbitrary units of macroscopic physics. So, if we wish to use the latter, we must include a conversion factor:

$$U = 1 + \mathrm{i}\frac{1}{\hbar}G\,, \quad G = \delta\boldsymbol{\epsilon}\cdot\boldsymbol{P} + \cdots + \hbar\delta\varphi\,, \tag{4.1.4}$$

where $\hbar$, the unit of action, is $(2\pi)^{-1}$ times Planck's* constant h. Experiment tells us that

$$\hbar = \frac{h}{2\pi} = 1.05457\times 10^{-27}\,\mathrm{erg\ sec} = 0.658\,212\,\mathrm{eV\ fs} \tag{4.1.5}$$

($1\,\mathrm{eV} = 1.602\,177\times 10^{-12}$ erg, electron-volt; $1\,\mathrm{fs} = 10^{-15}$s, femto-second).

It is important to recognize that the order in which these transformations, even infinitesimal ones, are made is important, in general. To use a familiar situation consider rotations. Compare 1,2:

$$\begin{aligned}\boldsymbol{r} \to \boldsymbol{r} - \delta_1\boldsymbol{\omega}\times\boldsymbol{r} &\to \boldsymbol{r} - \delta_1\boldsymbol{\omega}\times\boldsymbol{r} - \delta_2\boldsymbol{\omega}\times(\boldsymbol{r} - \delta_1\boldsymbol{\omega}\times\boldsymbol{r})\\ &= \boldsymbol{r} - \delta_1\boldsymbol{\omega}\times\boldsymbol{r} - \delta_2\boldsymbol{\omega}\times\boldsymbol{r} + \delta_2\boldsymbol{\omega}\times(\delta_1\boldsymbol{\omega}\times\boldsymbol{r})\end{aligned} \tag{4.1.6}$$

with 2,1:

$$\boldsymbol{r} \to \boldsymbol{r} - \delta_2\boldsymbol{\omega}\times\boldsymbol{r} - \delta_1\boldsymbol{\omega}\times\boldsymbol{r} + \delta_1\boldsymbol{\omega}\times(\delta_2\boldsymbol{\omega}\times\boldsymbol{r})\,. \tag{4.1.7}$$

The result of performing 1,2 and then the inverse of 2,1 is

$$\begin{aligned}\boldsymbol{r} \to \boldsymbol{r} &+ \delta_2\boldsymbol{\omega}\times(\delta_1\boldsymbol{\omega}\times\boldsymbol{r}) \underbrace{-\delta_1\boldsymbol{\omega}\times(\delta_2\boldsymbol{\omega}\times\boldsymbol{r})}_{=\delta_1\boldsymbol{\omega}\times(\boldsymbol{r}\times\delta_2\boldsymbol{\omega})}\\ &= \boldsymbol{r} - (\delta_1\boldsymbol{\omega}\times\delta_2\boldsymbol{\omega})\times\boldsymbol{r}\\ &= \boldsymbol{r} - \delta_{[12]}\boldsymbol{\omega}\times\boldsymbol{r}\,,\end{aligned} \tag{4.1.8}$$

i. e., another rotation described by

$$\delta_{[12]}\boldsymbol{\omega} = \delta_1\boldsymbol{\omega}\times\delta_2\boldsymbol{\omega} = -\delta_2\boldsymbol{\omega}\times\delta_1\boldsymbol{\omega}\,. \tag{4.1.9}$$

From the viewpoint of unitary transformations we are saying that $U_2U_1 \neq U_1U_2$ and

$$(U_1U_2)^{-1}U_2U_1 = U_{[12]}\,, \qquad U_{[12]}^{-1} = (U_2U_1)^{-1}U_1U_2 = U_{[21]} \tag{4.1.10}$$

which for infinitesimal transformations becomes

*Max Karl Ernst Ludwig PLANCK (1858–1947)

$$U_{[12]} = 1 + \frac{\mathrm{i}}{\hbar} G_{[12]} = 1 + \frac{1}{\hbar^2}(G_1 G_2 - G_2 G_1) \tag{4.1.11}$$

or

$$G_{[12]} = \frac{1}{\mathrm{i}\hbar}[G_1, G_2] , \tag{4.1.12}$$

since

$$\begin{aligned} U_2 U_1 &= \left(1 + \frac{\mathrm{i}}{\hbar} G_2\right)\left(1 + \frac{\mathrm{i}}{\hbar} G_1\right) \\ &= 1 + \frac{\mathrm{i}}{\hbar}(G_2 + G_1) - \frac{1}{\hbar^2} G_2 G_1 \end{aligned} \tag{4.1.13}$$

and

$$\begin{aligned} U_1 U_2 &= \left(1 + \frac{\mathrm{i}}{\hbar} G_1\right)\left(1 + \frac{\mathrm{i}}{\hbar} G_2\right) \\ &= 1 + \frac{\mathrm{i}}{\hbar}(G_1 + G_2) - \frac{1}{\hbar^2} G_1 G_2 . \end{aligned} \tag{4.1.14}$$

And so we have

$$\frac{1}{\mathrm{i}\hbar}[\delta_1 \boldsymbol{\omega} \cdot \boldsymbol{J}, \delta_2 \boldsymbol{\omega} \cdot \boldsymbol{J}] = (\delta_1 \boldsymbol{\omega} \times \delta_2 \boldsymbol{\omega}) \cdot \boldsymbol{J} + \hbar \delta_{[12]} \varphi . \tag{4.1.15}$$

Now the only possibility for the scalar $\delta_{[12]}\varphi$ is a multiple of $\delta_1\boldsymbol{\omega} \cdot \delta_2\boldsymbol{\omega}$, which is *symmetrical* in 1 and 2, not antisymmetrical. Hence $\delta_{[12]}\varphi = 0$. Then, written as

$$\frac{1}{\mathrm{i}\hbar}[\boldsymbol{J}, \delta\boldsymbol{\omega} \cdot \boldsymbol{J}] = \delta\boldsymbol{\omega} \times \boldsymbol{J} , \tag{4.1.16}$$

we recognize the characterization of a vector under rotations.

This immediately tells us that the analogous considerations for the vectors $\boldsymbol{P}$, $\boldsymbol{N}$, and $\boldsymbol{J}$ will yield

$$\begin{aligned} \frac{1}{\mathrm{i}\hbar}[\boldsymbol{P}, \delta\boldsymbol{\omega} \cdot \boldsymbol{J}] &= \delta\boldsymbol{\omega} \times \boldsymbol{P} , \\ \frac{1}{\mathrm{i}\hbar}[\boldsymbol{N}, \delta\boldsymbol{\omega} \cdot \boldsymbol{J}] &= \delta\boldsymbol{\omega} \times \boldsymbol{N} , \end{aligned} \tag{4.1.17}$$

whereas, for the scalar H,

$$\frac{1}{\mathrm{i}\hbar}[H, \delta\boldsymbol{\omega} \cdot \boldsymbol{J}] = 0 . \tag{4.1.18}$$

How about translations? As

$$\boldsymbol{r} \to \boldsymbol{r} - \delta_1\boldsymbol{\epsilon} \to \boldsymbol{r} - \delta_1\boldsymbol{\epsilon} - \delta_2\boldsymbol{\epsilon} \tag{4.1.19}$$

indicates, we have

$$\delta_{[12]}\boldsymbol{\epsilon} = 0 \tag{4.1.20}$$

and

$$\frac{1}{\mathrm{i}\hbar}[\delta_1\boldsymbol{\epsilon}\cdot\boldsymbol{P}, \delta_2\boldsymbol{\epsilon}\cdot\boldsymbol{P}] = \hbar\delta_{[12]}\varphi\ , \tag{4.1.21}$$

where the only possibility of $\delta_{[12]}\varphi \propto \delta_1\boldsymbol{\epsilon}\cdot\delta_2\boldsymbol{\epsilon}$ shows that

$$\frac{1}{\mathrm{i}\hbar}[\delta_1\boldsymbol{\epsilon}\cdot\boldsymbol{P}, \delta_2\boldsymbol{\epsilon}\cdot\boldsymbol{P}] = 0\ , \tag{4.1.22}$$

or

$$[P_k, P_l] = 0 \qquad \text{and} \qquad \boldsymbol{P}\times\boldsymbol{P} = 0\ . \tag{4.1.23}$$

Similarly,

$$[N_k, N_l] = 0\ , \quad \boldsymbol{N}\times\boldsymbol{N} = 0\ . \tag{4.1.24}$$

But when we come to

$$\frac{1}{\mathrm{i}\hbar}[\delta\boldsymbol{\epsilon}\cdot\boldsymbol{P}, \delta\boldsymbol{v}\cdot\boldsymbol{N}] = \hbar\delta\varphi = M\delta\boldsymbol{\epsilon}\cdot\delta\boldsymbol{v} \tag{4.1.25}$$

(dimension of M: mass) we can no longer conclude that $\delta\varphi = 0$ since two different vectors are involved. So

$$\frac{1}{\mathrm{i}\hbar}[P_k, N_l] = M\delta_{kl}\ . \tag{4.1.26}$$

With regard to transformations that include time displacement, consider

$$\begin{aligned} t &\to t - \delta_1 t \to t - \delta_1 t - \delta_2 t\ , \\ \boldsymbol{r} &\to \boldsymbol{r} - \delta_1\boldsymbol{v}\,t \to \boldsymbol{r} - \delta_1\boldsymbol{v}\,t - \delta_2\boldsymbol{v}(t - \delta_1 t)\ , \end{aligned} \tag{4.1.27}$$

so that $(1,2)\times(2,1)^{-1}$ leaves us with a net displacement

$$\delta_{[12]}\boldsymbol{\epsilon} = \delta_1\boldsymbol{v}\,\delta_2 t - \delta_2\boldsymbol{v}\,\delta_1 t \tag{4.1.28}$$

which will have no counterpart in displacements or rotations. So

$$\frac{1}{\mathrm{i}\hbar}[\delta\boldsymbol{v}\cdot\boldsymbol{N}, -\delta t\,H] = \delta\boldsymbol{v}\,\delta t\cdot\boldsymbol{P} + \underbrace{\hbar\delta\varphi}_{\to 0} \quad . \tag{4.1.29}$$

or

$$\frac{1}{\mathrm{i}\hbar}[\boldsymbol{N}, H] = -\boldsymbol{P}\ , \tag{4.1.30}$$

whereas

$$\frac{1}{i\hbar}[\delta\boldsymbol{\omega}\cdot\boldsymbol{J}, -\delta t\, H] = 0 \qquad \text{and} \qquad \frac{1}{i\hbar}[\delta\boldsymbol{\epsilon}\cdot\boldsymbol{P}, -\delta t\, H] = 0 \tag{4.1.31}$$

imply

$$[\boldsymbol{J}, H] = 0 \qquad \text{and} \qquad [\boldsymbol{P}, H] = 0\,. \tag{4.1.32}$$

The commutators involving $\boldsymbol{J}$ are the response to rotations, distinguishing vectors and scalars. Now let's look at the $\boldsymbol{P}$ commutators, the response to translations. From the $\boldsymbol{P}$ equation in (4.1.17) we get

$$\frac{1}{i\hbar}[\boldsymbol{J}, \delta\boldsymbol{\epsilon}\cdot\boldsymbol{P}] = \delta_\epsilon \boldsymbol{J} = \delta\boldsymbol{\epsilon}\times\boldsymbol{P}\,, \tag{4.1.33}$$

and since

$$\frac{1}{i\hbar}[\boldsymbol{P}, \delta\boldsymbol{\epsilon}\cdot\boldsymbol{P}] = \delta_\epsilon \boldsymbol{P} = 0\,, \tag{4.1.34}$$

also

$$\frac{1}{i\hbar}[\boldsymbol{N}, \delta\boldsymbol{\epsilon}\cdot\boldsymbol{P}] = \delta_\epsilon \boldsymbol{N} = -M\delta\boldsymbol{\epsilon} \tag{4.1.35}$$

and of course

$$\frac{1}{i\hbar}[H, \delta\boldsymbol{\epsilon}\cdot\boldsymbol{P}] = \delta_\epsilon H = 0\,. \tag{4.1.36}$$

Both $\boldsymbol{J}$ and $\boldsymbol{N}$ show a response to translation which can be expressed by a vector $\boldsymbol{R}$ such that

$$\delta_\epsilon \boldsymbol{R} = \frac{1}{i\hbar}[\boldsymbol{R}, \delta\boldsymbol{\epsilon}\cdot\boldsymbol{P}] = \delta\boldsymbol{\epsilon}\,,$$

$$\frac{1}{i\hbar}[R_k, P_l] = \delta_{kl}\,. \tag{4.1.37}$$

So

$$\delta_\epsilon(\boldsymbol{J} - \boldsymbol{R}\times\boldsymbol{P}) = 0\,, \quad \delta_\epsilon(\boldsymbol{N} + M\boldsymbol{R}) = 0 \tag{4.1.38}$$

and we write

$$\boldsymbol{J} = \boldsymbol{R}\times\boldsymbol{P} + \boldsymbol{S}\,, \tag{4.1.39}$$

where the components of $\boldsymbol{S}$ commute with those of $\boldsymbol{P}$,

$$[S_k, P_l] = 0\,. \tag{4.1.40}$$

Since $\boldsymbol{R}$ is a vector we must have

$$\begin{aligned}\frac{1}{\mathrm{i}\hbar}[\boldsymbol{R}, \delta\boldsymbol{\omega}\cdot\boldsymbol{J}] &= \delta\boldsymbol{\omega}\times\boldsymbol{R} \\ &= \frac{1}{\mathrm{i}\hbar}[\boldsymbol{R}, \delta\boldsymbol{\omega}\times\boldsymbol{R}\cdot\boldsymbol{P} + \delta\boldsymbol{\omega}\cdot\boldsymbol{S}]\end{aligned} \tag{4.1.41}$$

which is certainly satisfied if

$$[R_k, R_l] = 0 \quad \text{or} \quad \boldsymbol{R}\times\boldsymbol{R} = 0 \quad \text{and} \quad [R_k, S_l] = 0\,. \tag{4.1.42}$$

Also, since $\boldsymbol{N}$ generates a displacement proportional to t it must contain $\boldsymbol{P}t$, or

$$\boldsymbol{N} = \boldsymbol{P}t - M\boldsymbol{R}\,. \tag{4.1.43}$$

In particular, for $t = 0$, $\boldsymbol{N} = -M\boldsymbol{R}$, and $\boldsymbol{R}\times\boldsymbol{R} = 0$ follows from $\boldsymbol{N}\times\boldsymbol{N} = 0$.

Inasmuch as $\boldsymbol{R}$ and $\boldsymbol{P}$ are vectors, so is

$$\boldsymbol{L} = \boldsymbol{R}\times\boldsymbol{P} \tag{4.1.44}$$

and, in view of

$$\frac{1}{\mathrm{i}\hbar}[\boldsymbol{L}, \delta\boldsymbol{\omega}\cdot\boldsymbol{J}] = \frac{1}{\mathrm{i}\hbar}[\boldsymbol{L}, \delta\boldsymbol{\omega}\cdot\boldsymbol{L}] = \delta\boldsymbol{\omega}\times\boldsymbol{L} \tag{4.1.45}$$

one has

$$\boldsymbol{L}\times\boldsymbol{L} = \mathrm{i}\hbar\boldsymbol{L} \tag{4.1.46}$$

which implies that

$$\boldsymbol{S}\times\boldsymbol{S} = \mathrm{i}\hbar\boldsymbol{S}\,. \tag{4.1.47}$$

We see that

$$\boldsymbol{J} = \boldsymbol{L} + \boldsymbol{S} \tag{4.1.48}$$

is the decomposition into external or orbital angular momentum $\boldsymbol{L}$, and internal or spin angular momentum $\boldsymbol{S}$.

We have now recognized that the system as a whole is described by position vector $\boldsymbol{R}$, momentum vector $\boldsymbol{P}$, which for each direction in space constitute a q, p set of operators:

$$\frac{1}{\mathrm{i}\hbar}[R_k, P_l] = \delta_{kl}\,, \quad [R_k, R_l] = 0\,, \quad [P_k, P_l] = 0\,. \tag{4.1.49}$$

Accordingly all these operators have continuous spectra and have a classical limit.

Notice also that

$$\boldsymbol{R}\cdot\boldsymbol{L}=\boldsymbol{R}\cdot\boldsymbol{R}\times\boldsymbol{P}=\boldsymbol{R}\times\boldsymbol{R}\cdot\boldsymbol{P}=0 \tag{4.1.50}$$

which means that a rotation about the direction $\boldsymbol{R}$ has no effect, has zero quantum number,

$$\delta\langle\ |=\mathrm{i}\langle\ |\delta\boldsymbol{\omega}\cdot\boldsymbol{L}=0 \quad \text{if} \quad \delta\boldsymbol{\omega}\propto\boldsymbol{R}\,. \tag{4.1.51}$$

But zero is an integer and therefore all possible values of l in $\boldsymbol{L}^{2\prime}=l(l+1)\hbar^2$ are integers,

$$l=0,1,2,\ldots\,. \tag{4.1.52}$$

Now look at the information we have about H:

$$[\boldsymbol{J},H]=0\,,\qquad [\boldsymbol{P},H]=0\,,\qquad \frac{1}{\mathrm{i}\hbar}[\boldsymbol{N},H]=-\boldsymbol{P}\,. \tag{4.1.53}$$

The first says that H is a scalar, the second, according to

$$\langle\boldsymbol{R}'|\boldsymbol{P}=\frac{\hbar}{\mathrm{i}}\boldsymbol{\nabla}_{\boldsymbol{R}'}\langle\boldsymbol{R}'| \tag{4.1.54}$$

($\boldsymbol{R}$ components are compatible) says

$$[P,H]=\frac{\hbar}{\mathrm{i}}\boldsymbol{\nabla}_{\boldsymbol{R}}H=0\,, \tag{4.1.55}$$

H does not depend on $\boldsymbol{R}$; the third is

$$\frac{1}{\mathrm{i}\hbar}[\boldsymbol{P}t-M\boldsymbol{R},H]=-\boldsymbol{P} \tag{4.1.56}$$

or

$$\frac{1}{\mathrm{i}\hbar}[\boldsymbol{R},H]=\frac{1}{M}\boldsymbol{P}\,. \tag{4.1.57}$$

But, according to ($\boldsymbol{P}$ components are compatible, too)

$$\langle\boldsymbol{P}'|\boldsymbol{R}=\mathrm{i}\hbar\boldsymbol{\nabla}_{\boldsymbol{P}'}\langle\boldsymbol{P}'| \tag{4.1.58}$$

we have

$$\frac{1}{\mathrm{i}\hbar}[\boldsymbol{R},H]=\boldsymbol{\nabla}_{\boldsymbol{P}}H=\frac{1}{M}\boldsymbol{P} \tag{4.1.59}$$

or

$$H=\frac{\boldsymbol{P}^2}{2M}+H_{\text{int}} \qquad \text{with} \quad \boldsymbol{\nabla}_{\boldsymbol{P}}H_{\text{int}}=0\,. \tag{4.1.60}$$

4.2 Hamilton operator for a system of elementary particles

For us an elementary particle is defined as one without internal energy, or at least with inaccessible internal energy under the given circumstances. For atomic structure discussions the elementary particles are electrons and nuclei. For nuclear physics discussions, they are protons and neutrons, and so on.

Let each elementary particle be described by independent variables $\boldsymbol{r}_a$, $\boldsymbol{p}_a$, $\boldsymbol{s}_a$ and mass m_a. Then we construct $\boldsymbol{P}, \boldsymbol{J}, \boldsymbol{N}$ additively

$$\begin{aligned} \boldsymbol{P} &= \sum_a \boldsymbol{p}_a \,, \\ \boldsymbol{J} &= \sum_a (\boldsymbol{r}_a \times \boldsymbol{p}_a + \boldsymbol{s}_a) = \boldsymbol{R} \times \boldsymbol{P} + \boldsymbol{S} \,, \\ \boldsymbol{N} &= \sum_a (\boldsymbol{p}_a t - m_a \boldsymbol{r}_a) = \boldsymbol{P}t - M\boldsymbol{R} \,, \end{aligned} \tag{4.2.1}$$

where

$$M = \sum_a m_a \,, \quad \boldsymbol{R} = \sum_a \frac{m_a}{M} \boldsymbol{r}_a \tag{4.2.2}$$

and indeed

$$\begin{aligned} \frac{1}{\mathrm{i}\hbar}[R_k, P_l] &= \frac{1}{\mathrm{i}\hbar}\Big[\sum_a \frac{m_a}{M} r_{ak}, \sum_b p_{bl}\Big] \\ &= \underbrace{\sum_a \frac{m_a}{M}}_{=1} \underbrace{\frac{1}{\mathrm{i}\hbar}[r_{ak}, p_{al}]}_{=\delta_{kl}} = \delta_{kl} \,. \end{aligned} \tag{4.2.3}$$

We write

$$\begin{aligned} \sum_a \boldsymbol{r}_a \times \boldsymbol{p}_a &= \sum_a \Big[\boldsymbol{R} + (\boldsymbol{r}_a - \boldsymbol{R})\Big] \times \boldsymbol{p}_a \\ &= \boldsymbol{R} \times \boldsymbol{P} + \sum_a \underbrace{(\boldsymbol{r}_a - \boldsymbol{R}) \times (\boldsymbol{p}_a - \frac{m_a}{M}\boldsymbol{P})}_{\text{internal variables}} \,, \end{aligned} \tag{4.2.4}$$

since

$$\sum_a m_a (\boldsymbol{r}_a - \boldsymbol{R}) = 0 \quad \text{and} \quad \sum_a (\boldsymbol{p}_a - \frac{m_a}{M}\boldsymbol{P}) = 0 \,, \tag{4.2.5}$$

and get

$$\boldsymbol{S} = \sum_a \Big[(\boldsymbol{r}_a - \boldsymbol{R}) \times (\boldsymbol{p}_a - \frac{m_a}{M}\boldsymbol{P}) + \boldsymbol{s}_a\Big] \tag{4.2.6}$$

for the total internal angular momentum.

If the constituents were isolated from each other we would have

$$H = \sum_a \frac{\boldsymbol{p}_a^2}{2m_a} \,. \tag{4.2.7}$$

More general we write

$$H = \sum_a \frac{\boldsymbol{p}_a^2}{2m_a} + V = \frac{\boldsymbol{P}^2}{2M} + H_{\text{int}} \tag{4.2.8}$$

with

$$H_{\text{int}} = \sum_a \frac{1}{2m_a}\left(\boldsymbol{p}_a - \frac{m_a}{M}\boldsymbol{P}\right)^2 + V \,, \tag{4.2.9}$$

where V, the potential interaction energy, is a scalar function of the internal variables and the $\boldsymbol{s}_a$ and possibly others.

Problems

4-1 Verify explicitly that $\boldsymbol{L} = \boldsymbol{R} \times \boldsymbol{P}$ obeys the angular momentum commutation relations (4.1.46). Can you think of a reason, based on the vector structure of $\boldsymbol{L}$, for the fact that any component of $\boldsymbol{L}/\hbar$ has only integer values?

4-2 Show that $\boldsymbol{L}\cdot\boldsymbol{S}$ commutes with $\boldsymbol{J}$, $\boldsymbol{L}^2$, and $\boldsymbol{S}^2$. Then find the eigenvalues of $\boldsymbol{L}\cdot\boldsymbol{S}$.

4-3 Einsteinian* relativity: Replace the first line in (4.1.1) by

$$\bar{t} = t - \frac{1}{c}\delta\epsilon_0 - \frac{1}{c^2}\delta\boldsymbol{v}\cdot\boldsymbol{r} \,,$$

where c is the speed of light, and the Galilean form is formally recovered in the limit $c \to \infty$ if $(1/c)\delta\epsilon_0 \to \delta t$ is understood. Show that the commutators are the same, with two exceptions:

$$\frac{1}{i\hbar}\left[\delta\boldsymbol{\epsilon}\cdot\boldsymbol{P}, \delta\boldsymbol{v}\cdot\boldsymbol{N}\right] = \left(M + \frac{1}{c^2}H\right)\delta\boldsymbol{\epsilon}\cdot\delta\boldsymbol{v}$$

and

$$\frac{1}{i\hbar}\left[\delta_1\boldsymbol{v}\cdot\boldsymbol{N}, \delta_2\boldsymbol{v}\cdot\boldsymbol{N}\right] = -\frac{1}{c^2}\delta_1\boldsymbol{v}\times\delta_2\boldsymbol{v}\cdot\boldsymbol{J} \,.$$

4-4 In consequence of these modified commutation relations, what needs to be altered in the equations introducing $\boldsymbol{R}$ and $\boldsymbol{S}$?

---

*Albert EINSTEIN (1879–1955)

4-5 Photons have only spin angular momentum $+1$ or -1 along their direction of motion. (Incidentally, *helicity* is a more fitting term than spin under these circumstances.) A light beam is deflected through the angle θ. To what extent can you anticipate the dependence of the deflected beam's intensity on angle from the spin properties of a photon? [Hint: Recall Problem 3-5.]

Part B

Winter Quarter: Quantum Dynamics

5. Quantum Action Principle

5.1 Equations of motion

Consider infinitesimal displacements of the time origin,

$$\bar{t} = t - \delta t \,, \tag{5.1.1}$$

and the implied unitary transformation,

$$U = 1 + \frac{\mathrm{i}}{\hbar} G_t \qquad \text{with} \qquad G_t = -\delta t \, H \,, \tag{5.1.2}$$

where H, the Hamiltonian operator, depends upon a set of variables for the system, $v_\alpha(t)$, and possibly on t itself. When we shift the origin, the variables are redefined,

$$v_\alpha(t) = \bar{v}_\alpha(\bar{t}) \tag{5.1.3}$$

or

$$\begin{aligned} \bar{v}_\alpha(t) &= v_\alpha(t + \delta t) = v_\alpha(t) + \delta t \frac{\mathrm{d}}{\mathrm{d}t} v_\alpha(t) \\ &\equiv v_\alpha(t) - \delta v_\alpha(t) \,, \end{aligned} \tag{5.1.4}$$

where

$$\delta v_\alpha(t) = \frac{1}{\mathrm{i}\hbar} [v_\alpha(t), G_t] \,, \tag{5.1.5}$$

so that

$$-\delta t \frac{\mathrm{d}}{\mathrm{d}t} v_\alpha(t) = \frac{1}{\mathrm{i}\hbar} [v_\alpha(t), -\delta t \, H] \,. \tag{5.1.6}$$

This gives us the equations of motion

$$\frac{\mathrm{d}}{\mathrm{d}t} v_\alpha(t) = \frac{1}{\mathrm{i}\hbar} [v_\alpha(t), H] \,. \tag{5.1.7}$$

More generally, consider any $F\big(v(t), t\big)$, where $v(t)$ stands for the collection of all $v_\alpha(t)$'s. Then

$$U^{-1}FU = F - \delta F = F\big(U^{-1}v(t)U, t\big)$$
$$= F\big(\overline{v}(t), t\big) = F\big(v(t+\delta t), t\big) \,. \tag{5.1.8}$$

We write

$$F\big(v(t+\delta t), t\big) = F\big(v(t), t\big) + \delta t \left(\frac{\mathrm{d}}{\mathrm{d}t} - \frac{\partial}{\partial t}\right) F\big(v(t), t\big) \tag{5.1.9}$$

where $\mathrm{d}/\mathrm{d}t$ is the *total* time derivative and $\partial/\partial t$ refers to the *explicit* (or *parametric*) t dependence, so $\mathrm{d}/\mathrm{d}t - \partial/\partial t$ differentiates the time dependence *implicit* in the dynamical variables $v(t)$. This gives

$$\delta F = \frac{1}{\mathrm{i}\hbar}[F, -\delta t\, H] = -\delta t \left(\frac{\mathrm{d}}{\mathrm{d}t} - \frac{\partial}{\partial t}\right) F \tag{5.1.10}$$

or

$$\frac{\mathrm{d}}{\mathrm{d}t}F = \frac{\partial}{\partial t}F + \frac{1}{\mathrm{i}\hbar}[F, H] \,. \tag{5.1.11}$$

This is Heisenberg's equation of motion. It says that there are two contributions to the change in time of the arbitrary operator F: its explicit time dependence – the $\partial F/\partial t$ term – and its dynamical time dependence, given by $(\mathrm{i}\hbar)^{-1}$ times the commutator of F with the Hamilton operator.

The v equations of motion (5.1.7) are, of course, included since $\partial v/\partial t = 0$. By their nature, dynamical variables have no explicit time dependence.

For vectors $\langle \ldots, t|$, with the ellipsis indicating a time independent set of quantum numbers, we have

$$\overline{\langle \ldots, t|} = \langle \ldots, t| \left(1 + \frac{\mathrm{i}}{\hbar}G_t\right) = \langle \ldots, t| - \delta\langle \ldots, t| \tag{5.1.12}$$

or

$$\delta\langle \ldots, t| = \frac{1}{\mathrm{i}\hbar}\langle \ldots, t|G_t = \frac{\mathrm{i}}{\hbar}\langle \ldots, t|\delta t\, H \,. \tag{5.1.13}$$

To be more precise about $\langle \ldots, t|$, from the $v(t)$ select a complete set of commuting operators, $v_{c\alpha}(t)$, so that, at time t, they can all be assigned numerical values, collectively denoted by v'_c:

$$\langle v'_c, t|v_{c\alpha}(t) = v'_{c\alpha}\langle v'_c, t| \,. \tag{5.1.14}$$

The unitary operator (5.1.2) turns these equations into

$$\overline{\langle v'_c, t|}\,\underbrace{\overline{v}_{c\alpha}(t)}_{= v_{c\alpha}(t+\delta t)} = v'_{c\alpha}\overline{\langle v'_c, t|} \,, \tag{5.1.15}$$

so $\overline{\langle v'_c, t|}$ is the left eigenvector of the $v_{c\alpha}(t+\delta t)$ with eigenvalues $v'_{c\alpha}$,

$$\overline{\langle v_c', t|} = \langle v_c', t+\delta t| = \langle v_c', t| + \underbrace{\delta t \frac{\partial}{\partial t}\langle v_c', t|}_{= -\delta \langle v_c', t|} . \quad (5.1.16)$$

Accordingly,

$$-\delta t \frac{\partial}{\partial t}\langle v_c', t| = \frac{\mathrm{i}}{\hbar}\langle v_c', t|H(v(t), t)\delta t , \quad (5.1.17)$$

which is

$$\mathrm{i}\hbar \frac{\partial}{\partial t}\langle v_c', t| = \langle v_c', t|H , \quad (5.1.18)$$

Schrödinger's differential equation of motion. Its adjoint is

$$-\mathrm{i}\hbar \frac{\partial}{\partial t}|v_c', t\rangle = H|v_c', t\rangle . \quad (5.1.19)$$

These equations of motion come together in

$$\begin{aligned} &\frac{\mathrm{d}}{\mathrm{d}t}\langle v_c', t|F(v(t), t)|v_c'', t\rangle \\ &\quad = \langle v_c', t|\left(\frac{1}{\mathrm{i}\hbar}HF + \frac{\partial}{\partial t}F + \frac{1}{\mathrm{i}\hbar}[F, H] - \frac{1}{\mathrm{i}\hbar}FH\right)|v_c'', t\rangle \\ &\quad = \langle v_c', t|\frac{\partial F}{\partial t}(v(t), t)|v_c'', t\rangle . \end{aligned} \quad (5.1.20)$$

If F does not depend explicitly on t,

$$\frac{\mathrm{d}}{\mathrm{d}t}\langle v_c', t|F(v(t))|v_c'', t\rangle = 0 ; \quad (5.1.21)$$

the number $\langle v_c', t|F|v_c'', t\rangle$ is unchanged by the unitary transformation, applied to operators and vectors, that represents a change of t.

5.2 Conservation laws

Let's see some general consequences of Heisenberg's operator equation of motion (5.1.11). First, let $F = H$:

$$\frac{\mathrm{d}H}{\mathrm{d}t} = \frac{\partial H}{\partial t} + \frac{1}{\mathrm{i}\hbar}[H, H] = \frac{\partial H}{\partial t} . \quad (5.2.1)$$

In particular, if H does not depend explicitly on time, $\partial H/\partial t = 0$, that is, if H maintains its form under a time displacement, we have

$$\frac{\mathrm{d}H}{\mathrm{d}t} = 0 , \quad (5.2.2)$$

which expresses the conservation of energy: H is a constant of the motion.

Next, take $F = \delta\boldsymbol{\epsilon} \cdot \boldsymbol{P}$:

$$\frac{\mathrm{d}}{\mathrm{d}t}\delta\boldsymbol{\epsilon} \cdot \boldsymbol{P} = -\frac{1}{\mathrm{i}\hbar}[H, \delta\boldsymbol{\epsilon} \cdot \boldsymbol{P}] = -\delta_{\boldsymbol{\epsilon}} H \tag{5.2.3}$$

where $\delta_{\boldsymbol{\epsilon}} H$ is the change in H produced by a coordinate displacement. If H is unchanged by such a displacement, if it depends only on relative (internal) particle coordinates, the $\boldsymbol{r}_a - \boldsymbol{R}$ of Section 4.2, then the linear momentum is conserved,

$$\frac{\mathrm{d}}{\mathrm{d}t}\boldsymbol{P} = 0 \,. \tag{5.2.4}$$

Similarly, $F = \delta\boldsymbol{\omega} \cdot \boldsymbol{J}$ gives

$$\frac{\mathrm{d}}{\mathrm{d}t}\delta\boldsymbol{\omega} \cdot \boldsymbol{J} = -\delta_{\boldsymbol{\omega}} H \,, \tag{5.2.5}$$

and if H is unchanged by an infinitesimal rotation, if it is a scalar, the angular momentum is conserved,

$$\frac{\mathrm{d}}{\mathrm{d}t}\boldsymbol{J} = 0 \,. \tag{5.2.6}$$

Finally, consider $F = \delta\boldsymbol{v} \cdot \boldsymbol{N} = \delta\boldsymbol{v} \cdot \boldsymbol{P}t - \delta\boldsymbol{v} \cdot M\boldsymbol{R}$, for which

$$\begin{aligned} \frac{\mathrm{d}}{\mathrm{d}t}\delta\boldsymbol{v} \cdot \boldsymbol{N} &= \frac{\partial}{\partial t}(\delta\boldsymbol{v} \cdot \boldsymbol{P}t) + \frac{1}{\mathrm{i}\hbar}[\delta\boldsymbol{v} \cdot \boldsymbol{N}, H] \\ &= \delta\boldsymbol{v} \cdot \boldsymbol{P} - \delta_{\boldsymbol{v}} H \,. \end{aligned} \tag{5.2.7}$$

For $H = \boldsymbol{P}^2/(2M) + H_{\text{int}}$, as in (4.2.8), with an interaction Hamiltonian H_{int} that depends only on internal variables and is therefore unaffected by the transformation, whereas

$$\delta_{\boldsymbol{v}} \boldsymbol{P} = \frac{1}{\mathrm{i}\hbar}[\boldsymbol{P}, -\delta\boldsymbol{v} \cdot M\boldsymbol{R}] = \delta\boldsymbol{v}\, M \,, \tag{5.2.8}$$

as one would expect, one gets

$$\delta_{\boldsymbol{v}} H = \frac{\boldsymbol{P}}{M} \cdot M\delta\boldsymbol{v} = \delta\boldsymbol{v} \cdot \boldsymbol{P} \,, \tag{5.2.9}$$

so

$$\frac{\mathrm{d}}{\mathrm{d}t}\boldsymbol{N} = 0 \,. \tag{5.2.10}$$

The interpretation is the obvious one: In conjunction with the momentum conservation it establishes

$$\frac{\mathrm{d}}{\mathrm{d}t}\boldsymbol{N} = \frac{\mathrm{d}}{\mathrm{d}t}(\boldsymbol{P}t - M\boldsymbol{R}) = \boldsymbol{P} - M\frac{\mathrm{d}}{\mathrm{d}t}\boldsymbol{R} = 0 \,, \tag{5.2.11}$$

and tells us that the system moves with constant velocity $\boldsymbol{P}/M$.

These isolated statements come together in the recognition that they are all concerned with unitary transformations between equivalent frames of reference. Given some physical state, $|\ \rangle$, it is represented or described by the wave function $\langle v_c', t|\ \rangle$. Another description of the same state, in a different reference frame, is

$$\overline{\langle v_c', t|\ \rangle} = \langle v_c', t|U(t)|\ \rangle . \tag{5.2.12}$$

But this is also the description, in the original frame, of the different physical state $U(t)|\ \rangle$. Both states must obey the Schrödinger equation of motion

$$\mathrm{i}\hbar\frac{\partial}{\partial t}\langle v_c', t|\ \rangle = \langle v_c', t|H|\ \rangle \tag{5.2.13}$$

and

$$\mathrm{i}\hbar\frac{\partial}{\partial t}\langle v_c', t|U(t)|\ \rangle = \langle v_c', t|HU(t)|\ \rangle \tag{5.2.14}$$

which, in view of

$$\mathrm{i}\hbar\frac{\partial}{\partial t}\Big(\langle v_c', t|U(t)\Big) = \langle v_c', t|HU(t) + \mathrm{i}\hbar\langle v_c', t|\frac{\mathrm{d}}{\mathrm{d}t}U , \tag{5.2.15}$$

is only possible if

$$\frac{\mathrm{d}}{\mathrm{d}t}U(t) = 0 , \tag{5.2.16}$$

so that for $U = 1 + \frac{\mathrm{i}}{\hbar}G$ we get

$$\frac{\mathrm{d}}{\mathrm{d}t}G = 0 . \tag{5.2.17}$$

Here is the basis for all the conservation laws, of H, $\boldsymbol{P}$, $\boldsymbol{J}$, $\boldsymbol{N}$.

5.3 Sets of q, p pairs of variables

In $\boldsymbol{R}$ and $\boldsymbol{P}$,

$$[R_k, R_l] = 0 , \quad [P_k, P_l] = 0 , \quad \frac{1}{\mathrm{i}\hbar}[R_k, P_l] = \delta_{kl} \qquad (k, l = 1, 2, 3) \tag{5.3.1}$$

and, more generally, in the $\boldsymbol{r}_a$ and $\boldsymbol{p}_a$,

$$[r_{ka}, r_{lb}] = 0 , \quad [p_{ka}, p_{lb}] = 0 , \quad \frac{1}{\mathrm{i}\hbar}[r_{ka}, p_{lb}] = \delta_{ab}\delta_{kl} , \tag{5.3.2}$$

we see independent q, p pairs

$$[q_\alpha, q_\beta] = 0\,, \quad [p_\alpha, p_\beta] = 0\,, \quad \frac{1}{\mathrm{i}\hbar}[q_\alpha, p_\beta] = \delta_{\alpha\beta}\,. \tag{5.3.3}$$

More than this, particle spins $\boldsymbol{s}_a$, as with any angular momentum, can be represented as [recall (3.4.10), (3.4.21)],

$$\boldsymbol{s} = y^\dagger \frac{1}{2}\boldsymbol{\sigma} y = \sum_{\sigma,\sigma'} y^\dagger_\sigma \langle\sigma|\frac{1}{2}\boldsymbol{\sigma}|\sigma'\rangle y_{\sigma'}\,. \tag{5.3.4}$$

where, for each $y, y^\dagger$ pair,

$$y = \frac{1}{\sqrt{2}}\left(q + \frac{\mathrm{i}}{\hbar}p\right)\,, \quad y^\dagger = \frac{1}{\sqrt{2}}\left(q - \frac{\mathrm{i}}{\hbar}p\right) \tag{5.3.5}$$

expresses them in terms of a Hermitian q, p pair; or one can use y, $\mathrm{i}\hbar y^\dagger$ as a non-Hermitian q, p pair,

$$\left[y_{\sigma a}, y_{\sigma' b}\right] = 0\,, \quad \left[\mathrm{i}\hbar y^\dagger_{\sigma a}, \mathrm{i}\hbar y^\dagger_{\sigma' b}\right] = 0\,, \quad \frac{1}{\mathrm{i}\hbar}\left[y_{\sigma a}, \mathrm{i}\hbar y^\dagger_{\sigma' b}\right] = \delta_{ab}\delta_{\sigma\sigma'}\,. \tag{5.3.6}$$

So, with great generality, we consider sets of q, p variables, and a Hamiltonian operator that is a function of these sets: $H(q, p, t)$. We recall the lessons of Problems 1-54, 1-55, and 1-56 – which are immediately generalized to sets of q, p pairs:

$$\frac{1}{\mathrm{i}\hbar}[q_\alpha, F] = \frac{\partial F}{\partial p_\alpha}\,, \qquad \frac{1}{\mathrm{i}\hbar}[F, p_\beta] = \frac{\partial F}{\partial q_\beta} \tag{5.3.7}$$

[the basic commutation relations (5.3.3) are special cases] – and use them to get equations of motion for the $q_\alpha(t)$, $p_\alpha(t)$:

$$\begin{aligned} \frac{\mathrm{d}}{\mathrm{d}t}q_\alpha &= \frac{1}{\mathrm{i}\hbar}[q_\alpha, H] = \frac{\partial H}{\partial p_\alpha}\,, \\ \frac{\mathrm{d}}{\mathrm{d}t}p_\alpha &= \frac{1}{\mathrm{i}\hbar}[p_\alpha, H] = -\frac{\partial H}{\partial q_\alpha}\,. \end{aligned} \tag{5.3.8}$$

And now we can understand the origin of the classical Hamilton's equations of motion; they are already true at the fundamental quantal level!

Next, we give the symbolic Schrödinger equation

$$\mathrm{i}\hbar\frac{\partial}{\partial t}\langle v'_c, t|\ \rangle = \langle v'_c, t|H|\ \rangle \tag{5.3.9}$$

a more explicit form. As an example (the most usual one) of a complete set of commuting operators, at time t, we pick the set of $q_\alpha(t)$; any $p_\beta(t)$ must fail to commute with one of the operators (q_β, of course). So we have an equation of motion for the wave function

$$\langle q', t|\ \rangle \equiv \psi(q', t)\,. \tag{5.3.10}$$

Again recall that for a single q, p pair (one degree of freedom)

$$\langle q'| \mathrm{e}^{\frac{\mathrm{i}}{\hbar} p q''} = \langle q' + q''| \tag{5.3.11}$$

[cf. (1.16.19)] which, for $q'' \to \delta q$, and extended to any number of pairs, is

$$\langle q'| \frac{\mathrm{i}}{\hbar} G_q = \langle q'| \frac{\mathrm{i}}{\hbar} \sum_\alpha p_\alpha \delta q_\alpha = \sum_\alpha \delta q_\alpha \frac{\partial}{\partial q'_\alpha} \langle q'| \tag{5.3.12}$$

or [cf. (1.16.29)]

$$\langle q', t | p_\alpha(t) = \frac{\hbar}{\mathrm{i}} \frac{\partial}{\partial q'_\alpha} \langle q', t| \, , \tag{5.3.13}$$

which makes explicit that all operators and vectors here refer to a common time t. As before, we have the generalization [cf. (1.16.34)]

$$\langle q', t | F(q(t), p(t)) = F\left(q', \frac{\hbar}{\mathrm{i}} \frac{\partial}{\partial q'}\right) \langle q', t| \, , \tag{5.3.14}$$

if F depends algebraically on the p's. So, applied to the Hamiltonian, we get the numerical differential equation for the wave function:

$$\mathrm{i}\hbar \frac{\partial}{\partial t} \psi(q', t) = H\left(q', \frac{\hbar}{\mathrm{i}} \frac{\partial}{\partial q'}, t\right) \psi(q', t) \, . \tag{5.3.15}$$

This is Schrödinger's differential equation of motion for q wave functions (frequently just called *the* Schrödinger equation).

Equally well ($q \to p, p \to -q$) we have a Schrödinger equation for p wave functions,

$$\mathrm{i}\hbar \frac{\partial}{\partial t} \psi(p', t) = H\left(\mathrm{i}\hbar \frac{\partial}{\partial p'}, p', t\right) \psi(p', t) \, ; \tag{5.3.16}$$

as a rule this is useful only if $H(q, p, t)$ depends algebraically on the q's. But we can always construct one wave function from the other. Recall, for one degree of freedom, that the qp transformation function is [cf. (2.1.17)]

$$\langle q' | p' \rangle = \frac{1}{\sqrt{2\pi\hbar}} \mathrm{e}^{\frac{\mathrm{i}}{\hbar} q' p'} \, , \tag{5.3.17}$$

where we check the appropriateness of $1/\sqrt{\hbar}$ by verifying that

$$\begin{aligned} \int \langle q' | p' \rangle \, \mathrm{d}p' \, \langle p' | q'' \rangle &= \delta(q' - q'') \\ &= \int_{-\infty}^{\infty} \frac{\mathrm{d}p'}{2\pi\hbar} \mathrm{e}^{\mathrm{i}(p'/\hbar)(q' - q'')} \, , \end{aligned} \tag{5.3.18}$$

which is correct, the previous variable p' being replaced by $p'/\hbar$ in (2.1.18).

For n degrees of freedom,

$$\begin{aligned}\langle q'|p'\rangle &= \Big(\langle q_1'|\cdots\langle q_n'|\Big)\Big(|p_1'\rangle\cdots|p_n'\rangle\Big)\\ &= \prod_\alpha \langle q_\alpha'|p_\alpha'\rangle = \frac{1}{(2\pi\hbar)^{n/2}}\,\mathrm{e}^{\frac{\mathrm{i}}{\hbar}\sum_\alpha q_\alpha' p_\alpha'}\,,\end{aligned} \tag{5.3.19}$$

where, indeed, for example,

$$\int \langle q'|p'\rangle \underbrace{(\mathrm{d}p')}_{\equiv \mathrm{d}p_1'\cdots\mathrm{d}p_n'} \langle p'|q''\rangle = \prod_\alpha \underbrace{\int \langle q_\alpha'|p_\alpha'\rangle\,\mathrm{d}p_\alpha'\,\langle p_\alpha'|q_\alpha''\rangle}_{=\delta(q_\alpha'-q_\alpha'')} \equiv \delta(q'-q'')\,. \tag{5.3.20}$$

So,

$$\langle q',t|\ \rangle = \int \langle q'|p'\rangle\,(\mathrm{d}p')\,\langle p',t|\ \rangle\,, \tag{5.3.21}$$

or

$$\psi(q',t) = \int \frac{1}{(2\pi\hbar)^{n/2}}\,\mathrm{e}^{\frac{\mathrm{i}}{\hbar}\sum_\alpha q_\alpha' p_\alpha'}\,(\mathrm{d}p')\,\psi(p',t)\,, \tag{5.3.22}$$

and

$$\psi(p',t) = \int \frac{1}{(2\pi\hbar)^{n/2}}\,\mathrm{e}^{-\frac{\mathrm{i}}{\hbar}\sum_\alpha p_\alpha' q_\alpha'}\,(\mathrm{d}q')\,\psi(q',t)\,. \tag{5.3.23}$$

5.4 Wave functions for force-free motion

As the simplest illustration, consider a single particle of mass M (or system of particles without reference to internal motion),

$$H = \frac{\boldsymbol{p}^2}{2M}\,. \tag{5.4.1}$$

Here

$$\frac{\mathrm{d}\boldsymbol{p}}{\mathrm{d}t} = -\frac{\partial H}{\partial \boldsymbol{r}} = 0\,; \tag{5.4.2}$$

$\boldsymbol{p}$ is constant in time, and it is natural to be interested in the states with various $\boldsymbol{p}$ values, $|\boldsymbol{p}''\rangle$. In this situation we easily find the $\boldsymbol{p}$ wave functions:

$$\langle \boldsymbol{p}',t|\boldsymbol{p}''\rangle \equiv \psi_{\boldsymbol{p}''}(\boldsymbol{p}',t)\,, \tag{5.4.3}$$

where $|\boldsymbol{p}''\rangle$ refers to $t=0$. The Schrödinger equation (5.3.16) reads

$$\mathrm{i}\hbar\frac{\partial}{\partial t}\psi_{\boldsymbol{p}''}(\boldsymbol{p}',t) = \underbrace{\frac{\boldsymbol{p}'^2}{2M}}_{=E(\boldsymbol{p}')}\psi_{\boldsymbol{p}''}(\boldsymbol{p}',t)\,, \tag{5.4.4}$$

so

$$\psi_{\boldsymbol{p}''}(\boldsymbol{p}',t) = \mathrm{e}^{-\frac{\mathrm{i}}{\hbar}E(\boldsymbol{p}')t}\delta(\boldsymbol{p}'-\boldsymbol{p}'')\,, \tag{5.4.5}$$

which incorporates the initial condition

$$\psi_{\boldsymbol{p}''}(\boldsymbol{p}',0) = \langle\boldsymbol{p}'|\boldsymbol{p}''\rangle = \delta(\boldsymbol{p}'-\boldsymbol{p}'') \tag{5.4.6}$$

and obviously describes the fact that a $\boldsymbol{p}$ measurement at any time will certainly yield $\boldsymbol{p}''$, and the particle has the definite energy $E(\boldsymbol{p}') = E(\boldsymbol{p}'')$.

Now we construct [three degrees of freedom, $n = 3$ in (5.3.22)]

$$\begin{aligned}\psi_{\boldsymbol{p}''}(\boldsymbol{r}',t) &= \frac{1}{(2\pi\hbar)^{3/2}}\int \mathrm{e}^{\frac{\mathrm{i}}{\hbar}\boldsymbol{r}'\cdot\boldsymbol{p}'}\,(\mathrm{d}\boldsymbol{p}')\,\delta(\boldsymbol{p}'-\boldsymbol{p}'')\,\mathrm{e}^{-\frac{\mathrm{i}}{\hbar}E(\boldsymbol{p}')t}\\ &= \frac{1}{(2\pi\hbar)^{3/2}}\,\mathrm{e}^{\frac{\mathrm{i}}{\hbar}[\boldsymbol{p}''\cdot\boldsymbol{r}'-E(\boldsymbol{p}'')t]}\,;\end{aligned} \tag{5.4.7}$$

that's a wave function! Standard notation for plane waves is

$$\mathrm{e}^{\mathrm{i}(\boldsymbol{k}\cdot\boldsymbol{r}-\omega t)}\,, \tag{5.4.8}$$

and standard terminology calls $\boldsymbol{k}$ the propagation vector or wave vector and ω the angular frequency, related to the wavelength λ and the frequency ν by

$$|\boldsymbol{k}| = \frac{2\pi}{\lambda} = \frac{1}{\lambdabar}\,, \qquad \omega = 2\pi\nu\,. \tag{5.4.9}$$

We also note that $\lambdabar \equiv \lambda/(2\pi)$ is the reduced wavelength; the inverse wavelength $\lambda^{-1} = |\boldsymbol{k}|/(2\pi)$ is called wave number, and $|\boldsymbol{k}|$ is the reduced wave number. So (omitting primes) we have Planck's energy–frequency relation,

$$E = h\nu = \hbar\omega\,, \tag{5.4.10}$$

and de Broglie's momentum–wavelength relation,

$$\boldsymbol{p} = \hbar\boldsymbol{k}\,, \qquad |\boldsymbol{p}| = \frac{h}{\lambda} = \frac{\hbar}{\lambdabar}\,. \tag{5.4.11}$$

Naturally, the wave function (5.4.7) obeys the Schrödinger equation (5.3.15),

$$\underbrace{\mathrm{i}\hbar\frac{\partial}{\partial t}}_{\to E}\psi_{\boldsymbol{p}}(\boldsymbol{r},t) = \underbrace{\frac{1}{2M}\left(\frac{\hbar}{\mathrm{i}}\boldsymbol{\nabla}\right)^2}_{\to \boldsymbol{p}^2/(2M)}\psi_{\boldsymbol{p}}(\boldsymbol{r},t)\,, \tag{5.4.12}$$

and could have been found in this way also. While we're at it, let's find yet another kind of wave function for this simple system:

$$\langle \boldsymbol{r}', t | \boldsymbol{r}'' \rangle \equiv \langle \boldsymbol{r}', t | \boldsymbol{r}'', t = 0 \rangle , \tag{5.4.13}$$

which we get as

$$\begin{aligned}
\langle \boldsymbol{r}', t | \boldsymbol{r}'' \rangle &= \int \langle \boldsymbol{r}', t | \boldsymbol{p}' \rangle (\mathrm{d}\boldsymbol{p}') \langle \boldsymbol{p}' | \boldsymbol{r}'' \rangle \\
&= \int \frac{(\mathrm{d}\boldsymbol{p}')}{(2\pi\hbar)^3} \underbrace{\mathrm{e}^{\frac{\mathrm{i}}{\hbar}\boldsymbol{r}' \cdot \boldsymbol{p}'} \, \mathrm{e}^{-\frac{\mathrm{i}}{\hbar}[\boldsymbol{p}'^2/(2M)]t} \, \mathrm{e}^{-\frac{\mathrm{i}}{\hbar}\boldsymbol{p}' \cdot \boldsymbol{r}''}}_{= \exp\left(\frac{\mathrm{i}}{\hbar}\boldsymbol{p}' \cdot (\boldsymbol{r}' - \boldsymbol{r}'') - \frac{\mathrm{i}}{\hbar}\frac{t}{2M}\boldsymbol{p}'^2\right)} \\
&= \int \frac{(\mathrm{d}\boldsymbol{p}')}{(2\pi\hbar)^3} \, \mathrm{e}^{-\frac{\mathrm{i}}{\hbar}\frac{t}{2M}\left[\boldsymbol{p}' - (M/t)(\boldsymbol{r}' - \boldsymbol{r}'')\right]^2} \, \mathrm{e}^{\frac{\mathrm{i}}{\hbar}\frac{M}{2t}(\boldsymbol{r}' - \boldsymbol{r}'')^2} \\
&= \left(\frac{M}{2\pi\mathrm{i}\hbar t}\right)^{\frac{3}{2}} \mathrm{e}^{\frac{\mathrm{i}}{\hbar}\frac{M}{2t}(\boldsymbol{r}' - \boldsymbol{r}'')^2} .
\end{aligned} \tag{5.4.14}$$

This time transformation function is conceptually important but, as a wave function, it is too idealized for direct physical interpretation. For the following, notice that in all the above, based on

$$H = \frac{p_x^2}{2M} + \frac{p_y^2}{2M} + \frac{p_z^2}{2M} , \tag{5.4.15}$$

the x, y, and z motions are independent, as in

$$\langle \boldsymbol{r}', t | \boldsymbol{r}'' \rangle = \prod_{j=1}^{3} \sqrt{\frac{M}{2\pi\mathrm{i}\hbar t}} \, \mathrm{e}^{\frac{\mathrm{i}}{\hbar}\frac{M}{2t}(x_j' - x_j'')^2} , \tag{5.4.16}$$

for example. So consider just the $x_1 \equiv x$, $p_1 = p_x \equiv p$ motion in one dimension.

Suppose that, at $t = 0$, we have a minimum–uncertainty state $|\delta\rangle$, with $\langle x \rangle = x_0$, $\langle p \rangle = p_0$, and, of course, $\delta x \delta p = \frac{1}{2}\hbar$. According to (2.3.6) and (2.3.7), the initial wave functions are of Gaussian shape,

$$\begin{aligned}
\langle x, t = 0 | \delta \rangle = \psi_\delta(x) &= \frac{(2\pi)^{-\frac{1}{4}}}{\sqrt{\delta x}} \, \mathrm{e}^{\frac{\mathrm{i}}{\hbar}p_0(x - x_0)} \, \mathrm{e}^{-\left(\frac{x - x_0}{2\delta x}\right)^2} \, \mathrm{e}^{\frac{\mathrm{i}}{2\hbar}x_0 p_0} \\
&= \frac{(2/\pi)^{\frac{1}{4}}}{\sqrt{\hbar/\delta p}} \, \mathrm{e}^{\frac{\mathrm{i}}{\hbar}p_0(x - x_0)} \, \mathrm{e}^{-\left[\frac{1}{\hbar}(x - x_0)\delta p\right]^2} \, \mathrm{e}^{\frac{\mathrm{i}}{2\hbar}x_0 p_0} \\
&= \int_{-\infty}^{\infty} \frac{1}{\sqrt{2\pi\hbar}} \, \mathrm{e}^{\frac{\mathrm{i}}{\hbar}xp} \, \mathrm{d}p \, \psi_\delta(p)
\end{aligned} \tag{5.4.17}$$

and $(x \to p, \, p \to -x)$

$$\langle p,t=0|\delta\rangle = \psi_\delta(p) = \frac{(2\pi)^{-\frac{1}{4}}}{\sqrt{\delta p}}\,\mathrm{e}^{-\frac{\mathrm{i}}{\hbar}x_0(p-p_0)}\,\mathrm{e}^{-\left(\frac{p-p_0}{2\delta p}\right)^2}\,\mathrm{e}^{-\frac{\mathrm{i}}{2\hbar}p_0x_0}$$
$$= \frac{(2/\pi)^{\frac{1}{4}}}{\sqrt{\hbar/\delta x}}\,\mathrm{e}^{-\frac{\mathrm{i}}{\hbar}x_0(p-p_0)}\,\mathrm{e}^{-\left[\frac{1}{\hbar}(p-p_0)\delta x\right]^2}\,\mathrm{e}^{-\frac{\mathrm{i}}{2\hbar}p_0x_0}\,. \tag{5.4.18}$$

The time dependence of the latter wave function is immediate:

$$\psi_\delta(p,t) = \langle p,t|\delta\rangle = \mathrm{e}^{-\frac{\mathrm{i}}{\hbar}E(p)t}\psi_\delta(p)\,. \tag{5.4.19}$$

In a first step, we then get

$$\psi_\delta(x,t) = \langle x,t|\delta\rangle = \int \langle x|p\rangle\,\mathrm{d}p\,\langle p,t|\delta\rangle$$
$$= \int \frac{\mathrm{d}p}{\sqrt{2\pi\hbar}}\frac{(2\pi)^{-\frac{1}{4}}}{\sqrt{\delta p}}\,\mathrm{e}^{\frac{\mathrm{i}}{\hbar}p(x-x_0)}\,\mathrm{e}^{-\left[\frac{1}{\hbar}(p-p_0)\delta x\right]^2}\,\mathrm{e}^{-\frac{\mathrm{i}}{\hbar}\frac{p^2}{2M}t}\,\mathrm{e}^{\frac{\mathrm{i}}{2\hbar}x_0p_0}\,. \tag{5.4.20}$$

The exponent in this integrand is quadratic in p, with the p^2 term given by

$$-\left(\frac{p\delta x}{\hbar}\right)^2 - \frac{\mathrm{i}}{\hbar}\frac{p^2}{2M}t = -\frac{1}{\varepsilon(t)}\left(\frac{p\delta x}{\hbar}\right)^2, \tag{5.4.21}$$

where

$$\frac{1}{\varepsilon(t)} = 1 + \mathrm{i}\frac{\hbar t}{2M(\delta x)^2} = 1 + \mathrm{i}\frac{t}{M}\frac{\delta p}{\delta x} = 1 + 2\mathrm{i}\frac{t(\delta p)^2}{\hbar M}\,. \tag{5.4.22}$$

The identity

$$\varepsilon(t) = 1 - \mathrm{i}\frac{t}{M}\frac{\delta p}{\delta x}\varepsilon(t) \tag{5.4.23}$$

is useful when we complete a square to bring the whole exponent of (5.4.20) into the form

$$\text{exponent} = -\frac{1}{\varepsilon(t)}\left[\frac{p-\varepsilon(t)p_0}{\hbar/\delta x} - \frac{\mathrm{i}}{2}\varepsilon(t)\frac{x-x_0}{\delta x}\right]^2$$
$$+\frac{\mathrm{i}}{\hbar}p_0(x-x_0) - \frac{\mathrm{i}}{\hbar}\frac{p_0^2}{2M}t - \varepsilon(t)\left(\frac{x-x_0-p_0t/M}{2\delta x}\right)^2$$
$$+\frac{\mathrm{i}}{2\hbar}x_0p_0\,. \tag{5.4.24}$$

The p integration is now immediate and produces

$$\begin{aligned}\psi_\delta(x,t) &= \frac{(2\pi)^{-\frac{1}{4}}}{\sqrt{\delta x/\varepsilon(t)}}\, \mathrm{e}^{\frac{\mathrm{i}}{\hbar}p_0(x-x_0)}\, \mathrm{e}^{-\frac{\mathrm{i}}{\hbar}\frac{p_0^2}{2M}t}\, \mathrm{e}^{-\varepsilon(t)\left(\frac{x-x_0-p_0t/M}{2\delta x}\right)^2}\, \mathrm{e}^{\frac{\mathrm{i}}{2\hbar}x_0p_0} \\ &= (2/\pi)^{\frac{1}{4}}\left[\varepsilon(t)\delta p/\hbar\right]^{\frac{1}{2}}\, \mathrm{e}^{\frac{\mathrm{i}}{\hbar}p_0(x-x_0)}\, \mathrm{e}^{-\frac{\mathrm{i}}{\hbar}\frac{p_0^2}{2M}t} \\ &\qquad\times \mathrm{e}^{-\varepsilon(t)[\frac{1}{\hbar}(x-x_0-p_0t/M)\delta p]^2}\, \mathrm{e}^{\frac{\mathrm{i}}{2\hbar}x_0p_0}\,. \end{aligned} \tag{5.4.25}$$

The corresponding probability distribution

$$\left|\psi_\delta(x,t)\right|^2 = \frac{(2\pi)^{-\frac{1}{2}}}{\delta x(t)}\, \mathrm{e}^{-\frac{1}{2}\left(\frac{x-\langle x(t)\rangle}{\delta x(t)}\right)^2} \tag{5.4.26}$$

[make use of $\mathrm{Re}\,\varepsilon(t) = |\varepsilon(t)|^2$ to get it] is a Gaussian at all times and is, of course, properly normalized,

$$\int_{-\infty}^{\infty} \mathrm{d}x\, \left|\psi_\delta(x,t)\right|^2 = 1\,. \tag{5.4.27}$$

It exhibits the mean position

$$\langle x(t)\rangle = x_0 + \frac{p_0}{M}t\,, \tag{5.4.28}$$

which grows linearly in time and thus confirms that the particle moves with constant velocity p_0/M, and the t dependent spread in position,

$$\delta x(t) = |\varepsilon(t)|^{-1}\delta x = \sqrt{(\delta x)^2 + (t\delta p/M)^2}\,, \tag{5.4.29}$$

which shows that the Gaussian distribution (5.4.26) broadens in time. The spread in momentum is time independent,

$$\delta p(t) = \delta p\,, \tag{5.4.30}$$

of course, since

$$\left|\psi_\delta(p,t)\right|^2 = \left|\psi_\delta(p)\right|^2. \tag{5.4.31}$$

This means that

$$\left[\delta x(t)\,\delta p(t)\right]^2 = \left(\frac{1}{2}\hbar\right)^2 |\varepsilon|^{-2} = \left(\frac{1}{2}\hbar\right)^2 + \left(\frac{(\delta p)^2}{M}t\right)^2 \tag{5.4.32}$$

so the uncertainty product increases in time, eventually linearly,

$$\delta x(t)\,\delta p(t) \cong \frac{(\delta p)^2 t}{M} \qquad \text{if} \quad \frac{(\delta p)^2}{M}t \gg \hbar\,. \tag{5.4.33}$$

One should also observe that, at any $t \geq 0$, there is a minimum value possible for $\delta x(t)$:

$$
\begin{aligned}
\left[\delta x(t)\right]^2 &= (\delta x)^2 + \frac{1}{(\delta x)^2}\left(\frac{\hbar t}{2M}\right)^2 \\
&= \left(\delta x - \frac{1}{\delta x}\frac{\hbar t}{2M}\right)^2 + \frac{\hbar t}{M} \geq \frac{\hbar t}{M} .
\end{aligned}
\tag{5.4.34}
$$

The minimum of $\delta x(t)$ occurs at $t = T$ if $(\delta x)^2 = \frac{1}{2}\hbar T/M$ or $\delta x/\delta p = T/M$; in this optimal circumstance $\left[\delta x(t)\right]^2$ only doubles in time T. Not surprisingly, this time constant appears in $\varepsilon(t)$ of (5.4.22),

$$
\varepsilon = (1 + \mathrm{i}t/T)^{-1} \qquad \text{with} \quad T = M\frac{\delta x}{\delta p} , \tag{5.4.35}
$$

and therefore also in $\delta x(t)$,

$$
\delta x(t)/\delta x = \sqrt{1 + (t/T)^2} \geq \sqrt{2t/T} , \tag{5.4.36}
$$

which tells us that T sets the time scale for the spreading of the Gaussian probability distribution (5.4.26); the lower bound is the one of (5.4.34), the equal sign holding for $t = T$.

Relations such as (5.4.29), (5.4.30), (5.4.33), or (5.4.36) are more generally true than the particular initial state of minimum uncertainty suggests. This is the subject of Problem 5-8.

5.5 Quantum action principle

In classical mechanics, Hamilton's equations of motion are deduced from an action principle. Is there an action principle in quantum mechanics? Yes. Here is the derivation.

Consider the transformation function relating q states at infinitesimally different times:

$$
\langle q', t + \mathrm{d}t | q'', t\rangle = \langle q', t|\Bigl(1 - \frac{\mathrm{i}}{\hbar}\mathrm{d}t\, H\bigl(q(t), p(t), t\bigr)\Bigr)|q'', t\rangle \tag{5.5.1}
$$

and focus, not on what this equals, but how it *changes* when everything on which it depends is infinitesimally varied. Recalling

$$
\begin{aligned}
\frac{\partial}{\partial q'_\alpha}\langle q', t| &= \frac{\mathrm{i}}{\hbar}\langle q', t|p_\alpha(t) , \\
\frac{\partial}{\partial t}\langle q', t| &= -\frac{\mathrm{i}}{\hbar}\langle q', t|H\bigl(q(t), p(t), t\bigr) \equiv -\frac{\mathrm{i}}{\hbar}\langle q', t|H(t) ,
\end{aligned}
\tag{5.5.2}
$$

we begin with

$$
\begin{aligned}
&\delta_{q', t+\mathrm{d}t}\langle q', t + \mathrm{d}t| \\
&= \frac{\mathrm{i}}{\hbar}\langle q', t + \mathrm{d}t|\Bigl(\sum_\alpha p_\alpha(t + \mathrm{d}t)\delta q_\alpha(t + \mathrm{d}t) - H(t + \mathrm{d}t)\delta(t + \mathrm{d}t)\Bigr)
\end{aligned}
\tag{5.5.3}
$$

and

$$\delta_{q'',t}|q'',t\rangle = -\frac{\mathrm{i}}{\hbar}\Big(\sum_{\alpha} p_\alpha(t)\delta q_\alpha(t) - H(t)\delta t\Big)|q'',t\rangle \tag{5.5.4}$$

which are combined in

$$\begin{aligned}\delta'\langle q',t+\mathrm{d}t|q'',t\rangle = \frac{\mathrm{i}}{\hbar}\langle q',t+\mathrm{d}t|\Big(\sum_{\alpha}\big[p_\alpha(t+\mathrm{d}t)\delta q_\alpha(t+\mathrm{d}t) - p_\alpha(t)\delta q_\alpha(t)\big]\\ - \big[H(t+\mathrm{d}t)\delta(t+\mathrm{d}t) - H(t)\delta t\big]\Big)|q'',t\rangle\,,\end{aligned} \tag{5.5.5}$$

where $\delta' = \delta_{q',t+\mathrm{d}t} + \delta_{q'',t}$. Using

$$\begin{aligned}p_\alpha(t+\mathrm{d}t) &= p_\alpha(t) + \mathrm{d}t\frac{\mathrm{d}}{\mathrm{d}t}p_\alpha(t) = p_\alpha(t) - \mathrm{d}t\frac{\partial H}{\partial q_\alpha}\,,\\ H(t+\mathrm{d}t) &= H(t) + \mathrm{d}t\frac{\mathrm{d}H}{\mathrm{d}t} = H(t) + \mathrm{d}t\frac{\partial H}{\partial t}\,,\end{aligned} \tag{5.5.6}$$

and neglecting consistently all second-order changes, we rewrite $(\cdots)$ and get

$$\begin{aligned}\big(\cdots\big) =& \sum_{\alpha} p_\alpha(t)\big[\delta q_\alpha(t+\mathrm{d}t) - \delta q_\alpha(t)\big] - \mathrm{d}t\sum_{\alpha}\underbrace{\delta q_\alpha(t+\mathrm{d}t)}_{\to\,\delta q_\alpha(t)}\frac{\partial H}{\partial q_\alpha}\\ &- H\delta\mathrm{d}t - \mathrm{d}t\frac{\partial H}{\partial t}\delta t\\ =&\,\delta'\Big(\sum_{\alpha} p_\alpha(t)\big[q_\alpha(t+\mathrm{d}t) - q_\alpha(t)\big] - \mathrm{d}t\,H(t)\Big)\\ &-\sum_{\alpha}\delta p_\alpha(t)\underbrace{\big[q_\alpha(t+\mathrm{d}t) - q_\alpha(t)\big]}_{=\,\mathrm{d}q_\alpha\,=\,\mathrm{d}t\frac{\partial H}{\partial p_\alpha}} + \mathrm{d}t\sum_{\alpha}\frac{\partial H}{\partial p_\alpha}\delta p_\alpha\\ =&\,\delta'\Big(\sum_{\alpha} p_\alpha(t)\big[q_\alpha(t+\mathrm{d}t) - q_\alpha(t)\big] - \mathrm{d}t\,H(t)\Big)\,.\end{aligned} \tag{5.5.7}$$

Here, as in (5.5.5), δ' is the injunction to vary everything except the dynamics, that is: except the form of the Hamilton operator. As for changing the form of H, symbolized by δ'', we have

$$\begin{aligned}\delta''\langle q',t+\mathrm{d}t|q'',t\rangle &= \delta''\langle q',t|\left(1 - \frac{\mathrm{i}}{\hbar}\mathrm{d}t\,H\right)|q'',t\rangle\\ &= -\frac{\mathrm{i}}{\hbar}\langle q',t|\mathrm{d}t\,\delta''H|q'',t\rangle\\ &= -\frac{\mathrm{i}}{\hbar}\langle q',t+\mathrm{d}t|\mathrm{d}t\,\delta''H|q'',t\rangle\end{aligned} \tag{5.5.8}$$

which fits right into the structure of $\delta'(\cdots)$. So, with $\delta = \delta' + \delta''$,

$$\delta\langle q', t+\mathrm{d}t | q'', t\rangle = \frac{\mathrm{i}}{\hbar}\langle q', t+\mathrm{d}t | \delta W_{t+\mathrm{d}t,t} | q'', t\rangle \,, \tag{5.5.9}$$

where

$$W_{t+\mathrm{d}t,t} = \mathrm{d}t\, L(t) \quad \text{with} \quad L = \sum_\alpha p_\alpha \frac{\mathrm{d}q_\alpha}{\mathrm{d}t} - H \,. \tag{5.5.10}$$

Note that the order in which p_α and $\mathrm{d}q_\alpha/\mathrm{d}t$ are written does not matter after variation, for δp_α and δq_α are numbers (times the unit symbol 1), which commute with all operators.

Now we consider two consecutive infinitesimal time intervals: $t \to t+\mathrm{d}t \to t+2\mathrm{d}t$:

$$\langle q', t+2\mathrm{d}t | q'', t\rangle = \int \langle q', t+2\mathrm{d}t | \bar{q}, t+\mathrm{d}t\rangle (\mathrm{d}\bar{q}) \langle \bar{q}, t+\mathrm{d}t | q'', t\rangle \tag{5.5.11}$$

and compute $\delta\langle q', t+2\mathrm{d}t | q'', t\rangle$, using the previous result for each time interval:

$$\begin{aligned} \delta\langle q', t+2\mathrm{d}t | q'', t\rangle \\ &= \int \frac{\mathrm{i}}{\hbar}\langle q', t+2\mathrm{d}t | \delta W_{t+2\mathrm{d}t,t+\mathrm{d}t} | \bar{q}, t+\mathrm{d}t\rangle (\mathrm{d}\bar{q}) \langle \bar{q}, t+\mathrm{d}t | q'', t\rangle \\ &\quad + \int \langle q', t+2\mathrm{d}t | \bar{q}, t+\mathrm{d}t\rangle (\mathrm{d}\bar{q}) \frac{\mathrm{i}}{\hbar}\langle \bar{q}, t+\mathrm{d}t | \delta W_{t+\mathrm{d}t,t} | q'', t\rangle \\ &= \frac{\mathrm{i}}{\hbar}\langle q', t+2\mathrm{d}t | \left(\delta W_{t+2\mathrm{d}t,t+\mathrm{d}t} + \delta W_{t+\mathrm{d}t,t}\right) | q'', t\rangle \,, \end{aligned} \tag{5.5.12}$$

so that

$$\delta W_{t+2\mathrm{d}t,t} = \delta W_{t+2\mathrm{d}t,t+\mathrm{d}t} + \delta W_{t+\mathrm{d}t,t} \,. \tag{5.5.13}$$

We see that the structure is maintained, with the appropriate W produced additively from the constituents. The evident generality of this lets us jump immediately to the statement for any finite time interval, the *Quantum Action Principle*:

$$\delta\langle q', t_1 | q'', t_2\rangle = \frac{\mathrm{i}}{\hbar}\langle q', t_1 | \delta W_{12} | q'', t_2\rangle$$
$$\text{with} \quad W_{12} = \int_{t_1}^{t_2} \mathrm{d}t\, L(t) \,. \tag{5.5.14}$$

W, with the dimension of $\hbar$ – *action* – is the action operator; L is the Lagrangian, or Lagrange* operator, first met in (5.5.10).

---

*Joseph Louis de LAGRANGE (1736–1813)

5.6 Principle of stationary action

The time transformation function $\langle q', t_1 | q'', t_2 \rangle$ depends on the choice of final and initial state, specified by the vectors $\langle q', t_1 |$ and $| q'', t_2 \rangle$, and upon the form of the Hamiltonian operator that guides the time evolution. For a *given* Hamiltonian, the only freedom of change is of initial and final states, for which we write

$$\delta \langle q', t_1 | = \frac{\mathrm{i}}{\hbar} \langle q', t_1 | G_1 , \qquad \delta | q'', t_2 \rangle = -\frac{\mathrm{i}}{\hbar} G_2 | q'', t_2 \rangle , \tag{5.6.1}$$

where the generators G_1 and G_2 are infinitesimal Hermitian operators constructed form the physical variables at the respective times. This gives

$$\delta \langle q', t_1 | q'', t_2 \rangle = \frac{\mathrm{i}}{\hbar} \langle q', t_1 | (G_1 - G_2) | q'', t_2 \rangle \tag{5.6.2}$$

from which we conclude that

$$\delta W_{12} = G_1 - G_2 . \tag{5.6.3}$$

This is the *Principle of Stationary Action.* It asserts that the infinitesimal variation of W_{12} – which, according to (5.5.14) depends upon the variables at all values of t between t_1 and t_2 – in fact involves only variations at the end points, t_1 and t_2, and so is stationary with respect to variations at any intermediate time. We now want to recognize that, conversely, the equations of motion and the commutation relations follow from this single, fundamental dynamical principle.

In connection with δt_1 and δt_2, it is more convenient to regard t as a function of a variable τ, $t = t(\tau)$, where τ is not varied, but the form of the function is,

$$\delta t = \delta t(\tau) , \qquad \delta t_1 = \delta t(\tau_1) , \quad \delta t_2 = \delta t(\tau_2) . \tag{5.6.4}$$

So, with the reference to τ as integration variable left implicit, we have

$$W_{12} = \int_2^1 \Bigl(\sum_\alpha p_\alpha \mathrm{d}q_\alpha - H \mathrm{d}t \Bigr) . \tag{5.6.5}$$

Notice that, e. g.,

$$\begin{aligned} \delta \mathrm{d}q_\alpha(t) &= \delta [q_\alpha(t + \mathrm{d}t) - q_\alpha(t)] \\ &= \delta q_\alpha(t + \mathrm{d}t) - \delta q_\alpha(t) = \mathrm{d}\delta q_\alpha(t) \end{aligned} \tag{5.6.6}$$

so, symbolically,

$$\delta \mathrm{d} = \mathrm{d} \delta , \tag{5.6.7}$$

and

$$\begin{aligned}\delta W_{12} &= \int_2^1 \Big(\sum_\alpha (\delta p_\alpha \mathrm{d}q_\alpha + p_\alpha \mathrm{d}\delta q_\alpha) - \delta H \mathrm{d}t - H\mathrm{d}\delta t\Big) \\ &= \int_2^1 \Big(\sum_\alpha (\delta p_\alpha \mathrm{d}q_\alpha - \mathrm{d}p_\alpha \delta q_\alpha) - \delta H \mathrm{d}t + \mathrm{d}H\delta t\Big) \\ &\quad + \int_2^1 \mathrm{d}\Big(\sum_\alpha p_\alpha \delta q_\alpha - H\delta t\Big) . \end{aligned} \tag{5.6.8}$$

The last term refers to the end points only,

$$\int_2^1 \mathrm{d}\Big(\sum_\alpha p_\alpha \delta q_\alpha - H\delta t\Big) = \Big(\sum_\alpha p_\alpha \delta q_\alpha - H\delta t\Big)\Big|_2^1 . \tag{5.6.9}$$

The stationary action principle therefore requires that the first integrand on the right-hand side of (5.6.8) vanishes at all intermediate times:

$$\delta H = \frac{\mathrm{d}H}{\mathrm{d}t}\delta t + \sum_\alpha \left(\delta p_\alpha \frac{\mathrm{d}q_\alpha}{\mathrm{d}t} - \frac{\mathrm{d}p_\alpha}{\mathrm{d}t}\delta q_\alpha\right) . \tag{5.6.10}$$

The specific nature of the q's and p's is now introduced by asserting that the δq_α and δp_α are numerical multiples of the unit operator or, better, that they *commute* with all operators q and p. Then we can infer that

$$\frac{\mathrm{d}q_\alpha}{\mathrm{d}t} = \frac{\partial H}{\partial p_\alpha} , \qquad -\frac{\mathrm{d}p_\alpha}{\mathrm{d}t} = \frac{\partial H}{\partial q_\alpha} , \qquad \frac{\mathrm{d}H}{\mathrm{d}t} = \frac{\partial H}{\partial t} \tag{5.6.11}$$

and arrive at

$$\delta W_{12} = G_1 - G_2 \tag{5.6.12}$$

where

$$G = \sum_\alpha p_\alpha \delta q_\alpha - H\delta t \tag{5.6.13}$$

at each boundary.

From the interpretation of

$$G_t = -H\delta t \tag{5.6.14}$$

as the generator of infinitesimal time displacements, we infer the Heisenberg operator equation of motion and the Schrödinger vector equation of motion.

The interpretation of

$$G_q = \sum_\alpha p_\alpha \delta q_\alpha \tag{5.6.15}$$

as the generator of infinitesimal q displacements,

$$\delta_q F = \frac{1}{\mathrm{i}\hbar}[F, G_q] \qquad \text{or} \qquad \frac{\partial F}{\partial q_\beta} = \frac{1}{\mathrm{i}\hbar}[F, p_\beta] \,, \tag{5.6.16}$$

produces the basic commutators

$$\frac{1}{\mathrm{i}\hbar}[q_\alpha, p_\beta] = \frac{\partial q_\alpha}{\partial q_\beta} = \delta_{\alpha\beta} \,, \qquad \frac{1}{\mathrm{i}\hbar}[p_\alpha, p_\beta] = \frac{\partial p_\alpha}{\partial q_\beta} = 0 \,. \tag{5.6.17}$$

We do not, in this way, get $[q_\alpha, q_\beta] = 0$. But, of course, the latter is implicit in the choice of compatible q states, and the interpretation of the effect of G_q on these states:

$$\delta_q \langle q', t| = \langle q', t| \frac{\mathrm{i}}{\hbar} G_q \tag{5.6.18}$$

or

$$\frac{\hbar}{\mathrm{i}} \frac{\partial}{\partial q'_\alpha} \langle q', t| = \langle q', t| p_\alpha(t) \,. \tag{5.6.19}$$

On the other hand, we could have initially chosen as the complete set of compatible physical properties – commuting operators – the totality of the p's. Then we would have been led to the action principle

$$\delta \langle p', t_1 | p'', t_2 \rangle = \frac{\mathrm{i}}{\hbar} \langle p', t_1 | \delta \left(\int_2^1 \mathrm{d}t \, L_p \right) | p'', t_2 \rangle \tag{5.6.20}$$

where

$$L_p = -\sum_\alpha \frac{\mathrm{d}p_\alpha}{\mathrm{d}t} q_\alpha - H \tag{5.6.21}$$

is so labeled to distinguish it from

$$L_q = \sum_\alpha p_\alpha \frac{\mathrm{d}q_\alpha}{\mathrm{d}t} - H \,; \tag{5.6.22}$$

the time derivative term of L_p is produced from that of L_q by the substitution $q \to p$, $p \to -q$, and conversely. And since the operator equations of motion maintain their form under this substitution,

$$\frac{\mathrm{d}q_\alpha}{\mathrm{d}t} = \frac{\partial H}{\partial p_\alpha} \longleftrightarrow \frac{\mathrm{d}p_\alpha}{\mathrm{d}t} = -\frac{\partial H}{\partial q_\alpha} \,, \tag{5.6.23}$$

they are also produced by the new action principle. What is different is G now appearing as

$$G = G_p + G_t\,, \qquad G_p = -\sum_\alpha q_\alpha \delta p_\alpha\,. \tag{5.6.24}$$

The operator significance of G_p,

$$\delta_p F = \frac{1}{i\hbar}[F, G_p] \qquad \text{or} \qquad \frac{\partial F}{\partial p_\alpha} = \frac{1}{i\hbar}[q_\alpha, F]\,, \tag{5.6.25}$$

gives

$$\frac{1}{i\hbar}[q_\alpha, p_\beta] = \frac{\partial p_\beta}{\partial p_\alpha} = \delta_{\alpha\beta} \qquad \text{and} \qquad \frac{1}{i\hbar}[q_\alpha, q_\beta] = \frac{\partial q_\beta}{\partial p_\alpha} = 0\,. \tag{5.6.26}$$

Thus, all the fundamental commutators are produced when both Lagrangians, L_q and L_p, are employed.

5.7 Change of description

Although we used the $q \to p$, $p \to -q$ substitution to recognize that L_p produces the same equations of motion as L_q, the more fundamental observation is that

$$L_q - L_p = \frac{d}{dt}\sum_\alpha p_\alpha q_\alpha\,, \tag{5.7.1}$$

for, in general, if

$$L - \overline{L} = \frac{d}{dt}w \tag{5.7.2}$$

the Lagrangian $\overline{L}$ produces the same equations of motions as L:

$$W_{12} - \overline{W}_{12} = w_1 - w_2\,, \tag{5.7.3}$$

and the stationary action principle applied to $\overline{W}_{12}$ gives the same result as W_{12}, since the difference refers to the boundary, not the interior of the time interval. The remaining boundary terms are

$$(G_1 - G_2) - (\overline{G}_1 - \overline{G}_2) = \delta w_1 - \delta w_2 \tag{5.7.4}$$

or

$$G - \overline{G} = \delta w \tag{5.7.5}$$

at the terminal times t_1 and t_2.

In the example of L_q and L_p we have

$$G_q - G_p = \sum_\alpha p_\alpha \delta q_\alpha + \sum_\alpha q_\alpha \delta p_\alpha = \delta \sum_\alpha q_\alpha p_\alpha\,, \tag{5.7.6}$$

which leads directly to the transformation function connecting the q and p descriptions:

$$\delta\langle q'|p'\rangle = \frac{\mathrm{i}}{\hbar}\langle q'|(G_q - G_p)|p'\rangle = \frac{\mathrm{i}}{\hbar}\Big(\delta\sum_\alpha q'_\alpha p'_\alpha\Big)\langle q'|p'\rangle \tag{5.7.7}$$

or

$$\langle q'|p'\rangle = \frac{1}{(2\pi\hbar)^{n/2}}\,\mathrm{e}^{\frac{\mathrm{i}}{\hbar}\sum_\alpha q'_\alpha p'_\alpha}\,, \tag{5.7.8}$$

where the multiplicative constant emerges from the requirement that

$$\int \langle q'|p'\rangle\,(\mathrm{d}p')\,\langle p'|q''\rangle = \delta(q' - q'')\,, \tag{5.7.9}$$

as in (5.3.20).

5.8 Permissible variations

We began knowing all about p's and q's. Now let's turn it around and take as our starting point the quantum action principle:

$$\delta\langle 1|2\rangle = \frac{\mathrm{i}}{\hbar}\langle 1|\delta W_{12}|2\rangle \qquad \text{with} \qquad W_{12} = \int_2^1 \mathrm{d}t\, L\,, \tag{5.8.1}$$

where

$$L = \sum_\alpha p_\alpha \cdot \frac{\mathrm{d}}{\mathrm{d}t}q_\alpha - H(q,p,t) \qquad \Big[-\frac{\mathrm{d}}{\mathrm{d}t}w\Big] \tag{5.8.2}$$

with *no* a priori knowledge of operators p and q. For brevity we introduce the notation

$$A\,.\,B \equiv \frac{1}{2}\{A,B\} = \frac{1}{2}(AB + BA) \tag{5.8.3}$$

for the symmetrized product of two operators; the symmetrization ensures that the products $p_\alpha\,.\,\mathrm{d}q_\alpha$ are Hermitian. What additional input is required to infer the specific properties of the operators p and q? And how wide is the class of permissible variations?

For a given H we have the stationary action principle,

$$\begin{aligned} \delta W_{12} &= G_1 - G_2 \\ &= \int_2^1 \Big(\sum_\alpha [\delta p_\alpha\,.\,\mathrm{d}q_\alpha - \mathrm{d}p_\alpha\,.\,\delta q_\alpha] - \delta H \mathrm{d}t + \mathrm{d}H\delta t\Big) \\ &\quad + \int_2^1 \mathrm{d}\Big(\sum_\alpha p_\alpha\,.\,\delta q_\alpha - H\delta t\Big) \end{aligned} \tag{5.8.4}$$

[(5.6.3) and (5.6.8) with symmetrized products] with the inference that

$$\delta H = \sum_\alpha \left(\delta p_\alpha \cdot \frac{dq_\alpha}{dt} - \frac{dp_\alpha}{dt} \cdot \delta q_\alpha\right) + \frac{dH}{dt}\delta t . \tag{5.8.5}$$

To this point the δq_α and δp_α are unspecified operators. We now assume that *among* the possibilities are operators δq_α, δp_α that commute with the q's and p's. Then all is as before:

$$\delta H = \sum_\alpha \left(\delta p_\alpha \frac{dq_\alpha}{dt} - \frac{dp_\alpha}{dt}\delta q_\alpha\right) + \frac{dH}{dt}\delta t ,$$

$$\frac{\partial H}{\partial p_\alpha} = \frac{dq_\alpha}{dt}, \qquad \frac{\partial H}{\partial q_\alpha} = -\frac{dp_\alpha}{dt}, \qquad \frac{\partial H}{\partial t} = \frac{dH}{dt}, \tag{5.8.6}$$

and

$$G_q = \sum_\alpha p_\alpha \delta q_\alpha \tag{5.8.7}$$

as well as

$$G_p = -\sum_\alpha \delta p_\alpha q_\alpha \tag{5.8.8}$$

if $w = \sum_\alpha q_\alpha \cdot p_\alpha$ in (5.8.2).

Returning to the general expression for δH, we can now write

$$\delta H = \sum_\alpha \left(\delta p_\alpha \cdot \frac{\partial H}{\partial p_\alpha} + \delta q_\alpha \cdot \frac{\partial H}{\partial q_\alpha}\right) + \frac{\partial H}{\partial t}\delta t . \tag{5.8.9}$$

The commutation relations of the q's and p's are maintained if these variations appear as an infinitesimal unitary transformation,

$$\begin{aligned} \delta q_\alpha &= \frac{1}{i\hbar}[q_\alpha, G] = \frac{\partial G}{\partial p_\alpha} , \\ \delta p_\alpha &= \frac{1}{i\hbar}[p_\alpha, G] = -\frac{\partial G}{\partial q_\alpha} , \\ \delta H &= \frac{\partial H}{\partial t}\delta t + \frac{1}{i\hbar}[H, G] \end{aligned} \tag{5.8.10}$$

or

$$\frac{1}{i\hbar}[H, G] = \sum_\alpha \left(\frac{\partial H}{\partial q_\alpha} \cdot \frac{\partial G}{\partial p_\alpha} - \frac{\partial H}{\partial p_\alpha} \cdot \frac{\partial G}{\partial q_\alpha}\right) = (H, G) . \tag{5.8.11}$$

But we know from Problem 5-10a that this identity of commutator and symmetrized Poisson* bracket holds, for *arbitrary* H, only if G is less than cubic in the q's and p's. Another point to recognize is that one must have

*Siméon Denise Poisson (1781–1840)

$$G = \sum_\alpha p_\alpha \cdot \delta q_\alpha = \sum_\alpha p_\alpha \cdot \frac{\partial G}{\partial p_\alpha}, \tag{5.8.12}$$

for instance, which requires that the p dependent part of G – which produces δq_α – must be *linear* in the p's; likewise, G must be linear in the q's. Here we recall that, in general terms, a function f of a number of variables, say $x_1, \ldots, x_n$, is linear if

$$f(\lambda x_1, \ldots, \lambda x_n) = \lambda f(x_1, \ldots, x_2)\,; \tag{5.8.13}$$

we put $\lambda = 1 + \delta\lambda$,

$$\begin{aligned} f(x_1 + \delta\lambda\, x_1, \ldots, x_n + \delta\lambda\, x_n) &= f(x) + \delta\lambda \sum_k x_k \frac{\partial f}{\partial x_k} \\ &= f(x) + \delta\lambda\, f(x)\,, \end{aligned} \tag{5.8.14}$$

and conclude that

$$\sum_k x_k \frac{\partial f}{\partial x_k} = f \tag{5.8.15}$$

is characteristic of a linear function. Of course,

$$G_q = \sum_\alpha p_\alpha \delta q_\alpha \qquad \text{with numbers } \delta q_\alpha \tag{5.8.16}$$

is linear in the p's and q's and thus satisfies the linearity requirements on G. But now we see an additional possibility: a G that is linear in the p's and linear in the q's, therefore quadratic in the q's and p's,

$$G = \sum_{\alpha,\beta} p_\alpha \cdot g_{\alpha\beta} q_\beta\,,$$

$$\delta q_\alpha = \frac{\partial G}{\partial p_\alpha} = \sum_\beta g_{\alpha\beta} q_\beta\,, \qquad \delta p_\beta = -\frac{\partial G}{\partial q_\beta} = -\sum_\alpha p_\alpha g_{\alpha\beta}\,, \tag{5.8.17}$$

where the $g_{\alpha\beta}$'s are numbers. We shall make use of this possibility later, in Section 8.3.

Problems

5-1 Dynamical variables v_α have a dynamical, but no parametric time dependence. By contrast, probability operators (see Problems 1-30) do not change in time at all, $P(v(t_1), t_1) = P(v(t_2), t_2)$. Conclude that

$$\frac{\partial}{\partial t} P = \frac{\mathrm{i}}{\hbar}[P, H]\,,$$

which is frequently called the von Neumann* equation. Show that the dynamical variables can refer to any common time in this equation.

5-2a Descriptions at different times t and t' are related to each other by the unitary evolution operator U,

$$\langle\ldots,t| = \langle\ldots,t'|U_{t,t'}\bigl(v(t)\bigr)\,,$$

which has a parametric dependence on both times and a dynamical time dependence through the dynamical variables $v(t)$ of which it is a function. Show that U obeys the equations of motion

$$\mathrm{i}\hbar\frac{\partial}{\partial t}U_{t,t'} = HU\,, \qquad \mathrm{i}\hbar\frac{\mathrm{d}}{\mathrm{d}t}U_{t,t'} = UH\,,$$

where $H = H\bigl(v(t),t\bigr)$ is the Hamilton operator. Note the fundamental difference between taking the partial and the total time derivative of the evolution operator.

5-2b Show first that the replacement $v(t) \to v(t')$ does not change $U_{t,t'}$ as a whole, that is:

$$U_{t,t'}\bigl(v(t)\bigr) = U_{t,t'}\bigl(v(t')\bigr)\,,$$

and then verify the group property

$$U_{t,t''}\bigl(v(t)\bigr) = U_{t',t''}\bigl(v(t')\bigr)U_{t,t'}\bigl(v(t)\bigr) = U_{t,t'}\bigl(v(t)\bigr)U_{t',t''}\bigl(v(t)\bigr)$$

or tersely: $U_{t,t''} = U_{t,t'}U_{t',t''}$.

5-3 Show that the spread δA of observable A and the spread δH of the Hamilton operator H obey the inequality

$$\delta A\,\delta H \ge \frac{\hbar}{2}\left|\left\langle\frac{\mathrm{d}A}{\mathrm{d}t} - \frac{\partial A}{\partial t}\right\rangle\right|.$$

5-4 A slit is opened for the time interval δt and an emerging beam of particles moves along the x axis at the average speed

$$v = \frac{\partial E}{\partial p}\,.$$

What is the length δx of the beam? How large is δp, the unavoidable spread of momentum in the beam? Show that the related spread of energy, δE, is such that

$$\delta E\,\delta t \gtrsim \hbar\,.$$

*John (János) von Neumann (1903–1957)

5-5 Stern–Gerlach experiment, spin $\frac{1}{2}\hbar$: z magnetic moment, $\pm\mu$; field inhomogeneity, $(\mathrm{d}/\mathrm{d}z)B_z(z)$; time in field, t. For the experiment to work, the positive, or negative, z momentum acquired in time t must be large compared to (half) the inherent spread of z momenta in the beam, δp_z. Write out this condition. Now, Heisenberg's uncertainty relation requires a minimum spread of z values, δz. That implies a corresponding spread in ϕ, the angle of rotation of the angular momentum $\boldsymbol{s}$ in the xy plane, during time t. Write down ϕ, according to the torque equation based on $\boldsymbol{\mu} = \gamma\boldsymbol{s}$ [if this doesn't come to mind, derive it quantum mechanically from the Hamiltonian $H = -\gamma s_z B_z(z)$]. Evaluate $\delta\phi$. What known non-classical property of Stern-Gerlach measurements follows from the resulting inequality obeyed by $\delta\phi$?

5-6 For the Stern–Gerlach experiment on an atomic doublet, let E be the energy difference between the two states during the time interval δt that the atoms are in the magnetic field $B_z(z)$. What is the spread, δE, produced by the width of the beam, δz? By using the condition for the experiment to succeed, show that

$$\delta E\,\delta t \gtrsim \hbar\,.$$

5-7a One degree of freedom: Hamilton operator $H = p^2/(2M)$; minimum uncertainty state $|\delta\rangle$ at time $t = 0$. Solve the equations of motion and evaluate

$$[\delta x(t)]^2 = \left\langle \left[x(t) - \langle x(t)\rangle_\delta\right]^2\right\rangle_\delta \quad\text{and}\quad [\delta p(t)]^2\,.$$

5-7b According to (5.4.34), the inequality $[\delta x(t)]^2 \geq \hbar t/M$ holds. Give numerical values to this lower limit of $\delta x(t)$ for an electron at the times $t = 10^{-16}\,\mathrm{s}$; $1\,\mathrm{s}$. For an object of mass $1\,\mathrm{g}$, how large would t have to be in order that $\delta x(t) > 1\,\text{Å}$?

5-8a Reconsider the situation of Section 5.4, mass M moving freely along the x axis. But now do not assume that there is a minimum uncertainty state at $t = 0$; rather think of any arbitrary initial state. Solve the Heisenberg equations of motion to find the expectations values $\langle x(t)\rangle$, $\langle p(t)\rangle$, $\langle [x(t)]^2\rangle$, $\langle [p(t)]^2\rangle$, and $\frac{1}{2}\langle [x(t)p(t) + p(t)x(t)]\rangle$ in terms of their initial values. Use them to show that

$$\delta x(t) = \delta x\,\sqrt{1 - 2tt_0/T^2 + (t/T)^2}$$

with T as in (5.4.35) and

$$t_0 = -\frac{M}{(\delta p)^2}\left[\tfrac{1}{2}\langle xp + px\rangle - \langle x\rangle\langle p\rangle\right].$$

What is the physical significance of t_0?

5-8b How does the restriction in Problem 2-17a follow? [What is $\delta q\,\delta p$ there is $\delta x\,\delta p/\hbar$ here, of course.]

5-9a Recall Problem 2-8a and show that

$$U(q',p')U(q'',p'') = \mathrm{e}^{-\frac{1}{2}\mathrm{i}(q'p''-q''p')}U(q'+q'',p'+p'') ,$$
$$U(q'',p'')U(q',p') = \mathrm{e}^{\frac{1}{2}\mathrm{i}(q'p''-q''p')}U(q'+q'',p'+p'') .$$

5-9b Any $F(q,p)$ can be written as

$$F(q,p) = \int \frac{\mathrm{d}q'\,\mathrm{d}p'}{2\pi} f(q',p')U(q',p')$$

and similarly

$$G(q,p) = \int \frac{\mathrm{d}q''\,\mathrm{d}p''}{2\pi} g(q'',p'')U(q'',p'') .$$

To justify this assertion express $f(q',p')$ in terms of the normalized matrix elements of Problem 2-7a.

5-9c Prove that

$$\mathrm{e}^{-\frac{\mathrm{i}}{2}D}FG = \mathrm{e}^{\frac{\mathrm{i}}{2}D}GF$$

where (introducing $\hbar$)

$$D = \hbar\left(\frac{\partial}{\partial q_F}\frac{\partial}{\partial p_G} - \frac{\partial}{\partial p_F}\frac{\partial}{\partial q_G}\right) ,$$

in which $\partial/\partial q_F$, for example, means differentiation with respect to q in F only.

5-10a Introduce the notation [cf. (5.8.3)]

$$F\,.\,G \equiv \tfrac{1}{2}(FG+GF) = \tfrac{1}{2}\{F,G\}$$

for the symmetrized product of F and G, and use it to rewrite the result of Problem 5-9c as

$$\frac{1}{\mathrm{i}\hbar}[F,G] = \frac{\tan(\frac{1}{2}D)}{\frac{1}{2}D}(F,G)$$

where

$$(F,G) = \frac{D}{\hbar}F\,.\,G = \frac{\partial F}{\partial q}\cdot\frac{\partial G}{\partial p} - \frac{\partial F}{\partial p}\cdot\frac{\partial G}{\partial q}$$

is the symmetrized Poisson bracket operator. Recognize that

$$\frac{1}{i\hbar}[F,G] = (F,G)$$

if either F or G is less than cubic in q or p, or if both F and G are less than cubic in q, or in p.

5-10b Make a direct evaluation – that is $[F_1F_2,G] = F_1[F_2,G] + [F_1,G]F_2$ et cetera – of $(i\hbar)^{-1}[q^2,p^2]$ and $(i\hbar)^{-1}[q^3,p^3]$. Compare the results with the predictions of Problem 5-10a.

5-11 In Section 5.5 we found

$$\delta\langle q',t_1|q'',t_2\rangle = \frac{i}{\hbar}\langle q',t_1|\delta W_{12}^{(qq)}|q'',t_2\rangle$$

with the action operator

$$W_{12}^{(qq)} = \int_1^2 \left(\sum_\alpha p_\alpha \cdot \mathrm{d}q_\alpha - H\mathrm{d}t\right),$$

and in Section 5.6 we observed that

$$W_{12}^{(pp)} = \int_1^2 \left(-\sum_\alpha q_\alpha \cdot \mathrm{d}p_\alpha - H\mathrm{d}t\right),$$

is needed in

$$\delta\langle p',t_1|p'',t_2\rangle = \frac{i}{\hbar}\langle p',t_1|\delta W_{12}^{(pp)}|p'',t_2\rangle .$$

Now consider

$$\delta\langle q',t_1|p'',t_2\rangle = \frac{i}{\hbar}\langle q',t_1|\delta W_{12}^{(qp)}|p'',t_2\rangle$$

and

$$\delta\langle p',t_1|q'',t_2\rangle = \frac{i}{\hbar}\langle p',t_1|\delta W_{12}^{(pq)}|q'',t_2\rangle .$$

Show that the appropriate action operators are

$$\begin{aligned} W_{12}^{(qp)} &= W_{12}^{(qq)} + \sum_\alpha p_\alpha(t_2)\cdot q_\alpha(t_2) \\ &= W_{12}^{(pp)} + \sum_\alpha p_\alpha(t_1)\cdot q_\alpha(t_1) \end{aligned}$$

and

$$W_{12}^{(pq)} = W_{12}^{(qq)} - \sum_{\alpha} p_\alpha(t_1) \cdot q_\alpha(t_1)$$
$$= W_{12}^{(pp)} - \sum_{\alpha} p_\alpha(t_2) \cdot q_\alpha(t_2) \,,$$

respectively.

5-12 Show that $\delta\boldsymbol{\omega} \cdot \boldsymbol{J}$, the generator of infinitesimal rotations, is of the bilinear form in (5.8) if $\boldsymbol{J} = \boldsymbol{r} \times \boldsymbol{p} + \boldsymbol{s}$ is constructed from the position vector $\boldsymbol{r}$, the momentum vector $\boldsymbol{p}$, and the spin vector $\boldsymbol{s}$ of the quantum object considered.

6. Elementary Applications

Let's see the action principle at work solving problems. We'll consider the one-dimensional motion of a particle (position x, momentum p, mass M) without any force acting on it, exposed to a constant force, and under the influence of a linear restoring force.

6.1 Time transformation functions

6.1.1 Free particle

The Hamilton operator

$$H = \frac{p^2}{2M} \tag{6.1.1}$$

gives

$$\delta\langle x', t|x''\rangle = \frac{\mathrm{i}}{\hbar}\langle x', t|\Big[p(t)\delta x' - p\delta x'' - \delta t\frac{p^2}{2M}\Big]|x''\rangle \tag{6.1.2}$$

where

$$p(t) = p\,, \qquad x(t) = x + \frac{p}{M}t\,, \qquad \big[x, x(t)\big] = \frac{\mathrm{i}\hbar t}{M} \tag{6.1.3}$$

are immediate consequences of the Heisenberg equations of motion

$$\frac{\mathrm{d}x(t)}{\mathrm{d}t} = \frac{p(t)}{M}\,, \qquad \frac{\mathrm{d}p(t)}{\mathrm{d}t} = 0\,; \tag{6.1.4}$$

here and below we write $|x''\rangle \equiv |x'', t=0\rangle$, $p \equiv p(t=0)$, and $x \equiv x(t=0)$ for brevity. So

$$\delta\langle x', t|x''\rangle = \frac{\mathrm{i}}{\hbar}\langle x', t|\Big[\frac{M}{t}\big(x(t) - x\big)(\delta x' - \delta x'') - \delta t\frac{M}{2t^2}\big(x(t) - x\big)^2\Big]|x''\rangle \tag{6.1.5}$$

with

$$\begin{aligned}\big(x(t)-x\big)^2 &= \big(x(t)\big)^2 + x^2 - x(t)x - \underbrace{xx(t)}_{=x(t)x+\mathrm{i}\hbar t/M} \\ &= \big(x(t)\big)^2 - 2x(t)x + x^2 - \frac{\mathrm{i}\hbar t}{M} \,. \end{aligned} \tag{6.1.6}$$

Now that $[\cdots]$ of (6.1.5) is in a form where x and $x(t)$ stand next to their respective eigenvectors, we can introduce eigenvalues and get first

$$\begin{aligned}\frac{\delta\langle x',t|x''\rangle}{\langle x',t|x''\rangle} &= \frac{\mathrm{i}}{\hbar}\left[\frac{M}{t}(x'-x'')(\delta x'-\delta x'') - \frac{\delta t}{t^2}\frac{M}{2}(x'-x'')^2 + \mathrm{i}\hbar\frac{1}{2}\frac{\delta t}{t}\right] \\ &= \delta\left[\frac{\mathrm{i}}{\hbar}\frac{M}{2t}(x'-x'')^2 + \log\frac{1}{\sqrt{t}} + \log C\right] \end{aligned} \tag{6.1.7}$$

and then

$$\langle x',t|x''\rangle = \frac{C}{\sqrt{t}}\,\mathrm{e}^{\frac{\mathrm{i}}{\hbar}\frac{M}{2t}(x'-x'')^2} \,. \tag{6.1.8}$$

The value of the multiplicative integration constant C follows from the initial condition:

$$\langle x',t|x''\rangle \to \langle x'|x''\rangle = \delta(x'-x'') \qquad \text{for } t\to 0 \,. \tag{6.1.9}$$

Now

$$\int \mathrm{d}(x'-x'')\,\langle x',t|x''\rangle = C\sqrt{\frac{2\pi\mathrm{i}\hbar}{M}} \,, \tag{6.1.10}$$

is actually independent of t so that it must equal unity, and we get

$$C = \sqrt{\frac{M}{2\pi\mathrm{i}\hbar}} \,, \qquad \langle x',t|x''\rangle = \sqrt{\frac{M}{2\pi\mathrm{i}\hbar t}}\,\mathrm{e}^{\frac{\mathrm{i}}{\hbar}\frac{M}{2t}(x'-x'')^2} \,; \tag{6.1.11}$$

we have seen this time transformation function before as the factors in (5.4.16).

6.1.2 Constant force

Next consider the Hamilton operator

$$H = \frac{p^2}{2M} - Fx \qquad \text{with constant } F. \tag{6.1.12}$$

The Heisenberg equations of motion

$$\frac{\mathrm{d}x(t)}{\mathrm{d}t} = \frac{p(t)}{M} \,, \qquad \frac{\mathrm{d}p(t)}{\mathrm{d}t} = F \,, \tag{6.1.13}$$

are, of course, those of motion under the influence of constant force F. They are solved by

$$p(t) = p + Ft\,, \qquad x(t) = x + \frac{p}{M}t + \frac{1}{2}\frac{F}{M}t^2\,, \tag{6.1.14}$$

with the consequence

$$\bigl[x, x(t)\bigr] = \frac{\mathrm{i}\hbar t}{M}\,. \tag{6.1.15}$$

Note that this commutator is F independent.

The quantum action principle states here that

$$\delta\langle x', t|x''\rangle = \frac{\mathrm{i}}{\hbar}\langle x', t|\left[p(t)\delta x' - p\delta x'' - \delta t\left(\frac{p^2}{2M} - Fx\right)\right]|x''\rangle\,. \tag{6.1.16}$$

We bring it into a more useful form by expressing $[\cdots]$ in terms of x and $x(t)$, with $x(t)$ to the left of x in products so that they stand next to their respective eigenvectors. First, we use (6.1.14) to establish

$$\begin{aligned} p &= \frac{M}{t}\left(x(t) - x - \frac{Ft^2}{2M}\right),\\ p(t) &= \frac{M}{t}\left(x(t) - x + \frac{Ft^2}{2M}\right), \end{aligned} \tag{6.1.17}$$

where we note that initial and final quantities are interchanged,

$$x(t) \leftrightarrow x\,, \quad p(t) \leftrightarrow p \tag{6.1.18}$$

on letting $t \leftrightarrow -t$. That would be more evident had we continued to label the two times as t_1 and t_2, with $t = t_1 - t_2$. Then we are just talking about the interchange of labels 1 and 2.

Second, we then use (6.1.15) to get

$$\frac{p^2}{2M} = \frac{M}{2t^2}\left(\bigl[x(t)\bigr]^2 - 2x(t)x + x^2\right) - \frac{\mathrm{i}\hbar}{2t} - \frac{1}{2}F\bigl[x(t) - x\bigr] + \frac{F^2t^2}{8M}\,. \tag{6.1.19}$$

Therefore, the $[\cdots]$ of (6.1.16) turns into

$$\begin{aligned} \bigl[\cdots\bigr] \to{} & \frac{M}{t}\left(x' - x'' + \frac{Ft^2}{2M}\right)\delta x' - \frac{M}{t}\left(x' - x'' - \frac{Ft^2}{2M}\right)\delta x''\\ & - \delta t\left\{\frac{M}{2t^2}(x' - x'')^2 - \frac{\mathrm{i}\hbar}{2t} - \frac{1}{2}F(x' - x'') + \frac{F^2t^2}{8M} - Fx''\right\}\\ ={}& \delta\left\{\frac{M}{2t}(x' - x'')^2 - \mathrm{i}\hbar\log\frac{C}{\sqrt{t}} + \frac{1}{2}Ft(x' + x'') - \frac{F^2t^3}{24M}\right\}\\ ={}& \frac{\hbar}{\mathrm{i}}\frac{\delta\langle x', t|x''\rangle}{\langle x', t|x''\rangle} = \frac{\hbar}{\mathrm{i}}\delta\log\langle x', t|x''\rangle\,, \end{aligned} \tag{6.1.20}$$

and the time transformation function

$$\langle x', t | x'' \rangle = \sqrt{\frac{M}{2\pi \mathrm{i}\hbar t}}\, \mathrm{e}^{\frac{\mathrm{i}}{\hbar}\frac{M}{2t}(x' - x'')^2}\, \mathrm{e}^{\frac{\mathrm{i}}{\hbar}Ft\frac{x'+x''}{2}}\, \mathrm{e}^{-\frac{\mathrm{i}}{\hbar}\frac{F^2t^3}{24M}} \tag{6.1.21}$$

follows. The value of the integration constant C, as required by (6.1.9), is the one given in (6.1.11); it is accounted for in (6.1.21).

6.1.3 Linear restoring force: Harmonic oscillator

Now consider the Hamilton operator

$$H = \frac{p^2}{2M} + \frac{1}{2}M\omega^2 x^2 \qquad \text{with constant } \omega. \tag{6.1.22}$$

The Heisenberg equations of motion

$$\frac{\mathrm{d}x(t)}{\mathrm{d}t} = \frac{p(t)}{M}, \qquad \frac{\mathrm{d}p(t)}{\mathrm{d}t} = -M\omega^2 x(t), \tag{6.1.23}$$

are those of a harmonic oscillator; it's motion under the influence of a linear restoring force: proportional to the distance $x(t)$ and directed toward the equilibrium position $x = 0$. The familiar solutions of (6.1.23) are

$$\begin{aligned} x(t) &= x\cos(\omega t) + \frac{1}{M\omega}p\sin(\omega t), \\ p(t) &= p\cos(\omega t) - M\omega x\sin(\omega t), \end{aligned} \tag{6.1.24}$$

with the consequence

$$\bigl[x, x(t)\bigr] = \frac{\mathrm{i}\hbar}{M\omega}\sin(\omega t). \tag{6.1.25}$$

Here, the quantum action principle says

$$\delta\langle x', t | x'' \rangle = \frac{\mathrm{i}}{\hbar}\langle x', t| \left[p(t)\delta x' - p\delta x'' - \delta t\left(\frac{p^2}{2M} + \frac{M\omega^2}{2}x^2\right)\right] |x''\rangle, \tag{6.1.26}$$

and we express the operators in $[\cdots]$ in terms of x and $x(t)$: First we have

$$\begin{aligned} p &= \frac{M\omega}{\sin(\omega t)}\bigl[x(t) - x\cos(\omega t)\bigr], \\ p(t) &= \frac{M\omega}{\sin(\omega t)}\bigl[x(t)\cos(\omega t) - x\bigr], \end{aligned} \tag{6.1.27}$$

where we again note that $t \leftrightarrow -t$ interchanges initial and final quantities, and then

$$\frac{p^2}{2M} = \frac{M\omega^2}{2\sin^2(\omega t)}\Big([x(t)]^2 - 2x(t)x\cos\omega t + x^2\cos^2(\omega t)\Big) - \frac{\mathrm{i}\hbar\omega}{2}\frac{\cos(\omega t)}{\sin(\omega t)}\,, \tag{6.1.28}$$

so that in (6.1.26)

$$\begin{aligned}\Big[\cdots\Big] \to & \frac{M\omega}{\sin(\omega t)}[x'\cos(\omega t) - x'']\delta x' + \frac{M\omega}{\sin(\omega t)}[x''\cos(\omega t) - x']\delta x'' \\ & - \delta t\left\{\frac{M\omega^2}{2\sin^2(\omega t)}[x'^2 + x''^2 - 2x'x''\cos(\omega t)] - \frac{\mathrm{i}\hbar\omega}{2}\frac{\cos(\omega t)}{\sin(\omega t)}\right\} \\ = & \,\delta\left\{\frac{M\omega}{2\sin(\omega t)}[(x'^2 + x''^2)\cos(\omega t) - 2x'x''] - \mathrm{i}\hbar\log\sqrt{\frac{\omega}{\sin(\omega t)}}\right\} \\ = & \,\frac{\hbar}{\mathrm{i}}\delta\log\langle x', t|x''\rangle\,. \end{aligned} \tag{6.1.29}$$

Accordingly, we get

$$\langle x', t|x''\rangle = C\sqrt{\frac{\omega}{\sin(\omega t)}}\,\mathrm{e}^{\frac{\mathrm{i}}{\hbar}\frac{M\omega}{2\sin(\omega t)}[(x'^2 + x''^2)\cos(\omega t) - 2x'x'']}\,. \tag{6.1.30}$$

For $\omega t \ll 1$, this becomes (6.1.8), and therefore the $t \to 0$ limit of (6.1.9) requires that the integration constant C has the value given in (6.1.11), and so we arrive at the time transformation function

$$\langle x', t|x''\rangle = \sqrt{\frac{M\omega}{2\pi\mathrm{i}\hbar}\frac{1}{\sin(\omega t)}}\,\mathrm{e}^{\frac{\mathrm{i}}{\hbar}\frac{M\omega}{2\sin(\omega t)}[(x'^2 + x''^2)\cos(\omega t) - 2x'x'']}\,. \tag{6.1.31}$$

We note that the ambiguity of the square root of $\mathrm{i}\sin(\omega t)$ is only apparent; see Problem 6-12 for details.

6.2 Short times

We have just seen that we regain the free particle as $t \to 0$; there is no time for the force to act. Suppose now that t is small, $\omega t \ll 1$, but *not* approaching zero. It is helpful to first rewrite the exponent with the aid of the identities

$$\begin{aligned}(x'^2 + x''^2)\cos(\omega t) - 2x'x'' &= (x' - x'')^2\cos^2(\tfrac{1}{2}\omega t) - (x' + x'')^2\sin^2(\tfrac{1}{2}\omega t)\,, \\ \sin(\omega t) &= 2\sin(\tfrac{1}{2}\omega t)\cos(\tfrac{1}{2}\omega t)\,,\end{aligned} \tag{6.2.1}$$

so that

$$\begin{aligned}\langle x', t|x''\rangle = \sqrt{\frac{M\omega}{2\pi\mathrm{i}\hbar\sin(\omega t)}}\,&\mathrm{e}^{\frac{\mathrm{i}}{\hbar}\frac{M\omega}{4}(x' - x'')^2\cot(\frac{1}{2}\omega t)} \\ &\times \mathrm{e}^{-\frac{\mathrm{i}}{\hbar}\frac{M\omega}{4}(x' + x'')^2\tan(\frac{1}{2}\omega t)}\,.\end{aligned} \tag{6.2.2}$$

Now we use, for $\omega t \ll 1$,

$$\cot(\tfrac{1}{2}\omega t) \cong (\tfrac{1}{2}\omega t)^{-1} + \left[\text{terms of relative order } (\omega t)^2\right] ,$$
$$\tan(\tfrac{1}{2}\omega t) \cong \tfrac{1}{2}\omega t + \tfrac{1}{3}(\tfrac{1}{2}\omega t)^3 + \left[\text{terms of relative order } (\omega t)^4\right] . \tag{6.2.3}$$

We keep an extra term in the expansion for the tangent because it is multiplied by $(x'+x'')^2$, which can be large whereas $(x'-x'')^2$ is necessarily small, of order $\hbar t/M = [\hbar/(M\omega)]\omega t$. Indeed, we now shift the origin to a point $\overline{x}$,

$$x' \to \overline{x} + x' , \quad x'' \to \overline{x} + x'' \qquad \text{with} \quad |x'|, |x''| \ll |\overline{x}| ; \tag{6.2.4}$$

then

$$\langle x', t | x'' \rangle \cong \sqrt{\frac{M}{2\pi \mathrm{i}\hbar t}}\, \mathrm{e}^{\frac{\mathrm{i}}{\hbar}\frac{M}{2t}(x'-x'')^2}\, \mathrm{e}^{-\frac{\mathrm{i}}{\hbar}M\omega[\overline{x}^2 + \overline{x}(x'+x'')](\frac{\omega t}{2} + \frac{\omega^3 t^3}{24})} . \tag{6.2.5}$$

Recognize that the potential energy and the force at the point $x = \overline{x}$ are

$$\overline{V} = \tfrac{1}{2}M\omega^2\overline{x}^2 \quad \text{and} \quad \overline{F} = -M\omega^2\overline{x} , \tag{6.2.6}$$

respectively, so that the last exponent in (6.2.5) becomes

$$\begin{aligned} &\cong -\frac{\mathrm{i}}{\hbar}\frac{M\omega^2 t}{2}[\overline{x}^2 + \overline{x}(x'+x'')] - \frac{\mathrm{i}}{\hbar}\frac{M\omega^4 t^3}{24}\overline{x}^2 \\ &= -\frac{\mathrm{i}}{\hbar}\overline{V}t + \frac{\mathrm{i}}{\hbar}\overline{F}t\frac{x'+x''}{2} - \frac{\mathrm{i}}{\hbar}\frac{\overline{F}^2}{24M}t^3 \end{aligned} \tag{6.2.7}$$

and the approximation

$$\langle x', t | x'' \rangle \cong \sqrt{\frac{M}{2\pi \mathrm{i}\hbar t}}\, \mathrm{e}^{\frac{\mathrm{i}}{\hbar}\frac{M}{2t}(x'-x'')^2}\, \mathrm{e}^{\frac{\mathrm{i}}{\hbar}\overline{F}t\frac{x'+x''}{2}}\, \mathrm{e}^{-\frac{\mathrm{i}}{\hbar}\frac{\overline{F}^2}{24M}t^3}\, \mathrm{e}^{-\frac{\mathrm{i}}{\hbar}\overline{V}t} \tag{6.2.8}$$

obtains. Except for the last factor this is the transformation function (6.1.21) for constant force $\overline{F}$. In the latter discussion the potential energy $V = -Fx$ is zero at $x = 0$. Here, at $x = 0$, which is the physical point $\overline{x}$, the potential energy is $\overline{V}$ of (6.2.6).

Recall for a moment that the unitary operator for an infinitesimal time displacement is $1 + \frac{\mathrm{i}}{\hbar}(-\delta t\, H)$. If the system is conservative, H independent of t, then the repetition of this a number of times equal to $t/\delta t$ gives the unitary operator for a finite displacement of t,

$$\lim_{\delta t \to 0}\left(1 - \frac{\mathrm{i}}{\hbar}\delta t\, H\right)^{t/\delta t} = \mathrm{e}^{-\frac{\mathrm{i}}{\hbar}tH} . \tag{6.2.9}$$

This means that

$$\langle x', t | x'' \rangle = \langle x' | \mathrm{e}^{-\frac{\mathrm{i}}{\hbar} tH} | x'' \rangle . \tag{6.2.10}$$

So, if we change H by adding a constant $\overline{V}$, the transformation function changes by the phase factor $\exp\left(-\frac{\mathrm{i}}{\hbar}\overline{V}t\right)$ as observed. Incidentally, although we derived the small-t approximation (6.2.8) for the oscillator potential, all that enters is the potential and the force at the reference point $\overline{x}$,

$$\begin{aligned} V(x) &= \tfrac{1}{2}M\omega^2 x^2 = \tfrac{1}{2}M\omega^2 \left[\overline{x} - (x - \overline{x})\right]^2 \\ &= \overline{V} - \overline{F}(x - \overline{x}) + \tfrac{1}{2}M\omega^2 (x - \overline{x})^2 , \end{aligned} \tag{6.2.11}$$

where

$$\overline{V} = V(\overline{x}) , \qquad \overline{F} = -\frac{\mathrm{d}V}{\mathrm{d}x}(\overline{x}) , \qquad M\omega^2 = \frac{\mathrm{d}^2 V}{\mathrm{d}x^2}(\overline{x}) . \tag{6.2.12}$$

Upon using these equations to define $\overline{V}$, $\overline{F}$, ω^2 for an approximate description of an arbitrary potential energy in the vicinity of position $\overline{x}$, we get a "local-oscillator approximation" for any $V(x)$. Note that we have simplified the pre-factor in (6.2.2) by putting $\sin(\omega t) \to \omega t$. However, the next term in the expansion $\sin x = x - \frac{1}{6}x^3$ can be included, giving an additional factor of

$$1 + \frac{1}{12}\omega^2 t^2 = 1 + \frac{t^2}{12M}\frac{\mathrm{d}^2 V}{\mathrm{d}x^2} .$$

See Problems 6-16 for some more details.

6.3 Harmonic oscillator: Energy eigenvalues

Let's return to the oscillator result (6.1.31) and set $x' = x''$, and integrate. This is

$$\begin{aligned} \int_{-\infty}^{\infty} \mathrm{d}x' \, \langle x', t | x' \rangle &= \int_{-\infty}^{\infty} \mathrm{d}x' \, \langle x' | \mathrm{e}^{-\frac{\mathrm{i}}{\hbar} Ht} | x' \rangle = \mathrm{tr}\left\{ \mathrm{e}^{-\frac{\mathrm{i}}{\hbar} Ht} \right\} \\ &= \int_{-\infty}^{\infty} \mathrm{d}x' \sqrt{\frac{M\omega}{2\pi\mathrm{i}\hbar \sin(\omega t)}} \, \mathrm{e}^{-\frac{\mathrm{i}}{\hbar} M\omega x'^2 \tan(\frac{1}{2}\omega t)} . \end{aligned} \tag{6.3.1}$$

We have done such Gaussian integrals before, but here we have a situation in which the integral does not converge if $\tan(\frac{1}{2}\omega t) = 0$, that is $\omega t = 0, \pm 2\pi, \pm 4\pi, \ldots$. Therefore we regard it as a limit in which $t \to t - \mathrm{i}\epsilon$, with $\epsilon \to +0$. Indeed

$$\tan\left(\tfrac{1}{2}\omega(t - \mathrm{i}\epsilon)\right) = \tan(\tfrac{1}{2}\omega t) - \mathrm{i}\epsilon \frac{\frac{1}{2}\omega}{\cos^2(\frac{1}{2}\omega t)} \tag{6.3.2}$$

gives the convergence factor

$$e^{-\frac{1}{\hbar}\frac{M\omega^2\epsilon}{8}[\cos(\frac{1}{2}\omega t)]^{-2}x'^2} . \tag{6.3.3}$$

Now we get [the identity $2\sin(\omega t)\tan(\frac{1}{2}\omega t) = \left(2\sin(\frac{1}{2}\omega t)\right)^2$ is used]

$$\begin{aligned}\mathrm{tr}\left\{e^{-\frac{i}{\hbar}Ht}\right\} &= \sqrt{\frac{M\omega}{2\pi i\hbar\sin(\omega t)}}\sqrt{\frac{\pi\hbar}{iM\omega\tan(\frac{1}{2}\omega t)}} \\ &= \frac{1}{2i\sin(\frac{1}{2}\omega t)} = \frac{1}{e^{\frac{i}{2}\omega t} - e^{-\frac{i}{2}\omega t}} .\end{aligned} \tag{6.3.4}$$

At this point we must remember that t is actually $t - i\epsilon$, so $|\,e^{-i\omega(t - i\epsilon)}| = e^{-\epsilon\omega} < 1$, for which reason we write

$$\mathrm{tr}\left\{e^{-\frac{i}{\hbar}Ht}\right\} = \frac{e^{-\frac{i}{2}\omega t}}{1 - e^{-i\omega t}} = e^{-\frac{i}{2}\omega t}\sum_{n=0}^{\infty} e^{-in\omega t} = \sum_{n=0}^{\infty} e^{-\frac{i}{\hbar}(n+\frac{1}{2})\hbar\omega t} . \tag{6.3.5}$$

We know that [designating an eigenvalue of H by E, and its multiplicity by $m(E)$]

$$\mathrm{tr}\left\{e^{-\frac{i}{\hbar}Ht}\right\} = \sum_E m(E)\, e^{-\frac{i}{\hbar}Et} \tag{6.3.6}$$

and conclude

$$E = (n + \tfrac{1}{2})\hbar\omega , \qquad n = 0, 1, \ldots , \qquad m(E) = 1 . \tag{6.3.7}$$

Of course, we know all about the $n + \frac{1}{2}$ part. If we take

$$H = \frac{p_x^2}{2M} + \frac{M\omega^2}{2}x^2 \tag{6.3.8}$$

and introduce the dimensionless variables q, p,

$$x = \sqrt{\frac{\hbar}{M\omega}}\,q , \qquad p_x = \sqrt{m\hbar\omega}\,p , \qquad \frac{1}{i\hbar}[x, p_x] = \frac{1}{i}[q, p] = 1 , \tag{6.3.9}$$

into the Hamilton operator (6.1.22) we have

$$H = \hbar\omega\frac{p^2 + q^2}{2} , \qquad \text{where} \quad \left(\frac{p^2 + q^2}{2}\right)' = n + \tfrac{1}{2} \tag{6.3.10}$$

are the eigenvalues found in Section 2.4. Indeed, the stationary-uncertainty states $|n\rangle$ discussed there, then characterized on purely *kinematical* grounds, are now recognized to be the energy eigenstates of a harmonic oscillator, thus acquiring an equivalent *dynamical* characterization. We'll return to this in Section 6.5.

6.4 Free particle and constant force: State density

What happens when we carry out the trace calculation for the free particle and the constant force? If we set $x' = x''$ for the free particle we get from (6.1.11)

$$\langle x', t | x' \rangle = \sqrt{\frac{M}{2\pi i \hbar t}}\,, \tag{6.4.1}$$

which may not seem so informative. It helps to recall that

$$\langle x', t | x'' \rangle = \int_{-\infty}^{\infty} \frac{1}{\sqrt{2\pi\hbar}}\, \mathrm{e}^{\frac{i}{\hbar} x' p'}\, \mathrm{d}p'\, \mathrm{e}^{-\frac{i}{\hbar}\frac{p'^2}{2M} t}\, \frac{1}{\sqrt{2\pi\hbar}}\, \mathrm{e}^{-\frac{i}{\hbar} p' x''}\,, \tag{6.4.2}$$

so that

$$\langle x', t | x' \rangle = \int_{-\infty}^{\infty} \frac{\mathrm{d}p'}{2\pi\hbar}\, \mathrm{e}^{-\frac{i}{\hbar}\frac{p'^2}{2M} t} \tag{6.4.3}$$

which, on integration, gives back the stated result. Now (dropping primes)

$$\operatorname{tr}\left\{ \mathrm{e}^{-\frac{i}{\hbar} H t} \right\} = \int_{-\infty}^{\infty} \mathrm{d}x\, \langle x, t | x \rangle = \int_{-\infty}^{\infty} \frac{\mathrm{d}x\, \mathrm{d}p}{2\pi\hbar}\, \mathrm{e}^{-\frac{i}{\hbar} E t} \tag{6.4.4}$$

with

$$E = \frac{p^2}{2M}\,, \tag{6.4.5}$$

which makes evident that the spectrum of H is $0 \leq E < \infty$. We also see a correspondence between quantum states and classical phase space, an integrated range of $\mathrm{d}x\, \mathrm{d}p$ equal to $2\pi\hbar$ corresponding to one state. [In three dimensions it would be an integrated $(\mathrm{d}\boldsymbol{r})\,(\mathrm{d}\boldsymbol{p})$ equal to $(2\pi\hbar)^3$.] A more precise way to put it considers a large finite range of x ($\int \mathrm{d}x \to L$), and recognizes that, counting positive and negative p together,

$$\mathrm{d}p \to 2\mathrm{d}\sqrt{2ME} = \mathrm{d}E \sqrt{\frac{2M}{E}}\,, \tag{6.4.6}$$

so

$$\operatorname{tr}\left\{ \mathrm{e}^{-\frac{i}{\hbar} H t} \right\}_L = \int_0^{\infty} \mathrm{d}E\, \frac{L}{2\pi\hbar} \sqrt{\frac{2M}{E}}\, \mathrm{e}^{-\frac{i}{\hbar} E t}\,; \tag{6.4.7}$$

evidently,

$$m_L(E) = \frac{L}{2\pi\hbar} \sqrt{\frac{2M}{E}} \tag{6.4.8}$$

is the number of states per unit energy range over distance L, so

$$\frac{\partial}{\partial L} m_L(E) = \frac{1}{2\pi\hbar}\sqrt{\frac{2M}{E}} \tag{6.4.9}$$

is the ***state density***: the number of states per unit energy range and unit x range.

Similarly we rewrite the constant-force result (6.1.21),

$$\begin{aligned}\langle x,t|x\rangle &= \sqrt{\frac{M}{2\pi\mathrm{i}\hbar t}}\, \mathrm{e}^{\frac{\mathrm{i}}{\hbar}Fxt}\, \mathrm{e}^{-\frac{\mathrm{i}}{\hbar}\frac{F^2t^3}{24M}} \\ &= \int_{-\infty}^{\infty} \frac{\mathrm{d}p}{2\pi\hbar}\, \mathrm{e}^{-\frac{\mathrm{i}}{\hbar}\left(\frac{p^2}{2M} - Fx\right)t}\, \mathrm{e}^{-\frac{\mathrm{i}}{\hbar}\frac{F^2t^3}{24M}} \,.\end{aligned} \tag{6.4.10}$$

What do we do with

$$\mathrm{e}^{-\frac{\mathrm{i}}{\hbar}\frac{F^2t^3}{24M}} = \mathrm{e}^{-\frac{\mathrm{i}}{3}\tau^3}\,, \qquad \tau = \frac{1}{2}t\left(\frac{F^2}{M\hbar}\right)^{\frac{1}{3}}\ ? \tag{6.4.11}$$

Write it as a Fourier integral:

$$\mathrm{e}^{-\frac{\mathrm{i}}{3}\tau^3} = \int_{-\infty}^{\infty} \mathrm{d}\sigma\, \mathrm{e}^{\mathrm{i}\sigma\tau}\, \mathrm{Ai}(\sigma)\,, \tag{6.4.12}$$

where

$$\mathrm{Ai}(\sigma) = \int_{-\infty}^{\infty} \frac{\mathrm{d}\tau}{2\pi}\, \mathrm{e}^{-\mathrm{i}\sigma\tau}\, \mathrm{e}^{-\frac{\mathrm{i}}{3}\tau^3} = \frac{1}{\pi}\int_0^{\infty} \mathrm{d}\tau\, \cos\left(\sigma\tau + \tfrac{1}{3}\tau^3\right) \tag{6.4.13}$$

is Airy's* function. It obeys a simple differential equation, easily established by partial integration,

$$\begin{aligned}\frac{\mathrm{d}^2}{\mathrm{d}\sigma^2}\mathrm{Ai}(\sigma) &= -\int_{-\infty}^{\infty} \frac{\mathrm{d}\tau}{2\pi}\, \mathrm{e}^{-\mathrm{i}\sigma\tau}\, \tau^2\, \mathrm{e}^{-\frac{\mathrm{i}}{3}\tau^3} \\ &= -\int_{-\infty}^{\infty} \frac{\mathrm{d}\tau}{2\pi}\, \mathrm{e}^{-\mathrm{i}\sigma\tau}\, \mathrm{i}\frac{\partial}{\partial\tau}\, \mathrm{e}^{-\frac{\mathrm{i}}{3}\tau^3} \\ &= \sigma\, \mathrm{Ai}(\sigma)\,,\end{aligned} \tag{6.4.14}$$

and is normalized to unit integral,

$$\int_{-\infty}^{\infty} \mathrm{d}\sigma\, \mathrm{Ai}(\sigma) = 1\,, \tag{6.4.15}$$

as we see by putting $\tau = 0$ in (6.4.12). Now we find that

$$\begin{aligned}\mathrm{tr}\left\{\mathrm{e}^{-\frac{\mathrm{i}}{\hbar}Ht}\right\} &= \int_{-\infty}^{\infty} \mathrm{d}x\, \langle x,t|x\rangle \\ &= \int \frac{\mathrm{d}x\,\mathrm{d}p}{2\pi\hbar} \int \mathrm{d}\sigma\, \mathrm{Ai}(\sigma)\, \mathrm{e}^{-\frac{\mathrm{i}}{\hbar}\left(\frac{p^2}{2M} - Fx - \frac{\sigma}{2M}(\hbar M F)^{\frac{2}{3}}\right)t}\,.\end{aligned} \tag{6.4.16}$$

---

*Sir George Bidell AIRY (1801–1892)

With

$$E = \frac{p^2}{2M} - Fx - \frac{\sigma}{2M}(\hbar MF)^{\frac{2}{3}} \tag{6.4.17}$$

and varying p, we get

$$\frac{p\mathrm{d}p}{M} = \mathrm{d}E\,, \qquad p = \pm\sqrt{2ME + 2MFx + \sigma(\hbar MF)^{\frac{2}{3}}}\,. \tag{6.4.18}$$

So

$$\begin{aligned}\operatorname{tr}\left\{\mathrm{e}^{-\frac{\mathrm{i}}{\hbar}Ht}\right\} &= \int \mathrm{d}E\,\mathrm{e}^{-\frac{\mathrm{i}}{\hbar}Et}\int \mathrm{d}\sigma\,\mathrm{Ai}(\sigma)\\ &\quad\times\int\frac{\mathrm{d}x}{\pi\hbar}\frac{M}{\sqrt{2ME + 2MFx + \sigma(\hbar MF)^{\frac{2}{3}}}}\,,\end{aligned} \tag{6.4.19}$$

where the integration is over those values of E, σ, and x for which the argument of the square root is positive. The integral (6.4.15) ensures that, for $F = 0$, this is the free-particle trace of (6.4.7), as it must be.

Now for $F > 0$, we take the x integration from where the argument of the square root vanishes to a large positive value L,

$$\begin{aligned}&\int^L \frac{\mathrm{d}x}{\pi\hbar}\frac{M}{\sqrt{2ME + 2MFx + \sigma(\hbar MF)^{\frac{2}{3}}}}\\ &\quad= \int^L \frac{\mathrm{d}x}{\pi\hbar F}\frac{\mathrm{d}}{\mathrm{d}x}\mathrm{d}\sqrt{2ME + 2MFx + \sigma(\hbar MF)^{\frac{2}{3}}}\\ &\quad= \frac{1}{\pi\hbar}\sqrt{\frac{2ML}{F}}\sqrt{1 + \frac{E}{FL} + \frac{\sigma}{2L}\left(\frac{\hbar^2}{MF}\right)^{\frac{2}{3}}}\,.\end{aligned} \tag{6.4.20}$$

With L so large that

$$L \gg \frac{|E|}{F} \quad \text{and} \quad L \gg \left(\frac{\hbar^2}{MF}\right)^{\frac{1}{3}} \tag{6.4.21}$$

and noting that only σ values of order unity contribute significantly (see Problem 6-29 for a refined treatment), this gives

$$\begin{aligned}\operatorname{tr}\left\{\mathrm{e}^{-\frac{\mathrm{i}}{\hbar}Ht}\right\}_L &= \int_{-\infty}^{\infty}\mathrm{d}E\;m_L(E)\,\mathrm{e}^{-\frac{\mathrm{i}}{\hbar}Et}\\ \text{with}\quad m_L(E) &\cong \frac{1}{\pi\hbar}\sqrt{\frac{2ML}{F}}\end{aligned} \tag{6.4.22}$$

and the state density

$$\frac{\partial}{\partial L} m_L(E) \cong \frac{1}{\pi\hbar}\sqrt{\frac{M}{2FL}}\,, \tag{6.4.23}$$

where the range of E is $-\infty < E < \infty$, for unlike $F = 0$, where $1/\sqrt{E}$ appears in (6.4.8) to exclude $E < 0$, there is no restriction on E here.

The nature of the energy spectrum for a constant force is easily understood. Given the Hamiltonian (6.1.12), one gets an equivalent Hamiltonian by the displacement $x \to x + x_0$

$$H = \frac{p^2}{2M} - Fx - Fx_0\,. \tag{6.4.24}$$

Clearly the spectrum of $p^2/(2M) - Fx$ can have no lower limit, for if it did, by choosing $Fx_0 > 0$ one gets an even lower value. Similarly for an alleged upper value.

6.5 Harmonic oscillator: Energy eigenstates

The importance of the Hamiltonian in describing time evolution naturally directs attention not only to the eigenvalues of H but also its eigenvectors. All that information is in the time transformation function. Let the, possibly multiple, eigenvectors for energy E be labeled by E, γ, so that

$$\mathrm{e}^{-\frac{\mathrm{i}}{\hbar}Ht} = \sum_{E,\gamma} |E,\gamma\rangle\, \mathrm{e}^{-\frac{\mathrm{i}}{\hbar}Et} \langle E,\gamma|\,. \tag{6.5.1}$$

Then, for example,

$$\begin{aligned} \langle q', t|q''\rangle &= \langle q'|\,\mathrm{e}^{-\frac{\mathrm{i}}{\hbar}Ht}|q''\rangle \\ &= \sum_{E,\gamma} \underbrace{\langle q'|E,\gamma\rangle}_{=\psi_{E,\gamma}(q')}\, \mathrm{e}^{-\frac{\mathrm{i}}{\hbar}Et}\, \underbrace{\langle E,\gamma|q''\rangle}_{=\psi_{E,\gamma}(q'')^*} \end{aligned} \tag{6.5.2}$$

identifies their q wave functions. Hence knowledge of a time transformation function gives one all the eigenvalues and wave functions of H. As we have seen, the trace is a special application that focuses entirely on eigenvalues:

$$\begin{aligned} \mathrm{tr}\left\{\mathrm{e}^{-\frac{\mathrm{i}}{\hbar}Ht}\right\} &= \int \mathrm{d}q'\, \langle q'|\,\mathrm{e}^{-\frac{\mathrm{i}}{\hbar}Ht}|q'\rangle \\ &= \sum_{E,\gamma} \mathrm{e}^{-\frac{\mathrm{i}}{\hbar}Et} \int \mathrm{d}q'\, |\psi_{E,\gamma}(q')|^2 \\ &= \sum_{E,\gamma} \mathrm{e}^{-\frac{\mathrm{i}}{\hbar}Et} = \sum_{E} m(E)\, \mathrm{e}^{-\frac{\mathrm{i}}{\hbar}Et}\,. \end{aligned} \tag{6.5.3}$$

Let's illustrate this first with the discrete spectrum of the oscillator, starting with [dimensionless variable $q = \sqrt{M\omega/\hbar}\,x$]

$$
\begin{aligned}
\langle q',t|q''\rangle &\equiv \sqrt{\frac{\hbar}{M\omega}}\,\langle x',t|x''\rangle \\
&= \frac{1}{\sqrt{2\pi\mathrm{i}\sin(\omega t)}}\,\mathrm{e}^{\frac{\mathrm{i}}{2\sin(\omega t)}[(q'^2+q''^2)\cos(\omega t)-2q'q'']} \\
&= \frac{\pi^{-\frac{1}{2}}\,\mathrm{e}^{-\frac{\mathrm{i}}{2}\omega t}}{\sqrt{1-\mathrm{e}^{-2\mathrm{i}\omega t}}}\,\mathrm{e}^{\frac{1}{2}(q'^2+q''^2)} \\
&\qquad\times \mathrm{e}^{-(1-\mathrm{e}^{-2\mathrm{i}\omega t})^{-1}(q'^2+q''^2-2q'q''\,\mathrm{e}^{-\mathrm{i}\omega t})}\,.
\end{aligned}
\tag{6.5.4}
$$

The last factor, a Gaussian in two variables, is usefully presented as a double Fourier integral.

We stop for a moment to look at this n-dimensional Euclidean Fourier integral

$$
\int(\mathrm{d}u)\;\mathrm{e}^{-u^{\mathrm{T}}Au+2\mathrm{i}u^{\mathrm{T}}\phi} = \int(\mathrm{d}u)\;\mathrm{e}^{-u^{\mathrm{T}}Au+\mathrm{i}u^{\mathrm{T}}\phi+\mathrm{i}\phi^{\mathrm{T}}u} \tag{6.5.5}
$$

where u and ϕ are real n-component columns ($u_k = u_k^*$, $\phi_k = \phi_k^*$), A is a real, positive, symmetric $n\times n$ matrix ($A_{kl} = A_{lk} = A_{kl}^*$), and their products in the exponent stand for the obvious, namely

$$
u^{\mathrm{T}}Au = \sum_k u_k A_{kl} u_l\,, \qquad u^{\mathrm{T}}\phi = \sum_l u_l\phi_l = \phi^{\mathrm{T}}u\,. \tag{6.5.6}
$$

Complete the square in the exponent, then substitute $u \to u + \mathrm{i}A^{-1}\phi$ to turn the integral into

$$
\begin{aligned}
&\int(\mathrm{d}u)\;\mathrm{e}^{-(u-\mathrm{i}A^{-1}\phi)^{\mathrm{T}}A(u-\mathrm{i}A^{-1}\phi)}\,\mathrm{e}^{-\phi^{\mathrm{T}}A^{-1}\phi} \\
&\qquad = \int(\mathrm{d}u)\;\mathrm{e}^{-u^{\mathrm{T}}Au}\,\mathrm{e}^{-\phi^{\mathrm{T}}A^{-1}\phi}\,.
\end{aligned}
\tag{6.5.7}
$$

Now change the coordinate system to the eigencolumns of matrix A,

$$
\int(\mathrm{d}u)\;\mathrm{e}^{-u^{\mathrm{T}}Au} = \prod_k \int \mathrm{d}u_k\;\mathrm{e}^{-A_k u_k^2} = \prod_k \sqrt{\frac{\pi}{A_k}} = \frac{\pi^{\frac{1}{2}n}}{\sqrt{\det A}}\,. \tag{6.5.8}
$$

The result is

$$
\int(\mathrm{d}u)\;\mathrm{e}^{-u^{\mathrm{T}}Au+2\mathrm{i}u^{\mathrm{T}}\phi} = \frac{\pi^{\frac{1}{2}n}}{\sqrt{\det A}}\,\mathrm{e}^{-\phi^{\mathrm{T}}A^{-1}\phi}\,. \tag{6.5.9}
$$

The two-dimensional Gaussian we are presented with in (6.5.4) is

$$\phi^{\mathrm{T}} A^{-1} \phi = \begin{pmatrix} q' \\ q'' \end{pmatrix}^{\mathrm{T}} \begin{pmatrix} \dfrac{1}{1-\lambda^2} & -\dfrac{\lambda}{1-\lambda^2} \\ -\dfrac{\lambda}{1-\lambda^2} & \dfrac{1}{1-\lambda^2} \end{pmatrix} \begin{pmatrix} q' \\ q'' \end{pmatrix}$$

$$= (q', q'') \begin{pmatrix} 1 & \lambda \\ \lambda & 1 \end{pmatrix}^{-1} \begin{pmatrix} q' \\ q'' \end{pmatrix} \qquad \text{with} \quad \lambda \equiv \mathrm{e}^{-\mathrm{i}\omega t} \tag{6.5.10}$$

and we note that

$$\det \begin{pmatrix} 1 & \lambda \\ \lambda & 1 \end{pmatrix} = 1 - \lambda^2 \,. \tag{6.5.11}$$

Therefore

$$\frac{1}{\sqrt{1-\lambda^2}} \mathrm{e}^{-(q'^2 + q''^2 - 2\lambda q' q'')/(1-\lambda^2)}$$
$$= \frac{1}{\pi} \int_{-\infty}^{\infty} \mathrm{d}u_1 \mathrm{d}u_2 \, \mathrm{e}^{-u_1^2 - u_2^2 - 2\lambda u_1 u_2 + 2\mathrm{i}(u_1 q' + u_2 q'')} \,, \tag{6.5.12}$$

which we rearrange as

$$\frac{1}{\pi} \sum_{n=0}^{\infty} \frac{\left(\frac{1}{2}\lambda\right)^n}{n!} \int_{-\infty}^{\infty} \mathrm{d}u_1 \mathrm{d}u_2 \, (-2\mathrm{i}u_1)^n (-2\mathrm{i}u_2)^n \, \mathrm{e}^{-u_1^2 - u_2^2 + 2\mathrm{i}u_1 q' + 2\mathrm{i}u_2 q''}$$

$$= \sum_{n=0}^{\infty} \frac{\left(\frac{1}{2}\lambda\right)^n}{n!} \left(-\frac{\partial}{\partial q'}\right)^n \left(-\frac{\partial}{\partial q''}\right)^n \underbrace{\frac{1}{\pi} \int_{-\infty}^{\infty} \mathrm{d}u_1 \mathrm{d}u_2 \, \mathrm{e}^{-u_1^2 - u_2^2 + 2\mathrm{i}u_1 q' + 2\mathrm{i}u_2 q''}}_{= \mathrm{e}^{-q'^2 - q''^2}}$$

$$= \sum_{n=0}^{\infty} \lambda^n \frac{1}{2^n n!} \mathrm{H}_n(q') \mathrm{H}_n(q'') \, \mathrm{e}^{-q'^2 - q''^2} \,, \tag{6.5.13}$$

where we recognize the Hermite polynomials of (2.5.4). In summary,

$$\langle q', t | q'' \rangle = \sum_{n=0}^{\infty} \psi_n(q') \, \mathrm{e}^{-\frac{\mathrm{i}}{\hbar} E_n t} \psi_n(q'')^* \tag{6.5.14}$$

where

$$E_n = \hbar\omega \left(n + \tfrac{1}{2}\right) \tag{6.5.15}$$

are the energy eigenvalues of (6.3.7), and

$$\psi_n(q') = \frac{\pi^{-\frac{1}{4}}}{\sqrt{2^n n!}} \mathrm{H}_n(q') \, \mathrm{e}^{-\frac{1}{2} q'^2} \,, \qquad \psi_n(x') = \left(\frac{M\omega}{\hbar}\right)^{\frac{1}{4}} \psi_n(q') \tag{6.5.16}$$

are the wave functions of (2.5.5).

All this fuss to get back to a known result! Surely there are simpler ways? Indeed there are, not surprisingly involving non-Hermitian operators. But before going into that, in Section 7.1, let's look at the wave functions for a free particle and a particle under constant force.

6.6 Free particle and constant force: Energy eigenstates

For the free particle we know that [this is (6.4.2)]

$$\langle x,t|x'\rangle = \int \frac{\mathrm{e}^{\frac{\mathrm{i}}{\hbar}xp}}{\sqrt{2\pi\hbar}}\,\mathrm{d}p\,\mathrm{e}^{-\frac{\mathrm{i}}{\hbar}\frac{p^2}{2M}t}\,\frac{\mathrm{e}^{-\frac{\mathrm{i}}{\hbar}px'}}{\sqrt{2\pi\hbar}} \tag{6.6.1}$$

which we now rewrite in terms of $E = p^2/(2M)$, $\mathrm{d}E = (p/M)\mathrm{d}p$. We have only to recognize that for a given E, there are two values of p, $p = \pm\sqrt{2ME}$. (Henceforth we use p to mean the positive value.) That gives us the form

$$\langle x,t|x'\rangle = \sum_{\gamma=\pm}\int_0^\infty \mathrm{d}E\,\psi_{E,\gamma}(x)\,\mathrm{e}^{-\frac{\mathrm{i}}{\hbar}Et}\psi_{E,\gamma}(x')^* \tag{6.6.2}$$

where

$$\begin{aligned}\psi_{E,\pm}(x) &= \frac{1}{\sqrt{2\pi\hbar}}\sqrt{\frac{\mathrm{d}p}{\mathrm{d}E}}\,\mathrm{e}^{\pm\frac{\mathrm{i}}{\hbar}xp}\\ &= \frac{(2E/M)^{-\frac{1}{4}}}{\sqrt{2\pi\hbar}}\,\mathrm{e}^{\pm\frac{\mathrm{i}}{\hbar}x\sqrt{2ME}}\end{aligned} \tag{6.6.3}$$

represent the two states of the same energy, in which the particle is moving to the right (+) or the left (−).

From Problems 6-6 we know, for constant force $F > 0$, that

$$\langle p,t|p'\rangle = \mathrm{e}^{-\frac{\mathrm{i}}{\hbar}\frac{p^3}{6MF}}\,\delta(p-p'-Ft)\,\mathrm{e}^{\frac{\mathrm{i}}{\hbar}\frac{p'^3}{6MF}}\,. \tag{6.6.4}$$

Now

$$\delta(p-p'-Ft) = \int_{-\infty}^{\infty}\frac{\mathrm{d}E}{2\pi\hbar F}\,\mathrm{e}^{\frac{\mathrm{i}}{\hbar}\frac{E}{F}(p-p'-Ft)} \tag{6.6.5}$$

so that

$$\langle p,t|p'\rangle = \int_{-\infty}^{\infty}\mathrm{d}E\,\psi_E(p)\,\mathrm{e}^{-\frac{\mathrm{i}}{\hbar}Et}\,\psi_E(p')^* \tag{6.6.6}$$

where

$$\psi_E(p) = \frac{1}{\sqrt{2\pi\hbar F}}\,\mathrm{e}^{\frac{\mathrm{i}}{\hbar F}\left(Ep-\frac{p^3}{6M}\right)}\,. \tag{6.6.7}$$

What happens as $F \to 0$? Considered as a function of p, $\psi_E(p)$ oscillates infinitely rapidly in this limit, *except* if

$$\frac{\mathrm{d}}{\mathrm{d}p}\left(Ep - \frac{p^3}{6M}\right) = E - \frac{p^2}{2M} = 0\,, \tag{6.6.8}$$

which can only occur for $E > 0$. This gives us back the free particle spectrum. We leave to Problem 6-31 the task of verifying that the expected wave functions emerge.

Given $\psi_E(p)$, we get $\psi_E(x)$ as

$$\begin{aligned}\psi_E(x) &= \int_{-\infty}^{\infty} \frac{1}{\sqrt{2\pi\hbar}}\, \mathrm{e}^{\frac{\mathrm{i}}{\hbar}xp}\, \mathrm{d}p\, \frac{1}{\sqrt{2\pi\hbar F}}\, \mathrm{e}^{\frac{\mathrm{i}}{\hbar F}(Ep - \frac{p^3}{6M})} \\ &= \frac{1}{2\pi\hbar}\frac{1}{\sqrt{F}}\int_{-\infty}^{\infty} \mathrm{d}p\, \mathrm{e}^{\frac{\mathrm{i}}{\hbar F}(E+Fx)p}\, \mathrm{e}^{-\frac{\mathrm{i}}{\hbar F}\frac{p^3}{6M}}\,. \end{aligned}\tag{6.6.9}$$

The variable change

$$p = (2M\hbar F)^{\frac{1}{3}}\tau\,, \tag{6.6.10}$$

gives

$$\begin{aligned}\psi_E(x) &= \left(\frac{2M}{\hbar^2\sqrt{F}}\right)^{\frac{1}{3}} \int_{-\infty}^{\infty} \frac{\mathrm{d}\tau}{2\pi}\, \mathrm{e}^{-\mathrm{i}\,(2MF/\hbar^2)^{\frac{1}{3}}\,(-x - E/F)\,\tau}\, \mathrm{e}^{-\frac{\mathrm{i}}{3}\tau^3} \\ &= \left(\frac{2M}{\hbar^2\sqrt{F}}\right)^{\frac{1}{3}} \mathrm{Ai}\big((2MF/\hbar^2)^{\frac{1}{3}}(-x - E/F)\big)\,. \end{aligned}\tag{6.6.11}$$

They are orthonormal and complete, as they should be (see Problem 6-25).

Apart from the normalization constant, we could have recognized the appearance of the Airy function from the Schrödinger differential equation for an energy eigenstate,

$$\langle q'|H(q,p)|E,\gamma\rangle = H\Big(q', \frac{\hbar}{\mathrm{i}}\frac{\partial}{\partial q'}\Big)\langle q'|E,\gamma\rangle = \langle q'|E,\gamma\rangle E \tag{6.6.12}$$

(frequently called the time-independent Schrödinger equation) or

$$\left[E - H\Big(q', \frac{\hbar}{\mathrm{i}}\frac{\partial}{\partial q'}\Big)\right]\psi_{E,\gamma}(q') = 0\,, \tag{6.6.13}$$

which here is

$$\left(E + \frac{\hbar^2}{2M}\frac{\mathrm{d}^2}{\mathrm{d}x^2} + Fx\right)\psi_E(x) = 0\,. \tag{6.6.14}$$

Comparison with [this is (6.4.14)]

$$\left(\frac{\mathrm{d}^2}{\mathrm{d}\sigma^2} - \sigma\right) \mathrm{Ai}(\sigma) = 0 \tag{6.6.15}$$

indeed shows that

$$\sigma = \left(2MF/\hbar^2\right)^{\frac{1}{3}} (-x - E/F) \tag{6.6.16}$$

should be the argument of the Airy function in (6.6.11).

6.7 Constant force: Asymptotic wave functions

It is clear in a general way from the differential equation (6.6.15) that $\mathrm{Ai}(\sigma)$ will be oscillatory for σ negative and sufficiently large, but non-oscillatory and of exponential behavior if σ is positive and sufficiently large. We want to be a bit more precise about these asymptotic behaviors.

Consider first large negative σ values, $-\sigma \gg 1$. In

$$\mathrm{Ai}(\sigma) = \frac{1}{\pi} \operatorname{Re}\left(\int_0^\infty \mathrm{d}\tau \, \mathrm{e}^{\mathrm{i}(|\sigma|\tau - \frac{1}{3}\tau^3)}\right) \tag{6.7.1}$$

there is a stationary value of the phase:

$$0 = \frac{\mathrm{d}}{\mathrm{d}\tau}\left(|\sigma|\tau - \frac{1}{3}\tau^3\right) = |\sigma| - \tau^2 \,, \qquad \text{so that} \quad \tau = \sqrt{|\sigma|} = \sqrt{-\sigma} \,, \tag{6.7.2}$$

which in fact is a maximum value

$$\frac{1}{2}\frac{\mathrm{d}^2}{\mathrm{d}\tau^2}\left(|\sigma|\tau - \frac{1}{3}\tau^3\right) = -\tau \cong -\sqrt{|\sigma|} \,. \tag{6.7.3}$$

Expanding about this value up to quadratic terms, that is:

$$\begin{aligned} \tau \to \sqrt{-\sigma} + \tau \,, \quad |\sigma|\tau - \tfrac{1}{3}\tau^3 &\to (-\sigma)\left(\sqrt{-\sigma} + \tau\right) - \tfrac{1}{3}\left(\sqrt{-\sigma} + \tau\right)^3 \\ &\cong \tfrac{2}{3}(-\sigma)^{\frac{3}{2}} - \sqrt{-\sigma}\,\tau^2 \,, \end{aligned} \tag{6.7.4}$$

gives

$$\begin{aligned} \mathrm{Ai}(\sigma) &\cong \frac{1}{\pi} \operatorname{Re}\left(\int_{-\sqrt{-\sigma}\cong-\infty}^{\infty} \mathrm{d}\tau \, \mathrm{e}^{\mathrm{i}\frac{2}{3}(-\sigma)^{\frac{3}{2}}} \, \mathrm{e}^{-\mathrm{i}\sqrt{-\sigma}\tau^2}\right) \\ &\cong \frac{1}{\pi} \operatorname{Re}\left(\sqrt{\frac{\pi}{\mathrm{i}\sqrt{-\sigma}}} \, \mathrm{e}^{\mathrm{i}\frac{2}{3}(-\sigma)^{\frac{3}{2}}}\right) \\ &= \pi^{-\frac{1}{2}} (-\sigma)^{-\frac{1}{4}} \cos\left(\frac{2}{3}(-\sigma)^{\frac{3}{2}} - \frac{\pi}{4}\right) \,, \end{aligned} \tag{6.7.5}$$

which is the leading term in the asymptotic expansion for $-\sigma \gg 1$.

For $\sigma = 0$,

$$\mathrm{Ai}(0) = \frac{1}{\pi}\,\mathrm{Re}\left(\int_0^\infty d\tau\, e^{-\frac{i}{3}\tau^3}\right) \tag{6.7.6}$$

or, with the substitution $\tau = 3^{\frac{1}{3}}\, e^{-\frac{\pi i}{6}} y^{\frac{1}{3}}$,

$$\mathrm{Ai}(0) = \frac{1}{\pi}\mathrm{Re}\Bigl(3^{-\frac{2}{3}}\, e^{-\frac{\pi i}{6}} \underbrace{\int_0^\infty dy\, y^{-\frac{2}{3}}\, e^{-y}}_{=(-\frac{2}{3})! = 2.678\,939}\Bigr)$$

$$= \frac{3^{-\frac{1}{6}}}{2\pi}\left(-\frac{2}{3}\right)! = 0.3550\,. \tag{6.7.7}$$

Now we turn to large positive σ values, $\sigma \gg 1$. In

$$\mathrm{Ai}(\sigma) = \frac{1}{\pi}\,\mathrm{Re}\left(\int_0^\infty d\tau\; e^{-i(\sigma\tau + \frac{1}{3}\sigma^3)}\right) \tag{6.7.8}$$

it's well to notice that, for convergence at $\tau = \infty$, τ should have a negative imaginary part, i. e., the integration path should run below the real axis, such as

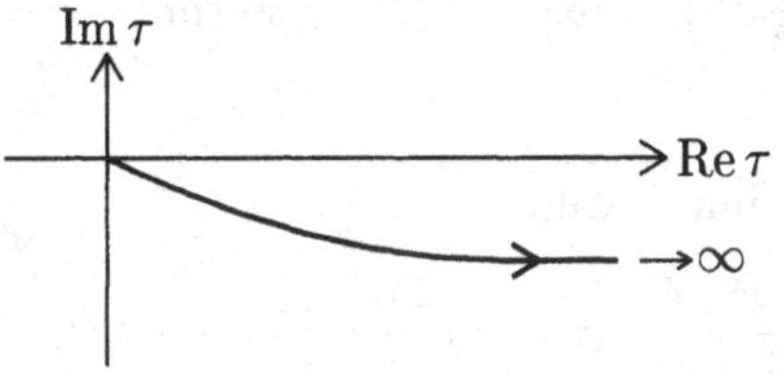

for example. Next, notice that the exponent still has a stationary point:

$$\frac{d}{d\tau}\left(\sigma\tau + \frac{1}{3}\tau^3\right) = \sigma + \tau^2 = 0 \quad \text{holds for} \quad \tau = -i\sqrt{\sigma}\,. \tag{6.7.9}$$

So let's choose this contour of integration:

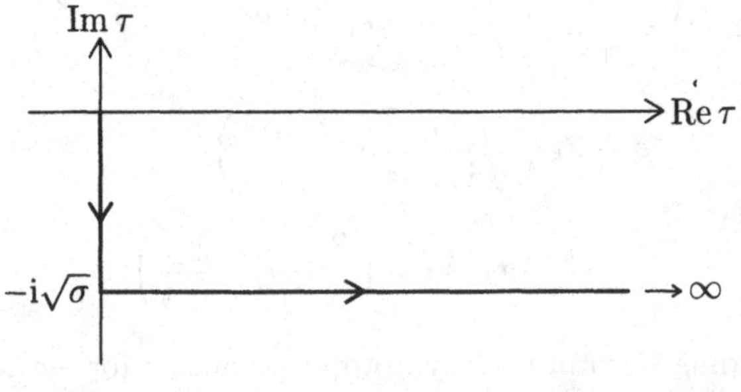

The first part, which we parameterize by $\tau = -\mathrm{i}v$ with $v : 0 \to \sqrt{\sigma}$, does not contribute:

$$\frac{1}{\pi}\,\mathrm{Re}\left(\int_0^{\sqrt{\sigma}} (-\mathrm{i})\mathrm{d}v\; \mathrm{e}^{-(\sigma v - \frac{1}{3}v^3)}\right) = 0\,. \tag{6.7.10}$$

The second part, where $\tau = -\mathrm{i}\sqrt{\sigma} + u$ with $u : 0 \to \infty$, is, for $\sigma \gg 1$,

$$\begin{aligned}\mathrm{Ai}(\sigma) &\cong \frac{1}{\pi}\,\mathrm{Re}\left(\int_0^\infty \mathrm{d}u\; \mathrm{e}^{-\frac{2}{3}\sigma^{\frac{3}{2}}}\,\mathrm{e}^{-\sqrt{\sigma}u^2}\right)\\ &= \frac{1}{\pi}\,\mathrm{e}^{-\frac{2}{3}\sigma^{\frac{3}{2}}}\underbrace{\int_0^\infty \mathrm{d}u\; \mathrm{e}^{-\sqrt{\sigma}u^2}}_{=\frac{1}{2}\sqrt{\pi/\sqrt{\sigma}}}\\ &= \frac{1}{2}\pi^{-\frac{1}{2}}\sigma^{-\frac{1}{4}}\,\mathrm{e}^{-\frac{2}{3}\sigma^{\frac{3}{2}}}\,,\end{aligned} \tag{6.7.11}$$

which is the leading term in the asymptotic expansion for $\sigma \gg 1$.

Now if we recall (6.6.16) and introduce the meaning of σ, the asymptotic forms of $\psi_E(x)$ become

$$\psi_E(x) \cong \pi^{-\frac{1}{2}}\frac{(2M/\hbar^2)^{\frac{1}{4}}}{(E+Fx)^{\frac{1}{4}}}\cos\Bigl(\frac{2}{3}(2M/\hbar^2)^{\frac{1}{2}}\frac{1}{F}(E+Fx)^{\frac{3}{2}} - \frac{\pi}{4}\Bigr)$$
$$\text{for}\quad E + Fx \gg \left[\hbar^2F^2/(2M)\right]^{\frac{1}{3}}\,, \tag{6.7.12}$$

and

$$\psi_E(x) \cong \frac{1}{2}\pi^{-\frac{1}{2}}\frac{(2M/\hbar^2)^{\frac{1}{4}}}{|E+Fx|^{\frac{1}{4}}}\,\mathrm{e}^{-\frac{2}{3}(2M/\hbar^2)^{\frac{1}{2}}\frac{1}{F}|E+Fx|^{\frac{3}{2}}}$$
$$\text{for}\quad -(E+Fx) \gg \left[\hbar^2F^2/(2M)\right]^{\frac{1}{3}}\,. \tag{6.7.13}$$

A simpler and more general presentation of these results comes from the recognition that $E+Fx = E-(-Fx)$ is the *classical* kinetic energy $p^2/(2M)$, so that

$$p(x) = \sqrt{2M(E+Fx)} \tag{6.7.14}$$

is the classical momentum at position x for energy E, and that

$$\frac{\mathrm{d}}{\mathrm{d}x}\left[\frac{2}{3F}(E+Fx)^{\frac{3}{2}}\right] = (E+Fx)^{\frac{1}{2}} = \frac{p(x)}{\sqrt{2M}}\,, \tag{6.7.15}$$

or

$$\sqrt{2M}\,\frac{2}{3F}(E+Fx)^{\frac{3}{2}} = \int_{-E/F}^{x} \mathrm{d}x'\,p(x')\,. \tag{6.7.16}$$

All this applies for $E + Fx > 0$, and

$$E + Fx > 0: \quad \psi_E(x) \cong \sqrt{\frac{2M}{\pi\hbar p(x)}} \cos\left(\frac{1}{\hbar}\int_{p=0}^{x} \mathrm{d}x'\, p(x') - \frac{\pi}{4}\right) \qquad (6.7.17)$$

repeats (6.7.12).

With $E + Fx < 0$, $p(x)$ is imaginary; this region is classically forbidden. But the wave function does not vanish, although it decreases rapidly as one penetrates into the region. With

$$|p(x)| = \sqrt{2M(-E - Fx)} \qquad (6.7.18)$$

we have

$$\int_{p=0}^{x} \mathrm{d}x'\, |p(x')| = (2M)^{\frac{1}{2}} \int_{-E/F}^{x} \mathrm{d}x'\, (-E - Fx)^{\frac{1}{2}}$$
$$= -(2M)^{\frac{1}{2}} \frac{2}{3F}(-E - Fx)^{\frac{3}{2}} \qquad (6.7.19)$$

or

$$E + Fx < 0: \quad \psi_E(x) \cong \frac{1}{2}\sqrt{\frac{2M}{\pi\hbar p(x)}}\, \mathrm{e}^{\frac{1}{\hbar}\int_{p=0}^{x} \mathrm{d}x'\, |p(x')|}\,, \qquad (6.7.20)$$

which repeats (6.7.13).

For the physical interpretation of these asymptotic wave functions, we first write

$$\psi_E \cong \frac{1}{2}\left(\frac{2M}{\pi\hbar p}\right)^{\frac{1}{2}} \Big[\underbrace{\mathrm{e}^{-\mathrm{i}\left(\frac{1}{\hbar}\int_{p=0}^{x} \mathrm{d}x'\, p(x') - \frac{\pi}{4}\right)}}_{\text{incident}} + \underbrace{\mathrm{e}^{\mathrm{i}\left(\frac{1}{\hbar}\int_{p=0}^{x} \mathrm{d}x'\, p(x') - \frac{\pi}{4}\right)}}_{\text{reflected}} \Big]$$

$$\text{for} \quad x > -E/F\,,$$

$$\psi_E \cong \frac{1}{2}\left(\frac{2M}{\pi\hbar p}\right)^{\frac{1}{2}} \underbrace{\mathrm{e}^{\frac{1}{\hbar}\int_{p=0}^{x} \mathrm{d}x'\, |p(x')|}}_{\text{transmitted}} \qquad \text{for} \quad x < -E/F\,. \qquad (6.7.21)$$

Matters are as sketched in this figure:

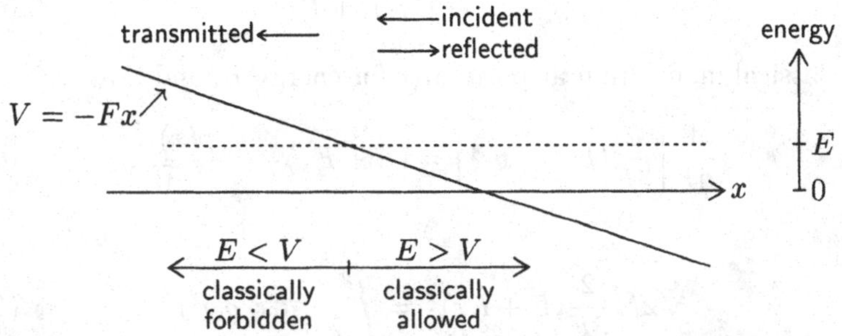

The incident wave moves to the left in the classically allowed region; it falls upon the boundary of the region, where $E - V(x) = E + Fx = 0$. There, a wave of equal amplitude is generated, moving back into the allowed region: the reflected wave. And, an exponentially attenuating wave moves on into the classically forbidden region: the transmitted wave.

The transmitted and reflected wave functions, $\psi_{\text{trans.}}$ and $\psi_{\text{refl.}}$, are both obtained from the incident wave function

$$\psi_{\text{inc.}}(x) = \frac{1}{2}\left(\frac{2M}{\pi\hbar p(x)}\right)^{\frac{1}{2}} \mathrm{e}^{-\mathrm{i}\left(\frac{1}{\hbar}\int_{p=0}^{x}\mathrm{d}x'\, p(x') - \frac{\pi}{4}\right)} \tag{6.7.22}$$

by simple changes of $p(x)$. The replacement

$$p(x) \to \mathrm{e}^{\frac{\pi\mathrm{i}}{2}}|p(x)| = \mathrm{i}|p(x)| \tag{6.7.23}$$

turns $\psi_{\text{inc.}}$ into $\psi_{\text{trans.}}$,

$$\psi_{\text{inc.}}(x) \to \frac{1}{2}\left(\frac{2M}{\pi\hbar p(x)}\right)^{\frac{1}{2}} \mathrm{e}^{\int_{p=0}^{x}\mathrm{d}x'\,|p(x')|} = \psi_{\text{trans.}}(x)\,, \tag{6.7.24}$$

and the replacement

$$p(x) \to \mathrm{e}^{\pi\mathrm{i}}p(x) = -p(x) \tag{6.7.25}$$

turns it into $\psi_{\text{refl.}}$,

$$\psi_{\text{inc.}}(x) \to \frac{1}{2}\left(\frac{2M}{\pi\hbar p(x)}\right)^{\frac{1}{2}} \mathrm{e}^{\mathrm{i}\left(\frac{1}{\hbar}\int_{p=0}^{x}\mathrm{d}x'\, p(x') - \frac{\pi}{4}\right)} = \psi_{\text{refl.}}\,. \tag{6.7.26}$$

Note that, owing to the phases $\pm\pi/4$ in the incident and reflected wave functions, no additional phase factors are needed. These rules for turning $\psi_{\text{inc.}}$ into $\psi_{\text{trans.}}$ and $\psi_{\text{refl.}}$ are known as connection formulas.

6.8 WKB approximation

The remark has been made that these results are more general than the particular potential $V(x) = -Fx$. To justify this, consider the Schrödinger equation

$$\left(E + \frac{\hbar^2}{2M}\frac{\mathrm{d}^2}{\mathrm{d}x^2} - V(x)\right)\psi_E(x) = 0\,, \tag{6.8.1}$$

or

$$\left(\frac{\mathrm{d}^2}{\mathrm{d}x^2} + \frac{1}{\hbar^2}\left[p(x)\right]^2\right)\psi_E(x) = 0\,, \qquad \left[p(x)\right]^2 = 2M\bigl(E - V(x)\bigr)\,; \tag{6.8.2}$$

$\pm p(x)$ being the classical momentum at x. In the light of what we have seen we look for a solution of the form

$$\psi_E(x) = A(x)\,\mathrm{e}^{\mathrm{i}\phi(x)} \tag{6.8.3}$$

with real amplitude $A(x)$ and real phase $\phi(x)$. For convenience, we denote differentiation with respect to x by primes and have

$$\begin{aligned} \frac{\mathrm{d}}{\mathrm{d}x}\psi_E(x) &\equiv \psi_E' = A'\,\mathrm{e}^{\mathrm{i}\phi} + A\mathrm{i}\phi'\,\mathrm{e}^{\mathrm{i}\phi}\,, \\ \frac{\mathrm{d}^2}{\mathrm{d}x^2}\psi_E(x) &\equiv \psi_E'' = A''\,\mathrm{e}^{\mathrm{i}\phi} + 2A'\mathrm{i}\phi'\,\mathrm{e}^{\mathrm{i}\phi} + A\mathrm{i}\phi''\,\mathrm{e}^{\mathrm{i}\phi} - A\phi'^2\,\mathrm{e}^{\mathrm{i}\phi}\,, \end{aligned} \tag{6.8.4}$$

so that

$$A'' - A\phi'^2 + \frac{1}{\hbar^2}p^2 A = 0\,, \qquad 2A'\phi' + A\phi'' = 0\,. \tag{6.8.5}$$

The second equation, multiplied by A, says that

$$\frac{\mathrm{d}}{\mathrm{d}x}\left(A^2\phi'\right) = 0 \qquad \text{or} \quad A = \frac{C}{\sqrt{\phi}} \tag{6.8.6}$$

with integration constant C [clearly, "constant" means "independent of x" here; C may (and will) depend on E]. Differentiation of

$$\frac{A'}{A} = -\frac{1}{2}\frac{\phi''}{\phi'} \tag{6.8.7}$$

gives

$$\frac{A''}{A} = \left(\frac{A'}{A}\right)' + \left(\frac{A'}{A}\right)^2 = -\frac{1}{2}\left(\frac{\phi''}{\phi'}\right)' + \frac{1}{4}\left(\frac{\phi''}{\phi'}\right)^2 . \tag{6.8.8}$$

and then we have that

$$\phi'^2 = \frac{1}{\hbar^2}p^2 - \frac{1}{2}\frac{\mathrm{d}}{\mathrm{d}x}\frac{\phi''}{\phi'} + \frac{1}{4}\left(\frac{\phi''}{\phi'}\right)^2 . \tag{6.8.9}$$

This is useful only as an approximation, beginning with $(p^2 > 0)$

$$\underline{0^{\text{th}}\ \text{order}}: \qquad \phi' = \pm\frac{1}{\hbar}p \tag{6.8.10}$$

and going on to

$$\begin{aligned} \underline{1^{\text{st}}\ \text{order}}: \qquad \phi' &= \pm\sqrt{\left(\frac{1}{\hbar}p\right)^2 - \frac{1}{2}\frac{\mathrm{d}}{\mathrm{d}x}\left(\frac{p'}{p}\right) + \frac{1}{4}\left(\frac{p'}{p}\right)^2} \\ &\cong \pm\left[\frac{1}{\hbar}p - \frac{1}{4}\frac{\hbar}{p}\frac{\mathrm{d}}{\mathrm{d}x}\frac{p'}{p} + \frac{1}{8}\frac{\hbar}{p}\left(\frac{p'}{p}\right)^2\right] . \end{aligned} \tag{6.8.11}$$

In terms of

$$\lambda(x) = \frac{\hbar}{p(x)}\,, \tag{6.8.12}$$

[sometimes called the local (reduced) de Broglie wavelength] this is

$$\underline{1^{\text{st}}\text{ order}}: \qquad \phi' \cong \pm\frac{1}{\hbar}\left[p - \frac{1}{4}\left(\lambda\frac{\mathrm{d}}{\mathrm{d}x}\right)^2 p + \frac{1}{8p}\left(\lambda\frac{\mathrm{d}p}{\mathrm{d}x}\right)^2\right]. \tag{6.8.13}$$

Thus the leading term is a good approximation wherever the momentum changes only by a small fraction in a distance of λ, that is: where

$$\left|\lambda(x)\frac{\mathrm{d}p(x)}{\mathrm{d}x}\right| \ll p(x) \tag{6.8.14}$$

is obeyed. That implies

$$\left|\lambda\frac{\mathrm{d}}{\mathrm{d}x}V\right| \ll |E - V| \tag{6.8.15}$$

or

$$\frac{\hbar^2}{2M}\frac{\left(\mathrm{d}V/\mathrm{d}x\right)^2}{|E-V|^3} \ll 1\,. \tag{6.8.16}$$

Obviously this fails in the vicinity of any $x = x_0$ for which $E - V = 0$. In the situation where, for small $x - x_0$,

$$E - V(x) \cong F(x - x_0) \quad \text{with } F = -\frac{\mathrm{d}V}{\mathrm{d}x}(x_0) \tag{6.8.17}$$

the approximation is good for

$$|x - x_0| \gg \left(\frac{\hbar^2}{MF}\right)^{\frac{1}{3}}\,, \tag{6.8.18}$$

($F > 0$ is assumed) provided of course that

$$|x - x_0|\left|\frac{\mathrm{d}^2V}{\mathrm{d}x^2}\right| \ll F\,, \tag{6.8.19}$$

which requires

$$\left|\frac{\mathrm{d}^2V}{\mathrm{d}x^2}(x_0)\right| \ll F\left(\frac{MF}{\hbar^2}\right)^{\frac{1}{3}}\,. \tag{6.8.20}$$

Under these circumstances we have

$$\psi \propto \frac{1}{\sqrt{p}} \mathrm{e}^{\pm \frac{\mathrm{i}}{\hbar} \int^x \mathrm{d}x'\, p(x')} \tag{6.8.21}$$

in the classically allowed region where $\left[p(x)\right]^2 > 0$, and the previous discussion, with the simple connection formulas among $\psi_{\text{inc.}}$, $\psi_{\text{refl.}}$, $\psi_{\text{trans.}}$, applies and gives back just the results found in the special example $V(x) = -Fx$, apart from the overall normalization constant.

In the quantum literature this approximation method is identified with the initials WKB, thereby referring to work of 1926 by Wentzel,* Kramers,† and Brillouin,‡ although it had already been given explicitly by Jeffreys§ in 1924, and had a long history stretching back to Green¶ and Carlini‖ (work of 1837 and 1817, respectively).

The asymptotic forms work only sufficiently far to the left and right of the point x_0 where $E - V = 0$. It would be nice to have a unified approximation that includes the region of transition between classically allowed [$x > x_0$, $p(x)$ positive] and forbidden [$x < x_0$, $p(x)$ imaginary] regions. For that, return to the Schrödinger equation (6.8.2) and define

$$\frac{1}{\hbar}\int_{x_0}^{x} \mathrm{d}x'\, p(x') = \frac{2}{3}(-\sigma)^{\frac{3}{2}} \qquad \text{with} \qquad \begin{cases} \sigma(x) < 0 & \text{if } x > x_0\,, \\ \sigma(x) > 0 & \text{if } x < x_0\,, \end{cases} \tag{6.8.22}$$

so

$$\frac{1}{\hbar}\mathrm{d}x\, p = -\sqrt{-\sigma}\, \mathrm{d}\sigma\,. \tag{6.8.23}$$

Now, introducing a scaled momentum $\chi(x)$ in accordance with

$$\chi \equiv p/\sqrt{-\sigma}\,, \qquad p^2 = -\sigma\chi^2\,, \tag{6.8.24}$$

we have

$$-\hbar\frac{\mathrm{d}}{\mathrm{d}x} = \chi\frac{\mathrm{d}}{\mathrm{d}\sigma}\,, \tag{6.8.25}$$

and the Schrödinger equation appears as

$$\left(\chi\frac{\mathrm{d}}{\mathrm{d}\sigma}\chi\frac{\mathrm{d}}{\mathrm{d}\sigma} - \sigma\chi^2\right)\psi = 0 \qquad \text{or} \qquad \left(\frac{1}{\chi}\frac{\mathrm{d}}{\mathrm{d}\sigma}\chi\frac{\mathrm{d}}{\mathrm{d}\sigma} - \sigma\right)\psi = 0\,. \tag{6.8.26}$$

Write

$$\psi = \chi^{-\frac{1}{2}}\overline{\psi} \tag{6.8.27}$$

---

*Gregor WENTZEL (1898–1978) †Hendrik Anthony KRAMERS (1894–1952) ‡Léon BRILLOUIN (1889–1969) §Sir Harold JEFFREYS (1891–1989) ¶George GREEN (1793–1841) ‖Francesco CARLINI (1783–1862)

and use (primes denote differentiation with respect to σ)

$$\chi^{\mp\frac{1}{2}}\frac{\mathrm{d}}{\mathrm{d}\sigma}\chi^{\pm\frac{1}{2}} = \frac{\mathrm{d}}{\mathrm{d}\sigma} \pm \frac{1}{2}\frac{\chi'}{\chi}\,,$$

$$\begin{aligned}\frac{1}{\chi}\frac{\mathrm{d}}{\mathrm{d}\sigma}\chi\frac{\mathrm{d}}{\mathrm{d}\sigma} &= \chi^{-\frac{1}{2}}\left(\frac{\mathrm{d}}{\mathrm{d}\sigma}+\frac{1}{2}\frac{\chi'}{\chi}\right)\left(\frac{\mathrm{d}}{\mathrm{d}\sigma}-\frac{1}{2}\frac{\chi'}{\chi}\right)\chi^{\frac{1}{2}} \\ &= \chi^{-\frac{1}{2}}\left[\left(\frac{\mathrm{d}}{\mathrm{d}\sigma}\right)^2 - \frac{1}{2}\left(\frac{\chi'}{\chi}\right)' - \frac{1}{4}\left(\frac{\chi'}{\chi}\right)^2\right]\chi^{\frac{1}{2}}\end{aligned} \tag{6.8.28}$$

to get

$$\left[\frac{\mathrm{d}^2}{\mathrm{d}\sigma^2} - \sigma - \frac{1}{2}\left(\frac{\chi'}{\chi}\right)' - \frac{1}{4}\left(\frac{\chi'}{\chi}\right)^2\right]\overline{\psi} = 0\,. \tag{6.8.29}$$

As before, $(\chi'/\chi)'$ and $(\chi'/\chi)^2$ will be relatively small (compared to σ) in the asymptotic regions. But now, near x_0, where

$$\begin{aligned} p(x) &\cong \begin{cases} \sqrt{2MF(x-x_0)} & \text{if } x \lesssim x_0\,, \\ \mathrm{i}\sqrt{2MF(x_0-x)} & \text{if } x \gtrsim x_0\,, \end{cases} \\ \sigma(x) &\cong (2MF/\hbar^2)^{\frac{1}{3}}(x_0 - x)\,, \\ \chi(x) &\cong (2\hbar MF)^{\frac{1}{3}}\,, \end{aligned} \tag{6.8.30}$$

the scaled momentum χ is a *constant* and those extra terms are negligible. Thus, we have an everywhere valid approximation

$$\left(\frac{\mathrm{d}^2}{\mathrm{d}\sigma^2} - \sigma\right)\overline{\psi} = 0\,, \tag{6.8.31}$$

solved by

$$\overline{\psi} \propto \mathrm{Ai}(\sigma)\,, \qquad \psi \propto \frac{(-\sigma)^{\frac{1}{4}}}{p^{\frac{1}{2}}}\,\mathrm{Ai}(\sigma)\,. \tag{6.8.32}$$

If, for convenience, we set the proportionality constant to $2\sqrt{\pi}$, and knowing the $|\sigma| \gg 1$ forms of the Airy function [cf. (6.7.5) and (6.7.11)], we now find the asymptotic form, for $-\sigma \gg 1$,

$$\psi \cong \frac{2}{\sqrt{p}}\cos\left(\frac{2}{3}(-\sigma)^{\frac{3}{2}} - \frac{\pi}{4}\right) = \frac{2}{\sqrt{p(x)}}\cos\left(\frac{1}{\hbar}\int_{x_0}^{x}\mathrm{d}x'\,p(x') - \frac{\pi}{4}\right) \tag{6.8.33}$$

as expected, and analogously, for $\sigma \gg 1$,

$$\psi \cong \frac{1}{\sqrt{|p|}}\,\mathrm{e}^{-\frac{2}{3}\sigma^{\frac{3}{2}}} = \frac{1}{\sqrt{|p(x)|}}\,\mathrm{e}^{\frac{1}{\hbar}\int_{x_0}^{x}\mathrm{d}x'\,|p(x')|}\,. \tag{6.8.34}$$

But now we also know the value of ψ at $x = x_0$, for example, to which the asymptotic forms incorrectly assign ∞ owing to the diverging $1/\sqrt{|p(x)|}$ amplitude factor. For that we need the limit of $(-\sigma)^{\frac{1}{4}}/p^{\frac{1}{2}} = \chi^{-\frac{1}{2}}$ as $x \to x_0$, when $\sigma \to 0$ and $p \to 0$. This is available in (6.8.30), and so

$$\psi(x_0) = 2\sqrt{\pi/\chi(x_0)}\,\mathrm{Ai}(0) = \frac{1.2585}{(2\hbar MF)^{\frac{1}{6}}}\,, \tag{6.8.35}$$

if we again take $2\sqrt{\pi}$ for the proportionality factor in (6.8.32).

Incidentally, we shouldn't fail to notice the physical significance of the factor $1/\sqrt{p(x)}$ in the asymptotic form, as given in

$$\psi_{\text{inc.}} \propto \frac{1}{\sqrt{p(x)}}\,\mathrm{e}^{-\mathrm{i}\left(\frac{1}{\hbar}\int_{x_0}^{x}\mathrm{d}x'\,p(x') - \frac{\pi}{4}\right)}\,, \tag{6.8.36}$$

with its implication

$$|\psi_{\text{inc.}}|^2 \propto \frac{1}{p(x)}\,, \tag{6.8.37}$$

and the associated (relative) probability of finding the incident particle in the interval $\mathrm{d}x$ about x,

$$\mathrm{d}x\,|\psi_{\text{inc.}}|^2 \propto \frac{\mathrm{d}x}{M\frac{\mathrm{d}x}{\mathrm{d}t}} \propto \mathrm{d}t\,. \tag{6.8.38}$$

Here is the very sensible result that the probability of being found in a given stretch is proportional to the time that the particle spends in that stretch. Notice however that we have considered only the incident wave. The same result would emerge if we had only the reflected wave. But in fact both are present and in reality

$$\mathrm{d}x\,|\psi|^2 \propto \frac{\mathrm{d}x}{p(x)}\,4\cos^2\left(\frac{1}{\hbar}\int \mathrm{d}x\,p - \frac{\pi}{4}\right)\,. \tag{6.8.39}$$

So, superimposed on the particle factor $[p(x)]^{-1}\mathrm{d}x \propto \mathrm{d}t$, is the wave factor describing the interference between incident and reflected wave. Indeed, at certain points, the probability per unit length is *zero*; at other points it is four times that of the incident wave alone. On the *average* it is twice that of $|\psi_{\text{inc.}}|^2$, representing the additive contribution of $|\psi_{\text{inc.}}|^2$ and $|\psi_{\text{refl.}}|^2$.

6.9 Zeros and extrema of the Airy function

The construction of ψ from $\mathrm{Ai}(\sigma)$ redirects attention to the latter function in its finer details. Let us specifically ask where $\mathrm{Ai}(\sigma)$ has zeros:

$$\mathrm{Ai}(\overline{\sigma}_n) = 0\,, \qquad n = 1,2,3,\dots\,, \tag{6.9.1}$$

and where it has extrema, maxima or minima:

$$\mathrm{Ai}'(\overline{\overline{\sigma}}_n) = 0\,, \qquad n = 1,2,3,\dots\,; \tag{6.9.2}$$

here and below primes denote differentiation with respect to σ, the argument of the Airy function.

For the purpose of determining the $\overline{\sigma}$'s and $\overline{\overline{\sigma}}$'s we return again to the amplitude–phase construction, this time in real form:

$$\begin{aligned} \mathrm{Ai}(\sigma) &= A\cos\phi\,, \\ \mathrm{Ai}' &= A'\cos\phi - A\phi'\sin\phi\,, \\ \mathrm{Ai}'' &= A''\cos\phi - 2A'\phi'\sin\phi - A\phi'^2\cos\phi - A\phi''\sin\phi\,. \end{aligned} \tag{6.9.3}$$

The differential equation (6.4.14), $\mathrm{Ai}'' = \sigma\,\mathrm{Ai}$, is satisfied by setting the coefficients of $\cos\phi$, $\sin\phi$ equal to zero,

$$A'' - A\phi'^2 = \sigma A\,, \qquad 2A'\phi' + A'\phi'' = 0\,. \tag{6.9.4}$$

The integration constant in

$$\frac{\mathrm{d}}{\mathrm{d}\sigma}(A^2\phi') = 0\,, \qquad A^2\phi' = -\frac{1}{\pi}\,, \tag{6.9.5}$$

follows from the asymptotic form (6.7.5), and approximations for $\phi(\sigma)$ are iteratively found from

$$\phi'^2 = -\sigma + \frac{A''}{A}\,. \tag{6.9.6}$$

The initial approximation is WKB: ignore A''/A, so that

<u>1st approximation:</u>

$$\phi' = -\sqrt{-\sigma}\,, \quad \phi = \frac{2}{3}(-\sigma)^{\frac{3}{2}} - \frac{\pi}{4}\,, \qquad A = \frac{\pi^{\;\frac{1}{2}}}{(-\sigma)^{\frac{1}{4}}}\,, \tag{6.9.7}$$

where, again, we make use of information available in (6.7.5) about the sign of ϕ' and the behavior of ϕ in the WKB regime ($-\sigma \gg 1$). In this approximation the zeros $\overline{\sigma}$ of $\mathrm{Ai}(\sigma)$ are given by

$$\cos\bigl(\phi(\overline{\sigma})\bigr) \cong \cos\Bigl(\frac{2}{3}(-\overline{\sigma})^{\frac{3}{2}} - \frac{\pi}{4}\Bigr) = 0\,, \tag{6.9.8}$$

so that

$$-\overline{\sigma}_n \cong \left[\frac{3\pi}{2}\left(n - \frac{1}{4}\right)\right]^{\frac{2}{3}} \qquad \text{for } n = 1,2,3,\dots\,, \tag{6.9.9}$$

and the comparison with the exact values,

$$\begin{aligned}-\bar{\sigma}_1 &\cong 2.3203 = 2.3381 \times 0.9924\,,\\ -\bar{\sigma}_2 &\cong 4.0818 = 4.0879 \times 0.9985\,,\\ -\bar{\sigma}_3 &\cong 5.5172 = 5.5205 \times 0.9994\,,\end{aligned} \tag{6.9.10}$$

shows that this simplest approximation is remarkably accurate.

In the approximation where A is slowly varying, the extrema of $A\cos\phi$ are given by $\phi(\bar{\bar{\sigma}}) = 0, \pi, 2\pi \dots$, so that

$$-\bar{\bar{\sigma}}_n \cong \left[\frac{3\pi}{2}\left(n - \frac{3}{4}\right)\right]^{\frac{2}{3}} \qquad \text{for } n = 1, 2, 3, \dots\,. \tag{6.9.11}$$

Here the comparison with exact values,

$$\begin{aligned}-\bar{\bar{\sigma}}_1 &\cong 1.1155 = 1.0188 \times 1.095\,,\\ -\bar{\bar{\sigma}}_2 &\cong 3.2616 = 3.2482 \times 1.0041\,,\\ -\bar{\bar{\sigma}}_3 &\cong 4.8263 = 4.8201 \times 1.0013\,,\end{aligned} \tag{6.9.12}$$

reveals very good agreement, except for $n = 1$.

We improve the approximation for $\phi(\sigma)$ by using the 1st value for A''/A in (6.9.6),

$$\left(\frac{A''}{A}\right)_{1^{\text{st}}} = \frac{5}{16}\frac{1}{(-\sigma)^2}\,, \tag{6.9.13}$$

giving

$$\phi'^2 = -\sigma\left[1 + \frac{5}{16}\frac{1}{(-\sigma)^3}\right]\,, \tag{6.9.14}$$

so

$$\phi' = -\sqrt{-\sigma}\left[1 + \frac{5}{32}\frac{1}{(-\sigma)^3}\right] \tag{6.9.15}$$

and

$$\underline{2^{\text{nd}}\text{ approximation:}} \qquad \phi = \frac{2}{3}(-\sigma)^{\frac{3}{2}} - \frac{\pi}{4} - \frac{5}{48}(-\sigma)^{-\frac{3}{2}}\,. \tag{6.9.16}$$

From $\phi(\bar{\sigma}) = \pi/2, 3\pi/2, 5\pi/2, \dots$, we then get a corresponding refinement of (6.9.9),

$$-\bar{\sigma}_n \cong \left[\frac{3\pi}{2}\left(n - \frac{1}{4}\right)\right]^{\frac{2}{3}} + \frac{5}{48}\left[\frac{3\pi}{2}\left(n - \frac{1}{4}\right)\right]^{-\frac{4}{3}}\,. \tag{6.9.17}$$

Not surprisingly, the comparison with the exact zeros of $\mathrm{Ai}(\sigma)$,

$$-\overline{\sigma}_1 \cong 2.320\,251 + 0.019\,349 = 2.339\,600 = 2.338\,107 \times 1.000\,638\,,$$
$$-\overline{\sigma}_2 \cong 4.081\,810 + 0.006\,252 = 4.088\,062 = 4.087\,949 \times 1.000\,028\,,$$
$$-\overline{\sigma}_3 \cong 5.517\,164 + 0.003\,422 = 5.520\,586 = 5.520\,560 \times 1.000\,005\,, \quad (6.9.18)$$

shows a substantial improvement.

Concerning the zeros of Ai$'$, we note that the extrema of $(-\phi')^{-\frac{1}{2}} \cos\phi$ are determined by

$$(-\phi')^{\frac{1}{2}} \left[\sin\phi + \frac{1}{2}\frac{\phi''}{(\phi')^2}\cos\phi\right] = 0 \qquad (6.9.19)$$

or, since

$$\frac{1}{2}\frac{\phi''}{(\phi')^2} = -\frac{1}{2}\frac{\mathrm{d}}{\mathrm{d}\sigma}\frac{1}{\phi'} \cong \frac{1}{2}\frac{\mathrm{d}}{\mathrm{d}\sigma}\frac{1}{\sqrt{-\sigma}} = \frac{1}{4}\frac{1}{(-\sigma)^{\frac{3}{2}}} \qquad (6.9.20)$$

is small,

$$\sin\left(\phi + \frac{1}{2}\frac{\phi''}{(\phi')^2}\right) = 0\,. \qquad (6.9.21)$$

With $\phi(\sigma)$ of (6.9.16), the requirement

$$\phi + \frac{1}{2}\frac{\phi''}{(\phi')^2} = 0, \pi, 2\pi, \ldots \qquad (6.9.22)$$

is met by

$$-\overline{\overline{\sigma}}_n \cong \left[\frac{3\pi}{2}\left(n - \frac{3}{4}\right)\right]^{\frac{2}{3}} - \frac{7}{48}\left[\frac{3\pi}{2}\left(n - \frac{3}{4}\right)\right]^{-\frac{4}{3}}, \qquad (6.9.23)$$

and the comparison with the exact zeros of Ai$'$,

$$-\overline{\overline{\sigma}}_1 \cong 1.115\,460 - 0.117\,206 = 0.998\,255 = 1.018\,793 \times 0.9980\,,$$
$$-\overline{\overline{\sigma}}_2 \cong 3.261\,626 - 0.013\,708 = 3.247\,917 = 3.248\,198 \times 0.999\,914\,,$$
$$-\overline{\overline{\sigma}}_3 \cong 4.826\,316 - 0.006\,261 = 4.820\,055 = 4.820\,099 \times 0.999\,991\,, \quad (6.9.24)$$

shows good agreement for $n = 1$ and very good agreement for $n > 1$.

If we now denote the zeros $\overline{\sigma}_1$, $\overline{\sigma}_2$, ... of the Airy function by σ_1, σ_3, σ_5, ... , and the extrema $\overline{\overline{\sigma}}_1$, $\overline{\overline{\sigma}}_2$, ... by σ_0, σ_2, σ_4, ... , the second-order approximations (6.9.17) and (6.9.23) are compactly presented as

$$-\sigma_m \cong \left[\frac{3\pi}{4}\left(m + \frac{1}{2}\right)\right]^{\frac{2}{3}} - \left(\frac{1}{48} + \frac{1}{8}(-1)^m\right)\left[\frac{3\pi}{4}\left(m + \frac{1}{2}\right)\right]^{-\frac{4}{3}},$$

$$\text{for } m = 0, 1, 2, \ldots \qquad (m \text{ odd: zeros; } m \text{ even: extrema}). \qquad (6.9.25)$$

The solid curve in

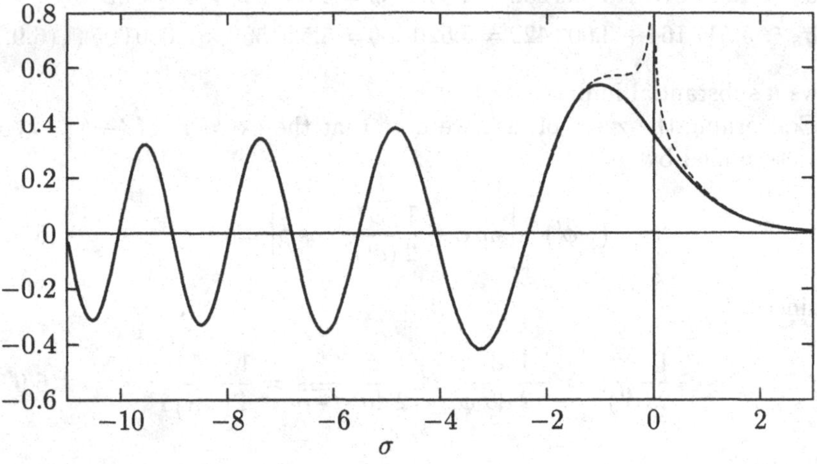

displays the function $\mathrm{Ai}(\sigma)$ that, for negative σ, has these zeros, maxima, minima, equals 0.3350 at $\sigma = 0$, and decreases rapidly for $\sigma > 0$. The dashed curve shows the leading asymptotic forms (6.7.5) and (6.7.11).

6.10 Constant restoring force

We get a direct application of these results if we ask for the energy values in the so-called linear potential:

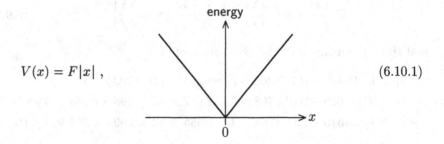

$$V(x) = F|x| \,, \tag{6.10.1}$$

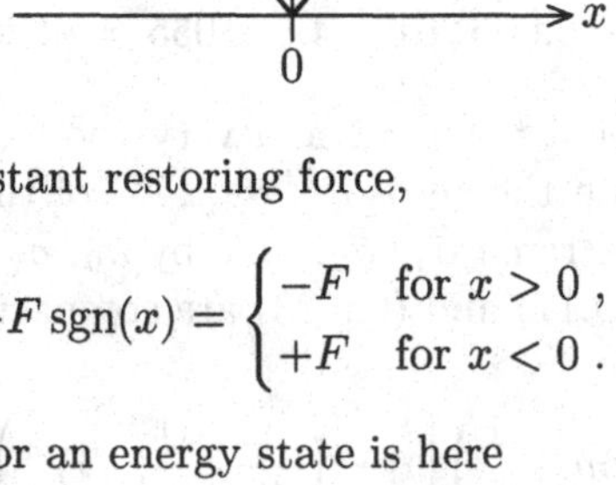

which corresponds to a constant restoring force,

$$-\frac{\partial V}{\partial x} = -F\,\mathrm{sgn}(x) = \begin{cases} -F & \text{for } x > 0\,, \\ +F & \text{for } x < 0\,. \end{cases} \tag{6.10.2}$$

The Schrödinger equation for an energy state is here

$$\left[E + \frac{\hbar^2}{2M}\frac{\mathrm{d}^2}{\mathrm{d}x^2} - F|x|\right]\psi(x) = 0\,. \tag{6.10.3}$$

This differential operator is even in x; it is unchanged by $x \to -x$:

$$\left[E + \frac{\hbar^2}{2M}\frac{\mathrm{d}^2}{\mathrm{d}x^2} - F|x|\right]\psi(-x) = 0\,. \tag{6.10.4}$$

If both $\psi(x)$ and $\psi(-x)$ are solutions, so also are their even and odd sums

$$\psi_{\text{even}}(x) = \frac{1}{2}\left[\psi(x) + \psi(-x)\right]\,, \qquad \psi_{\text{odd}}(x) = \frac{1}{2}\left[\psi(x) - \psi(-x)\right]\,, \tag{6.10.5}$$

which are characterized by

$$\psi_{\text{even}}(-x) = \psi_{\text{even}}(x)\,, \qquad \psi_{\text{odd}}(-x) = -\psi_{\text{odd}}(x)\,, \tag{6.10.6}$$

so that, taking the continuity of ψ and $\mathrm{d}\psi/\mathrm{d}x$ for granted,

$$\frac{\mathrm{d}}{\mathrm{d}x}\psi_{\text{even}}(0) = 0\,, \qquad \psi_{\text{odd}}(0) = 0\,. \tag{6.10.7}$$

For each kind of function it is sufficient to consider only the region $x > 0$, or $x < 0$. We use the latter:

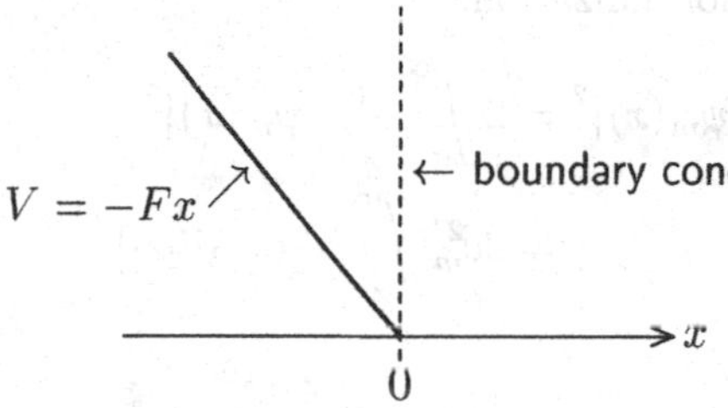

which gives us back the situation already studied, $V = -Fx$, now with a boundary condition at $x = 0$, appropriate to even or odd functions.

The solution, with $V = -Fx$, is

$$x < 0: \qquad \psi(x) \propto \mathrm{Ai}(\sigma) \quad \text{with} \quad -\sigma = \left(\frac{2MF}{\hbar^2}\right)^{\frac{1}{3}}\left(x + \frac{E}{F}\right)\,. \tag{6.10.8}$$

Therefore, in accordance with (6.10.6) we have

$$\begin{aligned}\psi_{\text{even}}(x) &\propto \mathrm{Ai}\Big(\left(2MF/\hbar^2\right)^{\frac{1}{3}}\left(|x| - E/F\right)\Big)\,,\\ \psi_{\text{odd}}(x) &\propto \operatorname{sgn}(x)\,\mathrm{Ai}\Big(\left(2MF/\hbar^2\right)^{\frac{1}{3}}\left(|x| - E/F\right)\Big)\,,\end{aligned} \tag{6.10.9}$$

and the boundary conditions (6.10.7) require

$$\begin{aligned}\left(\frac{2M}{\hbar^2F^2}\right)^{\frac{1}{3}} E &= \left\{\begin{matrix} -\bar{\bar{\sigma}}_1\,,\ -\bar{\bar{\sigma}}_2\,,\ \ldots \ \text{for } \psi_{\text{even}} \\ -\bar{\sigma}_1\,,\ -\bar{\sigma}_2\,,\ \ldots \ \text{for } \psi_{\text{odd}} \end{matrix}\right\}\\ &= \left\{\begin{matrix} -\sigma_0\,,\ -\sigma_2\,,\ \ldots \ \text{for } \psi_{\text{even}} \\ -\sigma_1\,,\ -\sigma_3\,,\ \ldots \ \text{for } \psi_{\text{odd}} \end{matrix}\right\}\end{aligned} \tag{6.10.10}$$

so that

$$E_m = (-\sigma_m)\left(\frac{\hbar^2 F^2}{2M}\right)^{\frac{1}{3}} \quad \text{for} \quad m = 0, 1, 2, \dots \tag{6.10.11}$$

are the energy eigenvalues for the linear potential (6.10.1). The second-order approximation for $-\sigma_m$ in (6.9.25) gives

$$\begin{aligned}\left(\frac{2M}{\hbar^2 F^2}\right)^{\frac{1}{3}} E_m \cong &\left[\frac{3\pi}{4}\left(m+\frac{1}{2}\right)\right]^{\frac{2}{3}} \\ &-\frac{1}{8}\left((-1)^m + \frac{1}{6}\right)\left[\frac{3\pi}{4}\left(m+\frac{1}{2}\right)\right]^{-\frac{4}{3}} .\end{aligned} \tag{6.10.12}$$

We write ψ_0, ψ_2, ψ_4, ... for the even solutions, and ψ_1, ψ_3, ψ_5, ... for the odd solutions. The (positive) constants c_m in

$$x > 0: \qquad \psi_m(x) = c_m (2MF/\hbar^2)^{\frac{1}{6}} \operatorname{Ai}\Big(\sigma_m + (2MF/\hbar^2)^{\frac{1}{3}} x\Big) \tag{6.10.13}$$

are determined by the normalization:

$$\begin{aligned}1 = \int_{-\infty}^{\infty} \mathrm{d}x \, |\psi_m(x)|^2 &= 2\int_0^{\infty} \mathrm{d}x \, |\psi_m(x)|^2 \\ &= 2c_m^2 \int_{\sigma_m}^{\infty} \mathrm{d}\sigma \left[\operatorname{Ai}(\sigma)\right]^2 ,\end{aligned} \tag{6.10.14}$$

so

$$c_m = \left(2\int_{\sigma_m}^{\infty} \mathrm{d}\sigma \left[\operatorname{Ai}(\sigma)\right]^2\right)^{-\frac{1}{2}} , \tag{6.10.15}$$

where, as we recall, $\operatorname{Ai}(\sigma_m) = 0$ for m odd, $\operatorname{Ai}'(\sigma_m) = 0$ for m even. Now, differentiation of the differential equation (6.4.14) obeyed by $\operatorname{Ai}(\sigma)$,

$$\frac{\mathrm{d}^2}{\mathrm{d}\sigma^2} \operatorname{Ai}(\sigma) = \sigma \operatorname{Ai}(\sigma) , \tag{6.10.16}$$

gives

$$\frac{\mathrm{d}^2}{\mathrm{d}\sigma^2} \operatorname{Ai}'(\sigma) = \sigma \operatorname{Ai}'(\sigma) + \operatorname{Ai}(\sigma) . \tag{6.10.17}$$

Therefore

$$\begin{aligned}\left[\operatorname{Ai}(\sigma)\right]^2 &= \operatorname{Ai}(\sigma)\frac{\mathrm{d}^2}{\mathrm{d}\sigma^2}\operatorname{Ai}'(\sigma) - \operatorname{Ai}'(\sigma)\frac{\mathrm{d}^2}{\mathrm{d}\sigma^2}\operatorname{Ai}(\sigma) \\ &= \frac{\mathrm{d}}{\mathrm{d}\sigma}\Big(\operatorname{Ai}(\sigma)\operatorname{Ai}''(\sigma) - \left[\operatorname{Ai}'(\sigma)\right]^2\Big) \\ &= -\frac{\mathrm{d}}{\mathrm{d}\sigma}\Big(\left[\operatorname{Ai}'(\sigma)\right]^2 - \sigma\left[\operatorname{Ai}(\sigma)\right]^2\Big)\end{aligned} \tag{6.10.18}$$

and

$$\frac{1}{2c_m^2} = \int_{\sigma_m}^{\infty} \mathrm{d}\sigma \left[\mathrm{Ai}(\sigma)\right]^2 = \left(\left[\mathrm{Ai}'(\sigma_m)\right]^2 - \sigma_m\left[\mathrm{Ai}(\sigma_m)\right]^2\right)$$

$$= \begin{cases} (-\sigma_m)\left[\mathrm{Ai}(\sigma_m)\right]^2 & \text{for } m \text{ even,} \\ \left[\mathrm{Ai}'(\sigma_m)\right]^2 & \text{for } m \text{ odd.} \end{cases} \tag{6.10.19}$$

Let's ask how well the leading (WKB) approximations represent the normalization constants c_m. Proceeding from

$$\mathrm{Ai}(\sigma) \cong \pi^{-\frac{1}{2}}(-\sigma)^{-\frac{1}{4}} \cos\left(\frac{2}{3}(-\sigma)^{-\frac{3}{2}} - \frac{\pi}{4}\right),$$

$$\mathrm{Ai}'(\sigma) \cong \pi^{-\frac{1}{2}}(-\sigma)^{\frac{1}{4}} \sin\left(\frac{2}{3}(-\sigma)^{-\frac{3}{2}} - \frac{\pi}{4}\right), \tag{6.10.20}$$

we first get

$$\left[\mathrm{Ai}'(\sigma)\right]^2 - \sigma\left[\mathrm{Ai}(\sigma)\right]^2 \cong \frac{1}{\pi}\sqrt{-\sigma} \tag{6.10.21}$$

and then with (6.9.25)

$$c_m \cong \left(\frac{\pi}{2}\right)^{\frac{1}{2}} \left[\frac{3\pi}{4}\left(m + \frac{1}{2}\right)\right]^{-\frac{1}{6}} = \pi^{\frac{1}{3}}(6m+3)^{-\frac{1}{6}}. \tag{6.10.22}$$

The comparison with exact values,

$$\begin{aligned} c_0 &\cong 1.21954 = 1.30784 \times 0.9325\,, & c_1 &\cong 1.01549 = 1.00841 \times 1.0070\,, \\ c_2 &\cong 0.93261 = 0.93634 \times 0.9960\,, & c_3 &\cong 0.88175 = 0.88046 \times 1.0015\,, \\ c_4 &\cong 0.84558 = 0.84666 \times 0.9987\,, & c_5 &\cong 0.81777 = 0.81727 \times 1.0006\,, \end{aligned} \tag{6.10.23}$$

shows that the error is well below 1% except for $m = 0$.

6.11 Rayleigh–Ritz variational method

The one state that did not fare too well with the WKB approximation is, not surprisingly, the lowest energy state, $m = 0$, which has a wave function without oscillations:

$$\psi_0(x) \propto \mathrm{Ai}(\sigma_0 + |q|)$$
with $q = (2MF/\hbar^2)^{\frac{1}{3}} x$.

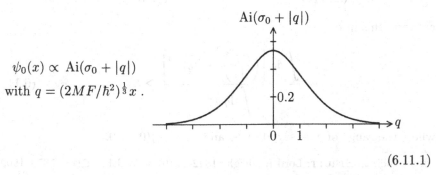

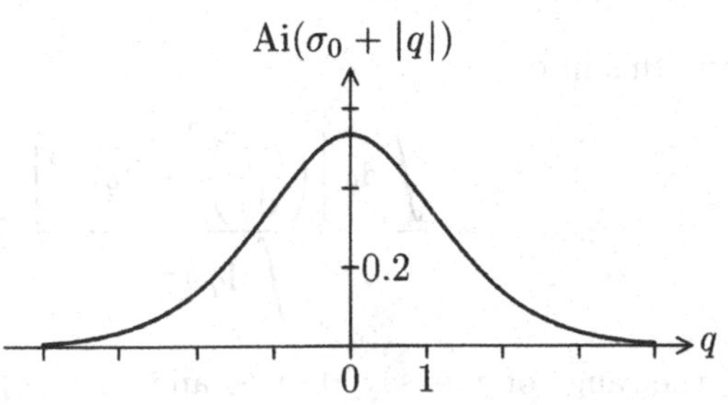

(6.11.1)

Here is a method directed specifically at that state. Consider any Hamiltonian for which the spectrum is bounded below:

$$H' = E \geq E_0 \,. \tag{6.11.2}$$

In the present circumstance, $H = p^2/(2M) + F|x|$, it is clear that $H' > 0$; there is a lowest energy state. Generally, we have

$$(H - E_0)' = E - E_0 \geq 0 \,, \tag{6.11.3}$$

so that, for any state $|\;\rangle$, the expectation value of $H - E_0$ is positive,

$$\langle (H - E_0) \rangle = \sum_E \underbrace{(E - E_0)}_{\geq 0} p(E) \geq 0 \,, \tag{6.11.4}$$

where the equal sign holds only if $|\;\rangle = |H' = E_0\rangle$. Equivalently,

$$\langle H \rangle \geq E_0 \,, \tag{6.11.5}$$

so that for any $|\;\rangle$, $\langle H \rangle$ provides an upper limit to E_0. One then tries to minimize $\langle H \rangle$ to get a good value. In the quantum literature, this is known as the Rayleigh*–Ritz† variational method.

It is often convenient to write a normalized (real) wave function as $\psi(x)/\sqrt{\int \mathrm{d}x' \, [\psi(x')]^2}$. Here then

$$\left\langle \frac{1}{2M} p^2 + F|x| \right\rangle = \frac{\displaystyle\int \mathrm{d}x \left[\frac{1}{2M} \left(-\frac{\hbar}{\mathrm{i}} \frac{\partial}{\partial x} \psi \right) \left(\frac{\hbar}{\mathrm{i}} \frac{\partial}{\partial x} \psi \right) + \psi F |x| \psi \right]}{\displaystyle\int \mathrm{d}x \, \psi^2} \geq E_0 \,. \tag{6.11.6}$$

Writing

$$x = \left(\frac{\hbar^2}{2MF} \right)^{\frac{1}{3}} q \quad \text{and} \quad E_0 = \left(\frac{\hbar^2 F^2}{2M} \right)^{\frac{1}{3}} \mathcal{E}_0 \tag{6.11.7}$$

converts this into

$$\frac{\displaystyle\int \mathrm{d}q \left[\left(\frac{\mathrm{d}\psi}{\mathrm{d}q} \right)^2 + |q| \psi^2 \right]}{\displaystyle\int \mathrm{d}q \, \psi^2} \geq \mathcal{E}_0 \tag{6.11.8}$$

where the range of q is, say, $0 \to \infty$ and $(\mathrm{d}\psi/\mathrm{d}q)(0) = 0$.

*John William STRUTT, Lord Rayleigh (1842–1919) †Walther RITZ (1878–1909)

Now we must pick a suitable trial wave function $\psi(q)$. It should be a maximum at $q = 0$, and it must decrease rapidly for large q. Suppose we try (having some knowledge of its shape)

$$\psi(q) = \mathrm{e}^{-\frac{2}{3}\lambda q^{\frac{3}{2}}}, \tag{6.11.9}$$

where λ is an adjustable parameter. Then we get

$$\frac{\int_0^\infty \mathrm{d}q\,(\lambda^2 q + q)\,\mathrm{e}^{-\frac{4}{3}\lambda q^{\frac{3}{2}}}}{\int_0^\infty \mathrm{d}q\,\mathrm{e}^{-\frac{4}{3}\lambda q^{\frac{3}{2}}}} \geq \mathcal{E}_0 \tag{6.11.10}$$

or, with

$$q = \left(\frac{3}{4\lambda}\right)^{\frac{2}{3}} s^{\frac{2}{3}}, \tag{6.11.11}$$

$$(\lambda^2+1)\left(\frac{3}{4\lambda}\right)^{\frac{2}{3}} \frac{\int_0^\infty \mathrm{d}s\, s^{\frac{1}{3}}\,\mathrm{e}^{-s}}{\int_0^\infty \mathrm{d}s\, s^{-\frac{1}{3}}\,\mathrm{e}^{-s}} = \left(\frac{3}{4}\right)^{\frac{2}{3}} \frac{(\frac{1}{3})!}{(-\frac{1}{3})!}\left(\lambda^{\frac{4}{3}} + \lambda^{-\frac{2}{3}}\right) \geq \mathcal{E}_0\,, \tag{6.11.12}$$

where $(\frac{1}{3})! = 0.892\,980$, $(-\frac{1}{3})! = 1.354\,118$, and $(\frac{1}{3})!\,(-\frac{1}{3})! = \frac{1}{3}\pi/\sin(\frac{1}{3}\pi) = 2\pi/3^{\frac{3}{2}}$ illustrates a property of the factorial function. We now pick λ to minimize this:

$$\frac{\mathrm{d}}{\mathrm{d}\lambda}\left(\lambda^{\frac{4}{3}} + \lambda^{-\frac{2}{3}}\right) = \frac{4}{3}\lambda^{\frac{1}{3}} - \frac{2}{3}\lambda^{-\frac{5}{3}} = 0 \tag{6.11.13}$$

or

$$\lambda^2 = \frac{1}{2}\,, \qquad \lambda^{\frac{4}{3}} + \lambda^{-\frac{2}{3}} = \frac{3}{2^{\frac{2}{3}}}\,. \tag{6.11.14}$$

Therefore

$$\mathcal{E}_0 \leq \frac{3^{\frac{5}{3}}}{4}\frac{(\frac{1}{3})!}{(-\frac{1}{3})!} = 1.0288 \quad \left[= 1.0188 \times 1.0098\right]. \tag{6.11.15}$$

The approximation is correctly in excess and remarkably close considering the simplicity of the trial wave function. Any more general choice will yield a lower and better answer.

Problems

6-1a One degree of freedom, Hamilton operator $H = p^2/(2M) + V(x)$ with arbitrary $V(x)$. The probability for finding the particle between x' and $x'+\mathrm{d}x'$ is $\mathrm{d}x'\,\rho(x',t)$ with the probability density

$$\rho(x',t) = \left\langle \delta\bigl(x(t) - x'\bigr) \right\rangle .$$

Show that the continuity equation

$$\frac{\partial}{\partial t}\rho(x',t) + \frac{\partial}{\partial x'}j(x',t) = 0$$

is obeyed by this ρ and the probability current density

$$j(x',t) = \frac{1}{M}\left\langle p(t) \,.\, \delta\bigl(x(t) - x'\bigr) \right\rangle .$$

Generalize this to motion in three dimensions.

6-1b Now consider the probability density for momentum p,

$$\rho(p',t) = \left\langle \delta\bigl(p(t) - p'\bigr) \right\rangle$$

and find the associated probability current density needed in

$$\frac{\partial}{\partial t}\rho(p',t) + \frac{\partial}{\partial p'}j(p',t) = 0 .$$

What is the three-dimensional analog?

6-2a One degree of freedom, Hamilton operator $H = p^2/(2M) - Fx$ with constant force F. State and solve the equations of motion. As in Problem 5-7a, again consider the minimum uncertainty state $|\delta\rangle$, at $t = 0$, and evaluate $[\delta x(t)]^2$, $[\delta p(t)]^2$.

6-2b Now consider an arbitrary initial state and repeat Problem 5-8a. Why could you have anticipated that T and t_0 do not depend on F? What changes when the force is time dependent, $F(t)$?

6-3a One degree of freedom, Hamilton operator $H = p^2/(2M) + \frac{1}{2}M\omega^2x^2$ with constant frequency ω. Same questions as in Problem 6-2a. In addition, use the solutions to verify that $x(t)$ and $p(t)$ obey the required commutation relations. Also prove that $\delta x(t)\,\delta p(t) \geq \frac{1}{2}\hbar$. Under what circumstances does the equality sign hold for all t?

6-3b Again lift the restriction of an initial minimum uncertainty state. Follow the strategy of Problem 5-8a and find corresponding expressions for $\delta x(t)$ and $\delta p(t)$.

6-4 Three degrees of freedom, Hamilton operator $H = \boldsymbol{p}^2/(2M)$. For $\boldsymbol{N} = \boldsymbol{p}(t)t - M\boldsymbol{r}(t)$ and constant $\boldsymbol{v}$, evaluate

$$\langle \boldsymbol{r}', t|\, e^{\frac{i}{\hbar}\boldsymbol{v}\cdot\boldsymbol{N}} |\boldsymbol{p}'\rangle$$

and interpret the result.

6-5a One degree of freedom: Prove that

$$e^{-\frac{i}{\hbar}f(p)}x\,e^{\frac{i}{\hbar}f(p)} = x - \frac{df(p)}{dp}$$

[Hint: Recall Problem 1-55.], and illustrate it with

$$\frac{p^2}{2M} - Fx = e^{-\frac{i}{\hbar}\frac{p^3}{6MF}}(-Fx)\,e^{\frac{i}{\hbar}\frac{p^3}{6MF}} .$$

6-5b Use this to show that

$$\begin{aligned} e^{-\frac{i}{\hbar}\left(\frac{p^2}{2M} - Fx\right)t} &= e^{-\frac{i}{\hbar}\frac{p^3}{6MF}}\,e^{\frac{i}{\hbar}Fxt}\,e^{\frac{i}{\hbar}\frac{p^3}{6MF}} \\ &= e^{\frac{i}{\hbar}Fxt}\,e^{-\frac{i}{\hbar}\frac{p^2}{2M}t}\,e^{-\frac{i}{\hbar}\frac{p}{M}\frac{1}{2}Ft^2}\,e^{-\frac{i}{\hbar}\frac{F^2t^3}{6M}} \\ &= e^{-\frac{i}{\hbar}\frac{F^2t^3}{6M}}\,e^{\frac{i}{\hbar}\frac{p}{M}\frac{1}{2}Ft^2}\,e^{-\frac{i}{\hbar}\frac{p^2}{2M}t}\,e^{\frac{i}{\hbar}Fxt} . \end{aligned}$$

Recognize in these results an example of Problem 2-11.

6-6a Apply a statement of Problem 6-5b to demonstrate that

$$\langle p',t|p''\rangle = e^{-\frac{i}{\hbar}\frac{p'^3}{6MF}}\,\delta(p'-p''-Ft)\,e^{\frac{i}{\hbar}\frac{p''^3}{6MF}} ,$$

for $H = p^2/(2M) - Fx$. What does this become in the limit $F \to 0$?

6-6b Verify that this $\langle p',t|p''\rangle$ is the solution of the appropriate p Schrödinger equation and its initial condition.

6-7 Apply another statement of Problem 6-5b and arrive at

$$\langle x',t|p'\rangle = \frac{1}{\sqrt{2\pi\hbar}}\,e^{\frac{i}{\hbar}x'(p'+Ft)}\,e^{-\frac{i}{\hbar}\frac{p'^2}{2M}t}\,e^{-\frac{i}{\hbar}\frac{p'}{M}\frac{1}{2}Ft^2}\,e^{-\frac{i}{\hbar}\frac{F^2t^3}{6M}}$$

for $H = p^2/(2M) - Fx$. Then use the action principle to produce another derivation of this result.

6-8 Apply the third result in Problem 6-5b to get $\langle p',t|x'\rangle$. Check that this is produced from $\langle x',t|p'\rangle$ by complex conjugation combined with the substitution $t \to -t$. Why should that be so?

6-9 Use the action principle, for $H = p^2/(2M)$ (one degree of freedom), to evaluate

$$\frac{\partial}{\partial M}\langle p',t|p''\rangle , \quad \frac{\partial}{\partial M}\langle x',t|p'\rangle , \quad \frac{\partial}{\partial M}\langle x',t|x''\rangle .$$

Are these results correct according to the known forms of the respective time transformation functions?

6-10 A question analogous to Problem 6-9, with $H = p^2/(2M) - Fx$, and

$$\frac{\partial}{\partial F}\langle x', t | p' \rangle = ? \,, \quad \frac{\partial}{\partial F}\langle x', t | x'' \rangle = ? \,.$$

6-11 One degree of freedom, Hamilton operator

$$H = \frac{1}{2M}\left[p - \frac{\partial \lambda(x,t)}{\partial x}\right]^2 - \frac{\partial \lambda(x,t)}{\partial t} \,.$$

Does the force $M(\mathrm{d}/\mathrm{d}t)^2 x$ depend on the "gauge" $\lambda(x,t)$? Use the quantum action principle to find the λ dependence of the time transformation function $\langle x', t_1 | x'', t_2 \rangle$.

6-12 Concerning the apparent ambiguity of the square root in (6.1.31): Follow the spirit of the discussion in Section 6.3 and write

$$\sqrt{\frac{1}{2\mathrm{i}\sin(\omega t)}} = \mathrm{e}^{-\frac{\mathrm{i}}{2}\omega t}\left(1 - \mathrm{e}^{-\epsilon}\mathrm{e}^{-2\mathrm{i}\omega t}\right)^{-\frac{1}{2}}\Big|_{\epsilon > 0,\ \epsilon \to 0} \,.$$

Then show that, for $\epsilon > 0$, the right-hand side is – in a natural way – a continuous function of ωt. Take the limit $\epsilon \to 0$ and state explicitly what you get for $k\pi < \omega t < (k+1)\pi$ with k integer.

6-13a One degree of freedom, Hamilton operator $H = p^2/(2M) - F(t)x$ with time dependent force $F(t)$ acting between $t = 0$ and $t = T$. Consider variations $\delta F(t)$ of the force and use the quantum action principle to find first $\delta_F \langle p', T | x', 0 \rangle$ and then $\langle p', T | x', 0 \rangle$.

6-13b Repeat for $\langle x', T | p', 0 \rangle$.

6-13c Finite time transformations are effected by the unitary evolution operator U,

$$\langle \ldots, T | = \langle \ldots, 0 | U \,;$$

see Problems 5-2. Regard U as a function of $x = x(t=0)$ and $p = p(t=0)$ and get the px-ordered form of U from $\langle p', T | x', 0 \rangle$ and its xp-ordered form from $\langle x', T | p', 0 \rangle$. Show that they can be written as

$$\begin{aligned} U &= \mathrm{e}^{-\frac{\mathrm{i}}{\hbar}\frac{p^2}{2M}T}\,\mathrm{e}^{-\frac{\mathrm{i}}{\hbar}p\Delta x}\,\mathrm{e}^{\frac{\mathrm{i}}{\hbar}x\Delta p}\,\mathrm{e}^{-\frac{\mathrm{i}}{2\hbar}\int_0^T \mathrm{d}t \int_0^T \mathrm{d}t'\, f(t)t_< F(t')/M} \\ &= \mathrm{e}^{\frac{\mathrm{i}}{\hbar}x\Delta p}\,\mathrm{e}^{-\frac{\mathrm{i}}{\hbar}p\Delta x}\,\mathrm{e}^{-\frac{\mathrm{i}}{\hbar}\frac{(p+\Delta p)^2}{2M}T}\,\mathrm{e}^{\frac{\mathrm{i}}{2\hbar}\int_0^T \mathrm{d}t \int_0^T \mathrm{d}t'\, f(t)t_> F(t')/M} \,, \end{aligned}$$

where $t_<$ and $t_>$ are the earlier and later one of the times t and t', respectively, and

$$\Delta p = \int_0^T \mathrm{d}t\, F(t)\,, \qquad \Delta x = -\int_0^T \mathrm{d}t\, \frac{t}{M} F(t)$$

are convenient abbreviations. What is the physical significance of Δp and Δx?

6-13d As a check, use either form for U to verify that $x(T) = U^\dagger x U$ and $p(T) = U^\dagger p U$.

6-13e Write U as a single exponential of a Hermitian operator, rather than a product of exponentials. [Hint: Problem 1-55.] Verify that you get the right answer for time-independent F.

6-14a Consider a Ag atom that passes through a succession of Stern–Gerlach magnets that are intended to first split a beam of atoms in two and then reunite it – a Stern–Gerlach interferometer. For simplicity, treat the longitudinal motion, along the y axis, as classical: $y \to vt$; ignore the x motion; and assume that only the z component of the magnetic field is relevant and that all z values of interest are sufficiently small to justify the approximation

$$\boldsymbol{\mu} \cdot \boldsymbol{B} = \mu \boldsymbol{\sigma} \cdot \boldsymbol{B} \cong \mu \sigma_z B_z(y,z) \cong \mu \sigma_z \left[B_z(y,0) + z \frac{\partial B_z}{\partial z}(y,0) \right]_{y \to vt} .$$

Show that the Hamilton operator $H = \boldsymbol{p}^2/(2M) - \boldsymbol{\mu} \cdot \boldsymbol{B}$ is then effectively reduced to

$$H = \frac{p_z^2}{2M} + \sigma_z \left[\hbar \Omega(t) - F(t) z \right]$$

where the precession frequency $\Omega(t)$ and the force $F(t)$ vanish before $t = 0$ and after $t = T$. Prior to entering the magnets at $t = 0$ the atom is in a $\sigma_x = 1$ spin state and its spatial properties are specified by a certain probability operator $P(z, p_z)$. Use the findings of Problems 6-13 to show that

$$\left| \langle \boldsymbol{\sigma}(T) \rangle \right|^2 = \left| \mathrm{tr} \left\{ P(z, p_z)\, \mathrm{e}^{2\mathrm{i}(p_z \Delta z - z \Delta p_z)/\hbar} \right\} \right|^2$$

where Δz and Δp_z are defined analogously.

6-14b Ideal reunification would be achieved for $\Delta z = 0$ and $\Delta p_z = 0$. Supposing now that the experimenter is satisfied by $\left| \langle \boldsymbol{\sigma}(T) \rangle \right| \geq 0.9$, say, how large are the tolerable uncontrolled errors in Δz and Δp_z? [Hint: Problem 2-16.] Conclusion?

6-15 One degree of freedom, Hamilton operator $H = p^2/(2M) + V_0 - Fx + \frac{1}{2}M\omega^2 x^2$ with V_0, F, ω^2 constant. Use the known transformation function for $V_0 = 0$, $F = 0$, $\omega^2 \neq 0$ to get

$$\begin{aligned}\langle x', t | x'' \rangle = \sqrt{\frac{M\omega}{2\pi i\hbar \sin(\omega t)}}\, & e^{\frac{i}{\hbar}\frac{M\xi^2}{2t}[\frac{1}{2}\omega t \cot(\frac{1}{2}\omega t)]} \\ & \times e^{-\frac{i}{\hbar}(V_0 - F\overline{x} + \frac{1}{2}M\omega^2\overline{x}^2)t} \\ & \times e^{-\frac{i}{\hbar}(M\omega^3)^{-1}(F - M\omega^2\overline{x})^2[\tan(\frac{1}{2}\omega t) - \frac{1}{2}\omega t]}\end{aligned}$$

where $\overline{x} = \frac{1}{2}(x' + x'')$, $\xi = x' - x''$. Check the limiting situations of vanishing F or vanishing ω^2. What happens for $\omega^2 < 0$?

6-16a One degree of freedom, Hamilton operator $H = p^2/(2M) + V(x)$ with (rather) arbitrary potential energy $V(x)$. The particle does not travel far during short time intervals. With this in mind, use the local-oscillator approximation for $V(x)$, that is

$$V(x) \cong V(\overline{x}) + (x - \overline{x})\frac{dV}{dx}(\overline{x}) + \frac{1}{2}(x - \overline{x})^2\frac{d^2V}{dx^2}(\overline{x})$$

with reference point $\overline{x} = \frac{1}{2}(x' + x'')$ half-way between the initial and final positions x'' and x', and the result of Problem 6-15 to obtain a short-time approximation for the time transformation function. You should get

$$\langle x', t | x'' \rangle \cong \underbrace{\sqrt{\frac{M}{2\pi i\hbar t}}\, e^{\frac{i}{\hbar}\frac{M\xi^2}{2t}}}_{\text{free particle}} \left(1 + \tfrac{1}{12}\overline{\omega^2}t^2\right) e^{-\frac{i}{\hbar}\frac{t}{24}M\overline{\omega^2}\xi^2}\, e^{-\frac{i}{\hbar}\overline{V}t}\, e^{-\frac{i}{\hbar}\frac{\overline{F}^2 t^3}{24M}}$$

where $\overline{V} = V(\overline{x})$, $\overline{F} = -\frac{dV}{dx}(\overline{x})$, $M\overline{\omega^2} = \frac{d^2V}{dx^2}(\overline{x})$, and $\xi = x' - x''$.

6-16b Generalize this to motion in three dimensions with Hamilton operator $H = \boldsymbol{p}^2/(2M) + V(\boldsymbol{r})$.

6-17a Put $x' = x'' = \overline{x}$ in Problem 6-16a and use the resulting approximation to $\langle \overline{x} | \exp(-\frac{i}{\hbar}Ht) | \overline{x} \rangle$ to show that

$$\operatorname{tr}\{f(H)\} \cong \int \frac{d\overline{x}\, dp}{2\pi\hbar} \int d\sigma\, \mathrm{Ai}(\sigma) f\Bigl(\frac{p^2}{2M} + \overline{V} - \frac{\sigma}{2}\bigl[(\hbar^2/M)\bigl(\overline{F}^2 - 2p^2\overline{\omega^2}\bigr)\bigr]^{\frac{1}{3}}\Bigr)$$

is the corresponding approximation for the trace of a function of H.

6-17b Now use properties of the Airy function and partial integrations to exhibit the leading quantum correction to the semiclassical phase space integral:

$$\operatorname{tr}\{f(H)\} \cong \underbrace{\int \frac{d\overline{x}\,dp}{2\pi\hbar} f\left(\frac{p^2}{2M}+\overline{V}\right)}_{\text{semiclassical value}} - \underbrace{\int \frac{d\overline{x}\,dp}{2\pi\hbar}\,\frac{\hbar^2\overline{\omega^2}}{24}\,f''\left(\frac{p^2}{2M}+\overline{V}\right)}_{\text{leading quantum correction}}$$

where primes denote differentiation with respect to the argument of $f(\ )$.

6-18 Reconsider the multiplicity of energy states for constant force. Concerning the phase space integral in (6.4.16): Suppose one integrates *first* over x, from $-\infty$ to ∞, and then limits the p integration to $|p| < P$. Demonstrate the equivalence of the resultant spectral density with that displayed in (6.4.22).

6-19 For function $f(H)$ of Hamilton operator $H = p^2/(2M) - Fx$ with constant force F show that

$$\operatorname{tr}\{f(H)\} = \int \frac{dx'\,dp'}{2\pi\hbar} f\left(\frac{p'^2}{2M} - Fx'\right).$$

Why did we not make use of this when evaluating the trace of (6.4.16)?

6-20 Prove that

$$\int_{-\infty}^{\infty} d\sigma\,\sigma\,\mathrm{Ai}(\sigma) = 0\,, \qquad \int_{-\infty}^{\infty} d\sigma\,\sigma^2\,\mathrm{Ai}(\sigma) = 0\,,$$

whereas

$$\int_{-\infty}^{\infty} d\sigma\,\sigma^3\,\mathrm{Ai}(\sigma) = 2\,.$$

6-21 Work out $\mathrm{Ai}'(0)$ analogously to $\mathrm{Ai}(0)$, and conclude that

$$-\,\mathrm{Ai}(0)\,\mathrm{Ai}'(0) = \frac{3^{-\frac{1}{2}}}{2\pi}\,.$$

[You will need a fundamental property of the factorial function.]

6-22 We know that

$$\int_{-\infty}^{\infty} d\sigma\ \mathrm{Ai}(\sigma) = 1\,.$$

What are the individual values of

$$\int_0^{\infty} d\sigma\ \mathrm{Ai}(\sigma) \quad \text{and} \quad \int_{-\infty}^{0} d\sigma\ \mathrm{Ai}(\sigma)\ ?$$

6-23 Here is a theorem about $\mathrm{Ai}(\sigma)$:

$$\left[\mathrm{Ai}(\sigma)\right]^2 = \frac{1}{2\pi}\int_{2^{2/3}\sigma}^{\infty} \mathrm{d}\tau\, \mathrm{Ai}(\tau)\frac{1}{\sqrt{2^{-2/3}\tau - \sigma}} .$$

In what sense does this make a true statement about $\left[\mathrm{Ai}(\sigma)\right]^2$ for $-\sigma \gg 1$, if one regards $-\sigma$ as large compared to the significant values of τ?

6-24 For a proof of the theorem in Problem 6-23 first show that

$$\int_{-\infty}^{\infty} \mathrm{d}y\, \delta(a - y^2) = \begin{cases} 0 & \text{if } a < 0 , \\ \frac{1}{\sqrt{a}} & \text{if } a > 0 , \end{cases}$$

and use this and (6.4.13) to establish

$$\frac{1}{2\pi}\int_{2^{2/3}\sigma}^{\infty} \mathrm{d}\tau \frac{\mathrm{Ai}(\tau)}{\sqrt{2^{-2/3}\tau - \sigma}} = \int_{-\infty}^{\infty}\frac{\mathrm{d}x}{2\pi}\int_{-\infty}^{\infty}\frac{\mathrm{d}y}{2\pi}\, \mathrm{e}^{-\frac{\mathrm{i}}{12}x^3 - \mathrm{i}xy^2 - \mathrm{i}\sigma x} .$$

Now introduce new integration variables in accordance with $x = \tau_1 + \tau_2$, $y = \frac{1}{2}(\tau_1 - \tau_2)$ and head home.

6-25 Show that both the orthonormality and the completeness of the $\psi_E(x)$ wave functions (6.6.11) imply

$$\int \mathrm{d}\sigma\, \mathrm{Ai}(\sigma - \sigma_1)\, \mathrm{Ai}(\sigma - \sigma_2) = \delta(\sigma_1 - \sigma_2) ,$$

which is, therefore, a completeness and orthogonality relation for the Airy function. Check it directly.

6-26a Use the defining integral representation of $\mathrm{Ai}(\sigma)$ to demonstrate that

$$\int_{-\infty}^{\infty} \mathrm{d}\sigma\, \mathrm{Ai}(\sigma)\, \mathrm{Ai}(-\sigma) = 2^{-\frac{1}{3}}\, \mathrm{Ai}(0) .$$

6-26b Extend the argument to establish

$$\int_{-\infty}^{\infty} \mathrm{d}\tau\, \mathrm{Ai}(\sigma + \tau)\, \mathrm{Ai}(\sigma - \tau)\, \mathrm{e}^{2\mathrm{i}k\tau} = 2^{-\frac{1}{3}}\, \mathrm{Ai}\big(2^{\frac{2}{3}}(\sigma + k^2)\big) .$$

6-26c As a check of consistency, derive a differential equation for this integral by first showing that

$$\left[\frac{\partial^2}{\partial\sigma^2} + \frac{\partial^2}{\partial\tau^2}\right]\left[\mathrm{Ai}(\sigma + \tau)\, \mathrm{Ai}(\sigma - \tau)\right] = 4\sigma\, \mathrm{Ai}(\sigma + \tau)\, \mathrm{Ai}(\sigma - \tau) .$$

6-27 According to Problem 6-23,

$$\left[\,\mathrm{Ai}(0)\right]^2 = \frac{2^{-2/3}}{\pi}\int_0^\infty \frac{\mathrm{d}\tau}{\sqrt{\tau}}\,\mathrm{Ai}(\tau)\,.$$

Evaluate this integral and arrive at

$$\left[\,\mathrm{Ai}(0)\right]^2 = \frac{1}{\pi^{3/2}}\frac{1}{2^{5/3}}\frac{1}{3^{5/6}}\left(-\tfrac{5}{6}\right)!\,.$$

Does this check out numerically?

6-28 Verify that another consequence of Problem 6-23 is

$$-\,\mathrm{Ai}(0)\,\mathrm{Ai}'(0) = -\frac{1}{2\pi}\int_0^\infty \frac{\mathrm{d}\tau}{\sqrt{\tau}}\,\mathrm{Ai}'(\tau)\,.$$

Evaluate this integral to recover the result in Problem 6-21.

6-29 Replace the unbounded x integral in (6.4.19) by the bounded one of (6.4.20), then perform the σ integration and the x integration (in this order) to arrive at

$$\mathrm{tr}\left\{\mathrm{e}^{-\frac{\mathrm{i}}{\hbar}Ht}\right\}_L = \int_{-\infty}^{\infty}\mathrm{d}E\;m_L(E)\,\mathrm{e}^{-\frac{\mathrm{i}}{\hbar}Et}$$

where

$$m_L(E) = \left(\frac{2M}{\hbar^2F^2}\right)^{\frac{1}{3}}\left(\left[\,\mathrm{Ai}'(\sigma)\right]^2 - \sigma\left[\,\mathrm{Ai}(\sigma)\right]^2\right)_{\sigma=(2MF/\hbar^2)^{\frac{1}{3}}(-E/F-L)}\,.$$

Show that (6.4.22) obtains under the circumstances of (6.4.21).

6-30 One degree of freedom, Hamilton operator $H = p^2/(2M) + \frac{1}{2}M\omega^2x^2$ with constant frequency ω. Use the action principle to find $\langle p',t|x'\rangle$. Can you identify the momentum wave functions $\psi_n(p')$? If you find it convenient, introduce dimensionless variables.

6-31 In order to perform, for $E > 0$, the limit $F \to 0$ in (6.6.7) first write $p = \pm\sqrt{2ME} + \overline{p}$, put aside all phase factors that do not depend on $\overline{p}$ (why is this allowed?), and get

$$\psi_{E,\pm}(p) = \frac{1}{\sqrt{\mp\mathrm{i}2\pi\hbar F}}\,\mathrm{e}^{\mp\frac{\mathrm{i}}{\hbar F}\sqrt{E/(2M)}\,\overline{p}^2}\,\mathrm{e}^{-\frac{\mathrm{i}}{\hbar F}\frac{\overline{p}^3}{6M}}\,.$$

Then show that

$$\frac{1}{\sqrt{\mp\mathrm{i}2\pi\hbar F}}\,\mathrm{e}^{\mp\frac{\mathrm{i}}{\hbar F}\sqrt{E/(2M)}\,\overline{p}^2} \to (2E/M)^{-\frac{1}{4}}\delta(\overline{p}) \quad \text{as } F \to 0\,,$$

and note that $\overline{p}^3/F \sim \sqrt{F}$ in this limit. Verify that your answer agrees with the $\psi_{E,\pm}(p)$ obtained from (6.6.3).

6-32 The nth oscillator state (dimensional variables used) has classical turning points at $q = \pm\sqrt{2n+1}$ (check this). Use the WKB approximation for the wave function in the classically forbidden region $q > \sqrt{2n+1}$, retaining the two leading terms in an expansion of the exponent for $q \gg \sqrt{2n+1}$. How does your result compare with the asymptotic limit of the known wave function?

6-33 You know the WKB approximation when the classical region is on the right. Find, in any manner, the WKB wave function for a classical region on the left. Consider a potential energy $V(x)$ with two classical turning points:

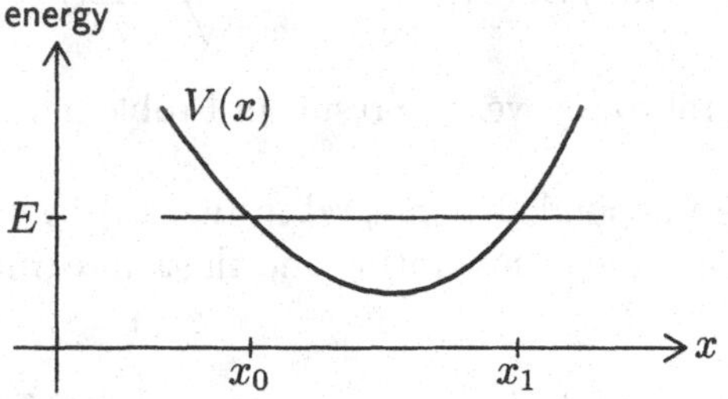

Write the WKB wave function for $x > x_0$, and for $x < x_1$. To within a possible minus sign these must be the same in the common region $x_0 < x < x_1$. Conclude that

$$\frac{1}{\pi\hbar}\int_{x_0}^{x_1} dx\,\sqrt{2M[E_n - V(x)]} = n + \tfrac{1}{2} \quad \text{for } n = 0, 1, 2, \ldots$$

determines the WKB approximations for the energy eigenvalues E_n.

6-34a Consider the oscillator potential $V(x) = \frac{1}{2}M\omega^2x^2$ and use the result of Problem 6-33 to work out the possible energy values. Compare with the known result.

6-34b Repeat for the linear potential $V(x) = F|x|$.

6-35a For a very different derivation of the WKB energies of Problem 6-33 return to the approximate trace evaluation in Problem 6-17b. Apply it to

$$f_E(H) = \begin{cases} 1 \text{ if } H < E\,, \\ 0 \text{ if } H > E\,, \end{cases}$$

so that tr $\{f_E(H)\}$ is the count of energy eigenstates below energy E. Equate the semiclassical value of tr $\{f_{E_n}(H)\}$ with $n + \frac{1}{2}$ (why is that reasonable?) and show that this reproduces the WKB quantization rule.

6-35b Now use the leading quantum correction of Problem 6-17b to improve upon the WKB rule. Find the implied corrections for the oscillator potential $V(x) = \frac{1}{2}M\omega^2x^2$ and for the linear potential $V(x) = F|x|$. Compare the latter with (6.10.12).

6-36 Consider a family $\psi_\lambda(q) = \psi(\lambda q)$ of scaled trial functions in (6.11.8). Which value of λ gives the lowest upper bound for $\mathcal{E}_0$? Use this to arrive at a scale-invariant version of (6.11.8). Try it for $\psi(q) = \exp(-q^\alpha)$ with $\alpha = \frac{3}{2}, \frac{7}{4}, 2$.

7. Harmonic Oscillators

7.1 Non-Hermitian operators

Now we turn to the treatment of the oscillator using non-Hermitian operators, with an eye toward more general dynamical circumstances. For simplicity we use dimensionless variables q, p, and the non-Hermitian variables $y, y^\dagger$ closely related to them, rather than dimensional x, p_x:

$$\begin{aligned} x &= \sqrt{\frac{\hbar}{M\omega}}\, q \longrightarrow q = \frac{1}{\sqrt{2}}(y^\dagger + y) , \\ p_x &= \sqrt{\hbar M \omega}\, p \longrightarrow p = \frac{\mathrm{i}}{\sqrt{2}}(y^\dagger - y) , \\ \frac{1}{\mathrm{i}\hbar}&[x, p_x] = \frac{1}{\mathrm{i}}[q, p] = 1 , \qquad [y, y^\dagger] = 1 , \end{aligned} \tag{7.1.1}$$

and express the energy in frequency units,

$$\begin{aligned} H = \frac{p_x^2}{2M} + \frac{1}{2}M\omega^2 x^2 &= \hbar\omega\left(\tfrac{1}{2}p^2 + \tfrac{1}{2}q^2\right) \\ &= \hbar\omega\left(y^\dagger y + \tfrac{1}{2}\right) \longrightarrow \omega y^\dagger y , \end{aligned} \tag{7.1.2}$$

where, in addition, the irrelevant constant $\frac{1}{2}\hbar\omega \mathrel{\widehat{=}} \frac{1}{2}\omega$ is subtracted, so that the eigenvalues of H are now ω, 2ω, 3ω, ...

Since $[y, \mathrm{i}y^\dagger] = \mathrm{i}$ analogous to $[q, p] = \mathrm{i}$, it must be possible to use, in addition to Lagrangian

$$L = p\frac{\mathrm{d}q}{\mathrm{d}t} - H(q, p, t) \tag{7.1.3}$$

the Lagrangian

$$\overline{L} = \mathrm{i}y^\dagger \frac{\mathrm{d}y}{\mathrm{d}t} - H(y, y^\dagger, t) , \tag{7.1.4}$$

and the generators

$$G_y = \mathrm{i}y^\dagger\, \delta y , \qquad G_{y^\dagger} = -\mathrm{i}y\, \delta y^\dagger \tag{7.1.5}$$

as the analogs of $G_q = p\,\delta q$ and $G_p = -q\,\delta p$. In accordance with Section 5.7, the condition is that

$$L - \overline{L} = \frac{\mathrm{d}}{\mathrm{d}t} w \,, \tag{7.1.6}$$

with, at any t,

$$\begin{aligned} \delta w &= (G_q + G_t) - (G_y + \overline{G}_t) \\ &= (p\delta q - H\delta t) - (\mathrm{i}y^\dagger \delta y - \overline{H}\delta t) \,. \end{aligned} \tag{7.1.7}$$

The transformation of interest here does not involve t:

$$\overline{H} = H \,, \qquad \text{so that} \quad G_t = -H\,\delta t = -\overline{H}\,\delta t = \overline{G}_t \,, \tag{7.1.8}$$

and we have

$$p\delta q - \mathrm{i}y^\dagger \delta y = \delta w(q, y) \,. \tag{7.1.9}$$

Now

$$p = \mathrm{i}(q - \sqrt{2}\,y);, \qquad y^\dagger = \sqrt{2}\,q - y \,, \tag{7.1.10}$$

so

$$\begin{aligned} \delta w &= \mathrm{i}(q - \sqrt{2}\,y)\delta q - \mathrm{i}(\sqrt{2}\,q - y)\delta y \\ &= \delta\left(\frac{\mathrm{i}}{2}q^2 + \frac{\mathrm{i}}{2}y^2 - \mathrm{i}\sqrt{2}\,qy\right) . \end{aligned} \tag{7.1.11}$$

This is used directly in finding the transformation function $\langle q' | y' \rangle$,

$$\begin{aligned} \delta\langle q' | y' \rangle &= \mathrm{i}\langle q' | (G_q - G_y) | y' \rangle = \mathrm{i}\langle q' | \delta w | y' \rangle \\ &= \mathrm{i}\langle q' | y' \rangle\,\delta\left(\frac{\mathrm{i}}{2}q'^2 + \frac{\mathrm{i}}{2}y'^2 - \mathrm{i}\sqrt{2}\,q'y'\right) , \end{aligned} \tag{7.1.12}$$

giving

$$\langle q' | y' \rangle = \pi^{-\frac{1}{4}}\,\mathrm{e}^{-\frac{1}{2}q'^2 - \frac{1}{2}y'^2 + \sqrt{2}\,q'y'} \tag{7.1.13}$$

which is the long known result (2.7.30) including the constant that normalizes the vector $|y' = 0\rangle$.

With the Hamilton operator of (7.1.2), the non-Hermitian equations of motion are

$$\frac{\mathrm{d}y}{\mathrm{d}t} = \frac{\partial H}{\partial \mathrm{i}y^\dagger} = -\mathrm{i}\omega y \tag{7.1.14}$$

and

$$\mathrm{i}\frac{\mathrm{d}y^\dagger}{\mathrm{d}t} = -\frac{\partial H}{\partial y} = -\omega y^\dagger . \tag{7.1.15}$$

The simplification here is evident, the equations of motion are solved immediately:

$$y(t) = \mathrm{e}^{-\mathrm{i}\omega t} y , \quad y^\dagger(t) = \mathrm{e}^{\mathrm{i}\omega t} y^\dagger . \tag{7.1.16}$$

Now consider the time transformation function $\langle y^{\dagger\prime}, t | y''\rangle$,

$$\begin{aligned}\delta\langle y^{\dagger\prime}, t | y''\rangle &= \mathrm{i}\langle y^{\dagger\prime}, t \big| \Big[(G_{y^\dagger} + G_t)(t) - G_y \Big] \big| y''\rangle \\ &= \mathrm{i}\langle y^{\dagger\prime}, t | [-\mathrm{i}\delta y^{\dagger\prime} y(t) - \delta t\, \omega y^\dagger y - \mathrm{i} y^\dagger \delta y''] | y''\rangle ,\end{aligned} \tag{7.1.17}$$

where

$$\begin{aligned}[\cdots] &= -\mathrm{i}\delta y^{\dagger\prime}\, \mathrm{e}^{-\mathrm{i}\omega t}\, \underset{\displaystyle y''}{\underset{\downarrow}{y}} - \delta t\, \omega\, \underset{\displaystyle y^{\dagger\prime}}{\underset{\downarrow}{y^\dagger(t)}}\, \mathrm{e}^{-\mathrm{i}\omega t}\, \underset{\displaystyle y''}{\underset{\downarrow}{y}} - \mathrm{i}\, \underset{\displaystyle y^{\dagger\prime}}{\underset{\downarrow}{y^\dagger(t)}}\, \mathrm{e}^{-\mathrm{i}\omega t} \delta y'' \\ &\to \delta \left[-\mathrm{i} y^{\dagger\prime}\, \mathrm{e}^{-\mathrm{i}\omega t} y'' \right] ,\end{aligned} \tag{7.1.18}$$

so

$$\langle y^{\dagger\prime}, t | y''\rangle = \mathrm{e}^{y^{\dagger\prime}\mathrm{e}^{-\mathrm{i}\omega t} y''} , \tag{7.1.19}$$

which satisfies the initial condition

$$\langle y^{\dagger\prime}, t | y''\rangle \to \langle y^{\dagger\prime} | y''\rangle = \mathrm{e}^{y^{\dagger\prime} y''} \qquad \text{for } t \to 0 . \tag{7.1.20}$$

Now let's use the time transformation function as

$$\langle y^{\dagger\prime}, t | y''\rangle = \langle y^{\dagger\prime} | \mathrm{e}^{-\mathrm{i}tH} | y''\rangle \tag{7.1.21}$$

which is immediate from the power series

$$\mathrm{e}^{y^{\dagger\prime}\mathrm{e}^{-\mathrm{i}\omega t} y''} = \sum_{n=0}^{\infty} \frac{(y^{\dagger\prime})^n}{\sqrt{n!}}\, \mathrm{e}^{-\mathrm{i}n\omega t}\, \frac{(y'')^n}{\sqrt{n!}} \tag{7.1.22}$$

telling us that

$$H' = n\omega \qquad \text{with} \quad n = 0, 1, 2, \ldots \tag{7.1.23}$$

are the energy eigenvalues and

$$\langle y^{\dagger\prime} | n\rangle = \frac{(y^{\dagger\prime})^n}{\sqrt{n!}} , \qquad \langle n | y''\rangle = \frac{(y'')^n}{\sqrt{n!}} \tag{7.1.24}$$

are the wave functions of the energy eigenvectors. This, of course, we already know. Indeed, we have seen the essential mathematical details in Section 2.9; that earlier treatment is recovered upon replacing ωt by $-\mathrm{i}\beta$.

Further, we know that the transformation function (7.1.13),

$$\langle q'|y'\rangle = \pi^{-\frac{1}{4}}\,\mathrm{e}^{-\frac{1}{2}q'^2 + \sqrt{2}\,q'y' - \frac{1}{2}y'^2} = \sum_{n=0}^{\infty}\psi_n(q')\frac{(y')^n}{\sqrt{n!}}\,, \tag{7.1.25}$$

produces the Hermite polynomial form for $\langle q'|n\rangle = \psi_n(q')$. It's also possible to use this generating function to arrive at asymptotic forms of the $\psi_n(q')$. We leave this matter to Problem 7-3, and turn to a study of more general dynamics.

7.2 Driven oscillator

Consider the time dependent Hamiltonian

$$H = \omega y^\dagger y + \kappa(t)^* y + \kappa(t) y^\dagger\,. \tag{7.2.1}$$

It gives the equations of motion

$$\mathrm{i}\frac{\mathrm{d}y}{\mathrm{d}t} = \frac{\partial H}{\partial y^\dagger} = \omega y + \kappa(t) \tag{7.2.2}$$

and

$$-\mathrm{i}\frac{\mathrm{d}y^\dagger}{\mathrm{d}t} = \omega y^\dagger + \kappa(t)^*\,, \tag{7.2.3}$$

which describe the system as driven by external forces. Clearly we can solve these first-order differential equations. But it is important to realize first the boundary conditions that accompany them. The time development of the system is given by

$$\langle y^{\dagger\prime}, t_1|y'', t_2\rangle^\kappa \tag{7.2.4}$$

where κ is written to recall the presence of the external forces, as distinguished from $(\kappa = 0)$

$$\langle y^{\dagger\prime}, t_1|y'', t_2\rangle = \mathrm{e}^{y^{\dagger\prime}\,\mathrm{e}^{-\mathrm{i}\omega(t_1 - t_2)}y''}\,. \tag{7.2.5}$$

Accordingly, it is natural to ask how $\langle 1|2\rangle^\kappa$ changes as we turn on the forces, as given by

$$\delta_\kappa\langle 1|2\rangle^\kappa = \mathrm{i}\langle 1|\Big[-\int_2^1 \mathrm{d}t\ (\delta\kappa^*\, y + \delta\kappa\, y^\dagger)\,(t)\Big]|2\rangle \tag{7.2.6}$$

where, of course, $\langle 1| = \langle y^{\dagger\prime}, t_1|$ and $|2\rangle = |y'', t_2\rangle$. Now we see that we should find $y(t)$ in terms of the given $y(t_2) \to y''$, and $y^\dagger(t)$ in terms of the given $y^\dagger(t_1) \to y^{\dagger\prime}$.

So, begin with (7.2.2) or, equivalently,

$$\frac{\mathrm{d}}{\mathrm{d}t}\Big(\mathrm{e}^{\mathrm{i}\omega t}y(t)\Big) = -\mathrm{i}\,\mathrm{e}^{\mathrm{i}\omega t}\kappa(t) \tag{7.2.7}$$

giving

$$\mathrm{e}^{\mathrm{i}\omega t}y(t) = \mathrm{e}^{\mathrm{i}\omega t_2}y(t_2) - \mathrm{i}\int_{t_2}^{t}\mathrm{d}t'\;\mathrm{e}^{\mathrm{i}\omega t'}\kappa(t')\,, \tag{7.2.8}$$

which is

$$y(t) = \mathrm{e}^{-\mathrm{i}\omega(t-t_2)}y(t_2) - \mathrm{i}\int_{t_2}^{t}\mathrm{d}t'\;\mathrm{e}^{-\mathrm{i}\omega(t-t')}\kappa(t')\,. \tag{7.2.9}$$

For the $y^\dagger(t)$ equation, it suffices to take the adjoint of the above, while replacing t_2 by t_1:

$$y^\dagger(t) = y^\dagger(t_1)\,\mathrm{e}^{-\mathrm{i}\omega(t_1-t)} - \mathrm{i}\int_{t}^{t_1}\mathrm{d}t'\,\kappa(t')^*\,\mathrm{e}^{-\mathrm{i}\omega(t'-t)}\,. \tag{7.2.10}$$

It will be helpful to introduce Heaviside's* unit step function

$$\eta(t-t') = \begin{cases} 1 & \text{for } t > t'\,, \\ 0 & \text{for } t < t'\,, \end{cases} \tag{7.2.11}$$

so that

$$\begin{aligned} \int_{t_2}^{t}\mathrm{d}t'\;\mathrm{e}^{-\mathrm{i}\omega(t-t')}\kappa(t') &= \int_{t_2}^{t_1}\mathrm{d}t'\,\eta(t-t')\,\mathrm{e}^{-\mathrm{i}\omega(t-t')}\kappa(t')\,, \\ \int_{t}^{t_1}\mathrm{d}t'\,\kappa(t')^*\,\mathrm{e}^{-\mathrm{i}\omega(t'-t)} &= \int_{t_2}^{t_1}\mathrm{d}t'\,\kappa(t')^*\eta(t'-t)\,\mathrm{e}^{-\mathrm{i}\omega(t'-t)}\,. \end{aligned} \tag{7.2.12}$$

Thus, the integral in (7.2.6) effectively becomes

$$\begin{aligned} &\int_{t_2}^{t_1}\mathrm{d}t\;\left(\delta\kappa^*\,y + \delta\kappa\,y^\dagger\right) \to \\ &\quad \int_{t_2}^{t_1}\mathrm{d}t\,\delta\kappa(t)^*\left[\mathrm{e}^{-\mathrm{i}\omega(t-t_2)}y'' - \mathrm{i}\int_{t_2}^{t_1}\mathrm{d}t'\,\eta(t-t')\,\mathrm{e}^{-\mathrm{i}\omega(t-t')}\kappa(t')\right] \\ &\quad + \int_{t_2}^{t_1}\mathrm{d}t\;\left[y^{\dagger\prime}\,\mathrm{e}^{-\mathrm{i}\omega(t_1-t)} - \mathrm{i}\int_{t_2}^{t_1}\mathrm{d}t'\,\kappa(t')^*\eta(t'-t)\,\mathrm{e}^{-\mathrm{i}\omega(t'-t)}\right]\delta\kappa(t) \end{aligned} \tag{7.2.13}$$

*Oliver HEAVISIDE (1850–1925)

or, on exchanging $t \leftrightarrow t'$ in the last (double) integral,

$$\int_{t_2}^{t_1} \mathrm{d}t\ (\delta\kappa^* y + \delta\kappa\, y^\dagger) \to$$
$$\delta\left[\int_2^1 \mathrm{d}t\, \kappa(t)^*\, \mathrm{e}^{-\mathrm{i}\omega t}\, \mathrm{e}^{\mathrm{i}\omega t_2} y'' + \mathrm{e}^{-\mathrm{i}\omega t_1} y^{\dagger\prime} \int_2^1 \mathrm{d}t\ \mathrm{e}^{\mathrm{i}\omega t} \kappa(t) \right.$$
$$\left. -\mathrm{i}\int_2^1 \mathrm{d}t\, \mathrm{d}t'\, \kappa(t)^* \eta(t-t')\, \mathrm{e}^{-\mathrm{i}\omega(t-t')} \kappa(t') \right] . \qquad (7.2.14)$$

This gives immediately the time transformation function

$$\begin{aligned}\langle y^{\dagger\prime}, t_1 | y'', t_2\rangle^\kappa = &\, \mathrm{e}^{y^{\dagger\prime}\, \mathrm{e}^{-\mathrm{i}\omega(t_1 - t_2)} y''} \\ &\times \mathrm{e}^{-\mathrm{i} y^{\dagger\prime}\, \mathrm{e}^{-\mathrm{i}\omega t_1} \int_2^1 \mathrm{d}t\ \mathrm{e}^{\mathrm{i}\omega t} \kappa(t)} \\ &\times \mathrm{e}^{-\mathrm{i}\int_2^1 \mathrm{d}t\, \kappa(t)^*\, \mathrm{e}^{-\mathrm{i}\omega t}\, \mathrm{e}^{\mathrm{i}\omega t_2} y''} \\ &\times \mathrm{e}^{-\int_2^1 \mathrm{d}t\, \mathrm{d}t'\, \kappa(t)^* \eta(t-t')\, \mathrm{e}^{-\mathrm{i}\omega(t-t')} \kappa(t')} . \end{aligned} \qquad (7.2.15)$$

We note that the first factor is $\langle y^{\dagger\prime}, t_1 | y'', t_2\rangle$, the $\kappa \equiv 0$ time transformation function of the not-driven oscillator, and the last factor is $\langle 0, t_1 | 0, t_2\rangle^\kappa$, the $y'' = 0 \to y^{\dagger\prime} = 0$ transition amplitude of the driven oscillator. The equivalent form

$$\langle y^{\dagger\prime}, t_1 | y'', t_2\rangle^\kappa = \langle y^{\dagger\prime}, t_1 | y'', t_2\rangle\, \mathrm{e}^{-\mathrm{i}[y^{\dagger\prime}\, \mathrm{e}^{-\mathrm{i}\omega t_1} \gamma + \gamma^*\, \mathrm{e}^{\mathrm{i}\omega t_2} y'']} \langle 0, t_1 | 0, t_2\rangle^\kappa \qquad (7.2.16)$$

with

$$\gamma = \int_2^1 \mathrm{d}t\ \mathrm{e}^{\mathrm{i}\omega t} \kappa(t) , \qquad \gamma^* = \int_2^1 \mathrm{d}t\, \kappa(t)^*\, \mathrm{e}^{-\mathrm{i}\omega t} \qquad (7.2.17)$$

reflects these observations.

7.2.1 Time-independent drive

First, let's run a check on this. We wrote $\kappa(t)$, but that includes κ independent of t:

$$\begin{aligned} H &= \omega y^\dagger y + \kappa^* y + \kappa y^\dagger \\ &= \omega\left(y^\dagger + \frac{\kappa}{\omega}\right)\left(y + \frac{\kappa}{\omega}\right) - \frac{|\kappa|^2}{\omega} \\ &= \omega(y^\dagger + \lambda^*)(y + \lambda) - \omega|\lambda|^2 \quad \text{with } \lambda \equiv \kappa/\omega . \end{aligned} \qquad (7.2.18)$$

The operators in the latter version are mutually adjoint and obey the commutation relation

$$[y+\lambda, y^\dagger+\lambda^*]=1\,. \tag{7.2.19}$$

Therefore we immediately see the spectrum:

$$E_n^\kappa = n\omega - \frac{|\kappa|^2}{\omega} = \left(n-|\lambda|^2\right)\omega\,, \tag{7.2.20}$$

just lowered by a constant. The eigenvectors are clearly given by

$$|n,\lambda\rangle = \frac{\left(y^\dagger+\lambda^*\right)^n}{\sqrt{n!}}|0,\lambda\rangle\,, \qquad (y+\lambda)\,|0,\lambda\rangle = 0\,. \tag{7.2.21}$$

Now

$$y+\lambda = \mathrm{e}^{-\lambda y^\dagger}\, y\, \mathrm{e}^{\lambda y^\dagger}\,, \tag{7.2.22}$$

so that

$$y\,\mathrm{e}^{\lambda y^\dagger}|0,\lambda\rangle = 0 \tag{7.2.23}$$

or

$$|0,\lambda\rangle = \frac{\mathrm{e}^{-\lambda y^\dagger}|0\rangle}{\sqrt{\langle 0|\,\mathrm{e}^{-\lambda^* y}\,\mathrm{e}^{-\lambda y^\dagger}|0\rangle}}\,, \tag{7.2.24}$$

where the denominator ensures proper normalization; it's explicit value is

$$\langle 0|\,\mathrm{e}^{-\lambda^* y}\,\mathrm{e}^{-\lambda y^\dagger}|0\rangle = \sum_{n=0}^{\infty}\frac{|\lambda|^{2n}}{n!} = \mathrm{e}^{|\lambda|^2}\,. \tag{7.2.25}$$

Accordingly,

$$|0,\lambda\rangle = \mathrm{e}^{-\frac{1}{2}|\lambda|^2}\,\mathrm{e}^{-\lambda y^\dagger}|0\rangle \tag{7.2.26}$$

which is conveyed, along with $|n,\lambda\rangle$, by the wave functions

$$\langle y^{\dagger\prime}|n,\lambda\rangle = \frac{\left(y^{\dagger\prime}+\lambda^*\right)^n}{\sqrt{n!}}\,\mathrm{e}^{-\lambda y^{\dagger\prime}}\,\mathrm{e}^{-\frac{1}{2}|\lambda|^2}\,. \tag{7.2.27}$$

Does the time transformation function (7.2.15) produce these results for $\kappa(t)=\kappa\;(=\omega\lambda)$? We observe that

$$\begin{aligned}\gamma &= \int_{t_2}^{t_1}\mathrm{d}t\;\mathrm{e}^{\mathrm{i}\omega t}\kappa = -\mathrm{i}\lambda\left(\mathrm{e}^{\mathrm{i}\omega t_1}-\mathrm{e}^{\mathrm{i}\omega t_2}\right)\,,\\ \gamma^* &= \int_{t_2}^{t_1}\mathrm{d}t\;\kappa^*\,\mathrm{e}^{-\mathrm{i}\omega t} = \mathrm{i}\lambda^*\left(\mathrm{e}^{-\mathrm{i}\omega t_1}-\mathrm{e}^{-\mathrm{i}\omega t_2}\right)\,,\end{aligned} \tag{7.2.28}$$

and

$$\int_{t_2}^{t_1} \mathrm{d}t\,\mathrm{d}t'\,\kappa^*\eta(t-t')\,\mathrm{e}^{-\mathrm{i}\omega(t-t')}\kappa = -\mathrm{i}\omega|\lambda|^2 \int_{t_2}^{t_1} \mathrm{d}t\,\left(1-\mathrm{e}^{-\mathrm{i}\omega(t-t_2)}\right)$$
$$= -\mathrm{i}\omega|\lambda|^2 \left[T + \frac{1}{\mathrm{i}\omega}\left(\mathrm{e}^{-\mathrm{i}\omega T}-1\right)\right] \tag{7.2.29}$$

where $T = t_1 - t_2$. Therefore,

$$\begin{aligned}\langle y^{\dagger\prime}, t_1 | y'', t_2\rangle^\kappa &= \mathrm{e}^{y^{\dagger\prime}\,\mathrm{e}^{-\mathrm{i}\omega T} y'' - \lambda(1-\mathrm{e}^{-\mathrm{i}\omega T})y^{\dagger\prime} - \lambda^*(\mathrm{e}^{-\mathrm{i}\omega T}-1)y''} \\ &\quad\times \mathrm{e}^{\mathrm{i}|\lambda|^2[\omega T - \mathrm{i}(\mathrm{e}^{-\mathrm{i}\omega T}-1)]} \\ &= \mathrm{e}^{\mathrm{i}\omega|\lambda|^2 T}\,\mathrm{e}^{-y^{\dagger\prime}\lambda}\,\mathrm{e}^{(y^{\dagger\prime}+\lambda^*)\,\mathrm{e}^{-\mathrm{i}\omega T}(y''+\lambda)}\,\mathrm{e}^{-\lambda^* y''}\,\mathrm{e}^{-|\lambda|^2} \\ &= \sum_{n=0}^{\infty} \langle y^{\dagger\prime}|n,\lambda\rangle\,\mathrm{e}^{-\mathrm{i}E_n^\kappa t}\langle n,\lambda|y''\rangle \end{aligned} \tag{7.2.30}$$

is the appropriate time transformation function where, indeed, E_n^κ is the energy of (7.2.20), $\langle y^{\dagger\prime}|n,\lambda\rangle$ are the wave functions of (7.2.27), and $\langle n,\lambda|y''\rangle$ are their adjoints.

7.2.2 Slowly varying drive

Next, suppose that $\kappa(t)$ changes *slowly* from 0 at t_2 to $\omega\lambda$ at t_1,

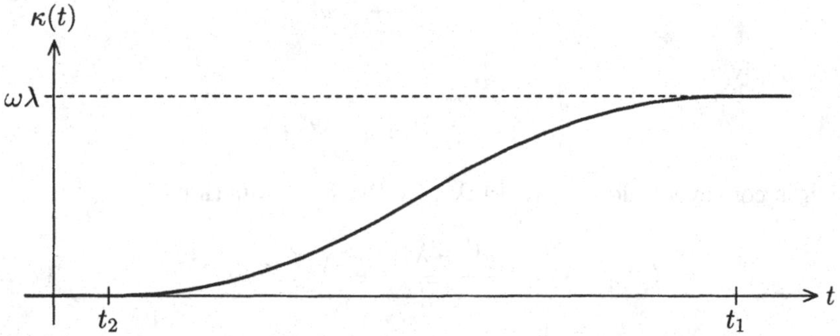

$$\omega(t_1 - t_2) \gg 1\,, \quad \left|\frac{1}{\omega^2}\frac{\mathrm{d}^2\kappa(t)}{\mathrm{d}t^2}\right| \ll |\kappa(t)|\,, \quad \frac{\mathrm{d}\kappa}{\mathrm{d}t} = 0 \text{ for } t < t_2 \text{ and } t > t_1\,. \tag{7.2.31}$$

What happens? We evaluate the integrals needed in (7.2.15) with the aid of partial integrations. First note that

$$\mathrm{e}^{\mathrm{i}\omega t}\kappa(t) = \frac{\mathrm{d}}{\mathrm{d}t}\left[\frac{\mathrm{e}^{\mathrm{i}\omega t}}{\mathrm{i}\omega}\kappa(t) - \frac{\mathrm{e}^{\mathrm{i}\omega t}}{(\mathrm{i}\omega)^2}\frac{\mathrm{d}\kappa}{\mathrm{d}t}(t)\right] + \frac{\mathrm{e}^{\mathrm{i}\omega t}}{(\mathrm{i}\omega)^2}\frac{\mathrm{d}^2}{\mathrm{d}t^2}\kappa(t)\,,$$

$$\cong \frac{\mathrm{d}}{\mathrm{d}t}\left[\frac{\mathrm{e}^{\mathrm{i}\omega t}}{\mathrm{i}\omega}\kappa(t) + \frac{\mathrm{e}^{\mathrm{i}\omega t}}{\omega^2}\frac{\mathrm{d}\kappa}{\mathrm{d}t}(t)\right] \tag{7.2.32}$$

so that

$$\gamma = \int_2^1 \mathrm{d}t\, \mathrm{e}^{\mathrm{i}\omega t}\kappa(t) \cong \frac{\mathrm{e}^{\mathrm{i}\omega t_1}}{\mathrm{i}\omega}\kappa(t_1) = -\mathrm{i}\lambda\,\mathrm{e}^{\mathrm{i}\omega t_1} \tag{7.2.33}$$

and

$$\gamma^* = \int_2^1 \mathrm{d}t\,\kappa(t)^*\,\mathrm{e}^{-\mathrm{i}\omega t} \cong \mathrm{i}\lambda^*\,\mathrm{e}^{-\mathrm{i}\omega t_1}\,. \tag{7.2.34}$$

The same approximation is used in

$$\begin{aligned}
&\int_2^1 \mathrm{d}t\,\mathrm{d}t'\,\kappa(t)^*\eta(t-t')\,\mathrm{e}^{-\mathrm{i}\omega(t-t')}\kappa(t') \\
&\cong \int_{t_2}^{t_1}\mathrm{d}t\,\kappa(t)^*\int_{t_2}^{t}\mathrm{d}t'\,\frac{\partial}{\partial t'}\left[\frac{\mathrm{e}^{-\mathrm{i}\omega(t-t')}}{\mathrm{i}\omega}\kappa(t) + \frac{\mathrm{e}^{-\mathrm{i}\omega(t-t')}}{\omega^2}\frac{\mathrm{d}\kappa}{\mathrm{d}t}(t)\right] \\
&= \frac{1}{\omega^2}\int_{t_2}^{t_1}\mathrm{d}t\left[-\mathrm{i}\omega|\kappa(t)|^2 + \kappa(t)^*\frac{\mathrm{d}}{\mathrm{d}t}\kappa(t)\right]. \qquad (7.2.35)
\end{aligned}$$

With

$$|\kappa(t)|^2 = \frac{\mathrm{d}}{\mathrm{d}t}\left[t|\kappa(t)|^2\right] - t\frac{\mathrm{d}}{\mathrm{d}t}|\kappa(t)|^2 \tag{7.2.36}$$

and

$$\kappa(t)^*\frac{\mathrm{d}}{\mathrm{d}t}\kappa(t) = \underbrace{\frac{1}{2}\frac{\mathrm{d}}{\mathrm{d}t}|\kappa(t)|^2}_{\text{real}} + \underbrace{\frac{1}{2}\left[\kappa(t)^*\frac{\mathrm{d}\kappa}{\mathrm{d}t}(t) - \frac{\mathrm{d}\kappa(t)^*}{\mathrm{d}t}\kappa(t)\right]}_{\text{imaginary}} \tag{7.2.37}$$

we write this as

$$\begin{aligned}
&\int_2^1 \mathrm{d}t\,\mathrm{d}t'\,\kappa(t)^*\eta(t-t')\,\mathrm{e}^{-\mathrm{i}\omega(t-t')}\kappa(t') \\
&\cong -\mathrm{i}\frac{t_1}{\omega}|\kappa(t_1)|^2 + \frac{1}{2\omega^2}|\kappa(t_1)|^2 - \mathrm{i}\phi \\
&= -\mathrm{i}\omega t_1|\lambda|^2 + \frac{1}{2}|\lambda|^2 - \mathrm{i}\phi\,, \qquad (7.2.38)
\end{aligned}$$

where the real phase ϕ is given by

$$\phi = \int_2^1 \mathrm{d}t \left[\frac{\mathrm{i}}{2\omega} \left(\kappa^* \frac{\mathrm{d}\kappa}{\mathrm{d}t} - \frac{\mathrm{d}\kappa^*}{\mathrm{d}t} \kappa \right) - \frac{t}{\omega} \left(\kappa^* \frac{\mathrm{d}\kappa}{\mathrm{d}t} + \frac{\mathrm{d}\kappa^*}{\mathrm{d}t} \kappa \right) \right] . \tag{7.2.39}$$

Therefore, the time transformation function (7.2.15) is here approximated by

$$\begin{aligned} \langle y^{\dagger\prime}, t_1 | y'', t_2 \rangle^\kappa &= \mathrm{e}^{\mathrm{i}\phi}\, \mathrm{e}^{\mathrm{i}\omega t_1 |\lambda|^2}\, \mathrm{e}^{-\frac{1}{2}|\lambda|^2}\, \mathrm{e}^{-y^{\dagger\prime}\lambda}\, \mathrm{e}^{(y^{\dagger\prime} + \lambda^*)\, \mathrm{e}^{-\mathrm{i}\omega(t_1 - t_2)} y''} \\ &= \mathrm{e}^{\mathrm{i}\phi} \sum_{n=0}^{\infty} \underbrace{\frac{(y^{\dagger\prime} + \lambda^*)^n}{\sqrt{n!}}\, \mathrm{e}^{-y^{\dagger\prime}\lambda}\, \mathrm{e}^{-\frac{1}{2}|\lambda|^2}\, \mathrm{e}^{-\mathrm{i}(n - |\lambda|^2)\omega t_1}}_{= \langle y^{\dagger\prime} | n, \lambda \rangle} \\ &\quad \times \mathrm{e}^{\mathrm{i}n\omega t_2} \underbrace{\frac{(y'')^n}{\sqrt{n!}}}_{= \langle n | y'' \rangle} \end{aligned} \tag{7.2.40}$$

or

$$\langle y^{\dagger\prime}, t_1 | y'', t_2 \rangle^\kappa = \mathrm{e}^{\mathrm{i}\phi} \sum_{n=0}^{\infty} \underbrace{\langle y^{\dagger\prime} | n, \lambda \rangle\, \mathrm{e}^{-\mathrm{i}E_n^\kappa t_1}}_{= \langle y^{\dagger\prime}, t_1 | n, \lambda \rangle} \times \underbrace{\mathrm{e}^{\mathrm{i}E_n t_2} \langle n | y'' \rangle}_{= \langle n | y'', t_2 \rangle} . \tag{7.2.41}$$

But, in general we can introduce energy states at t_2 and different energy states at t_1:

$$\langle y^{\dagger\prime}, t_1 | y'', t_2 \rangle^\kappa = \sum_{n,n'=0}^{\infty} \langle y^{\dagger\prime}, t_1 | n, \lambda \rangle \langle n, \lambda | n' \rangle \langle n' | y'', t_2 \rangle \tag{7.2.42}$$

from which we learn that

$$\langle n, \lambda | n' \rangle = \mathrm{e}^{\mathrm{i}\phi} \delta(n, n') \tag{7.2.43}$$

or

$$\left| \langle n, \lambda | n' \rangle \right|^2 = \delta(n, n') . \tag{7.2.44}$$

So the system is to be found, with certainty, in the state of the same quantum number n, although the energy changes by $-|\kappa(t_1)|^2/\omega = -\omega|\lambda|^2$ in going *slowly* from $\kappa = 0$ at time t_2 to $\kappa = \omega\lambda$ at time t_1.

7.2.3 Temporary drive

Now think of a situation in which κ is zero before t_1 and after t_2, being turned on and then off, as illustrated by

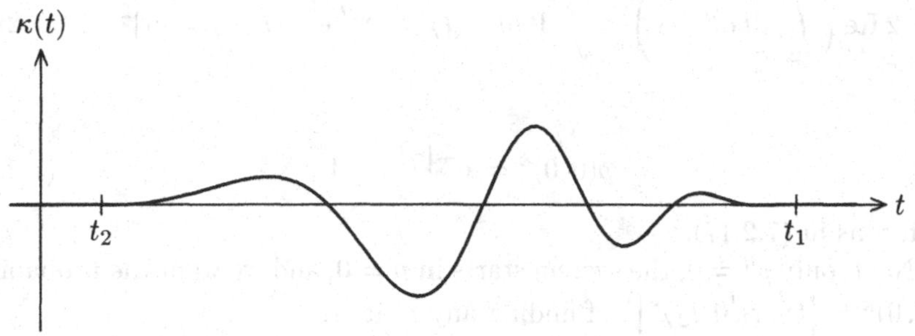

where

$$\kappa = 0 \quad \text{for } t < t_2 \text{ and } t > t_1 \tag{7.2.45}$$

but otherwise arbitrary. As a first example, put $y^{\dagger\prime} = 0$, $y'' = 0$, when (7.2.15) gives

$$\langle 0, t_1 | 0, t_2 \rangle^\kappa = \mathrm{e}^{-\int_2^1 \mathrm{d}t\, \mathrm{d}t'\, \kappa(t)^* \eta(t-t')\, \mathrm{e}^{-\mathrm{i}\omega(t-t')} \kappa(t')} \,. \tag{7.2.46}$$

Then the answer to the question, If the system is in the $n = 0$ state at t_2, what is the probability that it is still in the $n = 0$ state at t_1, despite the intervening κ disturbance?, is

$$\begin{aligned} p(0,0)^\kappa &= \left| \mathrm{e}^{-\int_2^1 \mathrm{d}t\, \mathrm{d}t'\, \kappa(t)^* \eta(t-t')\, \mathrm{e}^{-\mathrm{i}\omega(t-t')} \kappa(t')} \right|^2 \\ &= \mathrm{e}^{-2\mathrm{Re} \int_2^1 \mathrm{d}t\, \mathrm{d}t' \cdots} \,. \end{aligned} \tag{7.2.47}$$

In taking the complex conjugate, also interchange t and t':

$$\left[\int_2^1 \mathrm{d}t\, \mathrm{d}t' \cdots \right]^* = \int \mathrm{d}t\, \mathrm{d}t'\, \kappa(t)^* \eta(t'-t)\, \mathrm{e}^{-\mathrm{i}\omega(t-t')} \kappa(t') \,. \tag{7.2.48}$$

Then the exponent in (7.2.47) is

$$2\,\mathrm{Re} \left(\int_2^1 \mathrm{d}t\, \mathrm{d}t' \cdots \right) = \int \mathrm{d}t\, \mathrm{d}t'\, \kappa(t)^* \left[\eta(t-t') + \eta(t'-t) \right] \mathrm{e}^{-\mathrm{i}\omega(t-t')} \kappa(t') \,. \tag{7.2.49}$$

But

$$\eta(t-t') + \eta(t'-t) = \left\{ \begin{matrix} 1+0 \ \text{ for } t > t' \\ 0+1 \ \text{ for } t < t' \end{matrix} \right\} = 1 \,, \tag{7.2.50}$$

and we get

$$2\,\mathrm{Re}\left(\int_2^1 \mathrm{d}t\,\mathrm{d}t' \cdots\right) = \int \mathrm{d}t\,\mathrm{d}t'\,\kappa(t)^*\,\mathrm{e}^{-\mathrm{i}\omega t}\,\mathrm{e}^{\mathrm{i}\omega t'}\,\kappa(t') = |\gamma|^2 \tag{7.2.51}$$

and

$$p(0,0)^\kappa = \mathrm{e}^{-|\gamma|^2} \leq 1\,, \tag{7.2.52}$$

with γ as in (7.2.17).

Next, only $y'' = 0$; the system starts in $n = 0$, and we want the probability $p(n,0)^\kappa = \left|\langle n, t_1 | 0, t_2\rangle^\kappa\right|^2$ of finding any n at t_1:

$$\langle y^{\dagger\prime}, t_1 | 0, t_2\rangle^\kappa = \mathrm{e}^{-\mathrm{i}y^{\dagger\prime}\,\mathrm{e}^{-\mathrm{i}\omega t_1}\gamma}\langle 0, t_1 | 0, t_2\rangle^\kappa = \sum_{n=0}^{\infty} \underbrace{\langle y^{\dagger\prime}|n\rangle}_{=(y^{\dagger\prime})^n/\sqrt{n!}}\langle n, t_1 | 0, t_2\rangle^\kappa\,. \tag{7.2.53}$$

Therefore,

$$\langle n, t_1 | 0, t_2\rangle^\kappa = \mathrm{e}^{-\mathrm{i}n\omega t_1}\frac{(-\mathrm{i}\gamma)^n}{\sqrt{n!}}\langle 0, t_1 | 0, t_2\rangle^\kappa \tag{7.2.54}$$

and

$$p(n,0)^\kappa = \frac{\left(|\gamma|^2\right)^n}{n!}p(0,0)^\kappa = \frac{|\gamma|^{2n}}{n!}\,\mathrm{e}^{-|\gamma|^2} \tag{7.2.55}$$

where, indeed,

$$\sum_{n=0}^{\infty} p(n,0)^\kappa = \mathrm{e}^{|\gamma|^2}\,\mathrm{e}^{-|\gamma|^2} = 1\,. \tag{7.2.56}$$

Also note that $|\gamma|^2$ is the mean value of n,

$$\sum_{n=0}^{\infty} n p(n,0)^\kappa = \langle n\rangle^\kappa = \sum_{n=1}^{\infty}\frac{|\gamma|^{2n}}{(n-1)!}\,\mathrm{e}^{-|\gamma|^2} = |\gamma|^2\,, \tag{7.2.57}$$

so that one has a Poisson distribution of probabilities:

$$p(n,0)^\kappa = \frac{\left(\langle n\rangle^\kappa\right)^n}{n!}\,\mathrm{e}^{-\langle n\rangle^\kappa}\,. \tag{7.2.58}$$

Now we turn to the general situation and recall (7.2.16),

$$\begin{aligned}\langle y^{\dagger\prime}, t_1 | y'', t_2\rangle^\kappa &= \sum_{n,n'=0}^{\infty} \underbrace{\langle y^{\dagger\prime}|n\rangle}_{=(y^{\dagger\prime})^n/\sqrt{n!}} \times \langle n, t_1 | n', t_2\rangle^\kappa \times \underbrace{\langle n'|y''\rangle}_{=(y'')^n/\sqrt{n!}} \\ &= \mathrm{e}^{y^{\dagger\prime}\,\mathrm{e}^{-\mathrm{i}\omega t_1}\,\mathrm{e}^{\mathrm{i}\omega t_2}y''}\,\mathrm{e}^{-\mathrm{i}y^{\dagger\prime}\,\mathrm{e}^{-\mathrm{i}\omega t_1}\gamma}\,\mathrm{e}^{-\mathrm{i}\gamma^*\,\mathrm{e}^{\mathrm{i}\omega t_2}y''}\langle 0, t_1 | 0, t_2\rangle^\kappa\,.\end{aligned} \tag{7.2.59}$$

First we recognize that if we redefine

$$y^{\dagger\prime} \to e^{i\omega t_1} iu\,, \qquad y'' \to e^{-i\omega t_2} iv \tag{7.2.60}$$

and at the same time write

$$\langle n, t_1 | n', t_2 \rangle^{\kappa} = e^{-in\omega t_1} \langle n | n' \rangle^{\kappa} e^{in'\omega t_2}\,, \tag{7.2.61}$$

which is just making explicit the time dependences associated with the initial and final energies, we get, more simply, the generating function for the probability amplitudes $\langle n|n'\rangle^{\kappa}$,

$$e^{-uv + u\gamma + v\gamma^*} = \sum_{n,n'=0}^{\infty} \frac{u^n}{\sqrt{n!}} i^{n+n'} \frac{\langle n|n'\rangle^{\kappa}}{\langle 0|0\rangle^{\kappa}} \frac{v^{n'}}{\sqrt{n'!}}\,. \tag{7.2.62}$$

Then note a symmetry: The replacements

$$u \to \frac{\gamma^*}{\gamma} v\,, \qquad v \to \frac{\gamma}{\gamma^*} u \tag{7.2.63}$$

leave the left side unchanged. Therefore

$$\sum_{n,n'=0}^{\infty} \frac{u^n}{\sqrt{n!}} i^{n+n'} \frac{\langle n|n'\rangle^{\kappa}}{\langle 0|0\rangle^{\kappa}} \frac{v^{n'}}{\sqrt{n'!}} = \sum_{n,n'=0}^{\infty} \frac{\left[(\gamma^*/\gamma)v\right]^{n'}}{\sqrt{n'!}} i^{n+n'} \frac{\langle n'|n\rangle^{\kappa}}{\langle 0|0\rangle^{\kappa}} \frac{\left[(\gamma/\gamma^*)u\right]^{n}}{\sqrt{n!}} \tag{7.2.64}$$

or

$$\langle n'|n\rangle^{\kappa} = (\gamma^*/\gamma)^{n-n'} \langle n|n'\rangle^{\kappa} \tag{7.2.65}$$

with the consequence

$$p(n', n)^{\kappa} = \left|\langle n'|n\rangle^{\kappa}\right|^2 = p(n, n')^{\kappa}\,. \tag{7.2.66}$$

Now return to (7.2.62) and pick out the term proportional to $v^{n'}/\sqrt{n'!}$:

$$e^{u\gamma} \frac{(\gamma^* - u)^{n'}}{\sqrt{n'!}} = \sum_{n=0}^{\infty} \frac{u^n}{\sqrt{n!}} i^{n+n'} \frac{\langle n|n'\rangle^{\kappa}}{\langle 0|0\rangle^{\kappa}} \tag{7.2.67}$$

and, as a Taylor series in u,

$$i^{n+n'} \frac{\langle n|n'\rangle^{\kappa}}{\langle 0|0\rangle^{\kappa}} = \frac{1}{\sqrt{n!\,n'!}} \left(\frac{d}{du}\right)^n (\gamma^* - u)^{n'} e^{u\gamma}\Big|_{u=0}\,, \tag{7.2.68}$$

or ($x = |\gamma|^2 - u\gamma$; $u = 0 : x = |\gamma|^2$)

$$i^{n+n'} \frac{\langle n|n'\rangle^{\kappa}}{\langle 0|0\rangle^{\kappa}} = \frac{1}{\sqrt{n!\,n'!}} (-1)^n \gamma^{n-n'} e^{|\gamma|^2} \left(\frac{d}{dx}\right)^n x^{n'} e^{-x}\Big|_{x=|\gamma|^2}\,. \tag{7.2.69}$$

We meet the Laguerre* polynomial of order n (integer), index α (arbitrary)

$$\mathrm{L}_n^{(\alpha)}(x) = \frac{1}{n!} x^{-\alpha}\, \mathrm{e}^{x} \left(\frac{\mathrm{d}}{\mathrm{d}x}\right)^n x^{n+\alpha}\, \mathrm{e}^{-x} \tag{7.2.70}$$

so that

$$\begin{aligned}\frac{\langle n|n'\rangle^{\kappa}}{\langle 0|0\rangle^{\kappa}} &= \sqrt{\frac{n!}{n'!}}\,(-\mathrm{i}\gamma^*)^{n'-n}\,\mathrm{L}_n^{(n'-n)}(|\gamma|^2) \\ &= \sqrt{\frac{n'!}{n!}}\,(-\mathrm{i}\gamma)^{n-n'}\,\mathrm{L}_{n'}^{(n-n')}(|\gamma|^2)\,. \end{aligned} \tag{7.2.71}$$

A convenient way to write the implied probability is by means of the larger ($n_>$) and smaller ($n_<$) of n and n',

$$p(n,n')^{\kappa} = \frac{n_<!}{n_>!}(|\gamma|^2)^{n_>-n_<}\left[\mathrm{L}_{n_<}^{(n_>-n_<)}(|\gamma|^2)\right]^2 \mathrm{e}^{-|\gamma|^2} \tag{7.2.72}$$

and, in particular, for $n = n'$,

$$p(n,n)^{\kappa} = \left[\mathrm{L}_n^{(0)}(|\gamma|^2)\right]^2 \mathrm{e}^{-|\gamma|^2}\,. \tag{7.2.73}$$

As a polynomial, $\mathrm{L}_n^{(\alpha)}(x)$ is

$$\begin{aligned}\mathrm{L}_n^{(\alpha)}(x) &= \frac{1}{n!} x^{-\alpha}\left(\frac{\mathrm{d}}{\mathrm{d}x} - 1\right)^n x^{n+\alpha} \\ &= \sum_{k=0}^{n} (-1)^k \frac{(n+\alpha)!}{(n-k)!\,(k+\alpha)!}\,\frac{x^k}{k!}\,, \end{aligned} \tag{7.2.74}$$

so

$$\begin{aligned}\mathrm{L}_0^{(\alpha)} &= 1\,, \\ \mathrm{L}_1^{(\alpha)} &= 1+\alpha-x\,, \\ \mathrm{L}_2^{(\alpha)} &= \frac{1}{2}(1+\alpha)(2+\alpha) - (2+\alpha)x + \frac{1}{2}x^2\,, \end{aligned} \tag{7.2.75}$$

for example, and also

$$\mathrm{L}_n^{(\alpha)}(0) = \frac{(n+\alpha)!}{n!\,\alpha!}\,, \qquad \mathrm{L}_n^{(-n)}(x) = \frac{(-x)^n}{n!}\,, \tag{7.2.76}$$

which exhibit the first and last term of the series.

The situation is particularly simple for

$$|\gamma|^2 \ll 1 \quad \text{and} \quad \Delta n = n - n' \neq 0\,. \tag{7.2.77}$$

---

*Edmond LAGUERRE (1834–1886)

Then the leading factor is $\left(|\gamma|^2\right)^{|\Delta n|}$ and

$$p(n,n')^\kappa \cong \frac{n_<!}{n_>!}\left[\mathrm{L}_{n_<}^{(n_>-n_<)}(0)\right]^2\left(|\gamma|^2\right)^{|\Delta n|}$$
$$= \frac{n_>!}{n_<!}\frac{|\gamma|^{2|\Delta n|}}{\left(|\Delta n|!\right)^2} \qquad \text{if } \Delta n \neq 0\,, \tag{7.2.78}$$

for which

$$\begin{aligned}
p(n+1,n)^\kappa &\cong (n+1)|\gamma|^2\,,\\
p(n-1,n)^\kappa &\cong n|\gamma|^2\,,\\
p(n+2,n)^\kappa &\cong \frac{1}{4}(n+1)(n+2)|\gamma|^4\,,\\
p(n-2,n)^\kappa &\cong \frac{1}{4}n(n-1)|\gamma|^4
\end{aligned} \tag{7.2.79}$$

are simple examples, for $\Delta n = \pm 1$ and $\Delta n = \pm 2$. Note that $p(n-1,n) = p\big((n-1)+1,n-1\big)$ and $p(n-2,n) = p\big((n-2)+2,n-2\big)$ hold, as required by (7.2.66).

For the probability for $\Delta n = 0$, $p(n,n)^\kappa$, we must use the first terms of the power series,

$$\mathrm{L}_n^{(0)}(x) = 1 - nx + \frac{1}{4}n(n-1)x^2 + \cdots \tag{7.2.80}$$

To first order in $|\gamma|^2 \ll 1$, we get

$$p(n,n)^\kappa \cong \left(1-n|\gamma|^2\right)^2\left(1-|\gamma|^2\right) \cong 1-(2n+1)|\gamma|^2\,, \tag{7.2.81}$$

and we note that

$$p(n,n)^\kappa + p(n-1,n)^\kappa + p(n+1,n)^\kappa = 1 + \left[\text{terms of order } |\gamma|^4\right]\,, \tag{7.2.82}$$

a basic check of consistency.

These results for small $|\gamma|^2$ hold only if n and n' are not compensatingly large, that is: (7.2.81) assumes that $(2n+1)|\gamma|^2 \ll 1$, which is not true for $n|\gamma|^2 \gtrsim 1$. Let's return to (7.2.67) in the equivalent form

$$\mathrm{e}^{u\gamma}\left(1-\frac{\gamma^*}{u}\right)^{n'} = \sum_{n=0}^{\infty}\sqrt{\frac{n'!}{n!}}\,u^{n-n'}\mathrm{i}^{n-n'}\frac{\langle n|n'\rangle^\kappa}{\langle 0|0\rangle^\kappa} \tag{7.2.83}$$

and consider the regime $|\gamma| \ll 1$, $n' \gg 1$, for which

$$\left(1-\frac{\gamma^*}{u}\right)^{n'} = \mathrm{e}^{n'\log(1-\gamma^*/u)} \cong \mathrm{e}^{-n'\gamma^*/u}\,. \tag{7.2.84}$$

We also concentrate on the situation where we have $|\Delta n| \ll n'$. Then, for example,

$$\frac{n'!}{n!} = \frac{n'!}{(n'+\Delta n)!} \cong \frac{1}{(n')^{\Delta n}} \tag{7.2.85}$$

and

$$\sqrt{\frac{n'!}{n!}}\, u^{n-n'} \cong \left(\frac{u}{\sqrt{n'}}\right)^{\Delta n} \tag{7.2.86}$$

This suggests a variable change. Indeed, with

$$\gamma = |\gamma|\, \mathrm{e}^{\mathrm{i}\alpha} , \tag{7.2.87}$$

we put

$$\frac{u}{\sqrt{n'}} = \mathrm{e}^{-\mathrm{i}\alpha} t \tag{7.2.88}$$

and get

$$\mathrm{e}^{\sqrt{n'}|\gamma|\,(t-\frac{1}{t})} = \sum_{\Delta n} t^{\Delta n}\, \mathrm{e}^{-\mathrm{i}\alpha\Delta n}\, \mathrm{i}^{\Delta n} \frac{\langle n'+\Delta n | n'\rangle^{\kappa}}{\langle 0|0\rangle^{\kappa}} . \tag{7.2.89}$$

We recognize here the generating function of Bessel* coefficients, Bessel functions of integer order:

$$\mathrm{e}^{\frac{1}{2}z\,(t-\frac{1}{t})} = \sum_{m=-\infty}^{\infty} t^m \mathrm{J}_m(z) , \tag{7.2.90}$$

so, with account for $\langle 0|0\rangle^{\kappa} \cong 1$,

$$\langle n'+\Delta n | n'\rangle^{\kappa} \cong \left(-\mathrm{i}\,\mathrm{e}^{\mathrm{i}\alpha}\right)^{\Delta n} \mathrm{J}_{\Delta n}(2\sqrt{n'}|\gamma|) \tag{7.2.91}$$

and

$$p(n'+\Delta n, n')^{\kappa} \cong \left[\mathrm{J}_{\Delta n}(2\sqrt{n'}|\gamma|)\right]^2 . \tag{7.2.92}$$

For the essential probability check, we note that [$t \to 1/t$ in (7.2.90)]

$$\mathrm{e}^{-\frac{1}{2}z\,(t-\frac{1}{t})} = \sum_{m=-\infty}^{\infty} \frac{1}{t^m} \mathrm{J}_m(z) \tag{7.2.93}$$

and therefore

*Friedrich Wilhelm BESSEL (1784–1846)

$$1 = \sum_{m,m'=-\infty}^{\infty} t^{m-m'} \mathrm{J}_m(z)\mathrm{J}_{m'}(z) . \tag{7.2.94}$$

Picking out the t independent terms establishes the sum rule

$$1 = \sum_{m=-\infty}^{\infty} [\mathrm{J}_m(z)]^2 , \tag{7.2.95}$$

so that, indeed

$$\sum_n p(n,n')^\kappa = \sum_{\Delta n=-\infty}^{\infty} \left[\mathrm{J}_{\Delta n}\left(2\sqrt{n'}|\gamma|\right)\right]^2 = 1 . \tag{7.2.96}$$

The substitution $t \to -t$ in (7.2.90) tells us that

$$\mathrm{J}_m(z) = (-1)^m \mathrm{J}_{-m}(z) \tag{7.2.97}$$

or

$$p(n'+\Delta n, n')^\kappa \cong p(n'-\Delta n, n')^\kappa , \tag{7.2.98}$$

an approximate version for $n' \gg 1$ of (7.2.66).

We shall develop further properties of the J_m in Problems 7-11 and 7-12. For now let's notice that $n' \gg 1$ is not the classical limit, that requires also that $|\Delta n| \gg 1$ and $\sqrt{n'}|\gamma| \gg 1$. Let's put $t = \mathrm{e}^{\mathrm{i}\phi}$ in the generating function (7.2.90):

$$\mathrm{e}^{\frac{1}{2}z(\mathrm{e}^{\mathrm{i}\phi} - \mathrm{e}^{-\mathrm{i}\phi})} = \mathrm{e}^{\mathrm{i}z\sin\phi} = \sum_{m=-\infty}^{\infty} \mathrm{e}^{\mathrm{i}m\phi} \mathrm{J}_m(z) \tag{7.2.99}$$

or

$$\int_0^{2\pi} \frac{\mathrm{d}\phi}{2\pi}\, \mathrm{e}^{\mathrm{i}z\sin\phi}\, \mathrm{e}^{-\mathrm{i}m\phi} = \mathrm{J}_m(z) , \tag{7.2.100}$$

from which we get

$$\left[\mathrm{J}_{\Delta n}\left(2\sqrt{n'}|\gamma|\right)\right]^2 = \int \frac{\mathrm{d}\phi\, \mathrm{d}\phi'}{2\pi\ 2\pi}\, \mathrm{e}^{-\mathrm{i}\Delta n(\phi-\phi')}\, \mathrm{e}^{\mathrm{i}2\sqrt{n'}|\gamma|(\sin\phi - \sin\phi')} . \tag{7.2.101}$$

For $|\Delta n| \gg 1$ and $\sqrt{n'}|\gamma| \gg 1$, the important region is $\phi' \cong \phi$. So write $\phi' = \phi - \psi$, $|\psi| \ll 1$, and $\phi - \phi' = \psi$, $\sin\phi - \sin\phi' \cong \psi\cos\phi$ to get

$$p(n'+\Delta n, n')^\kappa \cong \frac{1}{(2\pi)^2} \int \mathrm{d}\phi\, \mathrm{d}\psi\, \mathrm{e}^{\mathrm{i}\psi\left[-\Delta n + 2\sqrt{n'}|\gamma|\cos\phi\right]} . \tag{7.2.102}$$

The ψ integration produces a delta function,

$$p(n'+\Delta n, n')^{\kappa} \cong \frac{1}{\pi}\int_0^{2\pi} \mathrm{d}\phi\, \delta\left(2\sqrt{n'}|\gamma|\cos\phi - \Delta n\right) , \qquad (7.2.103)$$

and the remaining ϕ integral is non-zero only if $|\Delta n| < 2\sqrt{n'}|\gamma|$. Substitute $x = \cos\phi$ and arrive at

$$\begin{aligned} p(n'+\Delta n, n')^{\kappa} &\cong \frac{2}{\pi}\int_{-1}^{1} \frac{\mathrm{d}x}{\sqrt{1-x^2}}\, \delta\left(2\sqrt{n'}|\gamma|x - \Delta n\right) \\ &= \begin{cases} 0 & \text{if } |\Delta n| > 2\sqrt{n'}|\gamma| , \\ \frac{1}{\pi}\left[4n'|\gamma|^2 - (\Delta n)^2\right]^{-\frac{1}{2}} & \text{if } |\Delta n| < 2\sqrt{n'}|\gamma| . \end{cases} \end{aligned} \qquad (7.2.104)$$

Consistent with the actual approximations, the total probability is unity:

$$\begin{aligned} \sum_{\Delta n} p(n'+\Delta n, n')^{\kappa} &\cong \sum_{\Delta n} \frac{1}{\pi} \frac{1}{\sqrt{4n'|\gamma|^2 - (\Delta n)^2}} \\ &\cong \frac{1}{\pi}\int_{-2\sqrt{n'}|\gamma|}^{2\sqrt{n'}|\gamma|} \frac{\mathrm{d}\nu}{\sqrt{4n'|\gamma|^2 - \nu^2}} = 1 . \end{aligned} \qquad (7.2.105)$$

A deeper understanding of this classical statement as the limit of a quantum statement is gained by solving Problem 7-10.

7.3 Remarks on Laguerre polynomials

A few words about the mathematical properties of $\mathrm{L}_n^{(\alpha)}(x)$ are in order. First, notice that

$$\begin{aligned} \mathrm{L}_n^{(\alpha)}(x) =& \frac{1}{n!}x^{-\alpha}\left(\frac{\mathrm{d}}{\mathrm{d}x} - 1\right)^n x^{\alpha+n} = \frac{1}{n!}x^{-\alpha}\left(\frac{\mathrm{d}}{\mathrm{d}x} - 1\right)^n x\, x^{\alpha-1+n} \\ =& \frac{1}{n!}x^{-(\alpha-1)}\left(\frac{\mathrm{d}}{\mathrm{d}x} - 1\right)^n x^{\alpha-1+n} \\ &+ \frac{1}{(n-1)!}x^{-\alpha}\left(\frac{\mathrm{d}}{\mathrm{d}x} - 1\right)^{n-1} x^{\alpha+n-1} \end{aligned} \qquad (7.3.1)$$

or

$$\mathrm{L}_n^{(\alpha)}(x) = \mathrm{L}_n^{(\alpha-1)}(x) + \mathrm{L}_{n-1}^{(\alpha)}(x) , \qquad (7.3.2)$$

an algebraic recurrence relation in n and α. Another version of $\mathrm{L}_n^{(\alpha)}(x)$ is

$$\mathrm{L}_n^{(\alpha)}(x) = \frac{1}{n!}x^{-\alpha}\left(\frac{\mathrm{d}}{\mathrm{d}x} - 1\right)^n x^\alpha x^n = \frac{1}{n!}\left(\frac{\mathrm{d}}{\mathrm{d}x} - 1 + \frac{\alpha}{x}\right)^n x^n \tag{7.3.3}$$

and then

$$\begin{aligned}\left(\frac{\mathrm{d}}{\mathrm{d}x} - 1 + \frac{\alpha}{x}\right)\mathrm{L}_n^{(\alpha)}(x) &= \frac{x^{-1}}{n!}x\left(\frac{\mathrm{d}}{\mathrm{d}x} - 1 + \frac{\alpha}{x}\right)^{n+1} x^{-1}x^{n+1} \\ &= \frac{x^{-1}}{n!}\left(\frac{\mathrm{d}}{\mathrm{d}x} - 1 + \frac{\alpha-1}{x}\right)^{n+1} x^{n+1}\end{aligned} \tag{7.3.4}$$

or

$$\left(x\frac{\mathrm{d}}{\mathrm{d}x} - x + \alpha\right)\mathrm{L}_n^{(\alpha)} = (n+1)\mathrm{L}_{n+1}^{(\alpha-1)} = (n+1)\left(\mathrm{L}_{n+1}^{(\alpha)} - \mathrm{L}_n^{(\alpha)}\right) \tag{7.3.5}$$

giving

$$\left(x\frac{\mathrm{d}}{\mathrm{d}x} - x + n + 1 + \alpha\right)\mathrm{L}_n^{(\alpha)} = (n+1)\mathrm{L}_{n+1}^{(\alpha)}\,, \tag{7.3.6}$$

a differential recurrence relation in n.

The equivalence of the two forms, for integer α,

$$\frac{n!}{(n+\alpha)!}(-x)^\alpha\,\mathrm{L}_n^{(\alpha)}(x) = \mathrm{L}_{n+\alpha}^{(-\alpha)}(x) = \frac{1}{(n+\alpha)!}x^\alpha\,\mathrm{e}^x\left(\frac{\mathrm{d}}{\mathrm{d}x}\right)^{n+\alpha} x^n\,\mathrm{e}^{-x} \tag{7.3.7}$$

tells us that

$$\begin{aligned}\alpha \text{ integer}: \qquad \mathrm{L}_n^{(\alpha)}(x) &= \frac{(-1)^\alpha}{n!}\,\mathrm{e}^x\left(\frac{\mathrm{d}}{\mathrm{d}x}\right)^{n+\alpha} x^n\,\mathrm{e}^{-x} \\ &= \frac{(-1)^\alpha}{n!}\left(\frac{\mathrm{d}}{\mathrm{d}x} - 1\right)^{n+\alpha} x^n\,.\end{aligned} \tag{7.3.8}$$

From the latter we learn that

$$\left(\frac{\mathrm{d}}{\mathrm{d}x} - 1\right)\mathrm{L}_n^{(\alpha)} = -\mathrm{L}_n^{(\alpha+1)}\,, \tag{7.3.9}$$

a differential recurrence relation in α, or with (7.3.2)

$$\frac{\mathrm{d}}{\mathrm{d}x}\mathrm{L}_n^{(\alpha)} = \mathrm{L}_n^{(\alpha)} - \mathrm{L}_n^{(\alpha+1)} = -\mathrm{L}_{n-1}^{(\alpha+1)} \tag{7.3.10}$$

which, although proved here for integer α, is generally true. The combination of the last result with [$n \to n-1$ and $\alpha \to \alpha+1$ in (7.3.5)]

$$\left(x\frac{\mathrm{d}}{\mathrm{d}x} - x + \alpha + 1\right)\mathrm{L}_{n-1}^{(\alpha+1)} = n\mathrm{L}_n^{(\alpha)} \tag{7.3.11}$$

is the differential equation

$$\left[x\frac{\mathrm{d}^2}{\mathrm{d}x^2} + (\alpha + 1 - x)\frac{\mathrm{d}}{\mathrm{d}x} + n\right]\mathrm{L}_n^{(\alpha)} = 0\,, \tag{7.3.12}$$

the Laguerre equation.

7.4 Two-dimensional oscillator

We now turn to the oscillator in two dimensions. The Hamilton operator is

$$\begin{aligned} H &= \omega\left(\tfrac{1}{2}p_1^2 + \tfrac{1}{2}q_1^2\right) - \tfrac{1}{2}\omega + \omega\left(\tfrac{1}{2}p_2^2 + \tfrac{1}{2}q_2^2\right) - \tfrac{1}{2}\omega \\ &= \omega\left(y_1^\dagger y_1 + y_2^\dagger y_2\right) , \end{aligned} \tag{7.4.1}$$

which is specialized in that

$$\omega_1 = \omega_2 = \omega ; \tag{7.4.2}$$

that is: this two-dimensional oscillator is isotropic. Clearly the energy eigenvalues are

$$E_{n_1,n_2} = n_1\omega + n_2\omega = N\omega \tag{7.4.3}$$

where

$$N = n_1 + n_2 = 0, 1, \ldots . \tag{7.4.4}$$

With the exception of $N = 0$, when $n_1 = n_2 = 0$, these eigenvalues are multiple, or degenerate: N is produced by $n_1 = N, n_2 = 0$; $n_1 = N-1, n_2 = 1$; $\ldots$; $n_1 = 0, n_2 = N$; there are $N + 1$ choices in all, so that

$$m(N) = N + 1 \tag{7.4.5}$$

is the multiplicity of energy $N\omega$.

Closely associated is the freedom to change the variables by a two-dimensional rotation:

$$\begin{aligned} \overline{y}_1 &= y_1 \cos\alpha + y_2 \sin\alpha , & \overline{y}_1^\dagger &= y_1^\dagger \cos\alpha + y_2^\dagger \sin\alpha , \\ \overline{y}_2 &= -y_1 \sin\alpha + y_2 \cos\alpha , & \overline{y}_2^\dagger &= -y_1^\dagger \sin\alpha + y_2^\dagger \cos\alpha . \end{aligned} \tag{7.4.6}$$

One checks immediately that the commutation relations are preserved, that is

$$\left[\overline{y}_k, \overline{y}_l\right] = \left[\overline{y}_k^\dagger, \overline{y}_l^\dagger\right] = 0 \quad \text{and} \quad \left[\overline{y}_k, \overline{y}_l^\dagger\right] = \delta_{kl} \quad \text{for} \quad k, l = 1, 2 , \tag{7.4.7}$$

or, more fundamentally, that the Lagrangian maintains its form:

$$L(y, y^\dagger) = \mathrm{i}\sum_k y_k^\dagger \frac{\mathrm{d}}{\mathrm{d}t} y_k - \omega \sum_k y_k^\dagger y_k = \overline{L} = L(\overline{y}, \overline{y}^\dagger) . \tag{7.4.8}$$

Two-dimensional rotations are conveniently expressed by

$$\begin{aligned} y_+ &= \frac{1}{\sqrt{2}}(y_1 - \mathrm{i}y_2) , & y_+^\dagger &= \frac{1}{\sqrt{2}}(y_1^\dagger + \mathrm{i}y_2^\dagger) , \\ y_- &= \frac{1}{\sqrt{2}}(y_1 + \mathrm{i}y_2) , & y_-^\dagger &= \frac{1}{\sqrt{2}}(y_1^\dagger - \mathrm{i}y_2^\dagger) , \end{aligned} \tag{7.4.9}$$

and again the form is maintained

$$L = \mathrm{i}\sum_{\sigma=\pm} y_\sigma^\dagger \frac{\mathrm{d}}{\mathrm{d}t} y_\sigma - \omega \sum_{\sigma=\pm} y_\sigma^\dagger y_\sigma \tag{7.4.10}$$

and, therefore, so are the commutation relations, as demonstrated by

$$\begin{aligned} \left[y_+, y_+^\dagger\right] &= \tfrac{1}{2}\left[y_1 - \mathrm{i}y_2, y_1^\dagger + \mathrm{i}y_2^\dagger\right] = \tfrac{1}{2}\left[y_1, y_1^\dagger\right] + \tfrac{1}{2}\left[y_2, y_2^\dagger\right] = 1\,, \\ \left[y_+, y_-^\dagger\right] &= \tfrac{1}{2}\left[y_1 - \mathrm{i}y_2, y_1^\dagger - \mathrm{i}y_2^\dagger\right] = \tfrac{1}{2}\left[y_1, y_1^\dagger\right] - \tfrac{1}{2}\left[y_2, y_2^\dagger\right] = 0\,, \end{aligned} \tag{7.4.11}$$

for example. The effect of a rotation is simply

$$\begin{aligned} \overline{y}_+ &= \mathrm{e}^{\mathrm{i}\alpha} y_+\,, & \overline{y}_+^\dagger &= \mathrm{e}^{-\mathrm{i}\alpha} y_+^\dagger\,, \\ \overline{y}_- &= \mathrm{e}^{-\mathrm{i}\alpha} y_-\,, & \overline{y}_-^\dagger &= \mathrm{e}^{\mathrm{i}\alpha} y_-^\dagger\,, \end{aligned} \tag{7.4.12}$$

and the maintainance of the commutation relations is transparent:

$$\left[\overline{y}_+, \overline{y}_+^\dagger\right] = \left[\mathrm{e}^{\mathrm{i}\alpha} y_+,\, \mathrm{e}^{-\mathrm{i}\alpha} y_+^\dagger\right] = \left[y_+, y_+^\dagger\right] = 1\,, \tag{7.4.13}$$

for example.

Consider an infinitesimal rotation,

$$\overline{y}_\pm = y_\pm - \delta y_\pm = (1 \pm \mathrm{i}\delta\alpha\, y_\pm) \tag{7.4.14}$$

with

$$\delta y_\pm = \mp \mathrm{i}\delta\alpha\, y_\pm = \frac{1}{\mathrm{i}}\left[y_\pm, G\right] = \frac{1}{\mathrm{i}}\frac{\partial G}{\partial y_\pm^\dagger} \tag{7.4.15}$$

so that the generator is

$$G = \delta\alpha\, L_3 \quad \text{where} \quad L_3 = y_+^\dagger y_+ - y_-^\dagger y_-\,. \tag{7.4.16}$$

Alternatively

$$L_3 = y_+^\dagger y_+ - y_-^\dagger y_- = \mathrm{i}\big(y_1 y_2^\dagger - y_2 y_1^\dagger\big) = q_1 p_2 - q_2 p_1 \tag{7.4.17}$$

as checked, for example, by

$$\begin{aligned} \delta q_1 &= \frac{1}{\mathrm{i}}\left[q_1, \delta\alpha\, L_3\right] = -\delta\alpha\, q_2\,, \\ \delta q_2 &= \frac{1}{\mathrm{i}}\left[q_2, \delta\alpha\, L_3\right] = \delta\alpha\, q_1\,. \end{aligned} \tag{7.4.18}$$

Evidently, these are infinitesimal two-dimensional rotations indeed, and we recognize in L_3 the two-dimensional version (or third component) of the three-dimensional orbital angular momentum vector $\boldsymbol{L} = \boldsymbol{q} \times \boldsymbol{p}$.

Using the $y_{\sigma=\pm}$ variables, the energy is $E_N = E_{n_+,n_-} = \omega(n_+ + n_-)$, still with multiplicity $m(N) = N+1$, of course. But now the different states of common N are labeled by the eigenvalue m of the angular momentum L_3,

$$L_3{}' \equiv m = n_+ - n_- = N - 2n_- = N, N-2, \dots, -N \,, \tag{7.4.19}$$

that is

$$\begin{aligned} N=1: &\quad m = 1,-1 &\quad \bigl[m(N) = N+1 = 2\bigr] \,, \\ N=2: &\quad m = 2,0,-2 &\quad \bigl[m(N) = N+1 = 3\bigr] \,, \\ N=3: &\quad m = 3,1,-1,-3 &\quad \bigl[m(N) = N+1 = 4\bigr] \,, \end{aligned} \tag{7.4.20}$$

the count of states is correct.

Question: What are the q wave functions for the state of definite energy and angular momentum? Of course, we know for one degree of freedom that

$$\langle q|n\rangle = \pi^{-\frac{1}{4}} \frac{1}{\sqrt{2^n n!}} \mathrm{H}_n(q)\,\mathrm{e}^{-\frac{1}{2}q^2} \tag{7.4.21}$$

and then for two degrees of freedom:

$$\begin{aligned} \langle q_1, q_2|n_1, n_2\rangle &= \langle q_1|n_1\rangle\langle q_2|n_2\rangle \\ &= \pi^{-\frac{1}{2}} \frac{1}{\sqrt{2^{n_1} n_1!\, 2^{n_2} n_2!}} \mathrm{H}_{n_1}(q_1)\mathrm{H}_{n_2}(q_2)\,\mathrm{e}^{-\frac{1}{2}(q_1^2+q_2^2)} \,, \end{aligned} \tag{7.4.22}$$

which are states of definite energy $E_N = \omega(n_1+n_2)$, but *not* states of definite angular momentum L_3. For that we go back to the generating function in one degree of freedom (7.1.25), so for two degrees of freedom (dropping primes) we have

$$\langle q_1, q_2|y_1, y_2\rangle = \pi^{-\frac{1}{2}}\,\mathrm{e}^{-\frac{1}{2}(q_1^2+q_2^2) + \sqrt{2}(q_1y_1+q_2y_2) - \frac{1}{2}(y_1^2+y_2^2)} \,, \tag{7.4.23}$$

which we proceed to rewrite in terms of $y_\pm$ (numbers!) and also introduce polar coordinates,

$$\begin{aligned} q_1 &= \rho\cos\phi \,, \\ q_2 &= \rho\sin\phi \,, \\ q_1 \pm \mathrm{i}q_2 &= \rho\,\mathrm{e}^{\pm\mathrm{i}\phi} \,. \end{aligned} \tag{7.4.24}$$

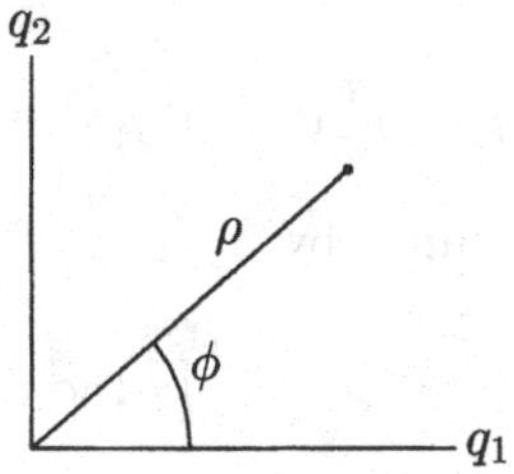

So

$$\begin{aligned} q_1^2+q_2^2 = \rho^2 \,, &\qquad \tfrac{1}{2}(y_1^2+y_2^2) = y_+y_- \,, \\ \sqrt{2}(q_1y_1+q_2y_2) &= \rho\bigl(y_+\,\mathrm{e}^{\mathrm{i}\phi} + y_-\,\mathrm{e}^{-\mathrm{i}\phi}\bigr) \,, \end{aligned} \tag{7.4.25}$$

and

$$\langle \rho, \phi | y_+, y_- \rangle = \pi^{-\frac{1}{2}} \mathrm{e}^{-y_+ y_- + \rho(y_+ \mathrm{e}^{\mathrm{i}\phi} + y_- \mathrm{e}^{-\mathrm{i}\phi})} \mathrm{e}^{-\frac{1}{2}\rho^2}$$
$$= \sum_{n_+, n_- = 0}^{\infty} \psi_{n_+, n_-}(\rho, \phi) \frac{y_+^{n_+}}{\sqrt{n_+!}} \frac{y_-^{n_-}}{\sqrt{n_-!}} . \tag{7.4.26}$$

Now we remember (7.2.62) and (7.2.71), which amount to

$$\mathrm{e}^{-uv + u\gamma + v\gamma^*} = \sum_{n,n'=0}^{\infty} \frac{u^n}{\sqrt{n!}} \frac{v^{n'}}{\sqrt{n'!}} \mathrm{i}^{n+n'} \sqrt{\frac{n'!}{n!}} (-\mathrm{i}\gamma)^{n-n'} \mathrm{L}_{n'}^{(n-n')}(|\gamma|^2) , \tag{7.4.27}$$

a generating function for Laguerre polynomials. So the substitutions

$$y_+ \to u , \qquad y_- \to v , \qquad \gamma \to \rho \, \mathrm{e}^{\mathrm{i}\phi} ,$$
$$n \to n_+ = \frac{N+m}{2} , \qquad n' \to n_- = \frac{N-m}{2} \tag{7.4.28}$$

give us the wave functions

$$\psi_{N,m}(\rho, \phi) = \frac{1}{\sqrt{\pi}} (-1)^{\frac{N-m}{2}} \mathrm{e}^{im\phi} \sqrt{\frac{(\frac{N-m}{2})!}{(\frac{N+m}{2})!}} \rho^m \mathrm{L}_{\frac{1}{2}(N-m)}^{(m)}(\rho^2) \, \mathrm{e}^{-\frac{1}{2}\rho^2}$$
$$= \frac{1}{\sqrt{2\pi}} \mathrm{e}^{im\phi} P_{N,m}(\rho) , \tag{7.4.29}$$

where the familiar azimuthal wave functions $(2\pi)^{-\frac{1}{2}} \mathrm{e}^{im\phi}$ are multiplied by the radial wave functions

$$P_{N,m}(\rho) = P_{N,-m}(\rho)$$
$$= (-1)^{\frac{N-|m|}{2}} \sqrt{2 \frac{(\frac{N-|m|}{2})!}{(\frac{N+|m|}{2})!}} \rho^{|m|} \mathrm{L}_{\frac{1}{2}(N-|m|)}^{(|m|)}(\rho^2) \, \mathrm{e}^{-\frac{1}{2}\rho^2} . \tag{7.4.30}$$

These wave functions are, of course, complete and orthonormal. Expressed in polar coordinates, the area element is

$$\mathrm{d}q_1 \, \mathrm{d}q_2 = \mathrm{d}\rho \, \rho \, \mathrm{d}\phi , \tag{7.4.31}$$

and orthonormality takes the form

$$\delta_{NN'} \delta_{mm'} = \int \mathrm{d}\rho \rho \, \mathrm{d}\phi \, \psi_{N,m}(\rho, \phi)^* \psi_{N',m'}(\rho, \phi)$$
$$= \int_0^\infty \mathrm{d}\rho \rho \, P_{N,m}(\rho) P_{N',m'}(\rho) \underbrace{\int_0^{2\pi} \frac{\mathrm{d}\phi}{2\pi} \mathrm{e}^{\mathrm{i}(m'-m)\phi}}_{=\delta_{mm'}} \tag{7.4.32}$$

which reduces to

$$\int_0^\infty \mathrm{d}\rho\,\rho\, P_{N,m}(\rho) P_{N',m}(\rho) = \delta_{NN'} \,. \tag{7.4.33}$$

To present this most simply as a property of Laguerre polynomials, we write

$$N = |m| + 2n_\rho \,, \qquad n_\rho = 0, 1, \ldots \tag{7.4.34}$$

(n_ρ is the radial quantum number) and

$$\rho^2 = x \,, \qquad |m| = \alpha = 0, 1, \ldots \tag{7.4.35}$$

to get ($n_\rho \to n$)

$$\int_0^\infty \mathrm{d}x\, x^\alpha\, \mathrm{e}^{-x} \mathrm{L}_n^{(\alpha)}(x) \mathrm{L}_{n'}^{(\alpha)}(x) = \delta_{nn'} \frac{(n+\alpha)!}{n!} \,. \tag{7.4.36}$$

This orthogonality statement about Laguerre polynomials is true for arbitrary values of α. The simplest example is $n = n' = 0$:

$$\int_0^\infty \mathrm{d}x\, x^\alpha\, \mathrm{e}^{-x} = \alpha! \,, \tag{7.4.37}$$

Euler's familiar integral representation of the factorial function.

What is the form of the Schrödinger equation we have solved? Here, the Hamilton operator is that in (7.4.1), and so a wave function $\psi(q_1, q_2)$ for the energy $E = \omega(|m| + 2n_\rho)$ obeys

$$\left[|m| + 2n_\rho + 1 + \frac{1}{2}\left(\frac{\partial^2}{\partial q_1^2} + \frac{\partial^2}{\partial q_2^2}\right) - \frac{1}{2}(q_1^2 + q_2^2)\right] \psi = 0 \,. \tag{7.4.38}$$

We know, however, that such a wave function appears naturally not in rectangular but polar coordinates.

That brings up the question of expressing the Laplacian differential operator in curvilinear coordinates, both in two and three dimensions. Consider a three-dimensional coordinate system, u_1, u_2, u_3 – that is: a parameterization of the position vector, $\boldsymbol{r} = \boldsymbol{r}(u_1, u_2, u_3)$. Infinitesimal changes of the coordinates,

$$\begin{aligned} \mathrm{d}\boldsymbol{r} &= \mathrm{d}u_1 \frac{\partial \boldsymbol{r}}{\partial u_1} + \mathrm{d}u_2 \frac{\partial \boldsymbol{r}}{\partial u_2} + \mathrm{d}u_3 \frac{\partial \boldsymbol{r}}{\partial u_3} \\ &= \mathrm{d}u_1\, h_1 \boldsymbol{e}_1 + \mathrm{d}u_2\, h_2 \boldsymbol{e}_2 + \mathrm{d}u_3\, h_3 \boldsymbol{e}_3 \,, \end{aligned} \tag{7.4.39}$$

define the (local) set of unit vectors $\boldsymbol{e}_k$, $k = 1, 2, 3$, and the positive metrical functions $h_k(u_1, u_2, u_3)$, $k = 1, 2, 3$. We take for granted that the coordinate system is orthogonal and right-handed, so that

$$\boldsymbol{e}_j \cdot \boldsymbol{e}_k = \delta_{jk} \quad \text{for } j, k = 1, 2, 3 \quad \text{and} \quad \boldsymbol{e}_1 \times \boldsymbol{e}_2 = \boldsymbol{e}_3 \,. \tag{7.4.40}$$

Then

$$ds^2 = dx^2 + dy^2 + dz^2 = h_1^2 du_1^2 + h_2^2 du_2^2 + h_3^2 du_3^2 \tag{7.4.41}$$

is the (squared) distance of neighboring points,

$$(d\boldsymbol{r}) = dx\,dy\,dz = \underbrace{h_1 h_2 h_3}_{\equiv h} du_1\,du_2\,du_3 \tag{7.4.42}$$

is the volume element, and

$$\boldsymbol{\nabla}\psi = \left(\boldsymbol{e_1}\frac{1}{h_1}\frac{\partial}{\partial u_1} + \boldsymbol{e_2}\frac{1}{h_2}\frac{\partial}{\partial u_2} + \boldsymbol{e_3}\frac{1}{h_3}\frac{\partial}{\partial u_3}\right)\psi \tag{7.4.43}$$

is the gradient of $\psi(\boldsymbol{r}) = \psi(u_1, u_2, u_3)$. To find $\boldsymbol{\nabla}^2\psi$, we apply Green's identity

$$\int_V (d\boldsymbol{r})\boldsymbol{\nabla}\cdot\boldsymbol{\nabla}\psi = \int_S d\boldsymbol{S}\cdot\boldsymbol{\nabla}\psi\,, \tag{7.4.44}$$

which relates the volume integral (over V) of the divergence of vector $\boldsymbol{\nabla}\psi$ to the surface integral (over the boundary S of volume V) of this vector field, to a small volume bounded by two surfaces each of constant u_1, constant u_2, and constant u_3, and we learn that the Laplacian differential operator is given by

$$\boldsymbol{\nabla}^2\psi = \frac{1}{h}\left[\frac{\partial}{\partial u_1}\underbrace{\frac{h}{h_1^2}}_{=h_2h_3/h_1}\frac{\partial}{\partial u_1} + \frac{\partial}{\partial u_2}\frac{h}{h_2^2}\frac{\partial}{\partial u_2} + \frac{\partial}{\partial u_3}\frac{h}{h_3^2}\frac{\partial}{\partial u_3}\right]\psi \tag{7.4.45}$$

with its obvious simplification in two dimensions.

For two-dimensional polar coordinates ρ, ϕ, for example, we have

$$h_\rho = 1\,,\quad h_\phi = \rho\,,\quad h = \rho\,,\qquad ds^2 = d\rho^2 + \rho^2 d\phi^2\,, \tag{7.4.46}$$

and

$$\begin{aligned}\boldsymbol{\nabla}^2 &= \frac{1}{\rho}\left[\frac{\partial}{\partial\rho}\rho\frac{\partial}{\partial\rho} + \frac{\partial}{\partial\phi}\frac{1}{\rho}\frac{\partial}{\partial\phi}\right] = \frac{\partial^2}{\partial\rho^2} + \frac{1}{\rho}\frac{\partial}{\partial\rho} + \frac{1}{\rho^2}\frac{\partial^2}{\partial\phi^2}\\ &= \frac{1}{\rho^2}\left[\left(\rho\frac{\partial}{\partial\rho}\right)^2 + \left(\frac{\partial}{\partial\phi}\right)^2\right]\,.\end{aligned} \tag{7.4.47}$$

Then, with

$$\psi = \frac{1}{\sqrt{2\pi}}e^{im\phi}P_{N,m}(\rho)\,,\qquad \frac{\partial^2}{\partial\phi^2}\psi = -m^2\psi \tag{7.4.48}$$

we get ($N = |m| + 2n_\rho$)

$$\left[\frac{\mathrm{d}^2}{\mathrm{d}\rho^2} + \frac{1}{\rho}\frac{\mathrm{d}}{\mathrm{d}\rho} - \frac{m^2}{\rho^2} + 2|m| + 4n_\rho + 2 - \rho^2\right] P(\rho) = 0 \tag{7.4.49}$$

in which $|m|$, freed from the relation to ϕ, is arbitrary; n_ρ is a non-negative integer.

The differential equations easily suggest the general character of the wave function. For large ρ the dominant terms in (7.4.49) are

$$\left(\frac{\mathrm{d}^2}{\mathrm{d}\rho^2} - \rho^2\right) P \cong 0 \tag{7.4.50}$$

and a suitable (bounded) asymptotic solution is

$$P(\rho) \propto \mathrm{e}^{-\frac{1}{2}\rho^2} , \tag{7.4.51}$$

for which

$$\frac{\mathrm{d}P}{\mathrm{d}\rho} = -\rho P , \qquad \frac{\mathrm{d}^2 P}{\mathrm{d}\rho^2} = (\rho^2 - 1)P \cong \rho^2 P . \tag{7.4.52}$$

For small distances the dominant terms are

$$\left(\frac{\mathrm{d}^2}{\mathrm{d}\rho^2} + \frac{1}{\rho}\frac{\mathrm{d}}{\mathrm{d}\rho} - \frac{m^2}{\rho^2}\right) P \cong 0 , \tag{7.4.53}$$

so that

$$P(\rho) \propto \rho^{\pm m} \qquad \left[\text{of which } P \propto \rho^{|m|} \text{ is bounded}\right] \tag{7.4.54}$$

since

$$\left(\rho^2 \frac{\mathrm{d}^2}{\mathrm{d}\rho^2} + \rho \frac{\mathrm{d}}{\mathrm{d}\rho}\right) \rho^{\pm m} = \left(\rho \frac{\partial}{\partial \rho}\right)^2 \rho^{\pm m} = (\pm m)^2 \rho^{\pm m} . \tag{7.4.55}$$

Then, on writing

$$P(\rho) \propto \rho^{|m|}\, \mathrm{e}^{-\frac{1}{2}\rho^2} \mathrm{L}(\rho^2) \tag{7.4.56}$$

we get a differential equation for $\mathrm{L}(x = \rho^2)$ that is the Laguerre equation (7.3.12).

Let us also note the form of the differential equation with the first derivative removed. For that, the identity

$$\begin{aligned} \rho^{-a} \frac{\mathrm{d}^2}{\mathrm{d}\rho^2} \rho^a &= \left(\rho^{-a} \frac{\mathrm{d}}{\mathrm{d}\rho} \rho^a\right)^2 = \left(\frac{\mathrm{d}}{\mathrm{d}\rho} + \frac{a}{\rho}\right)^2 \\ &= \frac{\mathrm{d}^2}{\mathrm{d}\rho^2} + \frac{2a}{\rho}\frac{\mathrm{d}}{\mathrm{d}\rho} + \frac{a^2 - a}{\rho^2} \end{aligned} \tag{7.4.57}$$

or

$$\rho^a\left(\frac{\mathrm{d}^2}{\mathrm{d}\rho^2}+\frac{2a}{\rho}\frac{\mathrm{d}}{\mathrm{d}\rho}+\frac{a^2-a}{\rho^2}\right)\rho^{-a}=\frac{\mathrm{d}^2}{\mathrm{d}\rho^2} \tag{7.4.58}$$

is useful. So, with $a=\frac{1}{2}$ and

$$P(\rho)=\frac{1}{\sqrt{\rho}}u(\rho) \tag{7.4.59}$$

in (7.4.49) we get

$$\left[\frac{\mathrm{d}^2}{\mathrm{d}\rho^2}-\frac{m^2-\frac{1}{4}}{\rho^2}+2|m|+4n_\rho+2-\rho^2\right]u(\rho)=0\,, \tag{7.4.60}$$

and (7.4.33) turns into the orthonormality statement

$$\int_0^\infty \mathrm{d}\rho\, u_{|m|,n_\rho}(\rho)u_{|m|,n'_\rho}(\rho)=\delta_{n_\rho n'_\rho} \tag{7.4.61}$$

for the radial wave functions

$$u_{|m|,n_\rho}(\rho)=(-1)^{n_\rho}\sqrt{2\frac{n_\rho!}{(n_\rho+|m|)!}}\,\rho^{|m|+\frac{1}{2}}\mathrm{L}_{n_\rho}^{(|m|)}(\rho^2)\,\mathrm{e}^{-\frac{1}{2}\rho^2}\,. \tag{7.4.62}$$

7.5 Three-dimensional oscillator

How about the isotropic oscillator in three dimensions? Its Hamilton operator is

$$H=\sum_{k=1}^{3}\omega\frac{1}{2}(p_k^2+q_k^2-1)=\omega\frac{1}{2}(\boldsymbol{p}^2+\boldsymbol{q}^2)-\frac{3}{2}\omega\,. \tag{7.5.1}$$

This time, for a change, we begin with the Schrödinger equation,

$$\left(\frac{E}{\omega}+\frac{1}{2}\boldsymbol{\nabla}^2-\frac{1}{2}\boldsymbol{q}^2+\frac{3}{2}\right)\psi(\boldsymbol{q})=0\,, \tag{7.5.2}$$

and, as a convenient way of exploiting the obvious freedom of three-dimensional rotations, we introduce spherical coordinates:

$$\begin{aligned} q_1&=\rho\sin\theta\,\cos\phi\,,\\ q_2&=\rho\sin\theta\,\sin\phi\,,\\ q_3&=\rho\cos\theta\,, \end{aligned} \tag{7.5.3}$$

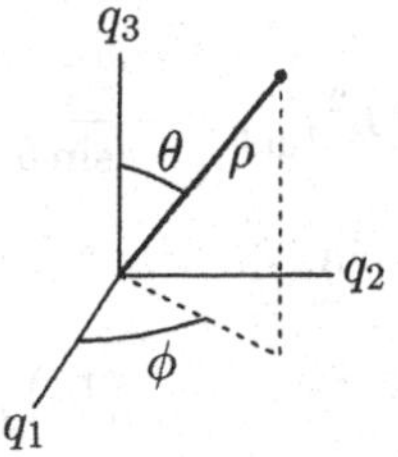

for which

$$h_\rho = 1\,,\ \ h_\theta = \rho\,,\ \ h_\phi = \rho\sin\theta\,,\quad h = \rho^2\sin\theta\,, \tag{7.5.4}$$

so that

$$\begin{aligned} ds^2 &= d\rho^2 + \rho^2 d\theta^2 + \rho^2\sin^2\theta d\phi^2\,,\\ (d\boldsymbol{q}) &= dq_1\,dq_2\,dq_3 = d\rho\,\rho^2\,d\theta\,\sin\theta\,d\phi\,, \end{aligned} \tag{7.5.5}$$

and

$$\begin{aligned} -\boldsymbol{p}^2 \to \boldsymbol{\nabla}^2 &= \frac{1}{\rho^2\sin\theta}\left[\frac{\partial}{\partial\rho}\rho^2\sin\theta\frac{\partial}{\partial\rho} + \frac{\partial}{\partial\theta}\frac{\rho^2\sin\theta}{\rho^2}\frac{\partial}{\partial\theta} + \frac{\partial}{\partial\phi}\frac{\rho^2\sin\theta}{\rho^2\sin\theta}\frac{\partial}{\partial\phi}\right]\\ &= \frac{\partial^2}{\partial\rho^2} + \frac{2}{\rho}\frac{\partial}{\partial\rho} + \frac{1}{\rho^2}\left[\frac{1}{\sin\theta}\frac{\partial}{\partial\theta}\sin\theta\frac{\partial}{\partial\theta} + \frac{1}{\sin^2\theta}\frac{\partial^2}{\partial\phi^2}\right]. \end{aligned} \tag{7.5.6}$$

Alternatively, it is clear that we must deal with angular momentum, three-dimensional angular momentum, and its differential-operator representation,

$$\boldsymbol{L} = \boldsymbol{q}\times\boldsymbol{p} \to \boldsymbol{q}\times\frac{1}{i}\boldsymbol{\nabla} = -\frac{1}{i}\boldsymbol{\nabla}\times\boldsymbol{q} \tag{7.5.7}$$

(primes omitted), and specifically

$$\begin{aligned} \boldsymbol{L}^2 &\to -(\boldsymbol{q}\times\boldsymbol{\nabla})\times(\boldsymbol{q}\times\boldsymbol{\nabla}) = \boldsymbol{\nabla}\times\boldsymbol{q}\cdot\boldsymbol{q}\times\boldsymbol{\nabla}\\ &= \boldsymbol{\nabla}\cdot\left[\boldsymbol{q}\times(\boldsymbol{q}\times\boldsymbol{\nabla})\right] = \boldsymbol{\nabla}\cdot(\boldsymbol{q}\,\boldsymbol{q}\cdot\boldsymbol{\nabla} - \boldsymbol{q}^2\boldsymbol{\nabla})\\ &= (\boldsymbol{q}\cdot\boldsymbol{\nabla}+3)\boldsymbol{q}\cdot\boldsymbol{\nabla} - \rho^2\boldsymbol{\nabla}^2 - 2\boldsymbol{q}\cdot\boldsymbol{\nabla} \end{aligned} \tag{7.5.8}$$

or, with $\boldsymbol{q}\cdot\boldsymbol{\nabla} = \rho\dfrac{\partial}{\partial\rho}$,

$$\boldsymbol{L}^2 \to \left(\rho\frac{\partial}{\partial\rho}\right)^2 + \rho\frac{\partial}{\partial\rho} - \rho^2\boldsymbol{\nabla}^2 = \rho^2\frac{\partial^2}{\partial\rho^2} + 2\rho\frac{\partial}{\partial\rho} - \rho^2\boldsymbol{\nabla}^2 \tag{7.5.9}$$

so that

$$-\boldsymbol{p}^2 + \frac{\boldsymbol{L}^2}{\rho^2} \to \frac{\partial^2}{\partial\rho^2} + \frac{2}{\rho}\frac{\partial}{\partial\rho} \tag{7.5.10}$$

which identifies the angular derivative term in $\left[-\rho^2\times\right]\boldsymbol{\nabla}^2$ of (7.5.6) as the differential operator representing $\boldsymbol{L}^2$. Symbolically, we write

$$\left(\boldsymbol{L}^2\right)_{\text{diff}} = -\frac{1}{\sin\theta}\frac{\partial}{\partial\theta}\sin\theta\frac{\partial}{\partial\theta} - \frac{1}{\sin^2\theta}\frac{\partial^2}{\partial\phi^2} \tag{7.5.11}$$

and, in the same vein,

$$\left(L_3\right)_{\text{diff}} = \frac{1}{i}\frac{\partial}{\partial\phi}\,. \tag{7.5.12}$$

In Section 3.7, we have already talked about solid harmonics, $\rho^l Y_{lm}(\theta, \phi)$, which satisfy Laplace's equation: $\nabla^2(\rho^l Y_{lm}) = 0$. But

$$\left(\frac{\partial^2}{\partial \rho^2} + \frac{2}{\rho}\frac{\partial}{\partial \rho}\right) \rho^l Y_{lm} = l(l+1)\frac{\rho^l Y_{lm}}{\rho^2} \tag{7.5.13}$$

so

$$(\boldsymbol{L}^2)_{\text{diff}} Y_{lm} = l(l+1) Y_{lm} \qquad \left[\text{and, of course, } (L_3)_{\text{diff}} Y_{lm} = m Y_{lm}\right] \tag{7.5.14}$$

and the $Y_{lm}(\theta, \phi)$ are the wave functions of orbital angular momentum with (as we already know) integral values of l. Accordingly [recall (7.4.58) with $a = 1$ here], we write

$$\psi(\rho, \theta, \phi) = \frac{1}{\rho} u(\rho) Y_{lm}(\theta, \phi) \tag{7.5.15}$$

and get the radial differential equation

$$\left[\frac{\mathrm{d}^2}{\mathrm{d}\rho^2} - \frac{l(l+1)}{\rho^2} + \frac{2E}{\omega} + 3 - \rho^2\right] u(\rho) = 0\,. \tag{7.5.16}$$

It has exactly the same form as the two-dimensional equation (7.4.60), with the correspondence

$$\begin{array}{cc} \underline{\text{two-dim. oscillator}} & \underline{\text{three-dim. oscillator}} \\ |m| & l + \frac{1}{2} \\ 2|m| + 4n_\rho & 2E/\omega + 1 \end{array} \tag{7.5.17}$$

and, therefore, with $|m| \to l + \frac{1}{2}$, we get

$$\frac{2E}{\omega} + 1 = 2\left(l + \tfrac{1}{2}\right) + 4n_\rho \tag{7.5.18}$$

or

$$E = N\omega \quad \text{with} \quad N = l + 2n_\rho = 0, 1, \ldots\,. \tag{7.5.19}$$

The energy states are degenerate (except for $N = 0$, where $l = n_\rho = 0$ is unique). The multiplicity is $m(N) = \frac{1}{2}(N+1)(N+2)$ since we have for even N

$$\begin{aligned} m(N) &= \sum_{l=0,2,\ldots}^{N} \sum_{m=-l}^{l} = \sum_{l=0,2,\ldots}^{N} (2l+1) \\ &= \underbrace{\frac{N+2}{2}}_{\text{number}} \times \underbrace{(N+1)}_{\text{average}} = \frac{1}{2}(N+1)(N+2)\,, \end{aligned} \tag{7.5.20}$$

and for odd N

$$m(N) = \sum_{l=1,3,\ldots}^{N} (2l+1) = \underbrace{\frac{N+1}{2}}_{\text{number}} \times \underbrace{(N+2)}_{\text{average}} = \frac{1}{2}(N+1)(N+2)\,. \quad (7.5.21)$$

In Problem 2-34 we got this from the alternative form

$$\frac{E}{\omega} = N = n_1 + n_2 + n_3\,. \quad (7.5.22)$$

In view of the orthonormality (3.7.45) of the spherical harmonics, the orthonormality of the wave functions

$$\int (\mathrm{d}\boldsymbol{q})\, \psi_{n_\rho lm}(\boldsymbol{q})^* \psi_{n_\rho lm}(\boldsymbol{q}) = \delta_{n_\rho n'_\rho}\delta_{ll'}\delta_{mm'}\,, \quad (7.5.23)$$

where $(\mathrm{d}\boldsymbol{q})$ as in (7.5.5) and

$$\psi_{n_\rho lm}(\boldsymbol{q}) = \frac{1}{\rho} u_{n_\rho l}(\rho) Y_{lm}(\theta,\phi)\,, \quad (7.5.24)$$

reduces to

$$\int_0^\infty \mathrm{d}\rho\, u_{n_\rho l}(\rho) u_{n'_\rho l}(\rho) = \delta_{n_\rho n'_\rho} \quad (7.5.25)$$

which is automatically satisfied by the two → three-dimensional construction

$$u_{n_\rho l}(\rho) = (-1)^{n_\rho} \sqrt{2\frac{n_\rho!}{(n_\rho + l + \frac{1}{2})!}}\, \rho^{l+1} \mathrm{L}_{n_\rho}^{(l+\frac{1}{2})}(\rho)\, \mathrm{e}^{-\frac{1}{2}\rho^2}\,. \quad (7.5.26)$$

We easily check the normalization for the simple situation $n_\rho = 0$:

$$\begin{aligned} \int_0^\infty \mathrm{d}\rho\, (u_{0l})^2 &= \frac{2}{(l+\frac{1}{2})!} \int_0^\infty \mathrm{d}\rho\, \rho\, (\rho^2)^{(l+\frac{1}{2})}\, \mathrm{e}^{-\rho^2} \\ &= \frac{1}{(l+\frac{1}{2})!} \int_0^\infty \mathrm{d}x\, x^{(l+\frac{1}{2})}\, \mathrm{e}^{-x} = 1\,, \end{aligned} \quad (7.5.27)$$

indeed.

Problems

7-1 Concerning the Lagrange operator (7.1.4): Find the additional term needed in the action operator

$$W_{12} = \int_2^1 [\mathrm{i}y^\dagger \mathrm{d}y - H(y, y^\dagger, t)\mathrm{d}t] + (?)$$

such that

$$\delta W_{12} = \left(G_{y^\dagger} + G_t\right)_1 - \left(G_y + G_t\right)_2$$

as required by the action principle applied to $\langle y^{\dagger\prime}, t_1 | y'', t_2\rangle$. [Hint: Problem 5-11.]

7-2 Dimensionless variables: $H = \frac{1}{2}p^2$, $[q,p] = \mathrm{i}$. Construct $\langle q', t | y''\rangle$, for which it is sufficient to consider time variations, in conjunction with the known initial value $\langle q' | y''\rangle$. Recall that the normalized minimum uncertainty state is $[y'' = (q'' + \mathrm{i}p'')/\sqrt{2}]$

$$|q'', p''\rangle = |y''\rangle \, \mathrm{e}^{-\frac{1}{2}|y''|^2} .$$

State $\langle q', t | q'', p''\rangle$ and work out the probability density $\left|\langle q', t | q'', p''\rangle\right|^2$. Check that it is normalized in q', give its physical meaning, and make contact with the $\psi_\delta(x,t)$ of (5.4.25).

7-3 Consider the generating function (7.1.25) for the oscillator wave functions $\psi_n(q)$. Show first that

$$\psi_n(q) = \sqrt{n!}\, t^{-n} \int \frac{\mathrm{d}\phi}{2\pi}\, \mathrm{e}^{-\mathrm{i}n\phi} \langle q | n\rangle \Big|_{y\,=\,t\,\mathrm{e}^{\mathrm{i}\phi}} \quad \text{with } t > 0 .$$

Then establish the circumstances under which the ϕ integral allows for a stationary-phase evaluation. Conclude that $\psi_n(q)$ is exponentially small for $q^2 > 2n$ and find the corresponding approximation for $q^2 < 2n$. Does the extreme semiclassical approximation

$$|\psi_n(q)|^2 \cong \begin{cases} \left(\pi\sqrt{2n - q^2}\right)^{-1} & \text{for } q^2 < 2n , \\ 0 & \text{for } q^2 > 2n , \end{cases}$$

emerge in the limit of very large n?

7-4 The driven oscillator: The driving force $\kappa(t)$ can be turned on, and off, smoothly as in

$$\kappa(t) = \mathrm{e}^{-\mathrm{i}\Omega t}\, \mathrm{e}^{-(t/\tau)^2} ,$$

or abruptly as in

$$\kappa(t) = \mathrm{e}^{-\mathrm{i}\Omega t} \eta\left(\tfrac{1}{2}\tau - |t|\right) .$$

Evaluate γ of (7.2.17) for both choices. Give a rough graphical comparison of the two $|\gamma|^2$, as a function of $\Omega - \omega$, for the situation

$$|\Omega - \omega| \cong \frac{1}{T}, \qquad \omega \gg \frac{1}{T}.$$

How significant is the change, in these circumstances, if complex $e^{-i\Omega t}$ is replaced by real $2\cos(\Omega t)$?

7-5a Dimensionless variables; Hamilton operator

$$H = \omega\tfrac{1}{2}(p^2 + q^2) - fq,$$

which describes an oscillator subjected to a constant force. Find the energy eigenvalues and q wave functions in terms of those for the undisturbed oscillator ($f = 0$). Then find the analogous p wave function relationship.

7-5b Write the ground state q wave function for the disturbed oscillator ($f \neq 0$). Then apply the $\langle q'|y'\rangle$ generating function of undisturbed oscillator wave functions to answer this question: If the system is in this $f \neq 0$ ground state, what is the probability that a measurement of $\omega\frac{1}{2}(p^2 + q^2)$ will have the outcome $n\omega$?

7-6 Dimensionless variables; Hamilton operator

$$H = \omega y^\dagger y + \frac{1}{2}\Omega\left(y^2 + y^{\dagger 2}\right)$$

with real parameter Ω. Apply perturbation theory to get the leading change in the energy values for small Ω. Can you find the exact eigenvalues? [Hint: Introduce q and p.]

7-7 Write the equations of motion for the Hamilton operator of Problem 7-5a. Evaluate

$$\int_0^t dt'\, q(t')$$

in terms of $p(t)$ and p. Then use the action principle to get

$$\frac{\partial}{\partial f}\langle p', t|p''\rangle^f$$

thereby obtaining $\langle p', t|p''\rangle^f$ in terms of $\langle p', t|p''\rangle^{f=0}$. Recognize here the same relations found in Problem 7-5a for energies and p wave functions.

7-8a Driven oscillator: The solution (7.2.9) of the equation of motion gives $y(t_1)$ in terms of $y(t_2)$ and κ. Construct the operator $N_1 = y(t_1)^\dagger y(t_1)$. What is its expectation value for a system initially in the nth state? Interpret this in terms of energy fed into the system by the driving force.

7-8b Evaluate the analogous expression for N_1^2 and interpret the result in terms of the dispersion of the energy transfer.

7-9 Dimensionless variables; Hamilton operator (7.2.1); minimum uncertainty states $|q',p'\rangle$ of (2.7.38). Write out $(y^{\dagger\prime} = y'^*)$

$$\langle q',p',t_1|q'',p'',t_2\rangle^\kappa = \mathrm{e}^{-\frac{1}{2}|y^{\dagger\prime}|^2}\langle y^{\dagger\prime},t_1|y'',t_2\rangle^\kappa\,\mathrm{e}^{-\frac{1}{2}|y''|^2}$$

and check that

$$\begin{aligned}|\langle q',p',t_1|q'',p'',t_2\rangle^\kappa|^2 &= \mathrm{e}^{-|y'\,\mathrm{e}^{\mathrm{i}\omega t_1} - y''\,\mathrm{e}^{\mathrm{i}\omega t_2} + \mathrm{i}\gamma|^2}\\ &= \mathrm{e}^{-|y' - y''\,\mathrm{e}^{-\mathrm{i}\omega(t_1-t_2)} + \mathrm{i}\gamma\,\mathrm{e}^{-\mathrm{i}\omega t_1}|^2}\,.\end{aligned}$$

Why does this combination of $y(t_1)$, $y(t_2)$ values look familiar? Verify the probability normalization

$$\int \frac{\mathrm{d}q'\,\mathrm{d}p'}{2\pi}\left|\langle q',p',t_1|q'',p'',t_2\rangle^\kappa\right|^2 = 1\,.$$

7-10 Driven oscillator: In the classical limit, the probability distribution becomes infinitely sharp. What is *the* final energy of the oscillator, in terms of the initial energy, $|\gamma|$, and the phase angle between the initial motion and the external force? If the phase angle is unknown, what probability distribution emerges for the final energy?

7-11 Use the generating function of Bessel coefficients to write out the infinite power series for $\mathrm{J}_m(z)$. [Although thus derived for integer m, it holds generally.] Write the analogous generating function and power series for the related function

$$\mathrm{I}_m(z) = \mathrm{i}^{-m}\mathrm{J}_m(\mathrm{i}z)\,.$$

Apply both generating function and power series to verify that

$$\mathrm{J}_{-m}(z) = (-1)^m\mathrm{J}_m(z) \quad \text{for integer } m.$$

What is the analogous statement for $\mathrm{I}_{-m}(z)$?

7-12 Take the Laguerre polynomial generating function (7.4.27) and, with the substitutions

$$u = \sqrt{\lambda}\,\mathrm{e}^{\mathrm{i}\phi}\,, \quad v = \sqrt{\lambda}\,\mathrm{e}^{-\mathrm{i}\phi}\,, \quad |\gamma| = \sqrt{x}\,,$$

arrive at the alternative generating function

$$\mathrm{e}^{-\lambda}\frac{\mathrm{I}_\alpha(2\sqrt{\lambda x})}{(\lambda x)^{\frac{1}{2}\alpha}} = \sum_{n=0}^{\infty}(-1)^n\frac{\lambda^n}{(n+\alpha)!}\mathrm{L}_n^{(\alpha)}(x)\,,$$

which, although thus proved for integer α, is true generally.

7-13a Concerning Laguerre polynomials: Return to Problem 2-26 with $f(y^\dagger) = (y^\dagger)^\alpha$, α arbitrary; $f(y) = y^\alpha$, $\alpha = 0, 1, 2, \dots$. Place $y^\dagger = x$, $y = \partial/\partial x$, and let the equivalent forms operate on e^{-x}. For arbitrary α, arrive thereby at yet another generating function:

$$\frac{1}{(1-\lambda)^{1+\alpha}}\,\mathrm{e}^{-\frac{x}{1-\lambda}} = \sum_{n=0}^{\infty}\lambda^n \mathrm{L}_n^{(\alpha)}(x)\,,$$

$$\text{with}\quad \mathrm{L}_n^{(\alpha)}(x) = \frac{1}{n!}x^{-\alpha}\,\mathrm{e}^{x}\left(\frac{\mathrm{d}}{\mathrm{d}x}\right)^n x^{n+\alpha}\,\mathrm{e}^{-x}$$

as in (7.2.70), and, for non-negative integral α, at the same expansion with the equivalent form

$$\mathrm{L}_n^{(\alpha)}(x) = \frac{(-1)^\alpha}{n!}\,\mathrm{e}^{x}\left(\frac{\mathrm{d}}{\mathrm{d}x}\right)^{n+\alpha} x^n\,\mathrm{e}^{-x}\,.$$

Recognize that the latter is equivalent to

$$\mathrm{L}_n^{(\alpha)}(x) = \frac{(-1)^\alpha}{n!}\left(\frac{\mathrm{d}}{\mathrm{d}x}-1\right)^{n+\alpha} x^n\,.$$

7-13b Derive the recurrence relations (7.3.9) and (7.3.10) directly from the generating function in Problem 7-13a.

7-13c Use the power series expansion for $\mathrm{L}_n^{(\alpha)}(x)$ to prove that (7.3.10) is true for arbitrary α.

8. Hydrogenic Atoms

8.1 Bound states

Now we are going to do a nice little trick: turn one kind of dynamical system into another one. Begin with the differential equation (7.4.60) that determines the energy eigenstates of the two-dimensional isotropic oscillator,

$$\left[\frac{\mathrm{d}^2}{\mathrm{d}\rho^2} - \frac{m^2 - \frac{1}{4}}{\rho^2} + 2|m| + 4n_\rho + 2 - \rho^2\right] u(\rho) = 0\,, \tag{8.1.1}$$

and put

$$\rho^2 = 2\lambda r \quad \text{with} \quad \lambda > 0\,. \tag{8.1.2}$$

Then

$$\begin{aligned}\left(\frac{\mathrm{d}}{\mathrm{d}\rho}\right)^2 &= 2\rho\frac{\mathrm{d}}{\mathrm{d}\rho^2}2\rho\frac{\mathrm{d}}{\mathrm{d}\rho^2} = \sqrt{2\lambda r}\frac{\mathrm{d}}{\mathrm{d}\lambda r}\sqrt{2\lambda r}\frac{\mathrm{d}}{\mathrm{d}\lambda r}\\ &= \frac{2}{\lambda}\sqrt{r}\frac{\mathrm{d}}{\mathrm{d}r}\sqrt{r}\frac{\mathrm{d}}{\mathrm{d}r} = \frac{2}{\lambda}\Bigl(r\frac{\mathrm{d}^2}{\mathrm{d}r^2} + \frac{1}{2}\frac{\mathrm{d}}{\mathrm{d}r}\Bigr)\,,\end{aligned} \tag{8.1.3}$$

and therefore

$$\left[\frac{\mathrm{d}^2}{\mathrm{d}r^2} + \frac{1}{2r}\frac{\mathrm{d}}{\mathrm{d}r} - \frac{m^2 - \frac{1}{4}}{4r^2} + \frac{\lambda}{r}(|m| + 2n_\rho + 1) - \lambda^2\right] u(\rho) = 0\,. \tag{8.1.4}$$

Here we have the $a = \frac{1}{4}$ case of (7.4.58), and the function change

$$u(\rho) = C(\lambda r)^{-\frac{1}{4}} u(r)\,, \tag{8.1.5}$$

with a (positive) proportionality constant C to be determined later, gives

$$\left[\frac{\mathrm{d}^2}{\mathrm{d}r^2} - \frac{m^2 - 1}{4r^2} + \frac{\lambda(|m| + 2n_r + 1)}{r} - \lambda^2\right] u(r) = 0\,, \tag{8.1.6}$$

where the radial quantum number n_ρ is renamed n_r.

Now, for any spherically symmetrical potential $V(r)$, the radial Schrödinger equation for a particle of mass M is

$$\left[E - \left(-\frac{\hbar^2}{2M}\nabla^2 + V(r)\right)\right]\frac{1}{r}u(r)Y_{lm}(\theta,\phi) = 0 \tag{8.1.7}$$

or

$$\left[\frac{\mathrm{d}^2}{\mathrm{d}r^2} - \frac{l(l+1)}{r^2} - \frac{2M}{\hbar^2}V(r) + \frac{2ME}{\hbar^2}\right]u(r) = 0\,. \tag{8.1.8}$$

The evident correspondence between the two equations tells us that we're now dealing with a $1/r$ potential which we identify with the attractive Coulomb* potential between the electron charge $-e$ and a nuclear charge Ze

$$V(r) = -\frac{Ze^2}{r}\,, \tag{8.1.9}$$

the potential energy of a hydrogenic atom. Then we have the correspondence

two-dim. oscillator	three-dim. Coulomb	
$\frac{1}{4}(m^2-1)$	$l(l+1)$	
$\lambda(\lvert m\rvert + 2n_r + 1)$	$\frac{2M}{\hbar^2}Ze^2$	
λ^2	$\frac{2M}{\hbar^2}(-E)$	(8.1.10)

and so first

$$|m|^2 \to 4l(l+1) + 1 = (2l+1)^2\,, \quad \text{that is} \quad |m| \to 2l+1 \tag{8.1.11}$$

and then

$$\lambda(2l + 2n_r + 2) \to 2\frac{Ze^2M}{\hbar^2}\,, \quad \text{that is} \quad \lambda \to \frac{Z}{na_0}\,, \tag{8.1.12}$$

where

$$n = n_r + l + 1 \tag{8.1.13}$$

is the principal quantum number (or energy quantum number), and

$$a_0 = \frac{\hbar^2}{Me^2} \tag{8.1.14}$$

is known as the (first) Bohr radius. Therefore

$$\frac{2M}{\hbar^2}(-E) = \frac{2}{e^2a_0}(-E) = \frac{Z^2}{n^2a_0^2} \tag{8.1.15}$$

or

*Charles–Augustin de COULOMB (1736–1806)

$$-E_n = \frac{Z^2e^2}{2n^2a_0} \quad \text{with} \quad n = 1, 2, 3, \ldots ; \tag{8.1.16}$$

these are the Bohr energies. Except for $n = 1$, when $l = 0$ and $n_r = 0$, the energy states of a hydrogenic atom are degenerate. In general their multiplicity is

$$m(n) = \sum_{l=0}^{n-1}(2l+1) = \sum_{l=0}^{n-1}\left[(l+1)^2 - l^2\right] = n^2 = 1, 4, 9, \ldots . \tag{8.1.17}$$

Here are some numbers pertinent to atomic physics. If M is the electron mass and e the elementary charge,

$$M = 9.10939 \times 10^{-28}\,\text{g}, \quad e = 4.80321 \times 10^{-10}\,\text{esu}, \tag{8.1.18}$$

then the Bohr radius is

$$a_0 = 0.5292\,\text{Å} \tag{8.1.19}$$

[1 Å = 10^{-8} cm, Ångström* unit]; it sets the atomic length scale, and

$$R_\infty \equiv \frac{Me^4}{2\hbar^2} = \frac{e^2}{2a_0} = 13.606\,\text{eV}, \tag{8.1.20}$$

called Rydberg† energy, sets the energy scale: $E_n = -R_\infty Z^2/n^2$. Corresponding atomic scales for frequency, wave number, time, and velocity are given by

$$\begin{aligned} R_\infty/(2\pi\hbar) &= 3.2898 \times 10^{15}\,\text{Hz}, \\ R_\infty/(2\pi\hbar c) &= 109\,737\,\text{cm}^{-1}, \\ \hbar/(2R_\infty) = \hbar a_0/e^2 &= 0.0242\,\text{fs}, \\ \sqrt{2R_\infty/M} = e^2/\hbar &= 2.188 \times 10^8\,\text{cm}\,\text{s}^{-1}, \end{aligned} \tag{8.1.21}$$

respectively.

Now, what about the wave functions of the hydrogenic system? We have the relation (8.1.5) between the oscillator $u(\rho)$ and the Coulomb $u(r)$, but need to determine the proportionality constant. Since

$$1 = \int_0^\infty d\rho\left[u(\rho)\right]^2 = C^2 \int_0^\infty \underbrace{d\sqrt{2\lambda r}\,(\lambda r)^{-\frac{1}{2}}}_{=2^{-\frac{1}{2}}dr/r}\left[u(r)\right]^2 \tag{8.1.22}$$

we find

$$C = 2^{\frac{1}{4}}\left(\int_0^\infty dr\,\frac{1}{r}\left[u(r)\right]^2\right)^{-\frac{1}{2}} = 2^{\frac{1}{4}}\left\langle\frac{1}{r}\right\rangle^{-\frac{1}{2}} \tag{8.1.23}$$

---

*Jonas Anders ÅNGSTRÖM (1814–1874) †Janne RYDBERG (1854–1919)

which gives C in terms of the average value of $1/r$ in the particular hydrogenic state, which, at the moment, we don't know. On the other hand, we could have turned it around:

$$1 = \int_0^\infty \mathrm{d}r\,[u(r)]^2 = \frac{1}{C^2}\int_0^\infty \frac{\mathrm{d}\rho\,\rho}{\lambda}\frac{\rho}{\sqrt{2}}[u(\rho)]^2 \tag{8.1.24}$$

gives

$$C = 2^{-\frac{1}{4}}\lambda^{-\frac{1}{2}}\left\langle\rho^2\right\rangle^{\frac{1}{2}} \quad \text{with} \quad \left\langle\rho^2\right\rangle = \int_0^\infty \mathrm{d}\rho\,\rho^2[u(\rho)]^2 . \tag{8.1.25}$$

Do we know, or can we easily find, $\langle\rho^2\rangle$? Sure! First, for one degree of freedom: In Section 6.3 we observed that the oscillator energy states are the stationary-uncertainty states of Section 2.4, and there we had found, in (2.4.22),

$$\left\langle p^2\right\rangle_n = \left\langle q^2\right\rangle_n = n + \tfrac{1}{2} , \tag{8.1.26}$$

giving back

$$\frac{E_n}{\omega} = \left\langle\frac{1}{\omega}H\right\rangle_n = \frac{1}{2}\left(\left\langle p^2\right\rangle_n + \left\langle q^2\right\rangle_n - 1\right)$$
$$= \frac{n+\frac{1}{2}+n+\frac{1}{2}-1}{2} = n . \tag{8.1.27}$$

For the two-dimensional oscillator, we have, then,

$$\left\langle\rho^2\right\rangle_n = \left\langle q_1^2 + q_2^2\right\rangle_n = \underbrace{n_1 + n_2}_{=N=|m|+2n_\rho} + 1 = |m| + 2n_\rho + 1 \tag{8.1.28}$$

so that with (8.1.11) and (8.1.13)

$$\left\langle\rho^2\right\rangle_n \to (2l+1) + (\underset{\displaystyle\downarrow \atop \displaystyle n_r}{2n_\rho} + 1) = 2(l + n_r + 1) = 2n . \tag{8.1.29}$$

Then, since $\lambda = Z/(na_0)$, we have

$$C = 2^{\frac{1}{4}}(a_0/Z)^{\frac{1}{2}}n , \tag{8.1.30}$$

giving

$$\left\langle\frac{1}{r}\right\rangle_n = \frac{Z}{n^2 a_0} ; \tag{8.1.31}$$

more about this in Section 8.2.

So now we are told that (8.1.2) and (8.1.5) with (8.1.12) and (8.1.30) turn the oscillator $u(\rho)$ of (7.4.30) into the hydrogenic wave function

$$u_{n,l}(r) = (-1)^{n-l-1}\sqrt{\frac{Z}{n^2 a_0}\frac{(n-l-1)!}{(n+l)!}}\left(\frac{2Zr}{na_0}\right)^{l+1} \mathrm{L}_{n-l-1}^{(2l+1)}\left(\frac{2Zr}{na_0}\right) \mathrm{e}^{-\frac{Zr}{na_0}} \tag{8.1.32}$$

or

$$R_{n,l}(r) = \frac{u_{n,l}(r)}{r} = (-1)^{n-l-1}\left(\frac{Z}{a_0}\right)^{\frac{3}{2}} \frac{2}{n^2}\sqrt{\frac{(n-l-1)!}{(n+l)!}}\, x^l \mathrm{L}_{n-l-1}^{(2l+1)}(x)\,\mathrm{e}^{-\frac{1}{2}x}$$
$$\text{with } x \equiv \frac{2Zr}{na_0}. \tag{8.1.33}$$

The simplest example is, of course, $n_r = 0$, $n = l+1$,

$$R_{n,n-1}(r) = \left(\frac{2Z}{na_0}\right)^{\frac{3}{2}}\sqrt{\frac{1}{(2n)!}}\left(\frac{2Zr}{na_0}\right)^{n-1} \mathrm{e}^{-\frac{Zr}{na_0}}, \tag{8.1.34}$$

in particular,

$$R_{10}(r) = \left(\frac{Z}{a_0}\right)^{\frac{3}{2}} 2\,\mathrm{e}^{-\frac{Zr}{a_0}} \tag{8.1.35}$$

for the lowest-energy state, the ground state in which $n = 1$, $l = 0$. As a check of consistency, we evaluate the normalization integral and find, indeed,

$$\int_0^\infty \mathrm{d}r\, r^2 \left[R_{n,n-1}(r)\right]^2 = \int_0^\infty \mathrm{d}x\, x^2 \frac{1}{(2n)!} x^{2n-2}\,\mathrm{e}^{-x} = 1\,. \tag{8.1.36}$$

8.2 Parameter dependence of energy eigenvalues

The incidental evaluation

$$\left\langle \frac{1}{r} \right\rangle_n = \frac{Z}{n^2 a_0} \tag{8.2.1}$$

directs attention to a general question: What information follows by knowing the dependence of energy eigenvalues E on various parameters λ? Consider the eigenvector equation

$$\big[E(\lambda) - H(\lambda)\big]\big|E,\gamma\big\rangle = 0 \tag{8.2.2}$$

and its adjoint

$$\big\langle E,\gamma\big|\big[E(\lambda) - H(\lambda)\big] = 0\,. \tag{8.2.3}$$

Differentiate with respect to λ:

$$\left(\frac{\partial E(\lambda)}{\partial\lambda} - \frac{\partial H(\lambda)}{\partial\lambda}\right)|E,\gamma\rangle + \left[E(\lambda) - H(\lambda)\right]\frac{\partial}{\partial\lambda}|E,\gamma\rangle = 0\,, \tag{8.2.4}$$

and multiply with $\langle E,\gamma|$ to get

$$\frac{\partial E(\lambda)}{\partial\lambda} = \left\langle\frac{\partial H(\lambda)}{\partial\lambda}\right\rangle\,. \tag{8.2.5}$$

This is frequently called the Hellmann*–Feynman† theorem. We'll reconsider these matters in somewhat more detail in Section 9.6.

As a first example, consider the one-dimensional oscillator for which

$$H = \frac{p^2}{2M} + \frac{M\omega^2}{2}x^2 = T + V\,, \tag{8.2.6}$$

where $T = p^2/(2M)$ is the kinetic energy and $V = \frac{1}{2}M\omega^2x^2$ is the potential energy. We recall that the dependence of the nth energy eigenvalue upon the mass M (no dependence) and the frequency ω (linear dependence) is given by

$$E = \langle T\rangle + \langle V\rangle = (n+\tfrac{1}{2})\hbar\omega\,. \tag{8.2.7}$$

Differentiation with respect to M gives

$$M\frac{\partial H}{\partial M} = -T + V\,, \quad M\frac{\partial E}{\partial M} = 0 \qquad \text{so that} \quad \langle T\rangle = \langle V\rangle\,, \tag{8.2.8}$$

and differentiation with respect to ω gives

$$\omega\frac{\partial H}{\partial\omega} = 2V\,, \quad \omega\frac{\partial E}{\partial\omega} = E \qquad \text{so that} \quad 2\langle V\rangle = E\,. \tag{8.2.9}$$

Together they say

$$\langle T\rangle = \langle V\rangle = \tfrac{1}{2}E\,, \tag{8.2.10}$$

which gives the correct sum of $\langle T\rangle$ and $\langle V\rangle$, as it should.

Now try three-dimensional hydrogenic atoms,

$$H = T + V \qquad \text{with } T = \frac{\boldsymbol{p}^2}{2M}\,, \quad V = -\frac{Ze^2}{r}\,, \tag{8.2.11}$$

and

$$E = -\frac{Z^2e^4M}{2n^2\hbar^2}\,. \tag{8.2.12}$$

---

*Hans HELLMANN (b. 1903) †Richard Phillips FEYNMAN (1918–1988)

We differentiate with respect to the mass M,

$$M\frac{\partial H}{\partial M} = -T\,, \quad M\frac{\partial E}{\partial M} = E \qquad \text{so that} \quad \langle T\rangle = -E\,, \tag{8.2.13}$$

and with respect to the nuclear charge Z,

$$Z\frac{\partial H}{\partial Z} = V\,, \quad Z\frac{\partial E}{\partial Z} = 2E \qquad \text{so that} \quad \langle V\rangle = 2E\,, \tag{8.2.14}$$

which give the correct sum,

$$\langle T\rangle + \langle V\rangle = -E + 2E = E\,. \tag{8.2.15}$$

The result in (8.2.14), presented as

$$-Ze^2\left\langle\frac{1}{r}\right\rangle = -\frac{Z^2e^2}{n^2a_0}\,, \tag{8.2.16}$$

is the known

$$\left\langle\frac{1}{r}\right\rangle = \frac{Z}{n^2a_0}\,. \tag{8.2.17}$$

8.3 Virial theorem

There is a related transformation in which we change the scale of the q's and p's. Consider a three-dimensional system

$$H = \frac{\boldsymbol{p}^2}{2M} + V(\boldsymbol{r}) = T + V\,, \qquad L = \boldsymbol{p}\cdot\frac{\mathrm{d}\boldsymbol{r}}{\mathrm{d}t} - H \tag{8.3.1}$$

and the infinitesimal transformation

$$\delta\boldsymbol{r} = \delta\lambda(t)\,\boldsymbol{r}\,, \quad \delta\boldsymbol{p} = -\delta\lambda(t)\,\boldsymbol{p}\,. \tag{8.3.2}$$

Then the induced change of the Lagrangian is

$$\begin{aligned}\delta L &= \boldsymbol{p}\cdot\boldsymbol{r}\frac{\mathrm{d}}{\mathrm{d}t}\delta\lambda + 2\delta\lambda T - \delta\lambda\,\boldsymbol{r}\cdot\boldsymbol{\nabla}V\\ &= \frac{\mathrm{d}}{\mathrm{d}t}(\delta\lambda\,\boldsymbol{p}\cdot\boldsymbol{r}) + \delta\lambda\left[-\frac{\mathrm{d}}{\mathrm{d}t}(\boldsymbol{p}\cdot\boldsymbol{r}) + 2T - \boldsymbol{r}\cdot\boldsymbol{\nabla}V\right],\end{aligned} \tag{8.3.3}$$

and the stationary action principle, $\delta W_{12} = G_1 - G_2$, applied to $\delta\lambda$ gives

$$G_\lambda = \delta\lambda\,\boldsymbol{p}\cdot\boldsymbol{r} \tag{8.3.4}$$

and

$$\frac{\mathrm{d}}{\mathrm{d}t}(\boldsymbol{p}\cdot\boldsymbol{r}) = 2T - \boldsymbol{r}\cdot\boldsymbol{\nabla}V\,, \tag{8.3.5}$$

which is known as the virial theorem. Note that G_λ is indeed a generator:

$$\begin{aligned}\delta\boldsymbol{r} &= \frac{1}{\mathrm{i}\hbar}[\boldsymbol{r}, G_\lambda] = \frac{\partial G_\lambda}{\partial \boldsymbol{p}} = \delta\lambda\,\boldsymbol{r}\,,\\ \delta\boldsymbol{p} &= \frac{1}{\mathrm{i}\hbar}[\boldsymbol{p}, G_\lambda] = -\frac{\partial G_\lambda}{\partial \boldsymbol{r}} = -\delta\lambda\,\boldsymbol{p}\,.\end{aligned} \tag{8.3.6}$$

The importance of the virial theorem lies in the remark that for a state of definite energy E, a ***stationary state***, the expectation value of the time derivative $\mathrm{d}F/\mathrm{d}t$ of an operator that has no parametric time dependence – that is $F = F(\boldsymbol{r},\boldsymbol{p})$ or $\partial F/\partial t = 0$ – vanishes:

$$\left\langle\frac{\mathrm{d}F}{\mathrm{d}t}\right\rangle_E = \langle E|\frac{1}{\mathrm{i}\hbar}(F\underset{E}{\underset{\downarrow}{H}} - \underset{E}{\underset{\downarrow}{H}}F)|E\rangle = 0\,. \tag{8.3.7}$$

When applied to $F = \boldsymbol{p}\cdot\boldsymbol{r}$, this implies

$$2\langle T\rangle = \langle \boldsymbol{r}\cdot\boldsymbol{\nabla}V\rangle \tag{8.3.8}$$

for these expectation values in a stationary state ($\equiv$ eigenstate of the Hamilton operator). If $V(\boldsymbol{r})$ is of degree n, that is $V \propto r^n$, then

$$\boldsymbol{r}\cdot\boldsymbol{\nabla}V = r\frac{\mathrm{d}}{\mathrm{d}r}V = nV \tag{8.3.9}$$

and

$$2\langle T\rangle = n\langle V\rangle\,. \tag{8.3.10}$$

In conjunction with $\langle T\rangle + \langle V\rangle = E$, then

$$\langle T\rangle = \frac{n}{n+2}E\,, \qquad \langle V\rangle = \frac{2}{n+2}E\,. \tag{8.3.11}$$

Thus, for the oscillator, $n = 2$ and

$$\langle T\rangle = \langle V\rangle = \tfrac{1}{2}E\,; \tag{8.3.12}$$

and for hydrogenic atoms, where the Coulomb potential has $n = -1$,

$$\langle T\rangle = -E\,, \qquad \langle V\rangle = 2E\,, \tag{8.3.13}$$

as seen in Section 8.2.

From the viewpoint of the one-dimensional radial motion described by $u(r)$, we have $p_r \mathrel{\widehat{=}} \frac{\hbar}{\mathrm{i}}\frac{\partial}{\partial r}$ and

$$p^2 \to p_r^2 + \frac{l(l+1)\hbar^2}{r^2} , \tag{8.3.14}$$

so that effectively

$$H \to H_l = \frac{p_r^2}{2M} + \frac{l(l+1)\hbar^2}{2Mr^2} + V(r) . \tag{8.3.15}$$

Then, with $\lambda = l$ in (8.2.5),

$$\frac{\partial E}{\partial l} = \frac{\hbar^2}{M} \left(l + \tfrac{1}{2}\right) \left\langle \frac{1}{r^2} \right\rangle . \tag{8.3.16}$$

For the Coulomb example, $E = -\frac{1}{2}(Z^2 e^2/a_0)(n_r + l + 1)^{-2}$, we get

$$\frac{\partial E}{\partial l} = \frac{Z^2 e^2}{a_0} \frac{1}{n^3} = \frac{\hbar^2}{M} \left(l + \tfrac{1}{2}\right) \left\langle \frac{1}{r^2} \right\rangle \tag{8.3.17}$$

or

$$\left\langle \frac{1}{r^2} \right\rangle = \frac{Z^2}{a_0^2} \frac{1}{n^3 \left(l + \frac{1}{2}\right)} , \tag{8.3.18}$$

whereas, for the three-dimensional isotropic oscillator, $E = \hbar\omega(l + 2n_r)$ and

$$\frac{\partial E}{\partial l} = \hbar\omega = \frac{\hbar^2}{M} \left(l + \tfrac{1}{2}\right) \left\langle \frac{1}{r^2} \right\rangle , \tag{8.3.19}$$

or

$$\left\langle \frac{1}{r^2} \right\rangle = \frac{M\omega}{\hbar} \frac{1}{l + \frac{1}{2}} . \tag{8.3.20}$$

A more subtle kind of average occurs when, beginning with the Schrödinger equation (8.1.8),

$$\left[\frac{\mathrm{d}^2}{\mathrm{d}r^2} - \frac{l(l+1)}{r^2} + \frac{2M}{\hbar^2}(E - V)\right] u(r) = 0 , \tag{8.3.21}$$

we differentiate,

$$\left[\frac{\mathrm{d}^2}{\mathrm{d}r^2} - \frac{l(l+1)}{r^2} + \frac{2M}{\hbar^2}(E - V)\right] u' = \left[-2\frac{l(l+1)}{r^3} + \frac{2M}{\hbar^2} V'\right] u \tag{8.3.22}$$

(primes denote r derivatives), then cross-multiply both equations to get

$$\left[-2\frac{l(l+1)}{r^3} + \frac{2M}{\hbar^2} V'\right] u^2 = \frac{\mathrm{d}}{\mathrm{d}r} \left(uu'' - u'^2\right) , \tag{8.3.23}$$

leading to

$$-2l(l+1)\left\langle\frac{1}{r^3}\right\rangle+\frac{2M}{\hbar^2}\left\langle V'\right\rangle=\left[u'(0)\right]^2=\left[R(0)\right]^2 . \tag{8.3.24}$$

Here we made use of $uu'' \to 0$ as $r \to 0$, which is an immediate consequence of $u(r) \propto r^{l+1}$ for $r \gtrsim 0$.

For $l \neq 0$, (8.3.24) is

$$\left\langle\frac{1}{r^3}\right\rangle=\frac{1}{l(l+1)}\frac{M}{\hbar^2}\left\langle V'\right\rangle ; \tag{8.3.25}$$

the three-dimensional isotropic oscillator, $V=\frac{1}{2}M\omega^2 r^2$, thus has

$$\left\langle\frac{1}{r^3}\right\rangle=\left(\frac{M\omega}{\hbar}\right)^2\frac{1}{l(l+1)}\langle r\rangle , \tag{8.3.26}$$

and for the Coulomb field we find ($l \neq 0$)

$$\left\langle\frac{1}{r^3}\right\rangle=\frac{1}{l(l+1)}\frac{M}{\hbar^2}Ze^2\left\langle\frac{1}{r^2}\right\rangle=\frac{Z^2}{a_0^3}\frac{1}{n^3}\frac{1}{l(l+\frac{1}{2})(l+1)} . \tag{8.3.27}$$

Now, for $l = 0$, we have

$$\frac{2M}{\hbar^2}\left\langle V'\right\rangle=\left[R(0)\right]^2 . \tag{8.3.28}$$

For the three-dimensional oscillator thus

$$2\left(\frac{M\omega}{\hbar}\right)^2\langle r\rangle=\left[R(0)\right]^2 , \tag{8.3.29}$$

and for the Coulomb potential

$$\left[R(0)\right]^2=\frac{2M}{\hbar^2}Ze^2\left\langle\frac{1}{r^2}\right\rangle=\frac{Z^3}{a_0^3}\frac{4}{n^3} . \tag{8.3.30}$$

The latter is quite simple; does it indeed follow from [$l = 0$ and $r = 0$ in (8.1.33)]

$$R_{n,0}(0)=(-1)^{n-1}\left(\frac{Z}{a_0}\right)^{\frac{3}{2}}\frac{2}{n^2}\sqrt{\frac{1}{n}}L_{n-1}^{(1)}(0)\ ? \tag{8.3.31}$$

Yes, since according to (7.2.76)

$$L_n^{(\alpha)}(0)=\frac{(n+\alpha)!}{n!\,\alpha!} \qquad \text{so that} \quad L_{n-1}^{(1)}(0)=\frac{n!}{(n-1)!}=n , \tag{8.3.32}$$

as required. As for the three-dimensional oscillator, relation (8.3.29) says [dimensionless variables and $l = 0$ in (7.5.26)]

$$\langle\rho\rangle = \frac{1}{2}\left[\frac{u(r)}{r}(r\to 0)\right]^2 = \frac{1}{2}2\frac{n_\rho!}{(n_\rho+\frac{1}{2})!}\left[L_{n_\rho}^{(\frac{1}{2})}(0)\right]^2$$
$$= \frac{n_\rho!}{(n_\rho+\frac{1}{2})!}\left[\frac{(n_\rho+\frac{1}{2})!}{n_\rho!\,\frac{1}{2}!}\right]^2 = \frac{(n_\rho+\frac{1}{2})!}{n_\rho!\,(\frac{1}{2}!)^2} . \qquad (8.3.33)$$

See Problem 8-16 for an explicit check.

8.4 Parabolic coordinates

The wonders of the Coulomb potential do not cease with the connection to the oscillator. Despite the evident spherical symmetry of the problem, there is another useful coordinate system, one with a preferred direction – parabolic coordinates ($\xi > 0$, $\eta > 0$, $0 \le \phi \le 2\pi$):

$$x = 2\sqrt{\xi\eta}\cos\phi\,,$$
$$y = 2\sqrt{\xi\eta}\sin\phi\,, \qquad (8.4.1)$$
$$z = \xi - \eta\,,$$

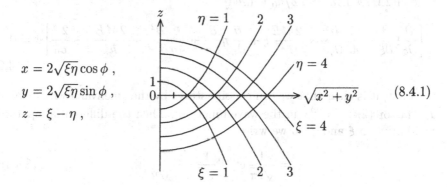

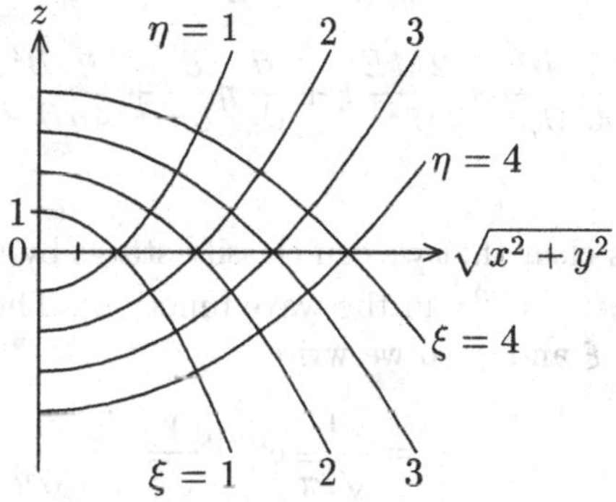

in terms of which the length r of the position vector $\boldsymbol{r}$ is

$$r = \sqrt{x^2+y^2+z^2} = \sqrt{4\xi\eta + (\xi-\eta)^2} = \xi + \eta\,, \qquad (8.4.2)$$

and

$$\xi = \tfrac{1}{2}(r+z)\,, \qquad \eta = \tfrac{1}{2}(r-z) \qquad (8.4.3)$$

emphasize the privileged role of the z direction. With

$$h_\xi = \sqrt{\frac{\xi+\eta}{\xi}}\,, \quad h_\eta = \sqrt{\frac{\xi+\eta}{\eta}}\,, \quad h_\phi = 2\sqrt{\xi\eta}\,, \qquad h = 2(\xi+\eta) \qquad (8.4.4)$$

in (7.4.39) and (7.4.42), we have [cf. (7.4.41)]

$$\mathrm{d}s^2 = \frac{\xi+\eta}{\xi}(\mathrm{d}\xi)^2 + \frac{\xi+\eta}{\eta}(\mathrm{d}\eta)^2 + 4\xi\eta(\mathrm{d}\phi)^2 \qquad (8.4.5)$$

and [cf. (7.4.45)]

$$\begin{aligned}\nabla^2 &= \frac{1}{2(\xi+\eta)}\left[\frac{\partial}{\partial\xi}2\xi\frac{\partial}{\partial\xi}+\frac{\partial}{\partial\eta}2\eta\frac{\partial}{\partial\eta}+\frac{\partial}{\partial\phi}\frac{\xi+\eta}{2\xi\eta}\frac{\partial}{\partial\phi}\right]\\ &= \frac{1}{\xi+\eta}\left[\frac{\partial}{\partial\xi}\xi\frac{\partial}{\partial\xi}+\frac{\partial}{\partial\eta}\eta\frac{\partial}{\partial\eta}\right]+\frac{1}{4\xi\eta}\frac{\partial^2}{\partial\phi^2}\,. \end{aligned} \tag{8.4.6}$$

Both this Laplacian differential operator and the Coulomb potential

$$V = -\frac{Ze^2}{r} = -\frac{Ze^2}{\xi+\eta} \tag{8.4.7}$$

have $\xi+\eta$ in the denominator and so, after multiplying with $(\xi+\eta)$, the Schrödinger equation

$$\left[\nabla^2+\frac{2M}{\hbar^2}(E-V)\right]\psi = 0 \tag{8.4.8}$$

reads $[(2M/\hbar^2)Ze^2 = 2Z/a_0$ is used]

$$\left[\frac{\partial}{\partial\xi}\xi\frac{\partial}{\partial\xi}+\frac{1}{4\xi}\frac{\partial^2}{\partial\phi^2}+\frac{2ME}{\hbar^2}\xi+\frac{\partial}{\partial\eta}\eta\frac{\partial}{\partial\eta}+\frac{1}{4\eta}\frac{\partial^2}{\partial\phi^2}+\frac{2ME}{\hbar^2}\eta+\frac{2Z}{a_0}\right]\psi = 0\,. \tag{8.4.9}$$

First, it is clear that we can classify states by the eigenvalues $L_z' = m\hbar$ of L_z [factor $(2\pi)^{-\frac{1}{2}}\,\mathrm{e}^{im\phi}$ in the wave function]. Then the differential operator is additive in ξ and η, so we write

$$\psi = \frac{1}{\sqrt{2\pi}}\,\mathrm{e}^{im\phi}\frac{1}{\sqrt{\xi}}u(\xi)\frac{1}{\sqrt{\eta}}u(\eta) \tag{8.4.10}$$

and get the pair of equations

$$\begin{aligned}&\left[\frac{\mathrm{d}^2}{\mathrm{d}\xi^2}-\frac{m^2-1}{4\xi^2}+\frac{2ME}{\hbar^2}+\frac{2Z_1}{a_0\xi}\right]u(\xi)=0\,,\\ &\left[\frac{\mathrm{d}^2}{\mathrm{d}\eta^2}-\frac{m^2-1}{4\eta^2}+\frac{2ME}{\hbar^2}+\frac{2Z_2}{a_0\eta}\right]u(\eta)=0\end{aligned} \tag{8.4.11}$$

with $Z_1+Z_2 = Z$. We set along side these the differential equation for $u(r)$, that is (8.1.8) with $(2M/\hbar^2)V = -2Z/(a_0 r)$,

$$\left[\frac{\mathrm{d}^2}{\mathrm{d}r^2}-\frac{l(l+1)}{r^2}+\frac{2ME}{\hbar^2}+\frac{2Z}{a_0 r}\right]u(r)=0\,, \tag{8.4.12}$$

where we know that

$$-\frac{2ME}{\hbar^2} = \frac{(Z/a_0)^2}{(n_r+l+1)^2}\,, \qquad n_r, l = 0,1,2,\ldots\,, \tag{8.4.13}$$

or, for our immediate purposes

$$n_r + l + 1 = \frac{Z/a_0}{\sqrt{-2ME/\hbar^2}} . \tag{8.4.14}$$

We see the correspondence: $|m| \leftrightarrow 2l+1$. And so, for the ξ and η equations, using $k_1 = 0, 1, \ldots$ and $k_2 = 0, 1, \ldots$ as analogs of $n_r = 0, 1, \ldots$, we get

$$k_1 + \tfrac{1}{2}\big(|m|+1\big) = \frac{Z_1/a_0}{\sqrt{-2ME/\hbar^2}} , \qquad k_2 + \tfrac{1}{2}\big(|m|+1\big) = \frac{Z_2/a_0}{\sqrt{-2ME/\hbar^2}} , \tag{8.4.15}$$

which, on addition ($Z_1 + Z_2 = Z$), give

$$k_1 + k_2 + |m| + 1 \equiv n = \frac{Z/a_0}{\sqrt{-2ME/\hbar^2}} , \tag{8.4.16}$$

so that

$$\frac{Z_1}{Z} = \frac{k_1 + \frac{1}{2}\big(|m|+1\big)}{n} , \qquad \frac{Z_2}{Z} = \frac{k_2 + \frac{1}{2}\big(|m|+1\big)}{n} , \tag{8.4.17}$$

and

$$E = -\frac{Z^2e^2}{2n^2a_0} , \qquad n = 1, 2, \ldots , \tag{8.4.18}$$

the familiar Bohr energies, of course.

What is the multiplicity of states of given n counted this way? Note that $k_1, k_2 = 0, 1, \ldots$, whereas $m = 0, \pm1, \pm2, \ldots$ From determining the multiplicities of oscillator energy states in two and three dimensions, we already know, for non-negative integers that

$$\begin{aligned} n_1 + n_2 = N \; &: \qquad m(N) = N + 1 , \\ n_1 + n_2 + n_3 = N \; &: \qquad m(N) = \frac{1}{2}(N+1)(N+2) , \end{aligned} \tag{8.4.19}$$

from which, if $n_3 = 0$ is excluded (n_3'), we get

$$\begin{aligned} n_1 + n_2 + n_3' = N \; : \qquad m(N) &= \frac{1}{2}(N+1)(N+2) - (N+1) \\ &= \frac{1}{2}N(N+1) . \end{aligned} \tag{8.4.20}$$

So, for

$$k_1 + k_2 + |m| = n - 1 \tag{8.4.21}$$

we have n states for $m = 0$ and twice $\frac{1}{2}n(n-1)$ states for $m \neq 0$, giving the multiplicity

$$m(n) = n + 2 \times \frac{1}{2}n(n-1) = n^2 , \tag{8.4.22}$$

which is, of course, the answer we found earlier in (8.1.17).

8.5 Weak external electric field

The utility of the parabolic coordinate system begins to appear when we ask for the effect on the energy levels of an applied homogeneous electric field $\boldsymbol{F}$, which we take to point along the z axis. Thus the potential energy of an electron (charge $-e$) is now

$$V = -\frac{Ze^2}{r} + eFz = -e\underbrace{\left(\frac{Ze}{r} - Fz\right)}_{\text{electrostatic potential}} \qquad (8.5.1)$$

which, in parabolic coordinates reads

$$V = -\frac{Ze^2}{\xi+\eta} + eF(\xi-\eta) = -\frac{1}{\xi+\eta}\left[Ze^2 - eF\left(\xi^2-\eta^2\right)\right] . \qquad (8.5.2)$$

Thus the factorization of the wave function persists, now with

$$\left[\frac{\mathrm{d}^2}{\mathrm{d}\xi^2} - \frac{m^2-1}{4\xi^2} + \frac{2ME}{\hbar^2} + \frac{2Z_1}{a_0\xi} - \frac{2F}{ea_0}\xi\right] u(\xi) = 0 ,$$
$$\left[\frac{\mathrm{d}^2}{\mathrm{d}\eta^2} - \frac{m^2-1}{4\eta^2} + \frac{2ME}{\hbar^2} + \frac{2Z_1}{a_0\eta} + \frac{2F}{ea_0}\eta\right] u(\eta) = 0 . \qquad (8.5.3)$$

Before continuing let's notice what plays the role of the effective potential V_{eff} for the two motions. Drawn for $m = 18$, $F = \frac{1}{2}e/a_0^2$ and $Z_1 = Z_2 = 42$ with abscissas linear in ξ^2 and η^2, respectively:

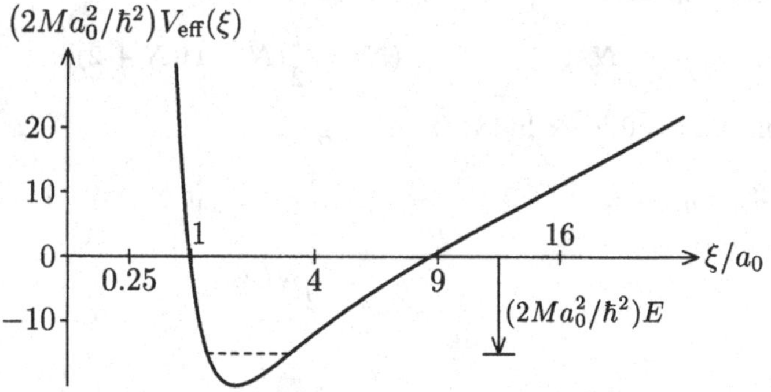

we see that

$$V_{\text{eff}}(\xi) = \frac{\hbar^2}{2M}\left[\frac{m^2-1}{4\xi^2} - \frac{2Z_1}{a_0\xi} + \frac{2F}{ea_0}\xi\right] \qquad (8.5.4)$$

binds the electron to the vicinity of the location of its minimum, whereas

$$V_{\text{eff}}(\eta) = \frac{\hbar^2}{2M}\left[\frac{m^2-1}{4\eta^2} - \frac{2Z_1}{a_0\eta} - \frac{2F}{ea_0}\eta\right] , \tag{8.5.5}$$

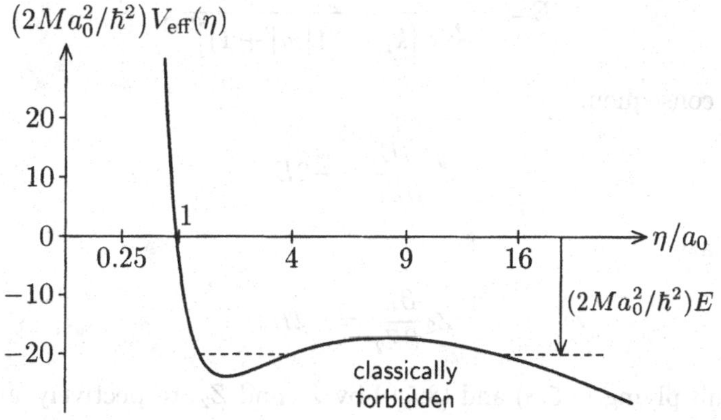

becomes arbitrarily large negative for large η values. In the latter situation there is a certain probability of finding the electron outside the atom, if ionizing the atom. That, however requires very strong electric fields to be effective, because one must overcome the exponential attenuation in the classically forbidden region (recall the discussion in Section 6.7); we are now interested only in weak fields.

Incidentally, we note that the ξ equation in (8.5.3) is known as the *up-hill equation*, and the η equation as the *down-hill equation* – names that are obviously suggested by the shapes of the effective potentials.

Think of the F in the $u(\xi)$ equation as a parameter and differentiate, anticipating that E and Z_1 will change (it is Z that is fixed)

$$\frac{2M}{\hbar^2}\frac{\partial E}{\partial F} = \frac{2}{ea_0}\langle\xi\rangle - \frac{2}{a_0}\frac{\partial Z_1}{\partial F}\left\langle\frac{1}{\xi}\right\rangle . \tag{8.5.6}$$

Of course, if we only change Z_1, we get

$$\frac{2M}{\hbar^2}\frac{\partial E}{\partial Z_1} = -\frac{2}{a_0}\left\langle\frac{1}{\xi}\right\rangle \tag{8.5.7}$$

so that

$$\frac{\partial E}{\partial F} - \frac{\partial E}{\partial Z_1}\frac{\partial Z_1}{\partial F} = e\langle\xi\rangle ; \tag{8.5.8}$$

we could have written the left side directly – it is the change of E produced by the *explicit* dependence on F. Similarly

$$\frac{\partial E}{\partial F} - \frac{\partial E}{\partial Z_2}\frac{\partial Z_2}{\partial F} = -e\langle\eta\rangle . \tag{8.5.9}$$

We are going to apply these results for $F = 0$, to get the linear dependence of E on F. First, for the ξ motion we known that

$$E = -\frac{Z_1^2}{2a_0}\frac{1}{\left[k_1 + \frac{1}{2}(|m|+1)\right]^2}, \tag{8.5.10}$$

with the consequence

$$Z_1 \frac{\partial E}{\partial Z_1} = -2E, \tag{8.5.11}$$

and similarly

$$Z_2 \frac{\partial E}{\partial Z_2} = -2E. \tag{8.5.12}$$

Then, multiplying (8.5.8) and (8.5.9) by Z_1 and Z_2, respectively, and adding ($F = 0$):

$$\underbrace{(Z_1 + Z_2)}_{=Z}\frac{\partial E}{\partial F} - 2E\underbrace{\frac{\partial}{\partial F}(Z_1 + Z_2)}_{=0} = eZ_1\langle\xi\rangle - eZ_2\langle\eta\rangle \tag{8.5.13}$$

so

$$\left.\frac{\partial E}{\partial F}\right|_{F=0} = \frac{e}{Z}\left[Z_1\langle\xi\rangle_{k_1} - Z_2\langle\eta\rangle_{k_2}\right]\Big|_{F=0}. \tag{8.5.14}$$

According to Problem 8-1d, for the Coulomb field in spherical coordinates,

$$\langle r\rangle_{nl} = \frac{a_0}{Z}\left[3n^2 - l(l+1)\right]. \tag{8.5.15}$$

So, with the correspondences

$$2l+1 \leftrightarrow |m|, \quad Z \leftrightarrow Z_{1,2}, \quad n \leftrightarrow k_{1,2} + \tfrac{1}{2}(|m|+1), \tag{8.5.16}$$

we get

$$\begin{aligned} Z_1\langle\eta\rangle_{k_1,|m|} &= \frac{a_0}{2}\left[3\left(k_1 + \frac{|m|+1}{2}\right)^2 - \frac{m^2-1}{4}\right], \\ Z_2\langle\eta\rangle_{k_2,|m|} &= \frac{a_0}{2}\left[3\left(k_2 + \frac{|m|+1}{2}\right)^2 - \frac{m^2-1}{4}\right], \end{aligned} \tag{8.5.17}$$

and

$$\left.\frac{\partial E}{\partial F}\right|_{F=0} = \frac{e}{Z}\frac{3a_0}{2}(k_1 - k_2)\underbrace{(k_1 + k_2 + |m| + 1)}_{=n} \tag{8.5.18}$$

follows.

This exact statement for $F = 0$ gives an *approximate* statement for the change of energy produced by a *weak* electric field:

$$\delta E_{n,k_1,k_2} = \frac{3}{2}\frac{ea_0}{Z}Fn(k_1 - k_2)\,. \tag{8.5.19}$$

This is an example of first-order perturbation theory, as applied to the linear Stark[*] effect.

The quantum numbers k_1 and k_2 range individually from 0 to $n - 1$, so that $k_1 - k_2$ goes from $n - 1$ to $-(n - 1)$, giving in all $2n - 1$ equally spaced values. For $n = 2$ with its $n^2 = 4$ states we get

$$
\begin{array}{lll}
k_1 = 1,\ k_2 = 0,\ m = 0: & \delta E = 3ea_0 F/Z\,, \\
k_1 = 0,\ k_2 = 0,\ m = \pm 1: & \delta E = 0\,, \\
k_1 = 0,\ k_2 = 1,\ m = 0: & \delta E = -3ea_0 F/Z\,.
\end{array}
$$

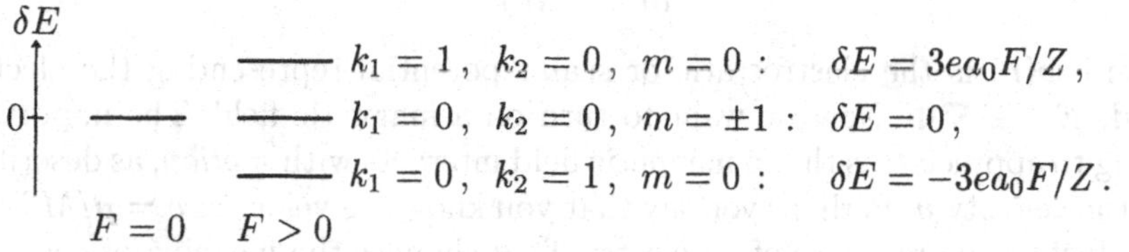

So the n^2-fold degeneracy is not removed; there are only $2n - 1$ different energy states. What in general is the multiplicity of a given energy, a given $k_1 - k_2$? Suppose $k_1 > k_2$. Then we get the same $k_1 - k_2$ by successively increasing k_1 and k_2 by 1, until we reach $k_1 = n - 1$ ($n - k_1$ values *including* the original) or by successively decreasing k_1 and k_2 by 1, until we reach $k_2 = 0$ (k_2 additional values), giving in all

$$m(n, k_1, k_2) = n - |k_1 - k_2| \tag{8.5.20}$$

for the multiplicity of the Stark-shifted energies. The total number of states is

$$\sum_{k=-(n-1)}^{n-1} \left(n - |k|\right) = n + 2\underbrace{\sum_{k=1}^{n-1}(n - k)}_{=(n-1)n} = n^2\,, \tag{8.5.21}$$

the multiplicity we know from (8.1.17) and (8.4.22).

8.6 Weak external magnetic field

This naturally raises a question: What would a homogeneous magnetic field do to the remaining degeneracy? It is time to appreciate generally how a charged particle is acted on by electric and magnetic fields that are given.

---

[*]Johannes STARK (1874–1957)

The Lagrangian of a free particle of mass M is

$$L = \boldsymbol{p} \cdot \frac{\mathrm{d}\boldsymbol{r}}{\mathrm{d}t} - \frac{1}{2M}\boldsymbol{p}^2 \,. \tag{8.6.1}$$

[Here and below we'll always understand products of potentially non-commuting operators as symmetrized, so $\boldsymbol{p} \cdot \mathrm{d}\boldsymbol{r}$ should be read as $\frac{1}{2}\boldsymbol{p} \cdot \mathrm{d}\boldsymbol{r} + \frac{1}{2}\mathrm{d}\boldsymbol{r} \cdot \boldsymbol{p}$.] We have already recognized in the context of Coulomb and constant electric fields that a particle of charge e (now either positive or negative) has the Lagrangian

$$L = \boldsymbol{p} \cdot \frac{\mathrm{d}\boldsymbol{r}}{\mathrm{d}t} - \frac{1}{2M}\boldsymbol{p}^2 - e\Phi(\boldsymbol{r}) \tag{8.6.2}$$

where $\Phi(\boldsymbol{r})$ is the electrostatic or scalar potential representing the electric field, $\boldsymbol{E} = -\boldsymbol{\nabla}\Phi$. Now we want to turn on a magnetic field. The important thing to appreciate is that a magnetic field interacts with *motion*, as described by the velocity $\boldsymbol{v}$. Perhaps you say that you know the velocity: $\boldsymbol{v} = \boldsymbol{p}/M$? But no. Just as the presence of an electric field changes the meaning of energy:

$$\frac{\boldsymbol{p}^2}{2M} \to \underbrace{\frac{\boldsymbol{p}^2}{2M}}_{\text{kinetic,}} + \underbrace{e\Phi}_{\text{potential energy}} \tag{8.6.3}$$

the presence of a magnetic field changes the meaning of momentum, $\boldsymbol{p}$ is no longer $M\boldsymbol{v}$. For that reason it behooves to introduce $\boldsymbol{v}$ as an independent variable. Consider this L:

$$\begin{aligned} L(\boldsymbol{r},\boldsymbol{p},\boldsymbol{v}) &= \boldsymbol{p} \cdot \left(\frac{\mathrm{d}\boldsymbol{r}}{\mathrm{d}t} - \boldsymbol{v}\right) + \frac{1}{2}M\boldsymbol{v}^2 - e\Phi(\boldsymbol{r}) \\ &= \boldsymbol{p} \cdot \frac{\mathrm{d}\boldsymbol{r}}{\mathrm{d}t} - H(\boldsymbol{r},\boldsymbol{p},\boldsymbol{v}) \,, \end{aligned} \tag{8.6.4}$$

with the Hamiltonian

$$H = \boldsymbol{p} \cdot \boldsymbol{v} - \frac{1}{2}M\boldsymbol{v}^2 + e\Phi \,. \tag{8.6.5}$$

The Hamilton–Heisenberg equations of motions are

$$\begin{aligned} \frac{\mathrm{d}\boldsymbol{r}}{\mathrm{d}t} &= \frac{\partial H}{\partial \boldsymbol{p}} = \boldsymbol{v} \,, \\ -\frac{\mathrm{d}\boldsymbol{p}}{\mathrm{d}t} &= \frac{\partial H}{\partial \boldsymbol{r}} = e\boldsymbol{\nabla}\Phi = -e\boldsymbol{E} \,, \end{aligned} \tag{8.6.6}$$

and, since there is no time derivative of $\boldsymbol{v}$, simply

$$0 = \frac{\partial H}{\partial \boldsymbol{v}} = \boldsymbol{p} - M\boldsymbol{v} \,. \tag{8.6.7}$$

So we identify $\boldsymbol{v}$ with velocity, $\mathrm{d}\boldsymbol{r}/\mathrm{d}t$, learn that momentum $\boldsymbol{p}$ is $M\mathrm{d}\boldsymbol{r}/\mathrm{d}t$ and get the force equation

$$\frac{\mathrm{d}\boldsymbol{p}}{\mathrm{d}t} = M\frac{\mathrm{d}^2\boldsymbol{r}}{\mathrm{d}t^2} = e\boldsymbol{E} . \tag{8.6.8}$$

If we wish we can simply accept that $\boldsymbol{v} = \boldsymbol{p}/M$ and come back to

$$H = \frac{\boldsymbol{p}^2}{2M} + e\Phi . \tag{8.6.9}$$

But now we want to add an interaction that depends on $\boldsymbol{v}$; the simplest possibility is $\boldsymbol{v}$ multiplied by a new vector. We shall in fact write

$$H = \boldsymbol{p}\cdot\boldsymbol{v} - \frac{1}{2}M\boldsymbol{v}^2 + e\Phi(\boldsymbol{r}) - \frac{e}{c}\boldsymbol{A}(\boldsymbol{r})\cdot\boldsymbol{v} , \tag{8.6.10}$$

where $c = 2.99792 \times 10^{10}$ cm/s is the speed of light. What are the new equations of motion? Rather than (8.6.6) and (8.6.7) we get now

$$\begin{aligned} \frac{\mathrm{d}\boldsymbol{r}}{\mathrm{d}t} &= \frac{\partial H}{\partial \boldsymbol{p}} = \boldsymbol{v} , \\ -\frac{\mathrm{d}\boldsymbol{p}}{\mathrm{d}t} &= \frac{\partial H}{\partial \boldsymbol{r}} = e\boldsymbol{\nabla}\Phi - \frac{e}{c}\boldsymbol{\nabla}\boldsymbol{A}(\boldsymbol{r})\cdot\boldsymbol{v} , \\ 0 &= \frac{\partial H}{\partial \boldsymbol{v}} = \boldsymbol{p} - M\boldsymbol{v} - \frac{e}{c}\boldsymbol{A}(\boldsymbol{r}) . \end{aligned} \tag{8.6.11}$$

In the second equation, $\boldsymbol{\nabla}$ does, of course, not differentiate the independent variable $\boldsymbol{v}$; nevertheless, since we'll shortly accept the third equation as defining $\boldsymbol{v}$ as a function of $\boldsymbol{p}$ and $\boldsymbol{r}$, it is expedient to use the vector identity

$$\boldsymbol{\nabla}\boldsymbol{A}(\boldsymbol{r})\cdot\boldsymbol{v} = \boldsymbol{v}\times(\boldsymbol{\nabla}\times\boldsymbol{A}) + \boldsymbol{v}\cdot\boldsymbol{\nabla}\boldsymbol{A} \tag{8.6.12}$$

and write the second equation in the equivalent form

$$-\frac{\mathrm{d}\boldsymbol{p}}{\mathrm{d}t} = e\boldsymbol{\nabla}\Phi - \frac{e}{c}\boldsymbol{v}\times(\boldsymbol{\nabla}\times\boldsymbol{A}) - \frac{e}{c}\boldsymbol{v}\cdot\boldsymbol{\nabla}\boldsymbol{A} . \tag{8.6.13}$$

So the first equation in (8.6.11) tells us that $\boldsymbol{v}$ is still the velocity $\mathrm{d}\boldsymbol{r}/\mathrm{d}t$, but the third says that $\boldsymbol{p}$ has changed:

$$\boldsymbol{p} = \underbrace{M\frac{\mathrm{d}\boldsymbol{r}}{\mathrm{d}t}}_{\text{kinetic,}} + \underbrace{\frac{e}{c}\boldsymbol{A}(\boldsymbol{r})}_{\text{potential momentum}} . \tag{8.6.14}$$

And the force equation? Differentiate,

$$\frac{\mathrm{d}\boldsymbol{p}}{\mathrm{d}t} = M\frac{\mathrm{d}^2\boldsymbol{r}}{\mathrm{d}t^2} + \frac{e}{c}\frac{\mathrm{d}}{\mathrm{d}t}\boldsymbol{A}(\boldsymbol{r}) , \tag{8.6.15}$$

and combine it with (8.6.13) to get

$$M\frac{\mathrm{d}^2\boldsymbol{r}}{\mathrm{d}t^2} = -e\boldsymbol{\nabla}\Phi + \frac{e}{c}\boldsymbol{v}\times\left[\boldsymbol{\nabla}\times\boldsymbol{A}(\boldsymbol{r})\right] + \frac{e}{c}\left[\boldsymbol{v}\cdot\boldsymbol{\nabla}\boldsymbol{A} - \frac{\mathrm{d}}{\mathrm{d}t}\boldsymbol{A}(\boldsymbol{r})\right]. \tag{8.6.16}$$

To evaluate $\frac{\mathrm{d}}{\mathrm{d}t}\boldsymbol{A}(\boldsymbol{r})$, we need H as a function of $\boldsymbol{p}$ and $\boldsymbol{r}$, which is

$$H = \underbrace{\left(\boldsymbol{p} - \frac{e}{c}\boldsymbol{A}\right)}_{=M\boldsymbol{v}}\cdot\boldsymbol{v} - \frac{1}{2}M\boldsymbol{v}^2 + e\Phi = \frac{1}{2}M\boldsymbol{v}^2 + e\Phi = \frac{1}{2M}\left(\boldsymbol{p} - \frac{e}{c}\boldsymbol{A}\right)^2 + e\Phi\,. \tag{8.6.17}$$

Now

$$\frac{\mathrm{d}}{\mathrm{d}t}\boldsymbol{A} = \frac{1}{\mathrm{i}\hbar}\left[\boldsymbol{A}(\boldsymbol{r}), H\right] = (\boldsymbol{A}, H)\,, \tag{8.6.18}$$

the symmetrized Poisson bracket, because $\boldsymbol{A}$ is independent of $\boldsymbol{p}$ and H is less than cubic in $\boldsymbol{p}$. So

$$\frac{\mathrm{d}}{\mathrm{d}t}\boldsymbol{A} = \frac{\partial\boldsymbol{A}}{\partial\boldsymbol{r}}\cdot\frac{\partial H}{\partial\boldsymbol{p}} = \boldsymbol{v}\cdot\boldsymbol{\nabla}\boldsymbol{A}\,, \tag{8.6.19}$$

the term in the second line of (8.6.16) vanishes, and we get

$$M\frac{\mathrm{d}^2\boldsymbol{r}}{\mathrm{d}t^2} = e\boldsymbol{E} + \frac{e}{c}\boldsymbol{v}\times\boldsymbol{B} \quad \text{with } \boldsymbol{E} = -\boldsymbol{\nabla}\Phi \ \text{ and } \boldsymbol{B} = \boldsymbol{\nabla}\times\boldsymbol{A} \tag{8.6.20}$$

which exhibits the dynamical action of the magnetic field $\boldsymbol{B}$, as it is constructed from the vector potential $\boldsymbol{A}$. The appearance of the Lorentz* force in (8.6.20) tells us that we have indeed found a way of incorporating magnetic fields. Thus the Lagrangian (8.6.4) with the Hamilton operator (8.6.10) is appropriate for a charged particle moving under the influence of both an electric and a magnetic field.

We are interested now in a homogeneous field $\boldsymbol{B}$. What is $\boldsymbol{A}$? The answer is not unique. If $\boldsymbol{A}$ is a possible potential, so is $\boldsymbol{A} + \boldsymbol{\nabla}\lambda$, λ arbitrary, for

$$\boldsymbol{\nabla}\times(\boldsymbol{A} + \boldsymbol{\nabla}\lambda) = \boldsymbol{\nabla}\times\boldsymbol{A} = \boldsymbol{B}\,. \tag{8.6.21}$$

(This is the freedom of gauge transformation.) Suppose for example that $\boldsymbol{B}$ points in the z direction, so

---

*Hendrik Antoon LORENTZ (1853–1929)

$$B_x = \frac{\partial}{\partial y}A_z - \frac{\partial}{\partial z}A_y = 0\,,$$
$$B_y = \frac{\partial}{\partial z}A_x - \frac{\partial}{\partial x}A_z = 0\,,$$
$$B_z = \frac{\partial}{\partial x}A_y - \frac{\partial}{\partial y}A_x = B\,. \tag{8.6.22}$$

One simple solution is

$$A_x = A_z = 0\,, \quad A_y = Bx\,. \tag{8.6.23}$$

Another is

$$A_y = A_z = 0\,, \quad A_x = -By\,. \tag{8.6.24}$$

A third is the average of these two

$$A_z = 0\,,\ A_x = -\frac{1}{2}By\,,\ A_y = \frac{1}{2}Bx\ :\quad \boldsymbol{A} = \frac{1}{2}\boldsymbol{B}\times\boldsymbol{r}\,; \tag{8.6.25}$$

it is the most natural form, being three-dimensional and not singling out arbitrary x or y directions. The first two appear as

$$\boldsymbol{A} = \frac{1}{2}\boldsymbol{B}\times\boldsymbol{r} + \boldsymbol{\nabla}\left(\pm\frac{1}{2}Bxy\right)\,. \tag{8.6.26}$$

If we use the natural form in the Hamiltonian (8.6.17), we get

$$\begin{aligned} H &= \frac{1}{2M}\left(\boldsymbol{p} - \frac{e}{2c}\boldsymbol{B}\times\boldsymbol{r}\right)^2 + e\Phi \\ &= \frac{\boldsymbol{p}^2}{2M} + e\Phi - \frac{e}{2Mc}\boldsymbol{B}\cdot\boldsymbol{r}\times\boldsymbol{p} + \frac{e^2}{8Mc^2}(\boldsymbol{B}\times\boldsymbol{r})^2\,. \end{aligned} \tag{8.6.27}$$

Note that the term linear in $\boldsymbol{B}$ is proportional to the orbital angular momentum $\boldsymbol{L} = \boldsymbol{r}\times\boldsymbol{p}$.

We now apply this to hydrogenic atoms. The charged particle is the electron, so $e \to -e$ and $e > 0$ denotes the elementary charge of (8.1.18). Then, in conjunction with the nuclear Coulomb field and a weak electric field along the z axis, consider a weak magnetic field along the same z axis. We omit the quadratic $\boldsymbol{B}$ term and get

$$H = \frac{\boldsymbol{p}^2}{2M} - \frac{Ze^2}{r} + eFz + \frac{e}{2Mc}BL_z\,. \tag{8.6.28}$$

This is easy! For any state specified by the *magnetic* quantum number m, $L'_z = m\hbar$, we simply get the additional energy (known as the energy shifts of the normal Zeeman* effect)

---

*Pieter ZEEMAN (1865–1943)

$$\frac{e\hbar}{2Mc}mB = m\mu_0 B \,, \tag{8.6.29}$$

where

$$\mu_0 = \frac{e\hbar}{2Mc} = 5.788\,382 \times 10^{-9}\,\mathrm{eV\,G^{-1}} \tag{8.6.30}$$

is called Bohr magneton. So the energy values in weak parallel electric and magnetic fields are

$$E = -\frac{Z^2e^2}{2a_0n^2} + \frac{3}{2}\frac{ea_0}{Z}Fn(k_1-k_2) + m\mu_0 B \,. \tag{8.6.31}$$

Electrically degenerate energy values, those of constant $k_1 - k_2$, have, according to (8.4.21), here written as

$$|m| = \underbrace{(n-1+k_1-k_2)}_{\text{given}} - 2k_1 \,, \tag{8.6.32}$$

different values of $|m|$, in general, or, for a given value of $|m|$, opposite values of m. The degeneracy is completely removed.

8.7 Insertion: Charge in a homogeneous magnetic field

Now consider just the magnetic field $\boldsymbol{B}$, so that $\Phi = 0$ in (8.6.17), and the Hamilton operator is

$$H = \frac{1}{2}M\boldsymbol{v}^2 = \frac{1}{2M}\left(\boldsymbol{p} - \frac{e}{c}\boldsymbol{A}\right)^2 \quad \text{with} \quad \boldsymbol{\nabla}\times\boldsymbol{A} = \boldsymbol{B} = B\boldsymbol{e}_z \,. \tag{8.7.1}$$

First, look at the commutator relations for $\boldsymbol{v} = M^{-1}\left(\boldsymbol{p} - \frac{e}{c}\boldsymbol{A}\right)$:

$$\begin{aligned} \boldsymbol{v}\times\boldsymbol{v} &= \frac{1}{M^2}\left(\boldsymbol{p} - \frac{e}{c}\boldsymbol{A}\right)\times\left(\boldsymbol{p} - \frac{e}{c}\boldsymbol{A}\right) \\ &= -\frac{1}{M^2}\frac{e}{c}\left(\boldsymbol{p}\times\boldsymbol{A} + \boldsymbol{A}\times\boldsymbol{p}\right) \\ &= \frac{\mathrm{i}\hbar}{M^2}\frac{e}{c}\boldsymbol{\nabla}\times\boldsymbol{A} = \frac{\mathrm{i}\hbar e}{M^2c}\boldsymbol{B} \,. \end{aligned} \tag{8.7.2}$$

That is

$$\frac{1}{\mathrm{i}\hbar}[v_x, v_y] = \frac{e}{M^2c}B \,, \quad \frac{1}{\mathrm{i}\hbar}[v_x, v_z] = \frac{1}{\mathrm{i}\hbar}[v_y, v_z] = 0 \,, \tag{8.7.3}$$

which shows a $(1+2)$-dimensional split,

$$H = \underbrace{\tfrac{1}{2}Mv_z^2}_{=p_z^2/(2M)} + \tfrac{1}{2}M\left(v_x^2 + v_y^2\right) \,. \tag{8.7.4}$$

For the electron, $e = -|e|$, the variables q, p introduced by

$$v_x = \sqrt{\frac{\hbar|e|B}{M^2c}}\,p\,, \qquad v_y = \sqrt{\frac{\hbar|e|B}{M^2c}}\,q \tag{8.7.5}$$

give

$$\frac{1}{\mathrm{i}}[q,p] = 1\,, \tag{8.7.6}$$

and then

$$H = \frac{p_z^2}{2M} + \frac{|e|\hbar B}{Mc}\frac{1}{2}\left(p^2+q^2\right) \tag{8.7.7}$$

gives us the energy spectrum

$$E = \frac{p_z'^{\,2}}{2M} + \underbrace{\frac{|e|\hbar B}{Mc}}_{=\hbar\omega_1}\left(n+\tfrac{1}{2}\right)$$

$$\text{with} \quad -\infty < p_z' < \infty \quad \text{and} \quad n = 0,1,2,\ldots\,. \tag{8.7.8}$$

Indeed $\omega_1 = |e|B/(Mc)$, called cyclotron frequency, is the rotation frequency of $\boldsymbol{v}$ in the x, y plane: the pair

$$\frac{\mathrm{d}v_x}{\mathrm{d}t} = -\omega_1 v_y\,, \qquad \frac{\mathrm{d}v_y}{\mathrm{d}t} = \omega_1 v_x \tag{8.7.9}$$

is just

$$\frac{\mathrm{d}q}{\mathrm{d}t} = \omega_1 p\,, \quad \frac{\mathrm{d}p}{\mathrm{d}t} = -\omega_1 q \tag{8.7.10}$$

and, of course, we knew this all along since (8.6.20) says

$$\frac{\mathrm{d}\boldsymbol{v}}{\mathrm{d}t} = \frac{e}{Mc}\boldsymbol{B}\times\boldsymbol{v} = \frac{|e|}{Mc}\boldsymbol{B}\times\boldsymbol{v}\,. \tag{8.7.11}$$

The energy (omitting the free z motion, or setting $p_z' = 0$) is that of a one-dimensional oscillator. What is the multiplicity of those energy states? We get a clear picture by, in contrast with the above, working with a particular gauge: $A_y = Bx$,

$$H = \frac{p_x^2}{2M} + \frac{1}{2M}\left(p_y - \frac{e}{c}Bx\right)^2. \tag{8.7.12}$$

Note that y does not appear, so

$$\frac{\mathrm{d}p_y}{\mathrm{d}t} = -\frac{\partial H}{\partial y} = 0\,; \tag{8.7.13}$$

a state can be specified by an eigenvalue of p_y, and

$$H \to \frac{p_x^2}{2M} + \frac{M\omega_1^2}{2}\left(x - \frac{c}{eB}p_y'\right)^2 \quad \text{with} \quad -\infty < p_y' < \infty \tag{8.7.14}$$

exhibits the one-dimensional oscillator. Clearly each energy state is *infinitely* degenerate, corresponding to the independence of the energy of p_y'.

How does this work out with $\boldsymbol{A} = \frac{1}{2}\boldsymbol{B}\times\boldsymbol{r}$? Here

$$\begin{aligned} H &= \frac{1}{2M}\left(\boldsymbol{p} - \frac{e}{2c}\boldsymbol{B}\times\boldsymbol{r}\right)^2 = \frac{\boldsymbol{p}^2}{2M} - \frac{e}{2Mc}\boldsymbol{B}\cdot\boldsymbol{L} + \frac{e^2}{8Mc^2}(\boldsymbol{B}\times\boldsymbol{r})^2 \\ &= \underbrace{\frac{1}{2M}}_{\uparrow\, p_z'=0}\left(p_x^2+p_y^2\right) + \underbrace{\frac{|e|}{2Mc}B}_{=\omega_2} L_z + \underbrace{\frac{e^2B^2}{8Mc^2}}_{=\frac{1}{2}M\omega_2^2}\left(x^2+y^2\right) \quad \text{with } \omega_2 = \tfrac{1}{2}\omega_1 \,. \end{aligned} \tag{8.7.15}$$

We recognize here the two-dimensional oscillator with an extra energy term $\omega_2 L_z$. It is convenient to use the quantum numbers $(y_+^\dagger y_+)' = n_+ = 0,1,2,\ldots$ and $(y_-^\dagger y_-)' = n_- = 0,1,2,\ldots$, for then $\omega_2 L_z' = \hbar\omega_2(n_+ - n_-)$ as we've seen in (7.4.17). So the eigenvalues are

$$\begin{aligned} E &= \hbar\omega_2(n_+ + n_- + 1) + \hbar\omega_2(n_+ - n_-) \\ &= \hbar\omega_2(2n_+ + 1) = \hbar\omega_1\left(n_+ + \tfrac{1}{2}\right) \,. \end{aligned} \tag{8.7.16}$$

This applies for $e = -|e| < 0$; for $e = |e| > 0$ we'd have $E = \hbar\omega_1(n_- + \frac{1}{2})$. The degeneracy comes from the dependence of the energy on only one of the integers n_+, n_-.

To describe the degeneracy more physically note that the equation of motion (8.7.11) is

$$\frac{\mathrm{d}\boldsymbol{v}}{\mathrm{d}t} = \frac{\mathrm{d}}{\mathrm{d}t}\left(-\frac{e}{Mc}\boldsymbol{B}\times\boldsymbol{r}\right) \,, \tag{8.7.17}$$

so

$$\boldsymbol{v} + \frac{e}{Mc}\boldsymbol{B}\times\boldsymbol{r} \equiv \frac{e}{Mc}\boldsymbol{B}\times\boldsymbol{r}_0 \tag{8.7.18}$$

obeys $\mathrm{d}\boldsymbol{r}_0/\mathrm{d}t = 0$. Indeed

$$\boldsymbol{v} = -\frac{e}{Mc}\boldsymbol{B}\times(\boldsymbol{r} - \boldsymbol{r}_0) \tag{8.7.19}$$

describes the rotational motion around the fixed point $\boldsymbol{r}_0$ with frequency ω_1. What are the properties of $\boldsymbol{r}_0 \mathrel{\widehat{=}} (x_0, y_0)$? First note

$$\frac{e}{c}\boldsymbol{B}\times\boldsymbol{r}_0 = \underbrace{M\boldsymbol{v}}_{\boldsymbol{p}-(e/c)\boldsymbol{A}=} + \frac{e}{c}\underbrace{\boldsymbol{B}\times\boldsymbol{r}}_{=2\boldsymbol{A}} = \boldsymbol{p} + \frac{e}{c}\boldsymbol{A} \tag{8.7.20}$$

so

$$p_x - \frac{e}{2c}By = -\frac{eB}{c}y_0 \, , \quad p_y + \frac{e}{2c}Bx = \frac{eB}{c}x_0 \, , \tag{8.7.21}$$

and conclude

$$\left(\frac{eB}{c}\right)^2 [x_0, y_0] = \left[p_x - \frac{e}{2c}By, p_y + \frac{e}{2c}Bx\right] = -i\hbar\frac{e}{c}B \, , \tag{8.7.22}$$

that is

$$[x_0, y_0] = -i\hbar\frac{c}{eB} \underbrace{\uparrow}_{e=-|e|} = \frac{i\hbar}{M\omega_1} \, ; \tag{8.7.23}$$

it is not possible to specify precisely the center of the motion. The coordinates x_0 and y_0 are subject to a Heisenberg uncertainty relation,

$$\delta x_0 \, \delta y_0 \geq \frac{1}{2}\frac{\hbar}{M\omega_1} \, . \tag{8.7.24}$$

Also observe that

$$\begin{aligned}\frac{1}{2M}\left(\boldsymbol{p} + \frac{e}{c}\boldsymbol{A}\right)^2 &= \underbrace{\frac{1}{2M}\left(\frac{eB}{c}\right)^2}_{=\frac{1}{2}M\omega_1^2} \underbrace{(x_0^2 + y_0^2)}_{=r_0^2} \\ &= \frac{1}{2M}(p_x^2 + p_y^2) - \underbrace{\frac{|e|}{2Mc}}_{=\omega_2}B\,L_z + \underbrace{\frac{e^2B^2}{8Mc^2}}_{=\frac{1}{2}M\omega_2^2}(x^2 + y^2) \, ,\end{aligned} \tag{8.7.25}$$

or

$$\frac{1}{2}M\omega_1^2 r_0^2 = \frac{1}{2M}(p_x^2 + p_y^2) + \frac{1}{2}M\omega_2^2(x^2 + y^2) - \omega_2 L_z \, , \tag{8.7.26}$$

so

$$\begin{aligned}\frac{1}{2}M\omega_1^2 r_0^{2\prime} &= \hbar\omega_2(n_+ + n_- + 1) - \hbar\omega_2(n_+ - n_-) \\ &= \hbar\omega_1\left(n_- + \tfrac{1}{2}\right) \, .\end{aligned} \tag{8.7.27}$$

Thus the second quantum number, the one that does not appear in the energy (8.7.16), specifies the required distance of the orbit center from the origin

$$r_0^{2\prime} = 2\frac{\hbar}{M\omega_1}\left(n_- + \tfrac{1}{2}\right) \, . \tag{8.7.28}$$

That, of course, was obvious once we saw (8.7.23); the eigenvalues of $x_0^2 + y_0^2$ are those of the oscillator, suitably scaled.

Incidentally, one can present E similarly:

$$H = \frac{1}{2}M\boldsymbol{v}^2 = \frac{M}{2}\left[-\frac{e}{Mc}\boldsymbol{B}\times(\boldsymbol{r}-\boldsymbol{r}_0)\right]^2 = \frac{M}{2}\omega_1^2(\boldsymbol{r}-\boldsymbol{r}_0)^2 \,, \tag{8.7.29}$$

and the comparison with (8.7.16) gives

$$(\boldsymbol{r}-\boldsymbol{r}_0)^{2\prime} = 2\frac{\hbar}{M\omega_1}\left(n_+ + \tfrac{1}{2}\right) . \tag{8.7.30}$$

It has now long been clear that the energy degeneracy simply means that the center of motion can be anywhere, although not precisely specifiable. One has the option of specifying r_0^2, but the location on that circle is unknown, or of, for example, specifying x_0 (that is the gauge $A_y = Bx$) but y_0 is unknown, or, most physically of using the minimum uncertainty states, the eigenvectors of $x_0 + \mathrm{i}y_0$. The latter gives us an easy way to answer the practical question: If there is only a finite (macroscopic) area A available, how many states are there of a given energy? We recall that

$$\int \frac{\mathrm{d}q'\,\mathrm{d}p'}{2\pi\hbar} \tag{8.7.31}$$

does that counting, where from (8.7.23) we have the correspondence

$$q \to \sqrt{M\omega_1}\, x_0 \,, \quad p \to \sqrt{M\omega_1}\, y_0 \,, \tag{8.7.32}$$

so

$$\int \frac{\mathrm{d}q'\,\mathrm{d}p'}{2\pi\hbar} \longrightarrow \frac{M\omega_1}{2\pi\hbar}\int \mathrm{d}x_0'\,\mathrm{d}y_0' = \frac{|e|B}{2\pi\hbar c}A \tag{8.7.33}$$

is the desired number. Another more special way takes the area A to be a circle. Then the largest radius and the largest n_- are such that

$$\frac{1}{\pi}A = r_0^{2\prime}\Big|_{\max} \cong 2\frac{\hbar}{M\omega_1}n_{-,\max} \,. \tag{8.7.34}$$

Therefore

$$\text{number of states} = n_{-,\max} = \frac{M\omega_1}{\hbar}\frac{A}{2\pi} \,. \tag{8.7.35}$$

8.8 Scattering states

Back to the unfinished Coulomb problem – unfinished because we have only considered the $E < 0$ states, the bound states. How about the $E > 0$ states?

They describe an electron, not bound, but coming from far away where it has positive kinetic energy, $\frac{1}{2}M\boldsymbol{v}^2 = \boldsymbol{p}^2/(2M)$, and negligible potential energy, $-Ze^2/r \to 0$. Such a particle approaching the nucleus will have its direction changed while receding from the nucleus – it is scattered.

Suppose that the particle, far away, is moving along the z axis with velocity $v = p/M$. Its wave function will be essentially the plane wave

$$e^{\frac{i}{\hbar}\boldsymbol{p}\cdot\boldsymbol{r}} = e^{\frac{i}{\hbar}pz} = e^{ikz} \quad \text{with } k = \frac{p}{\hbar} = \frac{Mv}{\hbar} . \tag{8.8.1}$$

The wave approaching the center of force at the origin will produce a new wave representing the scattered particle which will essentially emanate from the neighborhood of the origin, seen far away as a *spherical* wave e^{ikr}.

The two parts of the wave, crudely e^{ikz} and e^{ikr}, appear in parabolic coordinates as

$$e^{ik(\xi - \eta)} , \quad e^{ik(\xi + \eta)} \tag{8.8.2}$$

which suggests that the complete wave function may have the form

$$\psi = e^{ik\xi}G(\eta) , \tag{8.8.3}$$

certainly independent of ϕ because of the axial symmetry of the physical situation. Let's try it in (8.4.9) with $2ME/\hbar^2 = k^2$, $\partial\psi/\partial\phi = 0$:

$$\Big[\underbrace{\frac{\partial}{\partial\xi}\xi\frac{\partial}{\partial\xi} + k^2\xi}_{\to -k^2\xi + ik} + \frac{\partial}{\partial\eta}\eta\frac{\partial}{\partial\eta} + k^2\eta + \frac{2Z}{a_0}\Big] e^{ik\xi}G(\eta) = 0 , \tag{8.8.4}$$

so that

$$\left[\frac{d^2}{d\eta^2} + \frac{1}{\eta}\frac{d}{d\eta} + k^2 + \frac{2Z/a_0 + ik}{\eta}\right] G(\eta) = 0 \tag{8.8.5}$$

or, again, with $G = \sqrt{1/\eta}\,u(\eta)$

$$\left[\frac{d^2}{d\eta^2} + \frac{1}{4\eta^2} + k^2 + \frac{2Z/a_0 + ik}{\eta}\right] u(\eta) = 0 . \tag{8.8.6}$$

The analogous radial equation,

$$\left[\frac{d^2}{dr^2} + \frac{1}{4r^2} - \kappa^2 + \frac{2Z}{a_0 r}\right] u(r) = 0 , \tag{8.8.7}$$

is (8.1.8) with $l = -\frac{1}{2}$, $(2M/\hbar^2)V(r) = -2Z/(a_0 r)$, and $2ME/\hbar^2 = -\kappa^2$, $\kappa = Z/(na_0)$; it has the bound state solution [cf. (8.1.32)]

$$u(r) \propto r^{\frac{1}{2}} L^{(0)}_{Z/(\kappa a_0) - \frac{1}{2}}(2\kappa r)\, e^{-\kappa r} , \tag{8.8.8}$$

which indicates the correspondence

$$r \to \eta\,, \quad Z \to Z + \tfrac{1}{2}\mathrm{i}ka_0\,, \quad \kappa \to \mathrm{i}k\,. \tag{8.8.9}$$

So

$$u(\eta) \propto \eta^{\frac{1}{2}}\,\mathrm{L}^{(0)}_{-\mathrm{i}Z/(ka_0)}\,(2\mathrm{i}k\eta)\;\mathrm{e}^{-\mathrm{i}k\eta} \tag{8.8.10}$$

and

$$G(\eta) \propto \mathrm{L}_{-\mathrm{i}\beta}\,(2\mathrm{i}k\eta)\;\mathrm{e}^{-\mathrm{i}k\eta} \tag{8.8.11}$$

with

$$\beta = \frac{Z}{ka_0} = \frac{Ze^2M}{k\hbar^2} = \frac{Ze^2M}{\hbar p} = \frac{Ze^2}{Mv^2}k = \frac{Ze^2}{\hbar v}\,. \tag{8.8.12}$$

Now notice that the sign of i in $\kappa \to \mathrm{i}k$ is chosen so that

$$G(\eta) \propto \mathrm{e}^{-\mathrm{i}k\eta} \quad \text{for } Z \to 0 \quad \left[\text{when } \mathrm{L}^{(0)}_{-\mathrm{i}Z/(ka_0)} \to \mathrm{L}^{(0)}_0 = 1\right] \tag{8.8.13}$$

as it should. All very well but what is the complex extension of the Laguerre polynomial that appears here?

Recall that

$$\mathrm{L}^{(0)}_\nu(x) = \frac{1}{\nu!}\,\mathrm{e}^x\left(\frac{\mathrm{d}}{\mathrm{d}x}\right)^\nu x^\nu\,\mathrm{e}^{-x} \tag{8.8.14}$$

is the definition for integer ν. But we can extend it by using a contour integral:

$$\mathrm{L}^{(0)}_\nu(x) = \mathrm{e}^x \oint \frac{\mathrm{d}t}{2\pi\mathrm{i}}\,\frac{t^\nu\,\mathrm{e}^{-t}}{(t-x)^{\nu+1}}$$

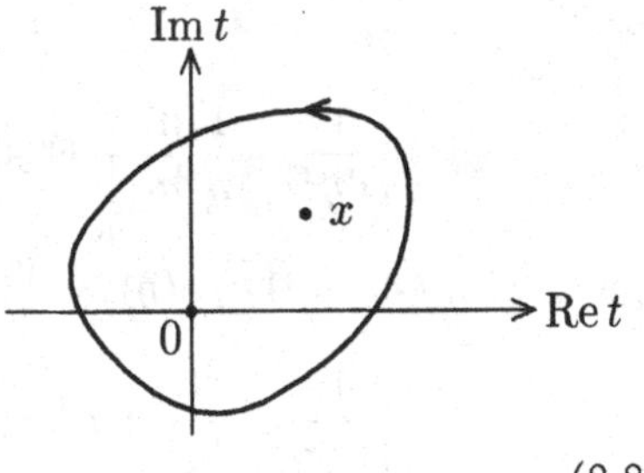

(8.8.15)

where a cut must connect $t = x$ and $t = 0$ for non-integer ν values. For $\nu = -\mathrm{i}\beta$ and $x = 2\mathrm{i}k\eta$, this gives us

$$\mathrm{L}^{(0)}_{-\mathrm{i}\beta}(2\mathrm{i}k\eta) = \mathrm{e}^{2\mathrm{i}k\eta} \oint \frac{\mathrm{d}t}{2\pi\mathrm{i}}\,\mathrm{e}^{-t}t^{-\mathrm{i}\beta}\,\frac{1}{(t-2\mathrm{i}k\eta)^{1-\mathrm{i}\beta}}\,. \tag{8.8.16}$$

A convenient way to take the contour is specified by having the cut extend from $t = 2\mathrm{i}k\eta$ to $t = 2\mathrm{i}k\eta + \infty$ and then from $t = \infty$ to $t = 0$. Then, integrating along the cut:

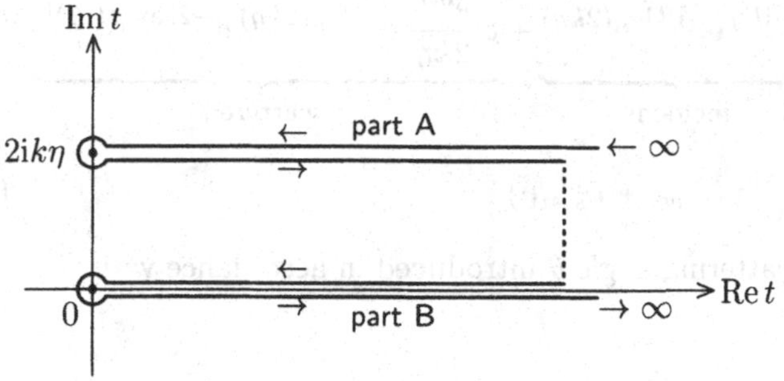

(8.8.17)

splits the integral into two parts. For part A we write $t = 2\mathrm{i}k\eta + u$ with $u : \infty \to 0 \to \infty$ and evaluate it for $k\eta \gg 1$, giving

$$\left[\mathrm{L}^{(0)}_{-\mathrm{i}\beta}(2\mathrm{i}k\eta)\right]_{\mathrm{A}} = \int \frac{\mathrm{d}u}{2\pi\mathrm{i}}\, \mathrm{e}^{-u}(2\mathrm{i}k\eta + u)^{-\mathrm{i}\beta} \frac{1}{u^{1-\mathrm{i}\beta}}$$

$$\underbrace{\overset{\uparrow}{\cong}}_{k\eta \gg 1} \mathrm{e}^{-\mathrm{i}\beta \log(2\mathrm{i}k\eta)} \frac{1}{2\pi\mathrm{i}} \left[\int_\infty^0 \mathrm{d}u\, \frac{\mathrm{e}^{-u}}{u^{1-\mathrm{i}\beta}} + \int_0^\infty \mathrm{d}u\, \frac{\mathrm{e}^{-u}}{\left(\mathrm{e}^{2\pi\mathrm{i}}u\right)^{1-\mathrm{i}\beta}}\right]$$

$$= \mathrm{e}^{-\mathrm{i}\beta[\frac{\mathrm{i}\pi}{2} + \log(2k\eta)]} \frac{1}{2\pi\mathrm{i}} \left(\mathrm{e}^{-2\pi\beta} - 1\right) \underbrace{\int_0^\infty \mathrm{d}u\, u^{-1+\mathrm{i}\beta}\, \mathrm{e}^{-u}}_{=(-1+\mathrm{i}\beta)!}$$

$$= \mathrm{e}^{\frac{\pi\beta}{2}}\, \mathrm{e}^{-\mathrm{i}\beta \log(2k\eta)}\, (\mathrm{i}\beta)! \frac{1 - \mathrm{e}^{-2\pi\beta}}{2\pi\beta}\,; \tag{8.8.18}$$

and for part B we find

$$\left[\mathrm{L}^{(0)}_{-\mathrm{i}\beta}(2\mathrm{i}k\eta)\right]_{\mathrm{B}} \cong \frac{\mathrm{e}^{2\mathrm{i}k\eta}}{\left(\mathrm{e}^{\frac{3\pi\mathrm{i}}{2}}\, 2k\eta\right)^{1-\mathrm{i}\beta}} \frac{1}{2\pi\mathrm{i}} \underbrace{\left[\int_\infty^0 \mathrm{d}t\, \mathrm{e}^{-t} t^{-\mathrm{i}\beta} + \mathrm{e}^{2\pi\beta} \int_0^\infty \mathrm{d}t\, \mathrm{e}^{-t} t^{-\mathrm{i}\beta}\right]}_{=\left(\mathrm{e}^{2\pi\beta} - 1\right)(-\mathrm{i}\beta)!}$$

$$= \frac{\beta\, \mathrm{e}^{2\mathrm{i}k\eta}}{2k\eta}\, \mathrm{e}^{-\frac{3\pi}{2}\beta}\, \mathrm{e}^{\mathrm{i}\beta \log(2k\eta)} \frac{\mathrm{e}^{2\pi\beta} - 1}{2\pi\beta} (-\mathrm{i}\beta)!\,. \tag{8.8.19}$$

So

$$\mathrm{L}^{(0)}_{-\mathrm{i}\beta}(2\mathrm{i}k\eta) = \left[\mathrm{L}^{(0)}_{-\mathrm{i}\beta}(2\mathrm{i}k\eta)\right]_{\mathrm{A}} + \left[\mathrm{L}^{(0)}_{-\mathrm{i}\beta}(2\mathrm{i}k\eta)\right]_{\mathrm{B}}$$

$$\cong \mathrm{e}^{-\frac{3\pi}{2}\beta} \frac{\mathrm{e}^{2\pi\beta} - 1}{2\pi\beta} (\mathrm{i}\beta)! \left[\mathrm{e}^{-\mathrm{i}\beta \log(2k\eta)} + \beta \frac{\mathrm{e}^{2\mathrm{i}k\eta}}{2k\eta}\, \mathrm{e}^{\mathrm{i}\beta \log(2k\eta)} \frac{(-\mathrm{i}\beta)!}{(\mathrm{i}\beta)!}\right] \tag{8.8.20}$$

If we omit the η independent pre-factor, we have, for $k\eta \gg 1$,

$$G \propto \underbrace{\mathrm{e}^{-\mathrm{i}k\eta}\,\mathrm{e}^{-\mathrm{i}\beta\log(2k\eta)}}_{\text{incident}} + \underbrace{\beta\frac{\mathrm{e}^{\mathrm{i}k\eta}}{2k\eta}\,\mathrm{e}^{\mathrm{i}\beta\log(2k\eta)}\,\mathrm{e}^{-2\mathrm{i}\arg((\mathrm{i}\beta)!)}}_{\text{scattered}}$$

$$= \mathrm{e}^{-\mathrm{i}k\xi}\left(\psi_{\text{inc.}} + \psi_{\text{scatt.}}\right) \tag{8.8.21}$$

and, with the scattering angle θ introduced in accordance with

$$z = r\cos\theta\,,$$
$$\eta = \tfrac{1}{2}(r - z) = r\sin^2\tfrac{1}{2}\theta\,, \tag{8.8.22}$$

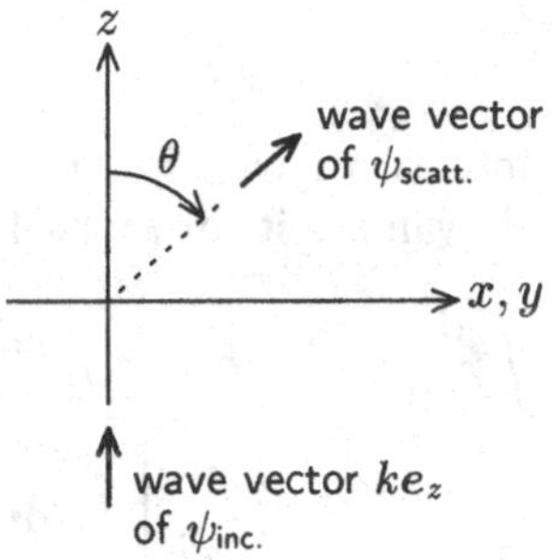

we write the incident and scattered amplitudes as

$$\psi_{\text{inc.}} = \mathrm{e}^{\mathrm{i}kz}\,\mathrm{e}^{-\mathrm{i}\beta\log\left(2kr\sin^2(\frac{1}{2}\theta)\right)}\,,$$
$$\psi_{\text{scatt.}} = \frac{\mathrm{e}^{\mathrm{i}kr}}{r}\frac{\beta}{2k\sin^2(\frac{1}{2}\theta)}\,\mathrm{e}^{\mathrm{i}\beta\log\left(2kr\sin^2(\frac{1}{2}\theta)\right)}\,\mathrm{e}^{-2\mathrm{i}\arg((\mathrm{i}\beta)!)}\,. \tag{8.8.23}$$

Note that although we anticipated that the incident particle is represented by just $\mathrm{e}^{\mathrm{i}kz}$, it actually feels the long-range effect of the slowly decreasing Coulomb potential; that is equally true of the outgoing spherical wave representing the scattered particle.

The asymptotic form $\psi = \psi_{\text{inc.}} + \psi_{\text{scatt.}}$ identifies the respective relative probability densities for the incident and scattered particles:

$$|\psi_{\text{inc.}}|^2 = 1\,,\quad |\psi_{\text{scatt.}}|^2 = \frac{1}{r^2}\left[\frac{\beta}{2k\sin^2(\frac{1}{2}\theta)}\right]^2\,. \tag{8.8.24}$$

These are also the relative fluxes – density times speed – because the asymptotic speed is the same; it is the direction that has changed. One speaks of the differential cross section per unit solid angle $\mathrm{d}\sigma/\mathrm{d}\Omega$, where $\mathrm{d}\sigma$ is the ratio of the scattered flux into the solid angle $\mathrm{d}\Omega$ to the incident flux

$$\mathrm{d}\sigma = \frac{|\psi_{\text{scatt.}}|^2 r^2\mathrm{d}\Omega}{|\psi_{\text{inc.}}|^2} = \mathrm{d}\Omega\left[\frac{\beta}{2k\sin^2(\frac{1}{2}\theta)}\right]^2 \tag{8.8.25}$$

or, with $\beta/k = Ze^2/(Mv^2)$,

$$\frac{d\sigma}{d\Omega} = \left(\frac{Ze^2}{2Mv^2}\right)^2 \frac{1}{\sin^4(\frac{1}{2}\theta)} . \tag{8.8.26}$$

It is independent of $\hbar$ and is identical with the classical Rutherford* cross section.

Problems

8-1a One-dimensional oscillator, dimensionless variables: Evaluate

$$\left\langle q^4\right\rangle_n , \quad \left\langle p^4\right\rangle_n , \quad \left\langle q^2 . p^2\right\rangle_n$$

by noting, for example, that $\left\langle q^4\right\rangle_n$ is the squared length of the vector $q^2|n\rangle$. Check these independent calculations by using them to find

$$\left\langle \left(\frac{q^2+p^2}{2}\right)^2 \right\rangle_n .$$

8-1b An example of a more general method for such calculations: Check that

$$e^{i\sqrt{2}\,\lambda q} = e^{i\lambda y^\dagger} e^{i\lambda y} e^{-\frac{1}{2}\lambda^2} ;$$

recall (or, better, show) that

$$y^k|n\rangle = |n-k\rangle \sqrt{\frac{n!}{(n-k)!}} ;$$

then deduce that

$$\left\langle e^{i\sqrt{2}\,\lambda q} \right\rangle_n = \mathrm{L}_n^{(0)}(\lambda^2)\, e^{-\frac{1}{2}\lambda^2} .$$

Use the initial terms of the λ expansion to recover $\langle q^2\rangle_n$ and $\langle q^4\rangle_n$. Anybody for $\langle q^6\rangle_n$?

8-1c For the two-dimensional oscillator, using dimensionless variables, evaluate $\langle \rho^4\rangle_{n_+,n_-}$ and express it in terms of N and m.

8-1d Use the correspondence between the three-dimensional Coulomb $u(r)$ and the two-dimensional oscillator $u(\rho)$ to show that

*Lord Ernest Rutherford, Baron of Nelson (1871–1937)

$$\langle r \rangle_{n,l} = \frac{na_0}{2Z} \frac{\langle \rho^4 \rangle_{n_+,n_-}}{\langle \rho^2 \rangle_{n_+,n_-}} .$$

Apply the known oscillator values and the connection between the two systems to arrive at

$$\langle r \rangle_{n,l} = \frac{a_0}{2Z} \left[3n^2 - l(l+1) \right] .$$

8-2 Two-dimensional oscillator: Express $\rho^2 = q_1^2 + q_2^2$ in terms of y_+ and y_-, introduced in accordance with $y_\pm = 2^{-\frac{1}{2}}(y_1 \mp \mathrm{i} y_2)$, and rederive (8.1.28).

8-3 The two-dimensional Coulomb problem is defined by the Hamilton operator

$$H = \frac{1}{2M}\left(p_x^2 + p_y^2\right) - \frac{Ze^2}{\sqrt{x^2+y^2}} .$$

What are the energy eigenvalues and their multiplicities?

8-4 The non-relativistic Hamiltonian for a spinning electron (charge $e = -|e|$) in a magnetic field is (very nearly)

$$H = \frac{1}{2M}\left(\boldsymbol{p} - \frac{e}{c}\boldsymbol{A}\right)^2 - \frac{e\hbar}{2Mc}\boldsymbol{\sigma}\cdot\boldsymbol{B} .$$

Verify that

$$H = \frac{1}{2M}\left[\boldsymbol{\sigma}\cdot\left(\boldsymbol{p} - \frac{e}{c}\boldsymbol{A}\right)\right]^2 .$$

Consider a homogeneous field along the z axis. What is the energy of the ground state in the circumstance $(p_z - \frac{e}{c}A_z)' = 0$? What are, more generally, the energy eigenvalues for $(p_z - \frac{e}{c}A_z)' = \hbar k$? [They are called Landau* levels.]

8-5 For a free particle, the relation between momentum $\boldsymbol{p}$ and the relativistic energy $W = Mc^2 + E$ is

$$W^2 = \boldsymbol{p}^2c^2 - (Mc^2)^2 ,$$

or

$$E = W - Mc^2 = \frac{\boldsymbol{p}^2c^2}{W + Mc^2} = \boldsymbol{\sigma}\cdot\boldsymbol{p}\frac{1}{2M + E/c^2}\boldsymbol{\sigma}\cdot\boldsymbol{p} .$$

[Note the connection with Problem 8-4, in the non-relativistic limit, when a magnetic field is introduced: $\boldsymbol{p} \to \boldsymbol{p} - \frac{e}{c}\boldsymbol{A}$.] Introduce the electrostatic

*Lev Davidovich LANDAU (1908–1968)

energy V (replacement $E \to E - V$) to arrive at this approximate relativistic correction to the Hamiltonian for a state of energy E:

$$H = \boldsymbol{\sigma} \cdot \boldsymbol{p} \frac{1}{2M + (E - V)/c^2} \boldsymbol{\sigma} \cdot \boldsymbol{p} + V$$

$$\cong \frac{\boldsymbol{p}^2}{2M} + V - \frac{1}{4M^2c^2} \boldsymbol{\sigma} \cdot \boldsymbol{p}(E - V) \boldsymbol{\sigma} \cdot \boldsymbol{p} .$$

What is the resulting energy shift for the ground state of a hydrogenic atom?

8-6a The three-dimensional relativistic Schrödinger equation for a spinless particle moving in the potential $V(r)$ is

$$\left[(E - V)^2 - c^2 \left(\frac{\hbar}{\mathrm{i}} \boldsymbol{\nabla} \right)^2 - (Mc^2)^2 \right] \psi = 0 .$$

Insert $V = -Ze^2/r$ and find the bound state energy values. [A comparison method is suggested.]

8-6b An approximate procedure for Problem 8-6a begins with $E = Mc^2 + \mathcal{E}$, $|\mathcal{E}| \ll Mc^2$, and arrives at the modified Schrödinger equation

$$\left[\mathcal{E} - V - \frac{1}{2M} \left(\frac{\hbar}{\mathrm{i}} \boldsymbol{\nabla} \right)^2 + \frac{(\mathcal{E} - V)^2}{2Mc^2} \right] \psi = 0 .$$

Use first-order perturbation theory to find the approximate energy shift. Check that your result is indeed an approximation of what you got in Problem 8-6a.

8-7a Regard $\boldsymbol{p}^2/(2M)$ as the leading term in the expansion of

$$c\sqrt{\boldsymbol{p}^2 + (Mc)^2} - Mc^2$$

in powers of $\boldsymbol{p}^2$. Show that the next term in this expansion gives a correction of

$$\Delta H = -\frac{1}{4} \alpha^2 R_\infty \left(\frac{\boldsymbol{p}^2}{(\hbar/a_0)} \right)^2$$

to the Hamilton operator, where $\alpha = e^2/(\hbar c) \cong 1/137$ is Sommerfeld's* fine structure constant.

8-7b Find the resulting shift of the ground-state energy in first-order perturbation theory. For this purpose evaluate the ground-state expectation value of $p^4 = (\boldsymbol{p}^2)^2$ by a variety of methods, indicated by

*Arnold SOMMERFELD (1868–1951)

$$\left\langle p^4\right\rangle = \int (\mathrm{d}\boldsymbol{p}')\, p'^4 \left|\psi(\boldsymbol{p}')\right|^2 = \int (\mathrm{d}\boldsymbol{r}')\, \psi(\boldsymbol{r}')^* \left(\hbar^2\boldsymbol{\nabla}'^2\right)^2 \psi(\boldsymbol{r}')$$
$$= \int (\mathrm{d}\boldsymbol{r}') \left|\hbar^2\boldsymbol{\nabla}'^2\psi(\boldsymbol{r}')\right|^2 = \left\langle \left[2M\left(Ze^2/r - Z^2R_\infty\right)\right]^2\right\rangle .$$

Justify these statements.

8-8 State the ground-state eigenvector equation as an integral equation obeyed by the $\psi(\boldsymbol{p}')$ of Problem 8-7b. Then verify that it is obeyed.

8-9a Since r is positive by its nature, one cannot apply the WKB quantization rule of Problem 6-33 to radial Schrödinger equations of the form (8.1.8) immediately. A suitable change to unrestricted variables has to be done first. Show that

$$r = r_0\,\mathrm{e}^{x/r_0}\,, \qquad u(r) = \mathrm{e}^{-\frac{1}{2}x/r_0}\psi(x)$$

(with arbitrary reference length $r_0 > 0$) turns the normalization integral for $u(r)$ into the one for $\psi(x)$, and that the resulting version of (8.1.8) corresponds to an effective one-dimensional Hamilton operator given by

$$H_l(x,p) = \frac{r_0}{r}\frac{p^2}{2M}\frac{r_0}{r} + \frac{\hbar^2(l+\frac{1}{2})^2}{2Mr^2} + V(r) \quad \text{with} \quad \frac{1}{\mathrm{i}\hbar}[x,p] = 1\,.$$

As in Problem 6-35a set the semiclassical value of $\operatorname{tr}\{\eta(E_{n_r,l} - H_l)\}$ equal to $n_r + \frac{1}{2}$ (n_r: radial quantum number) and arrive at

$$n_r + \tfrac{1}{2} = \frac{1}{\pi\hbar}\int \mathrm{d}r\,\sqrt{2ME_{n_r,l} - \hbar^2(l+\tfrac{1}{2})^2/r^2 - 2MV(r)}\,.$$

The transition from $l(l+1)$ in (8.1.8) to $(l+\frac{1}{2})^2$ here is known as Langer's* correction.

8-9b Apply this to the three-dimensional oscillator, $V(r) = \frac{1}{2}M\omega^2r^2$. How do the approximate WKB energy eigenvalues compare with the exact ones?

8-9c Repeat for the three-dimensional Coulomb problem, $V(r) = -Ze^2/r$.

8-9d These applications are easier if you first verify the integrals

$$\int_{r_1}^{r_2}\mathrm{d}r\,\frac{r}{\sqrt{(r_2-r)(r-r_1)}} = \frac{1}{2}\pi(r_1+r_2)\,,$$

$$\int_{r_1}^{r_2}\frac{\mathrm{d}r}{r}\,\frac{1}{\sqrt{(r_2-r)(r-r_1)}} = \frac{\pi}{\sqrt{r_1r_2}}\,,$$

$$\int_{r_1}^{r_2}\frac{\mathrm{d}r}{r}\sqrt{(r_2-r)(r-r_1)} = \frac{1}{2}\pi(r_1+r_2) - \pi\sqrt{r_1r_2}\,,$$

for $r_2 > r_1 > 0$.

*Rudolph Ernest LANGER (1894–1968)

8-10 Use arguments analogous to the ones in Problem 8-9a to find the WKB approximation to the energy eigenvalues for two-dimensional motion in a rotationally symmetric potential $V(x_1, x_2) = V(\rho)$, $\rho = (x_1^2 + x_2^2)^{\frac{1}{2}}$. What do you get for $V(\rho) = \frac{1}{2}M\omega^2\rho^2$?

8-11 The triton nucleus of a tritium atom (^{3}H, $Z = 1$) undergoes a β decay and we assume that the created electron (and also the neutrino) escapes very quickly. Before the decay, the atom is in its hydrogenic ground state. What is the probability that, after the decay, the resulting $^3\mathrm{He}^+$ ion ($Z = 2$) is found in its ground state as well?

8-12 Non-degenerate second-order perturbation theory. We know that

$$(E - H)\frac{\partial}{\partial\lambda}|E\rangle + \left(\frac{\partial E}{\partial\lambda} - \frac{\partial H}{\partial\lambda}\right)|E\rangle = 0\,,$$

from which follows

$$\frac{\partial E}{\partial\lambda} = \langle E|\frac{\partial H}{\partial\lambda}|E\rangle$$

and

$$E \neq E' : \qquad \langle E'|\frac{\partial}{\partial\lambda}|E\rangle = \frac{1}{E - E'}\langle E'|\frac{\partial H}{\partial\lambda}|E\rangle .$$

It is consistent with

$$\frac{\partial}{\partial\lambda}\langle E|E\rangle = 0$$

if one chooses

$$\langle E|\frac{\partial}{\partial\lambda}|E\rangle = 0\,.$$

Evaluate $\frac{\partial^2}{\partial\lambda^2}E$ and use it to write out the perturbation expansion for the eigenvalues of

$$H = H_0 + H_1 = H_0 + \lambda H_1\Big|_{\lambda=1}$$

as

$$E = E_0 + \frac{\partial E}{\partial\lambda}\Big|_{\lambda=0} + \frac{1}{2}\frac{\partial^2 E}{\partial\lambda^2}\Big|_{\lambda=0} + \cdots\,.$$

Apply this to the example of Problem 7-6.

8-13 Can you prove that, in general,

$$\langle r^{-1}\rangle \geq \langle r\rangle^{-1}\ ?$$

When does the equality hold? Is the inequality satisfied for hydrogenic atoms? Compare the two sides of the inequality for the $n_r = 0$ states, at large n. Conclusion?

8-14 The generator G_λ of (8.3.4) is not Hermitian as it stands. Why is this irrelevant as long as only infinitesimal scale changes are considered? After symmetrization we have

$$G_\lambda = \delta\lambda\,\tfrac{1}{2}(\boldsymbol{p}\cdot\boldsymbol{r}+\boldsymbol{r}\cdot\boldsymbol{p}) = \delta\lambda\,\Gamma$$

which *is* Hermitian. Find the explicit effect of finite scale transformations

$$\boldsymbol{r} \to \mathrm{e}^{\frac{\mathrm{i}}{\hbar}\lambda\Gamma}\boldsymbol{r}\,\mathrm{e}^{-\frac{\mathrm{i}}{\hbar}\lambda\Gamma}\,,$$

$$\boldsymbol{p} \to \mathrm{e}^{\frac{\mathrm{i}}{\hbar}\lambda\Gamma}\boldsymbol{p}\,\mathrm{e}^{-\frac{\mathrm{i}}{\hbar}\lambda\Gamma}\,.$$

How does a scale change affect the orbital angular momentum $\boldsymbol{L} = \boldsymbol{r}\times\boldsymbol{p}$?

8-15 Show that $r = |\boldsymbol{r}|$ and Γ constitute a pair of complementary observables. Which quantum degree of freedom are they associated with? [Hint: Consider the unitary operator $(r/r_0)^{\mathrm{i}\kappa}$, where $r_0 > 0$ is an arbitrary reference length and κ is any real number, and compare its product with $\mathrm{e}^{\frac{\mathrm{i}}{\hbar}\lambda\Gamma}$ with their product in reverse order.]

8-16 Three-dimensional oscillator: For $n_r = 0, 1$ and arbitrary l, calculate the expectation values $\langle r^{-2}\rangle$, $\langle r^{-3}\rangle$, $\langle r\rangle$, by direct integration using the radial wave functions (7.5.26). Compare with what is stated in (8.3.20), (8.3.26), and (8.3.33).

8-17 Hydrogenic atoms: Use the simple form of the radial function $u(r)$, $n_r = 0$, to evaluate $\langle r^k\rangle_n$, $n_r = 0$. Check that $\langle r\rangle$, $\langle r^{-1}\rangle$, $\langle r^{-2}\rangle$, $\langle r^{-3}\rangle$ are correctly reproduced. For what value of r is $[u(r)]^2$ a maximum? Can you give an approximate form for $[u(r)]^2$ near the maximum, when $n \gg 1$?

8-18 Parabolic coordinates:

$$\psi = C\frac{\mathrm{e}^{im\phi}}{\sqrt{2\pi}}\frac{1}{\sqrt{\xi}}u(\xi)\frac{1}{\sqrt{\eta}}u(\eta)\,.$$

With radial normalization:

$$\int_0^\infty \mathrm{d}\xi\,\left[u(\xi)\right]^2 = \int_0^\infty \mathrm{d}\eta\,\left[u(\eta)\right]^2 = 1\,,$$

and three-dimensional normalization:

$$\int (\mathrm{d}\boldsymbol{r})\, |\psi|^2 = 1\,,$$

what is C? Use $\frac{\partial E}{\partial F} = \langle ez \rangle$ to rederive (8.5.14).

8-19 In (8.6.10) the scalar potential and the vector potential are assumed to depend only on position $\boldsymbol{r}$ but not on time t. Now lift this simplifying restriction and consider $\Phi(\boldsymbol{r},t)$ and $\boldsymbol{A}(\boldsymbol{r},t)$, so that the Hamilton operator acquires a parametric time dependence. Show that the Lorentz force (8.6.20) emerges correctly with electric field $\boldsymbol{E} = -\partial \boldsymbol{A}/\partial t - \boldsymbol{\nabla}\Phi$ and, as before, magnetic field $\boldsymbol{B} = \boldsymbol{\nabla} \times \boldsymbol{A}$.

8-20 Three-dimensional Coulomb problem. Evaluate the expectation values $\langle r^2 \rangle_{nl}$. [Hint: Multiply the u differential equation (8.3.21) by $2r^3\frac{\mathrm{d}u}{\mathrm{d}r} - 3r^2 u$ and integrate.] Check your value for the ground state, $n = 1$, by using its wave function directly. What is the value of χ, the diamagnetic susceptibility [magnetic energy: $-\frac{1}{2}\chi B^2$; originates in the quadratic $\boldsymbol{B}$ term of (8.6.27)] for $n = 1$?

8-21 A particle of charge $-|e|$, mass M, moves in the homogeneous magnetic field B, which is directed along the z axis, and also in the electric field derived from the scalar potential

$$\Phi = -\frac{1}{2}Q\left(z^2 - \frac{x^2+y^2}{2}\right)\,, \qquad \nabla^2\Phi = 0$$

where $Q > 0$. What are the energy values of this system? Their multiplicities (excluding special relations)? [Hint: Look at Schrödinger's three-dimensional equation in the natural gauge. Do not take Q to be too large.]

8-22 Concerning the incident wave $\psi_{\text{inc.}}$ of (8.8.23): The surfaces of constant phase $kz - \beta\log(kr - kz)$ are not planes of constant z, as for $\beta = 0$. To get an impression of the mild distortion caused by the logarithmic term, consider a distant point on the z axis ($r = -z = R$, $kR \gg 1$, $Ze^2/R \ll Mv^2$) and find the points in the $z = 0$ plane of the same phase. How large is their common r value?

Part C

Spring Quarter: Interacting Particles

9. Two-Particle Coulomb Problem

9.1 Internal and external motion

In Chapter 8 we treated hydrogenic atoms as if their nuclei were infinitely massive. Let us now refine the analysis and consider two particles with masses m_1 and m_2 and charges Z_1e and Z_2e, respectively,

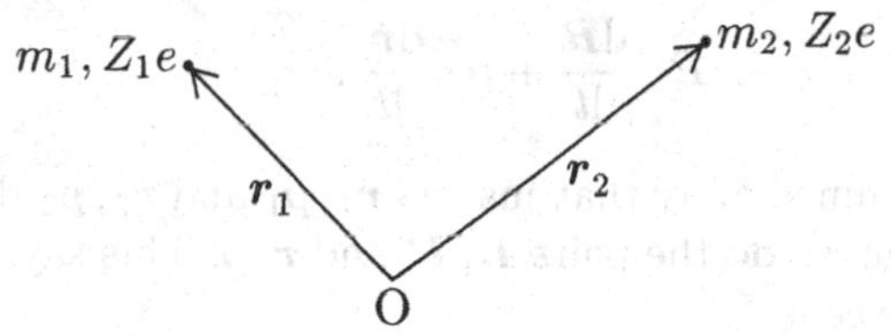

so that

$$H = \frac{\boldsymbol{p}_1^2}{2m_1} + \frac{\boldsymbol{p}_2^2}{2m_2} + \frac{Z_1Z_2e^2}{|\boldsymbol{r}_1 - \boldsymbol{r}_2|} \tag{9.1.1}$$

is the Hamilton operator. Introduce the position vector of the center of gravity, or center of mass:

$$\boldsymbol{R} = \frac{m_1\boldsymbol{r}_1 + m_2\boldsymbol{r}_2}{m_1 + m_2} , \tag{9.1.2}$$

the relative position vector:

$$\boldsymbol{r} = \boldsymbol{r}_1 - \boldsymbol{r}_2 , \tag{9.1.3}$$

the total momentum:

$$\boldsymbol{P} = \boldsymbol{p}_1 + \boldsymbol{p}_2 , \tag{9.1.4}$$

the relative momentum:

$$\boldsymbol{p} = \frac{m_2\boldsymbol{p}_1 - m_1\boldsymbol{p}_2}{m_1 + m_2} . \tag{9.1.5}$$

Then

$$r_1 = R + \frac{m_2}{m_1+m_2} r \,, \qquad r_2 = R - \frac{m_1}{m_1+m_2} r \,,$$
$$p_1 = \frac{m_1}{m_1+m_2} P + p \,, \qquad p_2 = \frac{m_2}{m_1+m_2} P - p \,. \tag{9.1.6}$$

The reason for the particular factors becomes evident from the Lagrangian

$$L = p_1 \cdot \frac{dr_1}{dt} + p_2 \cdot \frac{dr_2}{dt} - H \tag{9.1.7}$$

if we introduce the new variables:

$$\begin{aligned} p_1 \cdot \frac{dr_1}{dt} + p_2 \cdot \frac{dr_2}{dt} &= \left(\frac{m_1}{m_1+m_2} P + p \right) \cdot \left(\frac{dR}{dt} + \frac{m_2}{m_1+m_2} \frac{dr}{dt} \right) \\ &+ \left(\frac{m_2}{m_1+m_2} P - p \right) \cdot \left(\frac{dR}{dt} - \frac{m_1}{m_1+m_2} \frac{dr}{dt} \right) \\ &= P \cdot \frac{dR}{dt} + p \cdot \frac{dr}{dt} \,. \end{aligned} \tag{9.1.8}$$

This tells us immediately that just as r_1, p_1 and r_2, p_2 describe $3+3=6$ degrees of freedom, so do the pairs R, P and r, p. This says that all commutators are zero, except

$$\frac{1}{i\hbar}[R_k, P_k] = 1 \,, \quad \frac{1}{i\hbar}[r_k, p_k] = 1 \,. \tag{9.1.9}$$

In the Hamiltonian (9.1.1), we have the kinetic energy

$$\frac{p_1^2}{2m_1} + \frac{p_2^2}{2m_2} = \frac{P^2}{2M} + \frac{p^2}{2\mu} \,, \tag{9.1.10}$$

where

$$M = m_1 + m_2 \tag{9.1.11}$$

is the total mass and

$$\mu = \left(\frac{1}{m_1} + \frac{1}{m_2} \right)^{-1} = \frac{m_1 m_2}{m_1 + m_2} \tag{9.1.12}$$

is the reduced mass; and the electrostatic Coulomb potential energy

$$\frac{Z_1 Z_2 e^2}{|r_1 - r_2|} = \frac{Z_1 Z_2 e^2}{r} \tag{9.1.13}$$

depends only the distance $r = |r|$ between the particles. So, the Lagrangian splits in two,

$$L = L_{\text{ext}} + L_{\text{int}} \,, \tag{9.1.14}$$

where the external part

$$L_{\text{ext}} = \boldsymbol{P} \cdot \frac{\mathrm{d}\boldsymbol{R}}{\mathrm{d}t} - \frac{\boldsymbol{P}^2}{2M} \tag{9.1.15}$$

describes the whole system, which has mass M and moves as a free particle, and the internal part

$$L_{\text{int}} = \boldsymbol{p} \cdot \frac{\mathrm{d}\boldsymbol{r}}{\mathrm{d}t} - \left(\frac{\boldsymbol{p}^2}{2\mu} + \frac{Z_1 Z_2 e^2}{r} \right) = \boldsymbol{p} \cdot \frac{\mathrm{d}\boldsymbol{r}}{\mathrm{d}t} - H_{\text{int}} \tag{9.1.16}$$

describes the relative motion of the two particles.

For hydrogenic atoms, composed of an electron (particle 1) and the nucleus (particle 2), one has

$$\begin{aligned} &m_1 = m_{\text{el}} \,, \quad m_2 = m_{\text{nucl}} = M - m_{\text{el}} \,, \qquad M, m_{\text{nucl}} \gg m_{\text{el}} \,; \\ &\mu = \frac{m_{\text{el}}(M - m_{\text{el}})}{M} = m_{\text{el}} \left(1 - \frac{m_{\text{el}}}{M}\right) \cong m_{\text{el}} \left(1 - \frac{m_{\text{el}}}{m_{\text{nucl}}}\right) \,. \end{aligned} \tag{9.1.17}$$

Another example is positronium, electron and positron, which have opposite charge and equal mass, so

$$M = 2m_{\text{el}} \,, \quad \mu = \tfrac{1}{2} m_{\text{el}} \,. \tag{9.1.18}$$

The Bohr hydrogenic energy values, for $[m_{\text{nucl}} =]\, M = \infty$, are [put $Z \to -Z_1 Z_2$ and $M \to m_{\text{el}}$ in (8.1.16)]

$$E_\infty = -\frac{m_{\text{el}} Z^2 e^4}{2n^2 \hbar^2} \,. \tag{9.1.19}$$

We have only to replace the electron mass m_{el} by the reduced mass μ:

$$E = -\frac{\mu Z^2 e^4}{2n^2 \hbar^2} = (1 - \frac{m_{\text{el}}}{M}) E_\infty \,, \tag{9.1.20}$$

where E_∞ is the energy in the limit $m_{nucl} \to \infty$. For positronium, this means

$$E = \tfrac{1}{2} E_\infty \tag{9.1.21}$$

with $Z = 1$ in E_∞.

For scattering we found the cross section (8.8.26), for which the obvious extension is $[Z \to -Z_1 Z_2,\; M \to \mu,\; \theta \to \Theta]$

$$\frac{\mathrm{d}\sigma}{\mathrm{d}\Omega} = \left(\frac{Z_1 Z_2 e^2}{2\mu v^2} \right)^2 \frac{1}{\sin^4\left(\frac{1}{2}\Theta\right)} \,, \tag{9.1.22}$$

where v is (still) the relative speed of the two particles when they are far apart. Because this is the description of the relative motion, it is directly applicable only when there is *no* center-of-gravity motion, that is: when the total momentum vanishes, $\boldsymbol{P} = \boldsymbol{p}_1 + \boldsymbol{p}_2 = 0$. Under these conditions, $\boldsymbol{p}_1 = -\boldsymbol{p}_2 = \boldsymbol{p}$, the momentum relations of the collision are displayed as (all momentum vectors are of equal length)

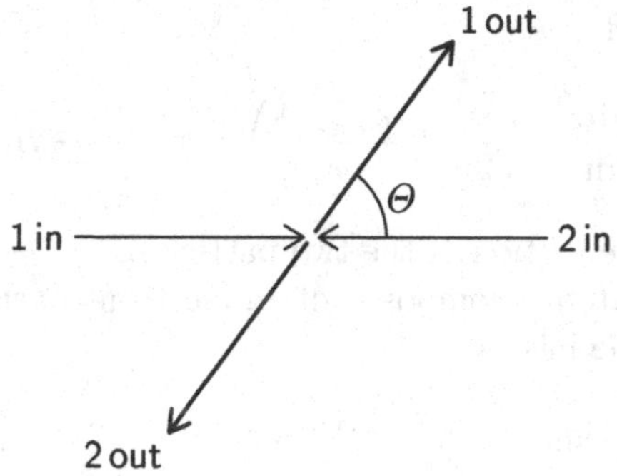

and Θ is the angle through which $\boldsymbol{p}$, the relative momentum, is turned.

9.2 Rutherford scattering revisited

The differential cross section involves just the ratio β/k [recall (8.8.25)], whereas the asymptotic form of the wave function refers to both β and k. Notice that β, initially $Ze^2/\hbar v$, is

$$\beta = -\frac{Z_1 Z_2 e^2}{\hbar v} \tag{9.2.1}$$

with v the asymptotic relative speed, and now

$$k = \frac{1}{\hbar} p = \frac{1}{\hbar} \mu v \,, \tag{9.2.2}$$

the relative momentum and the relative velocity being related by the reduced mass. We recall the asymptotic form [cf. (8.8.23)] of the wave function (θ is still used here)

$$\begin{aligned} \psi \propto\, & \mathrm{e}^{ikz}\, \mathrm{e}^{-\mathrm{i}\beta \log\left(2kr\sin^2(\frac{1}{2}\theta)\right)} \,, \\ & + \frac{\mathrm{e}^{ikr}}{r} \frac{\beta}{2k\sin^2(\frac{1}{2}\theta)}\, \mathrm{e}^{\mathrm{i}\beta \log\left(2kr\sin^2(\frac{1}{2}\theta)\right)}\, \mathrm{e}^{-2\mathrm{i}\arg\left((\mathrm{i}\beta)!\right)} \,. \end{aligned} \tag{9.2.3}$$

We used parabolic coordinates to get this; now we reexamine it using spherical coordinates, which are more useful when deviations from the

Coulomb potential come into play as, for example, in proton–proton scattering where, with increasing energy, nuclear forces begin to play a role.

Let's start by putting

$$e^{ikz} = e^{ikr\cos\theta} \tag{9.2.4}$$

into spherical coordinates. The azimuthal angle ϕ does not occur here, so the complete set of angle functions are just Legendre's polynomials $P_l(\zeta = \cos\theta)$; see Section 3.7. Accordingly, we write

$$e^{ikr\cos\theta} = \sum_{l=0}^{\infty}(2l+1)i^l j_l(kr)P_l(\cos\theta)\,, \tag{9.2.5}$$

where the factor i^l ensures that j_l is real. In fact, if we take the complex conjugate and at the same time let $\zeta \to -\zeta$, which leaves $e^{ikr\zeta}$ intact, $i^l P(\zeta)$ changes into

$$(-i)^l P_l(-\zeta) = i^l P_l(\zeta) \tag{9.2.6}$$

so j_l must be real. Upon recalling the orthogonality relation of Problem 3-15, we note that j_l is given explicitly by the integral

$$i^l j_l(kr) = \frac{1}{2}\int_{-1}^{1} d\zeta\, P_l(\zeta)\, e^{ikr\zeta}\,. \tag{9.2.7}$$

For our present purposes, the asymptotic form ($x = kr \gg 1$) will do:

$$\begin{aligned} i^l j_l(x) &= \frac{1}{2}\int_{-1}^{1} d\zeta\, P_l(\zeta)\frac{d}{d\zeta}\frac{e^{ix\zeta}}{ix} \\ &= \frac{1}{2}\left[P_l(1)\frac{e^{ix}}{ix} - P_l(-1)\frac{e^{-ix}}{ix}\right] - \frac{1}{2}\int_{-1}^{1} d\zeta\, \frac{e^{ix\zeta}}{ix}\frac{d}{d\zeta}P_l(\zeta) \\ &\cong \frac{1}{2ix}\left[e^{ix} - (-1)^l e^{-ix}\right] = i^l \frac{1}{2ix}\left(i^{-l}e^{ix} - i^l e^{-ix}\right) \end{aligned} \tag{9.2.8}$$

or

$$j_l(x) \cong \frac{\sin\left(x - \frac{1}{2}\pi l\right)}{x}\,. \tag{9.2.9}$$

So,

$$e^{ikr\cos\theta} \cong \sum_{l=0}^{\infty}(2l+1)i^l \frac{\sin\left(kr - \frac{1}{2}\pi l\right)}{kr}P_l(\cos\theta) \tag{9.2.10}$$

is an asymptotic expansion of the plane wave (9.2.4) in terms of Legendre's polynomials.

The wave function for the Coulomb potential in this axially symmetric situation can be expressed as

$$\psi(r,\theta) = \sum_{l=0}^{\infty}(2l+1)\mathrm{i}^l R_l(r)\mathrm{P}_l(\cos\theta)\,\mathrm{e}^{\mathrm{i}\delta_l}\,, \tag{9.2.11}$$

where the real radial function R_l obeys the Schrödinger equation

$$\left[\frac{\mathrm{d}^2}{\mathrm{d}r^2}+\frac{2}{r}\frac{\mathrm{d}}{\mathrm{d}r}-\frac{l(l+1)}{r^2}+\frac{2\mu}{\hbar^2}\left(E-\frac{Z_1Z_2e^2}{r}\right)\right]R_l(r)=0 \tag{9.2.12}$$

or, with

$$R_l(r)=\frac{u_l(r)}{r}\,,\quad E=\frac{p^2}{2\mu}=\frac{\hbar^2k^2}{2\mu}\,,\quad -\frac{Z_1Z_2e^2\mu}{\hbar^2}=\beta k\,, \tag{9.2.13}$$

more compactly

$$\left[\frac{\mathrm{d}^2}{\mathrm{d}r^2}-\frac{l(l+1)}{r^2}+k^2+\frac{2\beta k}{r}\right]u_l(r)=0\,. \tag{9.2.14}$$

The correspondence with our earlier discussion of hydrogenic bound states [(8.1.8) with $(2M/\hbar^2)V(r)=-2Z/(a_0r)$ and $2ME/\hbar^2=-Z^2/(na_0)^2$] is

$$k^2 \longleftrightarrow -\frac{Z^2}{n^2a_0^2}=-\frac{\beta^2k^2}{n^2}\,, \tag{9.2.15}$$

or

$$\beta k \longleftrightarrow \frac{Z}{a_0} \quad\text{with}\quad a_0 \longleftrightarrow \frac{\hbar^2}{\mu e^2} \quad\text{and}\quad Z \longleftrightarrow -Z_1Z_2\,, \tag{9.2.16}$$

so

$$n \longleftrightarrow -\mathrm{i}\beta \quad\text{and}\quad \frac{Z}{na_0} \longleftrightarrow \mathrm{i}k\,, \tag{9.2.17}$$

where either sign of i would do.

Recalling the bound-state wave functions (8.1.33) we thus note that

$$\begin{aligned} R_l &\propto \left(\frac{2Zr}{na_0}\right)^l \mathrm{L}_{n-l-1}^{(2l+1)}\left(\frac{2Zr}{na_0}\right)\mathrm{e}^{-\frac{Zr}{na_0}} \\ &\to (2kr)^l\mathrm{L}_{-\mathrm{i}\beta-l-1}^{(2l+1)}(2\mathrm{i}kr)\,\mathrm{e}^{-\mathrm{i}kr}\,. \end{aligned} \tag{9.2.18}$$

As in Section 8.8 we use a contour integral to generalize

$$\mathrm{L}_\nu^{(\alpha)}(x)=\frac{(-1)^\alpha}{\nu!}\,\mathrm{e}^x\left(\frac{\mathrm{d}}{\mathrm{d}x}\right)^{\nu+\alpha}x^\nu\,\mathrm{e}^{-x} \tag{9.2.19}$$

to non-integer ν values,

$$\mathrm{L}_\nu^{(\alpha)}(x) = (-1)^\alpha \frac{(\nu+\alpha)!}{\nu!}\, \mathrm{e}^x \oint \frac{\mathrm{d}t}{2\pi\mathrm{i}} \frac{t^\nu\, \mathrm{e}^{-t}}{(t-x)^{\nu+\alpha+1}} \tag{9.2.20}$$

with the contour as depicted in (8.8.15), so

$$R_l \propto \mathrm{e}^{ikr}(2kr)^l \oint \frac{\mathrm{d}t}{2\pi\mathrm{i}} t^{-\mathrm{i}\beta-l-1}(t-2\mathrm{i}kr)^{\mathrm{i}\beta-l-1}\,\mathrm{e}^{-t} \,. \tag{9.2.21}$$

This looks better with the substitution

$$t \to t + \mathrm{i}kr \,, \tag{9.2.22}$$

namely

$$R_l \propto (2kr)^l \oint \frac{\mathrm{d}t}{2\pi\mathrm{i}} (t-\mathrm{i}kr)^{\mathrm{i}\beta-l-1}(t+\mathrm{i}kr)^{-\mathrm{i}\beta-l-1}\,\mathrm{e}^{-t} \,. \tag{9.2.23}$$

First let's check that we have the physically acceptable solution as $r \to 0$, where, of r^l and r^{-l-1}, the first is chosen. Look at the integral for $r = 0$:

$$\begin{aligned} \oint \frac{\mathrm{d}t}{2\pi\mathrm{i}}\, t^{\mathrm{i}\beta-l-1} t^{-\mathrm{i}\beta-l-1}\,\mathrm{e}^{-t} &= \oint \frac{\mathrm{d}t}{2\pi\mathrm{i}} \frac{\mathrm{e}^{-t}}{t^{2l+2}} \\ &= \frac{1}{(2l+1)!}\left(\frac{\mathrm{d}}{\mathrm{d}t}\right)^{2l+1} \mathrm{e}^{-t}\Big|_{t=0} \\ &= -\frac{1}{(2l+1)!} \,. \end{aligned} \tag{9.2.24}$$

So indeed R_l behaves correctly, as r^l, for $r \to 0$.

Notice the symmetry of the integrand under $\mathrm{i} \to -\mathrm{i}$, indicating that it is intrinsically a real function. Now we want the asymptotic form, $kr \gg 1$. The singularities are at $\pm ikr$, so, with an eye on e^{-t}, we choose the closed contour

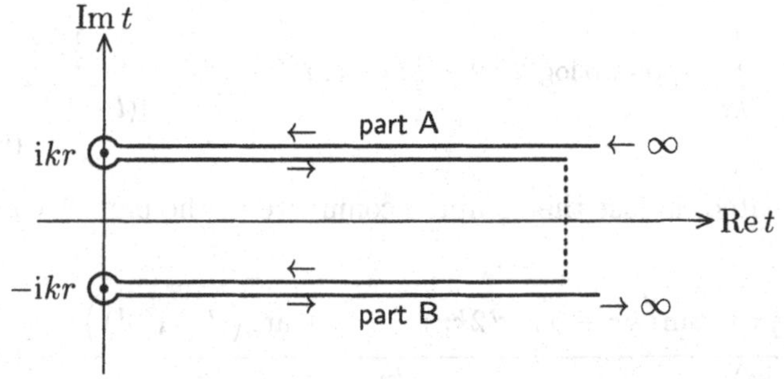

(9.2.25)

which is, of course, the contour of (8.8.17) after the shift (9.2.22).

For part A we write $t = \mathrm{i}kr + u$ with $u : \infty \to 0 \to \infty$ and get, for $kr \gg 1$,

$$\begin{aligned}\Big[R_l(r)\Big]_{\mathrm{A}} =&(2kr)^l(2\mathrm{i}kr)^{-\mathrm{i}\beta-l-1}\frac{\mathrm{e}^{-\mathrm{i}kr}}{2\pi\mathrm{i}}\bigg[\int_\infty^0 \mathrm{d}u\, u^{\mathrm{i}\beta-l-1}\,\mathrm{e}^{-u} \\ &+\int_0^\infty \mathrm{d}u\,(\mathrm{e}^{2\pi\mathrm{i}}u)^{\mathrm{i}\beta-l-1}\,\mathrm{e}^{-u}\bigg] \\ =&-\frac{\mathrm{e}^{-\mathrm{i}kr}}{2kr}\,\mathrm{e}^{-\mathrm{i}\beta\log(2kr)}\,\mathrm{e}^{\frac{1}{2}\pi\beta}\mathrm{i}^{-l}\frac{1}{2\pi}\underbrace{(\mathrm{e}^{-2\pi\beta}-1)}_{=-2\,\mathrm{e}^{-\pi\beta}\sinh(\pi\beta)}(\mathrm{i}\beta-l-1)! \\ =&\frac{1}{2kr}\,\mathrm{e}^{-\mathrm{i}\left[kr+\beta\log(2kr)+\frac{1}{2}\pi l\right]}\,\mathrm{e}^{-\frac{1}{2}\pi\beta}\frac{1}{\pi}\sinh(\pi\beta)\,(\mathrm{i}\beta-l-1)!\,.\end{aligned} \tag{9.2.26}$$

Let's notice here a property of the factorial function:

$$z!(-z)! = \frac{\pi z}{\sin(\pi z)} \quad \text{or} \quad (z-1)!(-z)! = \frac{\pi}{\sin(\pi z)}\,. \tag{9.2.27}$$

Then, with $z = -\mathrm{i}\beta + l + 1$, we get

$$(\mathrm{i}\beta-l-1)!\,(-\mathrm{i}\beta+l)! = \frac{\pi}{\sin\big(\pi(l+1-\mathrm{i}\beta)\big)} = (-1)^{l+1}\frac{\pi\mathrm{i}}{\sinh(\pi\beta)} \tag{9.2.28}$$

or

$$\sinh(\pi\beta)\,(\mathrm{i}\beta-l-1)! = (-1)^{l+1}\frac{\pi\mathrm{i}}{(-\mathrm{i}\beta+l)!}\,. \tag{9.2.29}$$

This gives

$$\Big[R_l(r)\Big]_{\mathrm{A}} = -\frac{\mathrm{i}}{2kr}\,\mathrm{e}^{-\mathrm{i}\left[kr+\beta\log(2kr)-\frac{1}{2}\pi l+\arg((l-\mathrm{i}\beta)!)\right]}\,\frac{\mathrm{e}^{-\frac{1}{2}\pi\beta}}{|(l-\mathrm{i}\beta)!|}\,. \tag{9.2.30}$$

Similarly, we parameterize part B by $t = -\mathrm{i}kr + u$ with $u : \infty \to 0 \to \infty$ and get

$$\Big[R_l(r)\Big]_{\mathrm{B}} = \frac{\mathrm{i}}{2kr}\,\mathrm{e}^{\mathrm{i}\left[kr+\beta\log(2kr)-\frac{1}{2}\pi l+\arg((l-\mathrm{i}\beta)!)\right]}\,\frac{\mathrm{e}^{-\frac{1}{2}\pi\beta}}{|(l-\mathrm{i}\beta)!|}\,, \tag{9.2.31}$$

which, as anticipated, is just the complex conjugate of the part-A contribution, so

$$R_l(r) \propto \frac{\mathrm{e}^{-\frac{1}{2}\pi\beta}}{|(l-\mathrm{i}\beta)!|}\frac{\sin\Big(kr+\beta\log(2kr)-\frac{1}{2}\pi l+\arg\big((l-\mathrm{i}\beta)!\big)\Big)}{kr}\,. \tag{9.2.32}$$

How do we find the factor of proportionality that is independent of r?

Let's return to the expansion (9.2.5) and consider large kr, where (9.2.9) applies,

$$\mathrm{i}^l j_l(kr) \cong \frac{\mathrm{e}^{ikr}}{2ikr} - (-1)^l \frac{\mathrm{e}^{-ikr}}{2ikr} \,. \tag{9.2.33}$$

From the point of view of radial motion, e^{-ikr} describes a spherical wave moving in, and e^{ikr} describes a spherical wave moving out. Were it not for the very long range nature of the Coulomb potential, we could argue that the effect of the interaction is to produce an additional *outgoing* wave; the incoming wave far away is still unaware of the presence of the interaction potential. While not entirely true, this is still the dominant aspect of the Coulomb interaction (see Problem 8-22).

For the incoming part of

$$\psi = \sum_l (2l+1)\mathrm{i}^l \, \mathrm{e}^{\mathrm{i}\delta_l} R_l(r) \mathrm{P}_l(\cos\theta) \tag{9.2.34}$$

to match that of e^{ikz}, where $\mathrm{e}^{\mathrm{i}\delta_l} R_l \to j_l(kr)$, it is clear that the factor in front of $(ikr)^{-1}\sin(kr+\cdots)$ in (9.2.32) must be unity, so that the looked-for proportionality factor is

$$-\,\mathrm{e}^{\frac{1}{2}\pi\beta}\,|(l-\mathrm{i}\beta)!| \,. \tag{9.2.35}$$

Then

$$\mathrm{i}^l R_l(r)_{\text{inc.}} = -(-1)^l \frac{\mathrm{e}^{-ikr}}{2ikr}\, \mathrm{e}^{-\mathrm{i}\beta\log(2kr)}\, \mathrm{e}^{-\mathrm{i}\arg((l-\mathrm{i}\beta)!)} \,. \tag{9.2.36}$$

The factor $\mathrm{e}^{-\mathrm{i}\beta\log(2kr)}$ is the long-range effect, it is simply a slowly varying phase that is independent of l. But we must remove $\mathrm{e}^{-\mathrm{i}\arg((l-\mathrm{i}\beta)!)}$, which is the purpose of $\mathrm{e}^{\mathrm{i}\delta_l}$:

$$\delta_l = \arg((l-\mathrm{i}\beta)!) \,. \tag{9.2.37}$$

Now, in fact, the incoming wave of the known solution does have the factor $\mathrm{e}^{-\mathrm{i}\beta\log(2kr)}$. But it also has $\mathrm{e}^{-\mathrm{i}\beta\log(\sin^2(\frac{1}{2}\theta))}$. That, however, must be implicit in our now completely determined wave function:

$$\psi = \sum_{l=0}^{\infty} (2l+1)\mathrm{i}^l \, \mathrm{e}^{\mathrm{i}\delta_l} R_l(r) \mathrm{P}_l(\cos\theta) \,,$$

$$\text{with} \quad R_l \cong \frac{1}{kr} \sin\left(kr + \beta\log(2kr) - \tfrac{1}{2}\pi l + \delta_l\right) , \tag{9.2.38}$$

being too subtle to show up in these crude asymptotic expansions.

Let's use this scattering wave function to answer the following question. The incident part of the wave is of unit amplitude; or the (relative) probability density of the incident beam is

$$|\psi_{\text{inc.}}|^2 = \left| e^{ikz}\, e^{-i\beta \log \ldots} \right|^2 = 1\,. \tag{9.2.39}$$

What is the probability density at the origin, where the particles are in contact? We know that only $l = 0$ survives for $r \to 0$ where $R_l \propto r^l$. So

$$\psi(0) = e^{i\delta_0} R_0(0) \tag{9.2.40}$$

where [this is the proportionality factor (9.2.35) with the sign change required by (9.2.24)]

$$R_0(0) = e^{\frac{1}{2}\pi\beta} |(l - i\beta)!| \tag{9.2.41}$$

or

$$|\psi(0)|^2 = e^{\pi\beta}(-i\beta)!(i\beta)! = e^{\pi\beta} \frac{\pi\beta}{\sinh(\pi\beta)}\,. \tag{9.2.42}$$

There are two situations here,

$$\begin{aligned} &\text{attractive interaction, } Z_1 Z_2 < 0\,,\ \beta > 0: \ |\psi(0)|^2 = \frac{2\pi\beta}{1 - e^{-2\pi\beta}} > 1\,, \\ &\text{repulsive interaction, } Z_1 Z_2 > 0\,,\ \beta < 0: \ |\psi(0)|^2 = \frac{2\pi|\beta|}{e^{2\pi|\beta|} - 1} < 1\,. \end{aligned} \tag{9.2.43}$$

Not surprising.

Particularly interesting is the situation $-\beta = Z_1 Z_2 e^2/(\hbar v) \gg 1$ (repulsion), or $v \ll Z_1 Z_2 e^2/\hbar$, where

$$|\psi(0)|^2 \cong 2\pi|\beta|\, e^{-2\pi|\beta|} \ll 1\,, \tag{9.2.44}$$

which is a semiclassical situation; the probability of penetrating the classically forbidden region is small, but not zero. One might wonder to what extent the semiclassical WKB description can reproduce this result, but we shall not explore this territory.

9.3 Additional short-range forces

Now suppose that when the colliding particles come quite close, additional forces come into play; for example, in proton–proton collisions nuclear forces, which are short range, become important at sufficiently high energies. This effect first begins with $l = 0$, where contact is possible. An example of such a potential is

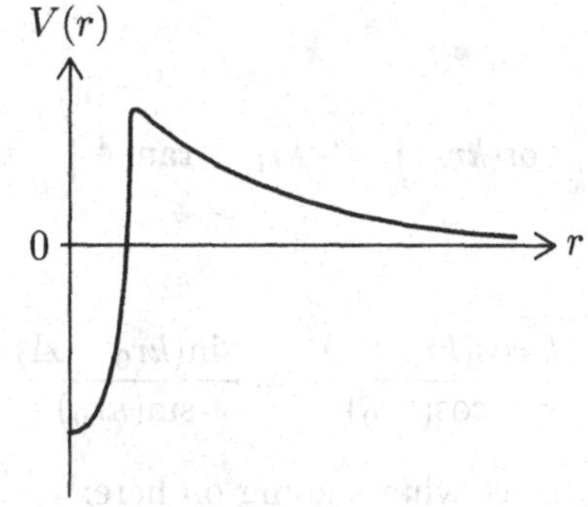

and, as a preparation, let's begin with a simpler model version that omits the Coulomb potential:

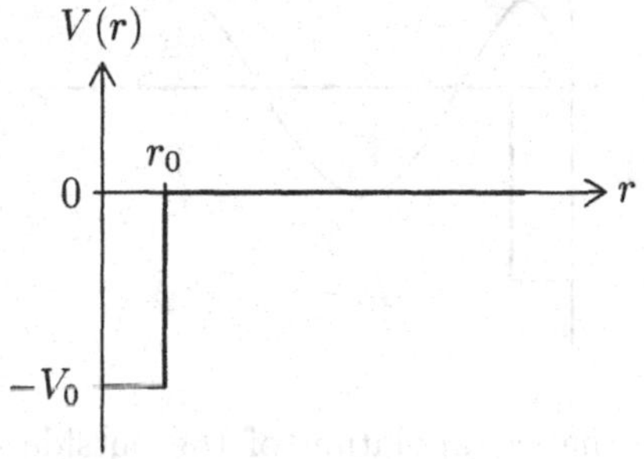

so that the Schrödinger equation is $[R(r) = (kr)^{-1}u(r)]$

$$\left(\frac{\mathrm{d}}{\mathrm{d}r}\right)^2 u = \begin{cases} [2\mu(E+V_0)/\hbar^2]u = \kappa^2 u & \text{for } r < r_0\,, \\ (2\mu E/\hbar^2)u = k^2 u & \text{for } r > r_0\,. \end{cases} \tag{9.3.1}$$

So, with attention to $u(0) = 0$,

$$u(r) = \begin{cases} C\sin(\kappa r) & \text{for } r < r_0\,, \\ \sin(kr + \Delta) & \text{for } r > r_0\,. \end{cases} \tag{9.3.2}$$

The real amplitude parameter C and the real phase parameter Δ are determined by the continuity condition at r_0 of the wave function:

$$C\sin(\kappa r_0) = \sin\left(kr_0 + \Delta\right) \tag{9.3.3}$$

and its derivative:

$$\kappa C\cos(\kappa r_0) = k\cos\left(kr_0 + \Delta\right)\,. \tag{9.3.4}$$

We get

$$\kappa \cot \kappa r_0 = k \cot(kr_0 + \Delta) \tag{9.3.5}$$

to determine Δ,

$$\Delta = -kr_0 + \cot^{-1}\left(\frac{\kappa}{k}\cot(kr_0)\right) = -kr_0 + \tan^{-1}\left(\frac{k}{\kappa}\tan(\kappa r_0)\right) , \tag{9.3.6}$$

and then C,

$$C = \frac{k}{\kappa}\frac{\cos(kr_0 + \Delta)}{\cos(\kappa r_0)} = \frac{\sin(kr_0 + \Delta)}{\sin(\kappa r_0)} . \tag{9.3.7}$$

A picture may help to see what's going on here:

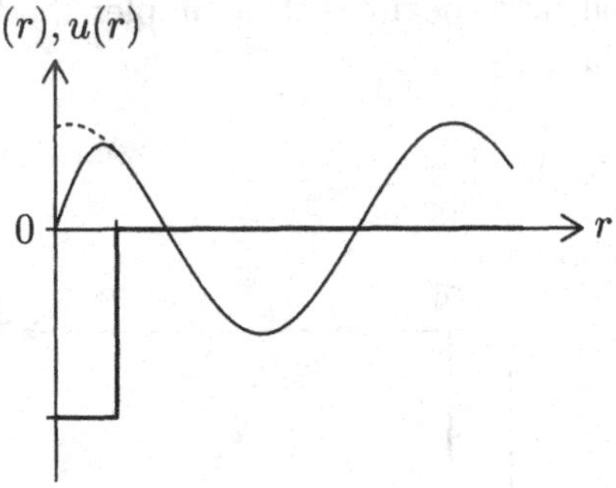

where the dashes are the extrapolation of the outside wave inside; the extrapolated wave does *not* vanish at the origin, which means that the phase of the outside wave is shifted (this is Δ, the *scattering phase shift*) relative to what it is for $V_0 = 0$.

The same thing happens for the Coulomb potential and short-range force, that is

$$\overline{R}_0(r) \propto \frac{1}{kr}\sin\left(kr + \beta\log(2kr) + \delta_0 + \Delta\right) , \tag{9.3.8}$$

as compared to pure Coulomb,

$$R_0(r) \sim \frac{1}{kr}\sin\left(kr + \beta\log(2kr) + \delta_0\right) . \tag{9.3.9}$$

So, to keep the given incoming spherical wave, we must delete the $l = 0$ contribution $\mathrm{e}^{\mathrm{i}\delta_0}R_0(r)$ and add $\mathrm{e}^{\mathrm{i}(\delta_0 + \Delta)}\overline{R}_0(r)$:

$$\psi = \psi_{\text{Coul.}} + \mathrm{e}^{\mathrm{i}(\delta_0 + \Delta)}\overline{R}_0(r) - \mathrm{e}^{\mathrm{i}\delta_0}R_0(r) \tag{9.3.10}$$

and then, recalling that $\delta_0 = \arg((-\mathrm{i}\beta)!) = -\arg((\mathrm{i}\beta)!)$,

$$\psi_{\text{scatt.}} \cong \frac{\mathrm{e}^{\mathrm{i}\left(kr+\beta\log(2kr)+2\delta_0\right)}}{r} f(\theta) \tag{9.3.11}$$

with

$$f(\theta) = \frac{\beta}{2k\sin^2(\frac{1}{2}\theta)}\,\mathrm{e}^{\mathrm{i}\beta\log\left(\sin^2(\frac{1}{2}\theta)\right)} + \frac{1}{k}\,\mathrm{e}^{\mathrm{i}\Delta}\sin\Delta\,. \tag{9.3.12}$$

Now the differential cross section in the center-of-mass frame, where Θ is the scattering angle, is

$$\begin{aligned}\frac{\mathrm{d}\sigma}{\mathrm{d}\Omega} &= r^2|\psi_{\text{scatt.}}|^2 = |f(\Theta)|^2 \\ &= \left(\frac{\beta}{2k}\right)^2 \frac{1}{\sin^4(\frac{1}{2}\Theta)} + \frac{\sin^2\Delta}{k^2} \\ &\quad + \frac{\beta}{k^2}\frac{\sin\Delta}{\sin^2(\frac{1}{2}\Theta)}\cos\left(\Delta - \beta\log\left(\sin^2(\tfrac{1}{2}\Theta)\right)\right)\,.\end{aligned} \tag{9.3.13}$$

The first term is the Rutherford cross section (9.1.22); the last, interference, term is purely quantum mechanical. In the classical limit, $\beta \to \infty$, it will oscillate very rapidly.

9.4 Scattering of identical particles

It is time to point out that the above discussion really applies only to two distinct particles, the proton and the deuteron, or the deuteron and the triton, for example, but not to two identical particles: proton and proton, for example. What's special about identical particles?

Think of the symbolic Schrödinger equation for particles labeled 1 and 2, with those numbers also used to represent analogous choices of physical quantities for each particle:

$$\mathrm{i}\hbar\frac{\partial}{\partial t}\langle 1,2,t| = \langle 1,2,t|H(1,2)\,. \tag{9.4.1}$$

The statement that they are identical particles means that the assignment of names is purely arbitrary; it makes no difference to the dynamics:

$$H(1,2) = H(2,1)\,. \tag{9.4.2}$$

Therefore, interchanging them we also have

$$\mathrm{i}\hbar\frac{\partial}{\partial t}\langle 2,1,t| = \langle 2,1,t|H(1,2)\,. \tag{9.4.3}$$

The clear inference is that the indistinguishable states $\langle 1,2,t|$ and $\langle 2,1,t|$ are really the same states, always remembering the phase freedom:

$$\langle 2,1,t| = \mathrm{e}^{\mathrm{i}\varphi}\langle 1,2,t| \tag{9.4.4}$$

with φ constant. Then interchanging $1 \leftrightarrow 2$ twice,

$$\langle 1,2,t| = \mathrm{e}^{\mathrm{i}\varphi}\langle 2,1,t| = (\mathrm{e}^{\mathrm{i}\varphi})^2\langle 1,2,t| \,, \tag{9.4.5}$$

tells us that $\mathrm{e}^{\mathrm{i}\varphi} = \pm 1$ are the actual possibilities. For $\mathrm{e}^{\mathrm{i}\varphi} = +1$ we have wave functions that are symmetrical in the particle labels, $\langle 1,2| = \langle 2,1|$, and one says that such particles obey Bose*–Einstein (BE) statistics; for $\mathrm{e}^{\mathrm{i}\varphi} = -1$ the situation is that of Fermi†–Dirac (FD) statistics with antisymmetrical wave functions, $\langle 1,2| = -\langle 2,1|$.

The importance of this for scattering is suggested in the center-of-mass diagram

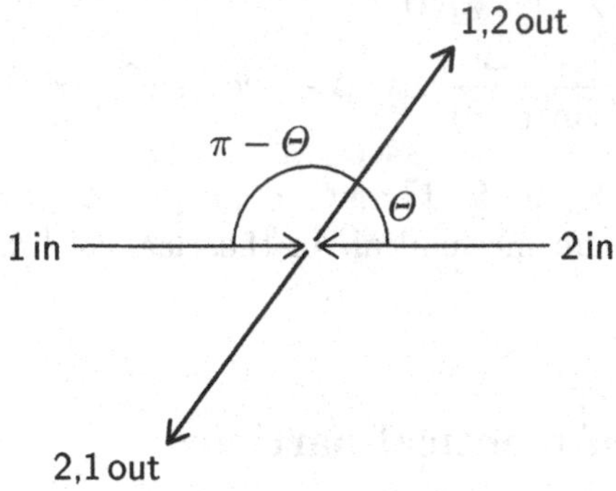

illustrating that after the collision there is no way of knowing whether the particle moving up is particle 1, deflected through angle Θ, or particle 2 deflected through angle $\pi - \Theta$; the wave functions for *both* contingencies must be used with due attention to the symmetry of the wave function, to the statistics of the particle. How do we find the statistics of a particular kind of particle?

Consider the scattering of two ^{4}He nuclei (α particles). The amplitude $f(\Theta)$ describes the scattering process in which particle 1 is detected moving up; $f(\pi - \Theta)$ describes the process in which it is particle 2 that is detected moving up. The combination of the two, produced by the interchange of labels 1 and 2 in the final state is $f(\Theta) \pm f(\pi - \Theta)$, respectively, that is

$$\frac{\mathrm{d}\sigma}{\mathrm{d}\Omega} = |f(\Theta) \pm f(\pi - \Theta)|^2 \quad \text{for} \quad \begin{cases} \text{BE statistics,} \\ \text{FD statistics.} \end{cases} \tag{9.4.6}$$

In the situation of pure Coulomb scattering the amplitudes $f(\Theta)$ and $f(\pi - \Theta)$ are available in (9.2.3), where the last factor is a θ independent, and therefore now irrelevant, phase factor. Accordingly, the modification of the Rutherford cross section is given by

*Satyendranath Bose (1894–1974) †Enrico Fermi (1901–1954)

$$\frac{\mathrm{d}\sigma}{\mathrm{d}\Omega} = \left(\frac{\beta}{2k} = -\frac{(2e)^2}{2\mu v^2}\right)^2 \left| \frac{\mathrm{e}^{\mathrm{i}\beta\log(\sin^2(\frac{1}{2}\Theta))}}{\sin^2(\frac{1}{2}\Theta)} \pm \frac{\mathrm{e}^{\mathrm{i}\beta\log(\cos^2(\frac{1}{2}\Theta))}}{\cos^2(\frac{1}{2}\Theta)} \right|^2$$
$$= \left(\frac{2e^2}{\mu v^2}\right)^2 \left[\frac{1}{\sin^4(\frac{1}{2}\Theta)} + \frac{1}{\cos^4(\frac{1}{2}\Theta)} \pm 2\frac{\cos\left(\beta\log(\tan^2(\frac{1}{2}\Theta))\right)}{\sin^2(\frac{1}{2}\Theta)\cos^2(\frac{1}{2}\Theta)}\right]. \tag{9.4.7}$$

In this so-called Mott* cross section, the first two terms would be the classical result for identical particles; the third is purely quantum mechanical. Notice what happens for right-angle scattering, that is: $\Theta = \frac{1}{2}\pi = \pi - \Theta$, where $f(\Theta) = f(\pi - \Theta)$:

$$\left.\frac{(\mathrm{d}\sigma/\mathrm{d}\Omega)_{\text{quant.}}}{(\mathrm{d}\sigma/\mathrm{d}\Omega)_{\text{class.}}}\right|_{\Theta = \frac{1}{2}\pi} = \frac{|f \pm f|^2}{2|f|^2} = \begin{cases} 2 & \text{for BE statistics,} \\ 0 & \text{for FD statistics.} \end{cases} \tag{9.4.8}$$

In fact, 2 is observed, at low enough energies that nuclear forces are ineffective: ^{4}He is a BE particle.

It's time to mention something we took for granted about ^{4}He; it has no spin. But other particles do have spin. And the requirement of symmetry or antisymmetry refers to *all* the degrees of freedom of a particle, position *and* spin. The spin states of the two particles, each of spin s, can be separated into symmetrical states and antisymmetrical states. We already know that for $s = \frac{1}{2}$, the $(2s+1)\times(2s+1) = 4$ states consist of three symmetrical states and one antisymmetrical one (see Section 3.5). In general, if you have two variables, each taking on n values, the number of antisymmetrical combinations is $\frac{1}{2}n(n-1)$, and the number of symmetrical ones is $\frac{1}{2}n(n-1)+n = \frac{1}{2}n(n+1)$ correctly adding to n^2. Thus the fraction of spin states that are symmetrical or antisymmetrical is $(n = 2s+1)$

$$\left.\begin{matrix}\text{symmetrical fraction} \\ \text{antisymmetrical fraction}\end{matrix}\right\} = \frac{\frac{1}{2}n(n\pm 1)}{n^2} = \frac{n \pm 1}{2n} = \begin{cases} \dfrac{s+1}{2s+1} > \dfrac{1}{2}, \\ \dfrac{s}{2s+1} < \dfrac{1}{2}; \end{cases} \tag{9.4.9}$$

as a check put $s = \frac{1}{2}$ and get the respective fractions of $\frac{3}{4}$ and $\frac{1}{4}$.

In a collision with all spin states equally probable, the fraction of symmetrical spin states will have the scattering amplitude $f(\Theta) \pm f(\pi - \Theta)$, for the respective BE/FD statistics, whereas the spin antisymmetrical fraction will have the scattering amplitude $f(\Theta) \mp f(\pi - \Theta)$. So

$$\frac{\mathrm{d}\sigma}{\mathrm{d}\Omega} = \frac{s+1}{2s+1}|f(\Theta) \pm f(\pi-\Theta)|^2 + \frac{s}{2s+1}|f(\Theta) \mp f(\pi-\Theta)|^2$$
$$= \underbrace{|f(\Theta)|^2 + |f(\pi-\Theta)|^2}_{\text{classical}} \pm \frac{2}{2s+1}\mathrm{Re}\big(f(\Theta)^* f(\pi-\Theta)\big) \tag{9.4.10}$$

*Sir Nevill Francis Mott (1905–1996)

and, for $\Theta = \frac{1}{2}\pi$,

$$\left.\frac{(\mathrm{d}\sigma/\mathrm{d}\Omega)_{\text{quant.}}}{(\mathrm{d}\sigma/\mathrm{d}\Omega)_{\text{class.}}}\right|_{\Theta=\frac{1}{2}\pi} = 1 \pm \frac{1}{2s+1} = \begin{cases} \dfrac{s+1}{s+\frac{1}{2}} > 1 & \text{for BE statistics,} \\ \dfrac{s}{s+\frac{1}{2}} < 1 & \text{for FD statistics,} \end{cases} \tag{9.4.11}$$

which shows how, in principle, the statistics and the spin can be determined. It is an empirical fact, one now understood theoretically, that there is a connection between spin and statistics:

$$\begin{aligned} &\text{BE statistics: } s = 0, 1, 2, \ldots \\ &\text{FD statistics: } s = \tfrac{1}{2}, \tfrac{3}{2}, \tfrac{5}{2}, \ldots, \end{aligned} \tag{9.4.12}$$

so, in fact the possibilities are

$$\left.\frac{(\mathrm{d}\sigma/\mathrm{d}\Omega)_{\text{quant.}}}{(\mathrm{d}\sigma/\mathrm{d}\Omega)_{\text{class.}}}\right|_{\Theta=\frac{1}{2}\pi} = \begin{cases} 2, \frac{4}{3}, \frac{6}{5} \ldots & \text{for BE statistics,} \\ \frac{1}{2}, \frac{3}{4}, \frac{5}{6}, \ldots & \text{for FD statistics.} \end{cases} \tag{9.4.13}$$

Note that these are reciprocal sequences.

9.5 Conserved axial vector

We can't leave the two-body Coulomb problem without presenting it in the following way, which uses the notation of the hydrogenic atom. We begin with recalling the basic dynamical properties,

$$H = \frac{\boldsymbol{p}^2}{2\mu} - \frac{Ze^2}{r}\,;$$

$$\frac{\mathrm{d}}{\mathrm{d}t}\boldsymbol{r} = \frac{1}{\mu}\boldsymbol{p}\,, \quad \frac{\mathrm{d}}{\mathrm{d}t}\boldsymbol{p} = -Ze^2\frac{\boldsymbol{r}}{r^3}\,;$$

$$\frac{\mathrm{d}}{\mathrm{d}t}\boldsymbol{L} = \frac{\mathrm{d}}{\mathrm{d}t}(\boldsymbol{r}\times\boldsymbol{p}) = \frac{1}{\mu}\boldsymbol{p}\times\boldsymbol{p} - \frac{Ze^2}{r^3}\boldsymbol{r}\times\boldsymbol{r} = 0\,. \tag{9.5.1}$$

We say that $\boldsymbol{L}$ is a constant of the motion. Can we find another vector constant of the motion? In this, we do not explicitly pay attention to the order of multiplication of operators, but verify its correctness later. Consider

$$\begin{aligned} \frac{\mathrm{d}}{\mathrm{d}t}(\boldsymbol{p}\times\boldsymbol{L}) &= -\frac{Ze^2}{r^3}\boldsymbol{r}\times(\boldsymbol{r}\times\boldsymbol{p}) = -\frac{\mu Ze^2}{r^3}\boldsymbol{r}\times\left(\boldsymbol{r}\times\frac{\mathrm{d}\boldsymbol{r}}{\mathrm{d}t}\right) \\ &= -\frac{\mu Ze^2}{r^3}\Big(\boldsymbol{r}\underbrace{\boldsymbol{r}\cdot\frac{\mathrm{d}\boldsymbol{r}}{\mathrm{d}t}}_{=r\mathrm{d}r/\mathrm{d}t} - r^2\frac{\mathrm{d}\boldsymbol{r}}{\mathrm{d}t}\Big) = \mu Ze^2\left(\frac{1}{r}\frac{\mathrm{d}\boldsymbol{r}}{\mathrm{d}t} - \frac{\boldsymbol{r}}{r^2}\frac{\mathrm{d}r}{\mathrm{d}t}\right) \\ &= \frac{\mathrm{d}}{\mathrm{d}t}\left(\mu Ze^2\frac{\boldsymbol{r}}{r}\right). \end{aligned} \tag{9.5.2}$$

Therefore,

$$\frac{\mathrm{d}}{\mathrm{d}t}\left(\boldsymbol{p}\times\boldsymbol{L}-\mu Ze^2\frac{\boldsymbol{r}}{r}\right)=0 \tag{9.5.3}$$

identifies a constant vector; understanding that $\boldsymbol{p}\times\boldsymbol{L}$ needs symmetrization, we write it as

$$\boldsymbol{A}=\frac{\boldsymbol{r}}{r}-\frac{1}{2\mu Ze^2}\left(\boldsymbol{p}\times\boldsymbol{L}-\boldsymbol{L}\times\boldsymbol{p}\right)\ . \tag{9.5.4}$$

[Although the same letter is used, there shouldn't be any danger of confusing this $\boldsymbol{A}$ with the vector potential of Section 8.6.] Notice that both H and $\boldsymbol{A}$, the *axial vector*, are less than cubic in $\boldsymbol{p}$. Hence,

$$\frac{\mathrm{d}\boldsymbol{A}}{\mathrm{d}t}=\frac{1}{\mathrm{i}\hbar}[\boldsymbol{A},H]=(\boldsymbol{A},H)=\frac{\partial\boldsymbol{A}}{\partial\boldsymbol{r}}\cdot\frac{\partial H}{\partial\boldsymbol{p}}-\frac{\partial\boldsymbol{A}}{\partial\boldsymbol{p}}\cdot\frac{\partial H}{\partial\boldsymbol{r}}\ , \tag{9.5.5}$$

where the symmetrization of the Poisson bracket is left implicit. One easily checks that the individual terms are only linear in $\boldsymbol{p}$ so that the symmetrization removes any contributions linear in $\hbar$, and $\hbar^2$ does not occur. Therefore if it is zero classically, it is zero.

For the next step we need the commutation relations between $\boldsymbol{p}$ and $\boldsymbol{L}$, the generator of rotations, for which

$$\frac{1}{\mathrm{i}\hbar}[p_x,L_y]=p_z\ ,\quad \frac{1}{\mathrm{i}\hbar}[p_y,L_x]=-p_z \tag{9.5.6}$$

are familiar examples. Now look at the z component of $\boldsymbol{p}\times\boldsymbol{L}+\boldsymbol{L}\times\boldsymbol{p}$:

$$\begin{aligned}(\boldsymbol{p}\times\boldsymbol{L}+\boldsymbol{L}\times\boldsymbol{p})_z&=p_xL_y-p_yL_x+L_xp_y-L_yp_x\\&=[p_x,L_y]-[p_y,L_x]=2\mathrm{i}\hbar p_z\end{aligned} \tag{9.5.7}$$

so that

$$\boldsymbol{p}\times\boldsymbol{L}+\boldsymbol{L}\times\boldsymbol{p}=2\mathrm{i}\hbar\boldsymbol{p} \tag{9.5.8}$$

summarizes the commutation relations (9.5.6) compactly, and

$$\begin{aligned}\boldsymbol{A}&=\frac{\boldsymbol{r}}{r}-\frac{1}{\mu Ze^2}(\boldsymbol{p}\times\boldsymbol{L}-\mathrm{i}\hbar\boldsymbol{p})\\&=\frac{\boldsymbol{r}}{r}-\frac{1}{\mu Ze^2}(-\boldsymbol{L}\times\boldsymbol{p}+\mathrm{i}\hbar\boldsymbol{p})\end{aligned} \tag{9.5.9}$$

are alternative ways of writing $\boldsymbol{A}$.

What is $\boldsymbol{A}^2$? Use both versions of (9.5.9) and write

$$\begin{aligned}
\boldsymbol{A}^2 =& \left[\frac{\boldsymbol{r}}{r} - \frac{1}{\mu Z e^2}(-\boldsymbol{L}\times\boldsymbol{p} + \mathrm{i}\hbar\boldsymbol{p})\right]\cdot\left[\frac{\boldsymbol{r}}{r} - \frac{1}{\mu Z e^2}(\boldsymbol{p}\times\boldsymbol{L} - \mathrm{i}\hbar\boldsymbol{p})\right] \\
=& 1 - \frac{1}{\mu Z e^2}\left(\frac{\boldsymbol{r}}{r}\cdot\boldsymbol{p}\times\boldsymbol{L} - \mathrm{i}\hbar\frac{\boldsymbol{r}}{r}\cdot\boldsymbol{p} - \boldsymbol{L}\times\boldsymbol{p}\cdot\frac{\boldsymbol{r}}{r} + \mathrm{i}\hbar\boldsymbol{p}\cdot\frac{\boldsymbol{r}}{r}\right) \\
&+ \left(\frac{1}{\mu Z e^2}\right)^2\Big(-(\boldsymbol{L}\times\boldsymbol{p})\cdot(\boldsymbol{p}\times\boldsymbol{L}) + \mathrm{i}\hbar\underbrace{\boldsymbol{L}\times\boldsymbol{p}\cdot\boldsymbol{p}}_{=0} + \mathrm{i}\hbar\underbrace{\boldsymbol{p}\cdot\boldsymbol{p}\times\boldsymbol{L}}_{=0} + \hbar^2 p^2\Big)
\end{aligned} \tag{9.5.10}$$

and note the identities

$$\begin{aligned}
\mathrm{i}\hbar\left(\boldsymbol{p}\cdot\frac{\boldsymbol{r}}{r} - \frac{\boldsymbol{r}}{r}\cdot\boldsymbol{p}\right) &= \mathrm{i}\hbar\frac{\hbar}{\mathrm{i}}\boldsymbol{\nabla}\cdot\frac{\boldsymbol{r}}{r} = \frac{2\hbar^2}{r}\,, \\
\frac{\boldsymbol{r}}{r}\cdot\boldsymbol{p}\times\boldsymbol{L} - \boldsymbol{L}\times\boldsymbol{p}\cdot\frac{\boldsymbol{r}}{r} &= \frac{1}{r}\boldsymbol{L}^2 + \boldsymbol{L}^2\frac{1}{r} = \frac{2}{r}\boldsymbol{L}^2\,, \\
(\boldsymbol{L}\times\boldsymbol{p})\cdot(\boldsymbol{p}\times\boldsymbol{L}) &= \boldsymbol{L}\cdot[\boldsymbol{p}\times(\boldsymbol{p}\times\boldsymbol{L})] \\
&= \boldsymbol{L}\cdot(\boldsymbol{p}\underbrace{\boldsymbol{p}\cdot\boldsymbol{L}}_{=0} - p^2\boldsymbol{L}) = -p^2\boldsymbol{L}^2\,.
\end{aligned} \tag{9.5.11}$$

So

$$\begin{aligned}
\boldsymbol{A}^2 &= 1 - \frac{2}{\mu Z e^2}\frac{1}{r}(\boldsymbol{L}^2 + \hbar^2) + \frac{1}{(\mu Z e^2)^2}p^2(\boldsymbol{L}^2 + \hbar^2) \\
&= 1 + \frac{2}{\mu Z^2 e^4}H(\boldsymbol{L}^2 + \hbar^2)\,,
\end{aligned} \tag{9.5.12}$$

or

$$H = -\frac{\mu Z^2 e^4}{2}\frac{1 - \boldsymbol{A}^2}{\boldsymbol{L}^2 + \hbar^2}\,, \tag{9.5.13}$$

a relation among the constants of the motion, H, $\boldsymbol{L}$, $\boldsymbol{A}$, which must exist since there are only six variables (count the components of $\boldsymbol{r}$ and $\boldsymbol{p}$, for instance).

Notice that bound states ($H' < 0$) have $\boldsymbol{A}^{2'} < 1$, whereas scattering states ($H' > 0$) have $\boldsymbol{A}^{2'} > 1$. This is consistent with the geometrical significance of the axial vector $\boldsymbol{A}$; see Problem 9-14.

What are the commutation relations among the six components of $\boldsymbol{L}$ and $\boldsymbol{A}$? We already know that

$$\frac{1}{\mathrm{i}\hbar}[L_x, L_y] = L_z\,,\ldots \quad\text{and}\quad \frac{1}{\mathrm{i}\hbar}[A_x, L_y] = A_z\,,\ldots \tag{9.5.14}$$

or, compactly,

$$\boldsymbol{L}\times\boldsymbol{L} = \mathrm{i}\hbar\boldsymbol{L} \quad\text{and}\quad \boldsymbol{A}\times\boldsymbol{L} + \boldsymbol{L}\times\boldsymbol{A} = 2\mathrm{i}\hbar\boldsymbol{A}\,. \tag{9.5.15}$$

What is $\boldsymbol{A}\times\boldsymbol{A}$? It is a vector that is a constant of the motion and therefore it must be a linear combination of $\boldsymbol{L}$, $\boldsymbol{A}$, and $\boldsymbol{L}\times\boldsymbol{A}$.

Now consider the transformation

$$\boldsymbol{r} \to -\boldsymbol{r}\,, \quad \boldsymbol{p} \to -\boldsymbol{p}\,, \tag{9.5.16}$$

which leaves the commutation relations and the Hamiltonian intact. (We have seen the one-dimensional version of this unitary reflection in Problems 2-9 and 2-24). Under this transformation

$$\boldsymbol{L} \to \boldsymbol{L}\,, \quad \boldsymbol{A} \to -\boldsymbol{A}\,, \quad \boldsymbol{A} \times \boldsymbol{A} \to \boldsymbol{A} \times \boldsymbol{A}\,, \tag{9.5.17}$$

therefore only $\boldsymbol{L}$ is possible on the right side:

$$\boldsymbol{A} \times \boldsymbol{A} = \mathrm{i}\hbar C \boldsymbol{L}\,, \tag{9.5.18}$$

where C is a scalar constant of motion. To determine its value, look at

$$\begin{aligned}\frac{1}{\mathrm{i}\hbar}[A_x, \boldsymbol{A}^2] &= \frac{1}{\mathrm{i}\hbar}[A_x, A_y^2] + \frac{1}{\mathrm{i}\hbar}[A_x, A_z^2] \\ &= A_y C L_z + C L_z A_y - A_z C L_y - C L_y A_z\end{aligned} \tag{9.5.19}$$

or, with (9.5.12),

$$\begin{aligned}\frac{1}{\mathrm{i}\hbar}[A_x, \boldsymbol{A}^2] &= (\boldsymbol{A} \times \boldsymbol{L})_x C - C(\boldsymbol{L} \times \boldsymbol{A})_x \\ &= \frac{1}{\mathrm{i}\hbar}\left[A_x, \frac{2}{\mu Z^2 e^4} H \boldsymbol{L}^2\right] \\ &= \frac{2H}{\mu Z^2 e^4} \frac{1}{\mathrm{i}\hbar}[A_x, \boldsymbol{L}^2]\end{aligned} \tag{9.5.20}$$

where

$$\begin{aligned}\frac{1}{\mathrm{i}\hbar}[A_x, \boldsymbol{L}^2] &= L_y A_z + A_z L_y - L_z A_y - A_y L_z \\ &= (\boldsymbol{L} \times \boldsymbol{A})_x - (\boldsymbol{A} \times \boldsymbol{L})_x\,.\end{aligned} \tag{9.5.21}$$

Therefore, $C = -2H/(\mu Z^2 e^4)$ in (9.5.18) and we have

$$\boldsymbol{A} \times \boldsymbol{A} = \mathrm{i}\hbar \frac{-2H}{\mu Z^2 e^4} \boldsymbol{L}\,. \tag{9.5.22}$$

Let's pretend that we don't know as yet the eigenenergies of H and let's write

$$H' = -\frac{\mu Z^2 e^4}{2\hbar^2} \frac{1}{\nu^2} \quad \text{with} \quad \nu > 0 \tag{9.5.23}$$

for the value of H in a subspace of given bound-state energy $H' < 0$. Then

$$\boldsymbol{A} \times \boldsymbol{A} = \frac{\mathrm{i}}{\nu^2} \frac{1}{\hbar} \boldsymbol{L} \quad \text{and} \quad \boldsymbol{A}^2 = 1 - \frac{1}{\nu^2}\left(\frac{1}{\hbar^2}\boldsymbol{L}^2 + 1\right)\,, \tag{9.5.24}$$

and we define

$$\boldsymbol{J}_{1,2} = \frac{1}{2}(\boldsymbol{L} \pm \hbar\nu\boldsymbol{A}) \ , \tag{9.5.25}$$

so that

$$\boldsymbol{L} = \boldsymbol{J}_1 + \boldsymbol{J}_2 \ , \quad \hbar\nu\boldsymbol{A} = \boldsymbol{J}_1 - \boldsymbol{J}_2 \ . \tag{9.5.26}$$

What are the commutation relations for $\boldsymbol{J}_1$ and $\boldsymbol{J}_2$? First,

$$\begin{aligned}\frac{1}{\mathrm{i}\hbar}[J_{1x}, J_{2y}] &= \frac{1}{4}\Big(\frac{1}{\mathrm{i}\hbar}[L_x, L_y] - \frac{1}{\mathrm{i}\hbar}[\hbar\nu A_x, \hbar\nu A_y] \\ &\qquad - \frac{1}{\mathrm{i}\hbar}[L_x, \hbar\nu A_y] + \frac{1}{\mathrm{i}\hbar}[\hbar\nu A_x, L_y]\Big) \\ &= \frac{1}{4}(L_z - L_z - \hbar\nu A_z + \hbar\nu A_z) = 0 \ ,\end{aligned} \tag{9.5.27}$$

so $\boldsymbol{J}_1$ and $\boldsymbol{J}_2$ *commute*. Then

$$\begin{aligned}\frac{1}{\mathrm{i}\hbar}[J_{1x}, J_{1y}] &= \frac{1}{4}\Big(\frac{1}{\mathrm{i}\hbar}[L_x, L_y] + \frac{1}{\mathrm{i}\hbar}[\hbar\nu A_x, \hbar\nu A_y] \\ &\qquad + \frac{1}{\mathrm{i}\hbar}[L_x, \hbar\nu A_y] + \frac{1}{\mathrm{i}\hbar}[\hbar\nu A_x, L_y]\Big) \\ &= \frac{1}{4}(L_z + L_z + \hbar\nu A_z + \hbar\nu A_z) = J_{1z} \ ,\end{aligned} \tag{9.5.28}$$

so

$$\boldsymbol{J}_1 \times \boldsymbol{J}_1 = \mathrm{i}\hbar\boldsymbol{J}_1 \quad \text{and similarly} \quad \boldsymbol{J}_2 \times \boldsymbol{J}_2 = \mathrm{i}\hbar\boldsymbol{J}_2 \ . \tag{9.5.29}$$

$\boldsymbol{J}_1$ and $\boldsymbol{J}_2$ are two independent angular momenta! There is one constraint, however. Notice that

$$\boldsymbol{A} \cdot \boldsymbol{L} = \frac{\boldsymbol{r}}{r} \cdot \boldsymbol{L} - \frac{1}{\mu Z e^2}(\boldsymbol{p} \times \boldsymbol{L} - \mathrm{i}\hbar\boldsymbol{p}) \cdot \boldsymbol{L} = 0 \ . \tag{9.5.30}$$

Therefore

$$\begin{aligned}\boldsymbol{J}_1^2 = \boldsymbol{J}_2^2 &= \frac{1}{4}(\boldsymbol{L}^2 + \underbrace{\hbar^2\nu^2\boldsymbol{A}^2}_{=\hbar^2\nu^2 - \boldsymbol{L}^2 - \hbar^2}) \\ &= \frac{1}{4}\hbar^2(\nu^2 - 1) \ .\end{aligned} \tag{9.5.31}$$

The eigenvalues of $\boldsymbol{J}_1^2 = \boldsymbol{J}_2^2$ are $j(j+1)\hbar^2$ where $j = 0, \frac{1}{2}, 1, \ldots$ or $2j + 1 = n = 1, 2, 3, \ldots$. So

$$\nu^2 = 4j(j+1) + 1 = (2j+1)^2 = n^2 \tag{9.5.32}$$

and we find, once again, the Bohr energies

$$H' = -\frac{\mu Z^2 e^4}{2n^2\hbar^2} . \tag{9.5.33}$$

It is clear that these energy eigenvalues are degenerate, corresponding to the $2j+1=n$ values for m_1 and m_2 independently: the multiplicity is n^2, as we know. Alternatively we can label the states by the eigenvalues of

$$\boldsymbol{L} = \boldsymbol{J}_1 + \boldsymbol{J}_2 : \quad \underbrace{|j_1 - j_2|}_{=0} \le l \le \underbrace{j_1 + j_2}_{=2j=n-1} . \tag{9.5.34}$$

The multiplicity computed this way, $\sum_{l=0}^{n-1}(2l+1) = n^2$, is, of course, the same.

Notice that $n=1$ is $j=j_1=j_2=0$ so that the ground-state vector $|\ \rangle$ obeys

$$\boldsymbol{J}_1|\ \rangle = 0 \quad \text{and} \quad \boldsymbol{J}_2|\ \rangle = 0 \tag{9.5.35}$$

or

$$\boldsymbol{L}|\ \rangle = 0 \quad \text{and} \quad \boldsymbol{A}|\ \rangle = 0 . \tag{9.5.36}$$

For the wave function $\psi(\boldsymbol{r})$ representing $|\ \rangle$, we have

$$\boldsymbol{r} \times \frac{\hbar}{\mathrm{i}}\boldsymbol{\nabla}\psi(\boldsymbol{r}) = 0 \quad \text{and} \quad \left(\frac{\boldsymbol{r}}{r} + \frac{\hbar^2}{\mu Z e^2}\boldsymbol{\nabla}\right)\psi(\boldsymbol{r}) = 0 \tag{9.5.37}$$

which say that

$$\psi(\boldsymbol{r}) = \psi(r) , \quad \left(\frac{\mathrm{d}}{\mathrm{d}r} + \frac{Z}{a_0}\right)\psi(r) = 0 , \tag{9.5.38}$$

and then

$$\psi(r) = \psi(0)\,\mathrm{e}^{-Zr/a_0} \quad \text{with} \quad a_0 = \frac{\hbar^2}{\mu e^2} . \tag{9.5.39}$$

The (positive) value of $\psi(0)$ is, of course, determined by the normalization,

$$1 = \int (\mathrm{d}\boldsymbol{r})\,|\psi|^2 = 4\pi\left[\psi(0)\right]^2 \underbrace{\int_0^\infty \mathrm{d}r\, r^2\,\mathrm{e}^{-2Zr/a_0}}_{=2(\frac{1}{2}a_0/Z)^3} = \frac{\pi a_0^3}{Z^3}\left[\psi(0)\right]^2, \tag{9.5.40}$$

and we arrive at the familiar ground-state wave function

$$\psi_{n=1}(\boldsymbol{r}) = \pi^{-\frac{1}{2}}\left(\frac{Z}{a_0}\right)^{\frac{3}{2}} \mathrm{e}^{-Zr/a_0} . \tag{9.5.41}$$

What is $\psi_{n=1}(\boldsymbol{p})$? We can find it by Fourier transformation – indeed, this is part of Problem 8-7b – but why not directly from

$$\left(\boldsymbol{A}+\frac{1}{\mu Ze^2}\boldsymbol{p}\times\boldsymbol{L}\right)|\,\rangle=\left(\frac{\boldsymbol{r}}{r}+\frac{\mathrm{i}\hbar}{\mu Ze^2}\boldsymbol{p}\right)|\,\rangle=0 \tag{9.5.42}$$

or

$$\left(\frac{\boldsymbol{r}}{r}+\mathrm{i}\frac{\boldsymbol{p}}{p_0}\right)|\,\rangle=0 \quad \text{with} \quad p_0=\hbar Z/a_0=\mu Ze^2/\hbar\,, \tag{9.5.43}$$

using the $\boldsymbol{p}$ description in which $\boldsymbol{r}\to \mathrm{i}\hbar\frac{\partial}{\partial \boldsymbol{p}}$? The problem is: how to handle $1/r$. But $|\,\rangle$ is the $n=1$ energy eigenstate:

$$\left(\frac{\boldsymbol{p}^2}{2\mu}-\frac{Ze^2}{r}\right)|\,\rangle=-\frac{1}{2\mu}\left(\frac{\hbar Z}{a_0}\right)^2|\,\rangle=-\frac{p_0^2}{2\mu}|\,\rangle\,, \tag{9.5.44}$$

so

$$\frac{1}{r}|\,\rangle=\frac{1}{2\mu Ze^2}(p^2+p_0^2)|\,\rangle=\frac{1}{2\hbar p_0}(p^2+p_0^2)|\,\rangle \tag{9.5.45}$$

which gives

$$[\boldsymbol{r}(p^2+p_0^2)+2\mathrm{i}\hbar\boldsymbol{p}]|\,\rangle=0\,. \tag{9.5.46}$$

And now we see that ($\boldsymbol{r}\to \mathrm{i}\hbar\frac{\partial}{\partial \boldsymbol{p}}$)

$$\left[\frac{\partial}{\partial \boldsymbol{p}}(p^2+p_0^2)+2\boldsymbol{p}\right]\psi_{n=1}(\boldsymbol{p})=0 \tag{9.5.47}$$

or

$$(p^2+p_0^2)^{-1}\frac{\partial}{\partial \boldsymbol{p}}(p^2+p_0^2)^2\psi_{n=1}(\boldsymbol{p})=0\,. \tag{9.5.48}$$

So,

$$\psi_{n=1}(\boldsymbol{p})\propto\frac{1}{(p^2+p_0^2)^2} \tag{9.5.49}$$

and, as expected for $l=0$, the wave function depends only on the length p of vector $\boldsymbol{p}$; there is no directional dependence. Normalization determines the modulus of the proportionality constant, but its phase cannot be chosen freely anymore because this choice was made earlier when we opted for a positive $\psi_{n=1}(\boldsymbol{r})$. Therefore, we find the $\boldsymbol{p}=0$ value from

$$\psi(\boldsymbol{p}=0)=\int\frac{(\mathrm{d}\boldsymbol{r})}{(2\pi\hbar)^{\frac{3}{2}}}\,\psi(\boldsymbol{r})\,, \tag{9.5.50}$$

here:

$$\psi_{n=1}(\boldsymbol{p}=0) = 4\pi(2\pi\hbar)^{-\frac{3}{2}}\pi^{-\frac{1}{2}}\left(\frac{Z}{a_0}\right)^{\frac{3}{2}} \underbrace{\int_0^\infty \mathrm{d}r\, r^2\, \mathrm{e}^{-\frac{Zr}{a_0}}}_{=2(a_0/Z)^3}$$

$$= \frac{1}{\pi}\left(\frac{2a_0}{\hbar Z}\right)^{\frac{3}{2}} = \frac{1}{\pi}\left(\frac{2}{p_0}\right)^{\frac{3}{2}}, \tag{9.5.51}$$

with the consequence

$$\psi_{n=1}(\boldsymbol{p}) = \frac{2}{\pi}\sqrt{2p_0^5}\,\frac{1}{(p^2+p_0^2)^2}\,. \tag{9.5.52}$$

When verifying, as a check, that the normalization is correct,

$$1 = \int (\mathrm{d}\boldsymbol{p})\,\left|\psi_{n=1}(\boldsymbol{p})\right|^2 = 4\pi\int_0^\infty \mathrm{d}p\,p^2\,\frac{4}{\pi^2}\frac{2p_0^5}{(p^2+p_0^2)^4}$$

$$\underset{\uparrow\ \underbrace{\ }_{p^2+p_0^2=p_0^2/t}}{=} \frac{16}{\pi}\underbrace{\int_0^1 \mathrm{d}t\,t^{\frac{3}{2}}(1-t)^{\frac{1}{2}}}_{=\frac{3}{2}!\frac{1}{2}!/3!} = \frac{16}{\pi}\times\frac{\pi}{16}, \tag{9.5.53}$$

we meet, in a typical context, Euler's beta function integral

$$\int_0^1 \mathrm{d}t\;t^n(1-t)^m = \frac{n!\,m!}{(n+m+1)!}, \tag{9.5.54}$$

here for $n=\frac{3}{2}$, $m=\frac{1}{2}$; and $(-\frac{1}{2})! = \sqrt{\pi}$ is worth memorizing.

9.6 Weak external fields

Now let's impose a weak electric and a weak magnetic field (homogeneous in space and time). The small change in the internal Hamiltonian is

$$\delta H = -e'\boldsymbol{r}\cdot\boldsymbol{E} - \frac{(e/M)'}{2c}\boldsymbol{L}\cdot\boldsymbol{B}\,, \quad e=-|e|\,. \tag{9.6.1}$$

This is essentially the leading correction of (8.6.27), for $e=|e|$, except that effective coupling strengths – e' for the electric interaction, $(e/M)'$ for the magnetic interaction – appear; see Problem 9-18 for details.

First we review, in a little more detail than in Section 8.2, the basics of first-order perturbation theory. We recall that

$$\bigl[E(\lambda)-H(\lambda)\bigr]\bigl|E,\gamma\bigr\rangle = 0\,, \tag{9.6.2}$$

implies

$$\langle E,\gamma|\frac{\partial H(\lambda)}{\partial\lambda}|E,\gamma'\rangle = \frac{\partial E(\lambda)}{\partial\lambda}\delta(\gamma,\gamma')\,, \tag{9.6.3}$$

and note that, on multiplication by a small $\delta\lambda$, this becomes

$$\langle E,\gamma|\delta H|E,\gamma'\rangle = \delta E(E,\gamma')\delta(\gamma,\gamma')\,. \tag{9.6.4}$$

As told by the appearance of the Kronecker symbol $\delta(\gamma,\gamma')$, the correct choice of degenerate states is that which diagonalizes the matrix of δH for these states. In short, they are the eigenvectors of δH in this space of degenerate states.

Finding the eigenvector of $\boldsymbol{L}\cdot\boldsymbol{B}$ is easy; what to do about $\boldsymbol{r}\cdot\boldsymbol{E}$? Here, we want to remember something, namely that

$$\langle E,\gamma|\frac{\mathrm{d}F}{\mathrm{d}t}|E,\gamma'\rangle = \langle E,\gamma|\frac{1}{\mathrm{i}\hbar}(FH-HF)|E,\gamma'\rangle = 0 \tag{9.6.5}$$

for any operator F that has no parametric time dependence, $\partial F/\partial t = 0$. Now look at the axial vector:

$$\begin{aligned}\boldsymbol{A} &= \frac{\boldsymbol{r}}{r} - \frac{1}{2\mu Ze^2}(\boldsymbol{p}\times\boldsymbol{L}-\boldsymbol{L}\times\boldsymbol{p}) \\ &= \frac{\boldsymbol{r}}{r} - \frac{\mathrm{d}}{\mathrm{d}t}\frac{1}{2Ze^2}(\boldsymbol{r}\times\boldsymbol{L}-\boldsymbol{L}\times\boldsymbol{r}) \\ &\doteq \frac{\boldsymbol{r}}{r}\,,\end{aligned} \tag{9.6.6}$$

where the dotted equal sign states equality up to a total time derivative, and at the unperturbed Hamiltonian:

$$H = \frac{\boldsymbol{p}^2}{2\mu} - \frac{Ze^2}{r} = T+V \tag{9.6.7}$$

so that

$$\frac{\boldsymbol{r}}{r} = -\frac{1}{Ze^2}\boldsymbol{r}V\,. \tag{9.6.8}$$

If we can effectively replace V by H, we shall exhibit $\boldsymbol{r}$ in terms of $\boldsymbol{A}/H$ and time derivatives that do not contribute. For this we use the virial theorem:

$$\frac{\mathrm{d}}{\mathrm{d}t}\boldsymbol{r}\cdot\boldsymbol{p} = 2T+V = 2H-V \tag{9.6.9}$$

where it should be remembered that (although it does not matter) $\boldsymbol{r}\cdot\boldsymbol{p}$ is really $\frac{1}{2}(\boldsymbol{r}\cdot\boldsymbol{p}+\boldsymbol{p}\cdot\boldsymbol{r})$, and

$$\frac{1}{2}(\boldsymbol{r}\cdot\boldsymbol{p}+\boldsymbol{p}\cdot\boldsymbol{r}) = \mu\frac{1}{2}\left(\boldsymbol{r}\cdot\frac{\mathrm{d}\boldsymbol{r}}{\mathrm{d}t}+\frac{\mathrm{d}\boldsymbol{r}}{\mathrm{d}t}\cdot\boldsymbol{r}\right) = \frac{\mathrm{d}}{\mathrm{d}t}\frac{1}{2}\mu r^2\,; \tag{9.6.10}$$

so

$$-V = -2H + \frac{\mathrm{d}^2}{\mathrm{d}t^2}\frac{1}{2}\mu r^2 \,. \tag{9.6.11}$$

Now, then

$$\begin{aligned}\frac{\boldsymbol{r}}{r} = \frac{1}{Ze^2}\boldsymbol{r}(-V) &= -\frac{2}{Ze^2}\boldsymbol{r}H + \frac{\boldsymbol{r}}{Ze^2}\frac{\mathrm{d}^2}{\mathrm{d}t^2}\frac{1}{2}\mu r^2 \\ &\doteq -\frac{2}{Ze^2}\boldsymbol{r}H + \frac{1}{2Ze^2}\underbrace{\mu\frac{\mathrm{d}^2\boldsymbol{r}}{\mathrm{d}t^2}}_{=\frac{\mathrm{d}}{\mathrm{d}t}\boldsymbol{p} = -Ze^2\boldsymbol{r}/r^3} r^2 \\ &= -\frac{2}{Ze^2}\boldsymbol{r}H - \frac{1}{2}\frac{\boldsymbol{r}}{r}\end{aligned} \tag{9.6.12}$$

or, in conjuction with (9.6.6),

$$\frac{3}{2}\boldsymbol{A} \doteq -\frac{2}{Ze^2}\boldsymbol{r}H \,. \tag{9.6.13}$$

So

$$\boldsymbol{r} \doteq \frac{3}{4}\frac{Ze^2}{-H}\boldsymbol{A} \tag{9.6.14}$$

and

$$\delta H \to -\frac{3}{4}\frac{Ze^2}{-H}e'\boldsymbol{A}\cdot\boldsymbol{E} - \frac{(e/M)'}{2c}\boldsymbol{L}\cdot\boldsymbol{B} \,. \tag{9.6.15}$$

In effect, we have replaced the position vector $\boldsymbol{r}$ in (9.6.1) by a suitable multiple of the axial vector $\boldsymbol{A}$. Since $\boldsymbol{A}$ commutes with the unperturbed Hamiltonian (9.6.7), this equivalent version of δH is fit better for a perturbative evaluation.

For the degenerate states of principal quantum number n, $Ze^2/(-H') = 2n^2 a_0/Z$, this becomes

$$\begin{aligned}\delta H &\to -\frac{3}{2}\frac{a_0}{Z}n^2 e'\boldsymbol{A}\cdot\boldsymbol{E} - \frac{(e/M)'}{2c}\boldsymbol{L}\cdot\boldsymbol{B} \\ &= -\frac{3ne'}{2p_0}n\hbar\boldsymbol{A}\cdot\boldsymbol{E} - \frac{(e/M)'}{2c}\boldsymbol{L}\cdot\boldsymbol{B} \,,\end{aligned} \tag{9.6.16}$$

where we must also observe that, for such states,

$$n\hbar\boldsymbol{A} = \boldsymbol{J}_1 - \boldsymbol{J}_2 \,, \quad \boldsymbol{L} = \boldsymbol{J}_1 + \boldsymbol{J}_2 \tag{9.6.17}$$

[$\nu \to n$ in (9.5.26)]. So

$$\begin{aligned}\delta H \to & -\frac{3ne'}{2p_0}\boldsymbol{E}\cdot(\boldsymbol{J}_1-\boldsymbol{J}_2)-\frac{(e/M)'}{2c}\boldsymbol{B}\cdot(\boldsymbol{J}_1+\boldsymbol{J}_2)\\ &= -\left(\frac{3ne'}{2p_0}\boldsymbol{E}+\frac{(e/M)'}{2c}\boldsymbol{B}\right)\cdot\boldsymbol{J}_1\\ &+\left(\frac{3ne'}{2p_0}\boldsymbol{E}-\frac{(e/M)'}{2c}\boldsymbol{B}\right)\cdot\boldsymbol{J}_2 .\end{aligned} \tag{9.6.18}$$

We see that the physically significant directions are of these two combinations of $\boldsymbol{E}$ and $\boldsymbol{B}$, not of $\boldsymbol{E}$ and $\boldsymbol{B}$ individually. The eigenvalues of $\boldsymbol{J}_1$ and $\boldsymbol{J}_2$ along these directions are $m_1\hbar$ and $m_2\hbar$, where both m_1 and m_2 range independently from $j=\frac{1}{2}(n-1)$ to $-j=-\frac{1}{2}(n-1)$, which are $(2j+1)^2=n^2$ states:

$$\begin{aligned}\delta E_{n,m_1,m_2} &= -\left|\frac{3ne'}{2p_0}\boldsymbol{E}+\frac{(e/M)'}{2c}\boldsymbol{B}\right|\hbar m_1\\ &+\left|\frac{3ne'}{2p_0}\boldsymbol{E}-\frac{(e/M)'}{2c}\boldsymbol{B}\right|\hbar m_2 .\end{aligned} \tag{9.6.19}$$

In general, the degeneracy is removed, the exception occurs where $\boldsymbol{E}$ and $\boldsymbol{B}$ are perpendicular, when the two combinations have the same length, and only the difference m_1-m_2 appears. Note that for $\boldsymbol{B}=0$ and $F=|\boldsymbol{E}|$ we get

$$\delta E = -\frac{3ne'}{2p_0}F\hbar(m_1-m_2) = -\frac{3}{2}\frac{ne'a_0}{Z}F(m_1-m_2)\ ; \tag{9.6.20}$$

this agrees with the parabolic coordinate result (8.5.19) – there e was $|e|$ – which was stated in terms of $k_{1,2}=m_{1,2}+\frac{1}{2}(n-1)=0,1,\dots,n-1$.

Problems

9-1 Verify directly from the definition that $\boldsymbol{p}=\mu\boldsymbol{v}$ with $\boldsymbol{v}=\boldsymbol{v}_1-\boldsymbol{v}_2$.

9-2 Verify explicitly the commutation relations between all components of $\boldsymbol{R}$, $\boldsymbol{r}$, $\boldsymbol{P}$, $\boldsymbol{p}$.

9-3a The binding potential of some two-atomic molecules may be approximated by

$$V(r)=V_0\left[\left(\frac{r_0}{r}\right)^2-2\frac{r_0}{r}\right] ,$$

where V_0 and r_0 are phenomenological constants. Find the energy eigenvalues (relative motion) associated with this potential.

9-3b For many large molecules one has

$$\mu V_0 r_0^2 \gg \hbar^2 .$$

For the corresponding small parameter, give the three leading terms of the energy eigenvalues.

9-4 Particle 2 is initially at rest. Particle 1, of equal mass, and initial momentum $\boldsymbol{p}_1''$, collides with it. What are the initial values of $\boldsymbol{P}$ and $\boldsymbol{p}$? The collision turns the relative momentum through the angle Θ. What is the angle θ between the initial momentum $\boldsymbol{p}_1''$ and the final momentum $\boldsymbol{p}_1'$, which is the scattering angle when particle 2 is initially at rest? Supplement your analytical derivation by a diagram that makes the result evident.

9-5a Use the construction (3.7.34) of Legendre's polynomials $\mathrm{P}_l(\zeta)$ to show that

$$\mathrm{j}_l(x) = \frac{x^l}{2^l\, l!}\left(1 + \frac{\mathrm{d}^2}{\mathrm{d}x^2}\right)^l \frac{\sin x}{x}$$

for the radial wave function defined in (9.2.5). State, in particular, the explicit forms of $\mathrm{j}_0(x)$ and $\mathrm{j}_1(x)$.

9-5b Note that (why?)

$$\left(\boldsymbol{\nabla}^2 + k^2\right) \mathrm{e}^{ikr\cos\theta} = 0 .$$

What differential equation does $\mathrm{j}_l(kr)$ obey as a function of r; as a function of $x = kr$? What would you conclude from the differential equation about the behavior of $\mathrm{j}_l(x)$ for small x; for large x?

9-5c Through an appropriate transformation of a differential equation and comparison of asymptotic forms, recognize that

$$\mathrm{j}_l(x) = \sqrt{\frac{\pi}{2x}}\, \mathrm{J}_{l+\frac{1}{2}}(x)$$

expresses $\mathrm{j}_l(x)$ in terms of a closely related Bessel function. [The j_l are known as spherical Bessel functions.]

9-6a One of the definitions of the factorial function is

$$\frac{1}{z!} = \mathrm{e}^{\gamma z} \prod_{n=1}^{\infty} \left(1 + \frac{z}{n}\right) \mathrm{e}^{-\frac{z}{n}} ,$$

where Euler's constant $\gamma = 0.5772\cdots$ is fixed by $1! = 1$. Check that the relation $z! = z\,(z-1)!$ is obeyed.

9-6b Use this construction of $z!$ to prove the relation stated in (9.2.27).

9-7 The spherically symmetrical potential $V(r)$ falls off faster than $1/r$ for large r; the radial wave function for positive energy E, $R_l(r)$, is asymptotically

$$R_l(r) \cong \frac{1}{kr}\sin(kr - \tfrac{1}{2}\pi l + \delta_l)\,.$$

Explain why the correct wave function for a particle (relative motion) incident along the z axis, with wave function e^{ikz} is

$$\psi(r,\theta) = \sum_{l=0}^{\infty}(2l+1)\mathrm{i}^l\,\mathrm{e}^{\mathrm{i}\delta_l}R_l(r)\mathrm{P}_l(\cos\theta)\,.$$

Find the scattering amplitude $f(\theta)$:

$$\psi_{\text{scatt.}} \cong \frac{\mathrm{e}^{ikr}}{r}f(\theta)\,.$$

What is the differential scattering cross section per unit solid angle? What is the total cross section?

9-8a Wave function $\psi(\boldsymbol{r},t)$ obeys the Schrödinger equation

$$\mathrm{i}\hbar\frac{\partial}{\partial t}\psi = \left[-\frac{\hbar^2}{2M}\boldsymbol{\nabla}^2 + V(\boldsymbol{r})\right]\psi\,.$$

The probability density ρ and the probability current density $\boldsymbol{j}$ of Problem 6-1a are then given by (check this)

$$\rho = \psi^*\psi\,, \qquad \boldsymbol{j} = \frac{1}{M}\,\mathrm{Re}\left(\psi^*\frac{\hbar}{\mathrm{i}}\boldsymbol{\nabla}\psi\right)\,.$$

They satisfy the continuity equation

$$\frac{\partial}{\partial t}\rho + \boldsymbol{\nabla}\cdot\boldsymbol{j} = 0\,,$$

which justifies the interpretation of ρ and $\boldsymbol{j}$ as particle density and particle flux vector, respectively.

9-8b In a state of definite energy the flux vector of Problem 9-8a obeys

$$\int_S \mathrm{d}\boldsymbol{S}\cdot\boldsymbol{j} = 0\,,$$

for any closed surface S. Why? Consider the asymptotic scattering wave function (time factor omitted)

$$\psi \propto \mathrm{e}^{ikz} + \frac{\mathrm{e}^{ikr}}{r} f(\theta) \quad \text{with} \quad z = r\cos\theta \,.$$

By integrating over a sphere of large radius ($kr \gg 1$), conclude that the total cross section is given by

$$\sigma = \frac{4\pi}{k} \operatorname{Im}\left(f(\theta = \pi)\right) \,.$$

Verify this so-called *optical theorem* for the phase construction of $f(\theta)$ and σ you found in Problem 9-7.

9-9a Use the operator equation of motion to show, for any operator F not explicitly dependent on time, that

$$\langle E', \gamma' | F(t) | E'', \gamma'' \rangle = \mathrm{e}^{\frac{\mathrm{i}}{\hbar}(E' - E'')t} \langle E', \gamma' | F | E'', \gamma'' \rangle \,.$$

Is this consistent with the fact that $\langle E', \gamma'; t | F(t) | E'', \gamma''; t \rangle$ is independent of t?

9-9b Perturbation theory: $H = H_0 + H_1$, H_1 small. Use the action principle to show that the small change that H_1 produces in the time transformation function is

$$\delta\langle 1|2\rangle = -\frac{\mathrm{i}}{\hbar}\langle 1| \int_2^1 \mathrm{d}t\, H_1(t) |2\rangle \,.$$

9-9c With H_1 not an explicit function of t, and the initial and final states eigenstates of H_0, that is

$$\langle 1|2\rangle = \langle E', \gamma'; t_1 | E'', \gamma''; t_2 \rangle$$

show that the time transformation function in Problem 9-9b produces the factor

$$-\mathrm{i}\hbar \frac{\mathrm{e}^{\frac{\mathrm{i}}{\hbar}(E' - E'')t_1} - \mathrm{e}^{\frac{\mathrm{i}}{\hbar}(E' - E'')t_2}}{E' - E''} \,.$$

9-9d Demonstrate that in the limit of large $T = t_1 - t_2$,

$$\left| \frac{\mathrm{e}^{\frac{\mathrm{i}}{\hbar}(E' - E'')t_1} - \mathrm{e}^{\frac{\mathrm{i}}{\hbar}(E' - E'')t_2}}{E' - E''} \right|^2 \to \frac{2\pi}{\hbar} T\, \delta(E' - E'') \,.$$

At this point you should conclude that the probability per unit time for $E'', \gamma'' \to E', \gamma'$ is

$$\frac{2\pi}{\hbar}\,\delta(E'-E'')\big|\langle E',\gamma'|H_1|E'',\gamma''\rangle\big|^2 .$$

In the quantum literature this is known as Fermi's golden rule.

9-10a Scattering: Let $H_0 = \boldsymbol{p}^2/(2M)$, $H_1 = V(\boldsymbol{r})$. The wave function describing a momentum state [compare with $\phi_{\boldsymbol{p}}$ of (10.8.12) below] is

$$\psi_{\boldsymbol{p}}(\boldsymbol{r}) = \sqrt{\frac{(\mathrm{d}\boldsymbol{p})}{(2\pi)^3}}\,\mathrm{e}^{\frac{\mathrm{i}}{\hbar}\boldsymbol{p}\cdot\boldsymbol{r}} \quad \text{with} \quad E = \frac{\boldsymbol{p}^2}{2M} .$$

The average flux of particles in the initial state is

$$F = v|\psi_{\boldsymbol{p}''}|^2 \quad \text{with} \quad v = |\boldsymbol{p}''|/M .$$

Introduce momentum-space spherical coordinates to write $(\mathrm{d}\boldsymbol{p}')$ in terms of $\mathrm{d}E'$ and the solid angle $\mathrm{d}\Omega$. The differential cross section is defined by

$$\mathrm{d}\sigma = \frac{\text{probability per unit time for all deflections into } \mathrm{d}\Omega}{\text{flux } F} .$$

You should end up with

$$\frac{\mathrm{d}\sigma}{\mathrm{d}\Omega} = \left(\frac{M}{2\pi\hbar}\right)^2 \left|\int (\mathrm{d}\boldsymbol{r})\,\mathrm{e}^{-\frac{\mathrm{i}}{\hbar}(\boldsymbol{p}'-\boldsymbol{p}'')\cdot\boldsymbol{r}}V(\boldsymbol{r})\right|^2 .$$

This is known as the (first-order) Born approximation.

9-10b Let V be the screened Coulomb potential

$$V(\boldsymbol{r}) = \frac{Ze^2}{r}\,\mathrm{e}^{-r/a} ;$$

what is $\mathrm{d}\sigma/\mathrm{d}\Omega$? What do you get in the limit $a \to \infty$?

9-11 The differential cross section (center-of-mass frame) for the scattering of two particles that are distinguishable in principle, but not in practice, is

$$|f(\Theta)|^2 + |f(\pi-\Theta)|^2 ,$$

the "classical" part in (9.4.10). Consider the scattering of two spin-$\frac{1}{2}$ FD particles. Apply the above to the scattering of particles with opposite spins; use antisymmetrical wave functions for parallel spins. Compare your result with that of Section 9.4.

9-12 Carry out the Fourier transformation required to get the hydrogenic ground-state wave function $\psi_{n=1}(\boldsymbol{p})$ from the known $\psi_{n=1}(\boldsymbol{r})$.

9-13 Verify the statement about linear $\boldsymbol{p}$ dependence at (9.5.5).

9-14 The polar equation of a conic of eccentricity ε is

$$r = \frac{a(1-\varepsilon^2)}{1-\varepsilon\cos\theta}\,.$$

State the geometrical significance of a and θ. Then use the axial vector $\boldsymbol{A}$ to arrive at this form of the classical Kepler* orbit. Draw a Kepler ellipse and indicate $\boldsymbol{A}$. So, why is $\boldsymbol{A}$ called *axial* vector?

9-15 Axial vector: Use the result for the time average of $\boldsymbol{r}$, along with the structure of $\boldsymbol{A}$, to rederive the expression for $\langle r\rangle_{n,l}$ given in Problem 8-1d.

9-16 The hydrogenic state $m_1 = m_2 = \frac{1}{2}(n-1)$ obeys (why?)

$$(J_{1x} + \mathrm{i}J_{1y})|\,\rangle = 0\,, \qquad (J_{2x} + \mathrm{i}J_{2y})|\,\rangle = 0\,,$$

or

$$(A_x + \mathrm{i}A_y)|\,\rangle = 0\,, \qquad (L_x + \mathrm{i}L_y)|\,\rangle = 0\,.$$

Check that, to within a numerical factor, the solid harmonic with $m = l$, a solution of Laplace's equation, is $(x + \mathrm{i}y)^l$, so the required wave function is proportional to $(x + \mathrm{i}y)^{n-1}f(r)$. Use the axial vector condition to find $f(r)$. Exhibit the normalized wave function. [These states are known as circular Rydberg states, in particular for large principal quantum number n.]

9-17 The $n = 2$ hydrogenic state, of multiplicity 4, is described by two spin-$\frac{1}{2}$'s. We know that two spins have three symmetrical states ($l = 1$) and one antisymmetrical state ($l = 0$). Explain why (to within a constant) $A_z|n = 2, l = 1, m = 0\rangle$ is $|n = 2, l = 0, m = 0\rangle$.

9-18 The two-particle system of: (1) charge e, mass m; (2) charge $-Ze$, mass $M - m$; is acted on by weak, homogeneous, electric and magnetic fields. Consider the internal motion and arrive at the effective electric charge e' for the electric dipole moment, and the effective ratio $(e/M)'$ for the magnetic dipole moment. Apply your results to the situations: (a) $m/M \ll 1$ (to first order in m/M) as in a hydrogenic atom; (b) $M = 2m$, $Z = 1$, as in positronium.

9-19a Consider the four degenerate hydrogenic states for $n = 2$ that are labeled by angular momentum. Construct the 4×4 matrix of $-e'Fz$. What are its eigenvalues and eigencolumns? Compare with the known result for the linear Stark effect.

*Johannes KEPLER (1571–1630)

9-19b Repeat the above for the matrix of $-e'Fx$. Were your eigenvalue answers to be expected?

9-19c Extend the last calculation to the matrix of $-e'Fx - \frac{1}{2c}(e/M)'L_z B$. Does it work out correctly?

10. Identical Particles

10.1 Modes. Creation and annihilation operators

We now begin the construction of an algebraic theory of identical particles known (inaccurately) as second quantization. A brief reference, in the context of angular momentum, already appears on page 158.

Recall that, for a single system, physical property f is displayed relative to the complete set of measurements A,

$$\sum_{a'} |a'\rangle\langle a'| = 1\,, \tag{10.1.1}$$

as

$$f = \sum_{a',a''} |a'\rangle\langle a'|f|a''\rangle\langle a''| = \sum_{a',a''} \langle a'|f|a''\rangle |a'a''|\,. \tag{10.1.2}$$

Now consider a collection of identical systems, henceforth called *particles*, which are labeled $k = 1, 2, \ldots, n$. Then

$$f_k = \sum_{a',a''} \underbrace{\langle a'_k|f_k|a''_k\rangle}_{=\langle a'|f|a''\rangle} |a'a''|_k\,, \tag{10.1.3}$$

inasmuch as the relation between f and A is the same for each of the identical particles. Linear momentum, angular momentum, kinetic energy, are examples of a *one*-particle physical property, described by a one-particle operator that is constructed additively:

$$F = \sum_{k=1}^{n} f_k = \sum_{a',a''} \langle a'|f|a''\rangle \underbrace{\sum_{k=1}^{n} |a'a''|_k}_{\equiv |a'a''| = |a''a'|^\dagger}\,. \tag{10.1.4}$$

This generalization of the measurement symbol $|a'a''|$ is distinguished from the original by the property

$$\sum_{a'}|a'a'| = \sum_{a'}\left[\sum_{k=1}^{n}|a'a'|_k\right] = \sum_{k=1}^{n}\underbrace{\sum_{a'}|a'a'|_k}_{=1} = n\,. \tag{10.1.5}$$

We also note the product

$$\begin{aligned}|a'a''||a'''a^{\text{v}}| &= \sum_{k=1}^{n}\sum_{l=1}^{n}|a'a''|_k|a'''a^{\text{v}}|_l \\ &= \sum_{k\neq l}|a'a''|_k|a'''a^{\text{v}}|_l + \sum_{k}\underbrace{|a'a''|_k|a'''a^{\text{v}}|_k}_{=\delta(a'',a''')|a'a^{\text{v}}|_k} \\ &= \sum_{k\neq l}|a'a''|_k|a'''a^{\text{v}}|_l + \delta(a'',a''')|a'a^{\text{v}}|\end{aligned} \tag{10.1.6}$$

and, similarly $[a' \leftrightarrow a''', a'' \leftrightarrow a^{\text{v}}, k \leftrightarrow l]$

$$|a'''a^{\text{v}}||a'a''| = \sum_{k\neq l}|a'''a^{\text{v}}|_l|a'a''|_k + \delta(a',a^{\text{v}})|a'''a''|\,. \tag{10.1.7}$$

The two $k \neq l$ terms are the same, because operators associated with different particles, different degrees of freedom, commute. Therefore, on subtracting the two, we get

$$\Big[|a'a''|, |a'''a^{\text{v}}|\Big] = \delta(a'',a''')|a'a^{\text{v}}| - \delta(a',a^{\text{v}})|a'''a''|\,. \tag{10.1.8}$$

Note that on putting $a' = a''$ and summing, we get, consistently:

$$\Big[n, |a'''a^{\text{v}}|\Big] = |a'''a^{\text{v}}| - |a'''a^{\text{v}}| = 0\,. \tag{10.1.9}$$

Also,

$$\Big[|a'a'|, |a''a''|\Big] = \delta(a',a'')|a'a''| - \delta(a',a'')|a''a'| = 0\,. \tag{10.1.10}$$

In order to distinguish between the states of the n particles, and those of a single particle, we henceforth call the latter *modes*. Accordingly, we describe

$$|a'a''| = \sum_{k=1}^{n}|a'a''|_k \tag{10.1.11}$$

as symbolizing a measurement in which *one* of the n particles is removed from the a'' mode and put into the a' mode (this is sinistral reading – right to left) or, in factored form:

$$|a'a''| = \begin{pmatrix}\text{create an}\\ a'\text{ particle}\end{pmatrix} \times \begin{pmatrix}\text{annihilate}\\ \text{an } a''\text{ particle}\end{pmatrix} = \psi(a')^{\dagger}\psi(a'')\,, \tag{10.1.12}$$

where $\psi(a')^\dagger$ and $\psi(a'')$ are generalizations of $|a'\rangle$ and $\langle a''|$, respectively, as introduced in Section 1.3 (reading was dextral then). One calls the $\psi(a')^\dagger$'s creation operators, the $\psi(a'')$'s annihilation operators, and either kind ladder operators. Note that the construction (10.1.12) is consistent with the relation

$$|a'a''|^\dagger = |a''a'| . \tag{10.1.13}$$

The meaning we give to $\psi(a'')$, that, applied to a state with n particles, $|n\rangle$, it produces a state with $n-1$ particles, and that $\psi(a')^\dagger$, applied to a state with $n-1$ particles produces a state with n particles, frees us of the restriction to a specific number of particles. From now on, n is any non-negative integer, and

$$\sum_{a'} |a'a'| = n \to \sum_{a'} \psi(a')^\dagger \psi(a') = N \tag{10.1.14}$$

where N is the total number operator, with eigenvalues

$$N' = n = 0, 1, 2, \ldots . \tag{10.1.15}$$

It is natural to extend this to

$$\psi(a')^\dagger \psi(a') = N(a') , \tag{10.1.16}$$

the operator for the number of particles of mode a'. That is consistent since

$$[N(a'), N(a'')] = 0 \tag{10.1.17}$$

as a consequence of (10.1.10), so that it is possible to specify simultaneously all the eigenvalues

$$N(a')' = n(a') = 0, 1, 2, \ldots . \tag{10.1.18}$$

Indeed, a state is characterized by that collection of eigenvalues:

$$|\{n(a')\}\rangle = |n_1, n_2, \ldots\rangle \tag{10.1.19}$$

where $n_1, n_2, \ldots$ represent some ordered labeling of the modes and the number of particles occupying them. Such states with a definite number of particles in each mode are frequently called Fock states.

The physical meaning of $\psi(a')$ and $N(a'')$ is conveyed by

$$\begin{aligned} N(a'')\psi(a')|\{n\}\rangle &= \left\{ \begin{matrix} n(a'')\psi(a')|\{n\}\rangle & \text{for } a'' \neq a' \\ [n(a') - 1]\psi(a')|\{n\}\rangle & \text{for } a'' = a' \end{matrix} \right\} \\ &= \psi(a')[n(a'') - \delta(a', a'')]|\{n\}\rangle \\ &= \psi(a')[N(a'') - \delta(a', a'')]|\{n\}\rangle \end{aligned} \tag{10.1.20}$$

which is the operator relation

$$N(a'')\psi(a') = \psi(a')\big[N(a'') - \delta(a',a'')\big] \tag{10.1.21}$$

or

$$\big[\psi(a'), N(a'')\big] = \delta(a',a'')\psi(a') . \tag{10.1.22}$$

The analogous treatment of the creation operator $\psi(a')^\dagger$ gives

$$N(a'')\psi(a')^\dagger = \psi(a')^\dagger\big[N(a'') + \delta(a',a'')\big] \tag{10.1.23}$$

or

$$\big[N(a''), \psi(a')^\dagger\big] = \delta(a',a'')\psi(a')^\dagger \tag{10.1.24}$$

which, as it should be, is the adjoint of the ψ relation (10.1.22).

We can check this against the general commutation relation (10.1.8), now written as

$$\begin{aligned}\big[\psi(a')^\dagger\psi(a''), \psi(a''')^\dagger\psi(a^{\mathrm{v}})\big] =& \delta(a'',a''')\psi(a')^\dagger\psi(a^{\mathrm{v}}) \\ &- \delta(a',a^{\mathrm{v}})\psi(a''')^\dagger\psi(a'') .\end{aligned} \tag{10.1.25}$$

Putting $a''' = a^{\mathrm{v}}$ specializes it to

$$\big[\psi(a')^\dagger\psi(a''), N(a''')\big] = \delta(a'',a''')\psi(a')^\dagger\psi(a'') - \delta(a',a''')\psi(a')^\dagger\psi(a'') \tag{10.1.26}$$

where, indeed,

$$\begin{aligned}\big[\psi(a')^\dagger\psi(a''), N(a''')\big] &= \psi(a')^\dagger\big[\psi(a''), N(a''')\big] + \big[\psi(a')^\dagger, N(a''')\big]\psi(a'') \\ &= \delta(a'',a''')\psi(a')^\dagger\psi(a'') - \delta(a',a''')\psi(a')^\dagger\psi(a'') .\end{aligned} \tag{10.1.27}$$

For the decisive step consider the creation of two additional particles in either order

$$\psi(a')^\dagger\psi(a'')^\dagger|\{n\}\rangle ; \qquad \psi(a'')^\dagger\psi(a')^\dagger|\{n\}\rangle . \tag{10.1.28}$$

The states are physically indistinguishable, the vectors can differ only by a phase factor,

$$\psi(a')^\dagger\psi(a'')^\dagger = C(a',a'')\psi(a'')^\dagger\psi(a')^\dagger \tag{10.1.29}$$

with

$$|C(a',a'')|^2 = 1 , \quad C(a',a'')^* = \frac{1}{C(a',a'')} . \tag{10.1.30}$$

Reversing the order twice is no change at all, so that

$$C(a',a'')C(a'',a') = 1 . \tag{10.1.31}$$

Then taking the adjoint of (10.1.29) and interchanging a' and a'' gives

$$\psi(a')\psi(a'') = C(a'',a')^* \psi(a'')\psi(a') . \tag{10.1.32}$$

Now look at

$$\begin{aligned}[\psi(a'), N(a'')] &= \psi(a')\psi(a'')^\dagger \psi(a'') - \psi(a'')^\dagger \psi(a'')\psi(a') \\ &= \left(\psi(a')\psi(a'')^\dagger - C(a'',a')\psi(a'')^\dagger \psi(a')\right)\psi(a'') \\ &= \delta(a',a'')\psi(a'') \end{aligned} \tag{10.1.33}$$

from which we infer that

$$\psi(a')\psi(a'')^\dagger - C(a'',a')\psi(a'')^\dagger\psi(a') = \delta(a',a'') . \tag{10.1.34}$$

Next, put this in the general commutation relation (10.1.25), requiring

$$\begin{aligned}\psi(a')^\dagger\psi(a'')\psi(a''')^\dagger\psi(a^{\mathrm{v}}) &= \psi(a')^\dagger\left[C(a''',a'')\psi(a''')^\dagger\psi(a'')\right. \\ &\qquad \left. + \delta(a'',a''')\right]\psi(a^{\mathrm{v}}) \\ &= \delta(a'',a''')\psi(a')^\dagger\psi(a^{\mathrm{v}}) \\ &\quad + C(a',a''')C(a''',a'')\psi(a''')^\dagger\psi(a')^\dagger\psi(a'')\psi(a^{\mathrm{v}}) \end{aligned} \tag{10.1.35}$$

and likewise

$$\begin{aligned}\psi(a''')^\dagger\psi(a^{\mathrm{v}})\psi(a')^\dagger\psi(a'') &= \delta(a',a^{\mathrm{v}})\psi(a''')^\dagger\psi(a'') \\ &\quad + C(a',a^{\mathrm{v}})C(a^{\mathrm{v}},a'')\psi(a''')^\dagger\psi(a')^\dagger\psi(a'')\psi(a^{\mathrm{v}}) . \end{aligned} \tag{10.1.36}$$

So, (10.1.25) is satisfied provided that

$$C(a',a''')C(a''',a'') = C(a',a^{\mathrm{v}})C(a^{\mathrm{v}},a'') . \tag{10.1.37}$$

This relation and the ones in (10.1.30) and (10.1.31) can only hold if $C(a',a'')$ is of the form

$$C(a',a'') = \pm\, \mathrm{e}^{\mathrm{i}\varphi(a') - \mathrm{i}\varphi(a'')} . \tag{10.1.38}$$

Only the sign is relevant, however, because, if $\varphi(a')$ is not zero, it can be made so by a redefinition of the annihilation and creation operators:

$${}'\psi(a') = \mathrm{e}^{\mathrm{i}\varphi(a')N}\psi(a') , \qquad {}'\psi(a')^\dagger = \psi(a')^\dagger\,\mathrm{e}^{-\mathrm{i}\varphi(a')N} . \tag{10.1.39}$$

Note the commutation relation expressing the significance of ψ, $\psi^\dagger$:

$$\psi(a')\,\mathrm{e}^{\mathrm{i}\varphi N} = \mathrm{e}^{\mathrm{i}\varphi(N+1)}\psi(a')\,,$$
$$\mathrm{e}^{-\mathrm{i}\varphi N}\psi(a')^\dagger = \psi(a')^\dagger\,\mathrm{e}^{-\mathrm{i}\varphi(N+1)} \tag{10.1.40}$$

which generalize (10.1.22) and (10.1.24). So

$$\begin{aligned}
{}'\psi(a')\,{}'\psi(a'') &= \mathrm{e}^{\mathrm{i}\varphi(a')N}\psi(a')\,\mathrm{e}^{\mathrm{i}\varphi(a'')N}\psi(a'') \\
&= \mathrm{e}^{\mathrm{i}\varphi(a')+\mathrm{i}\varphi(a'')N}\,\mathrm{e}^{\mathrm{i}\varphi(a'')}\psi(a')\psi(a'') \\
&= \pm\,\mathrm{e}^{\mathrm{i}\varphi(a')+\mathrm{i}\varphi(a'')N}\,\mathrm{e}^{\mathrm{i}\varphi(a')}\psi(a'')\psi(a') \\
&= \pm{}'\psi(a'')\,{}'\psi(a')
\end{aligned} \tag{10.1.41}$$

and

$$\begin{aligned}
&{}'\psi(a')\,{}'\psi(a'')^\dagger \mp {}'\psi(a'')^\dagger\,{}'\psi(a') \\
&\quad= \mathrm{e}^{\mathrm{i}\varphi(a')N}\psi(a')\psi(a'')^\dagger\,\mathrm{e}^{-\mathrm{i}\varphi(a'')N} \\
&\quad\mp \mathrm{e}^{\mathrm{i}\varphi(a')N}\,\mathrm{e}^{-\mathrm{i}\varphi(a')}\psi(a'')^\dagger\psi(a')\,\mathrm{e}^{\mathrm{i}\varphi(a'')}\,\mathrm{e}^{-\mathrm{i}\varphi(a'')N}
\end{aligned} \tag{10.1.42}$$

or

$$\begin{aligned}
&{}'\psi(a')\,{}'\psi(a'')^\dagger \mp {}'\psi(a'')^\dagger\,{}'\psi(a') \\
&\quad= \mathrm{e}^{\mathrm{i}\varphi(a')N}\Big[\underbrace{\psi(a')\psi(a'')^\dagger - C(a'',a')\psi(a'')^\dagger\psi(a')}_{=\,\delta(a',a'')}\Big]\,\mathrm{e}^{-\mathrm{i}\varphi(a'')N} \\
&\quad= \delta(a',a'')\,.
\end{aligned} \tag{10.1.43}$$

In summary, there are two types of identical particles: those with BE statistics (ψ's commute), called *bosons* – and those with FD statistics (ψ's anticommute), called *fermions*. The table

BE statistics	FD statistics
$[\psi(a'),\psi(a'')] = 0$	$\{\psi(a'),\psi(a'')\} = 0$
$[\psi(a')^\dagger,\psi(a'')^\dagger] = 0$	$\{\psi(a')^\dagger,\psi(a'')^\dagger\} = 0$
$[\psi(a'),\psi(a'')^\dagger] = \delta(a',a'')$	$\{\psi(a'),\psi(a'')^\dagger\} = \delta(a',a'')$ (10.1.44)

lists corresponding commutation relations. For later use we introduce the notation

$$[\{A,B\}] = \begin{cases} [A,B] = AB - BA & \text{for BE symbols,} \\ \{A,B\} = AB + BA & \text{for FD symbols,} \end{cases} \tag{10.1.45}$$

then

$$[\{\psi(a'),\psi(a'')\}] = 0\,, \qquad [\{\psi(a')^\dagger,\psi(a'')^\dagger\}] = 0\,,$$
$$[\{\psi(a'),\psi(a'')^\dagger\}] = \delta(a',a'') \tag{10.1.46}$$

replace the three pairs of (10.1.44).

10.2 One-particle and two-particle operators

The BE system we recognize as the non-Hermitian operator $(y, y^\dagger)$ description of a number of degrees of freedom, and consistently, the eigenvalues of $N(a')$ are indeed

$$N(a')' = \left(\psi(a')^\dagger \psi(a')\right)' = n(a') = 0, 1, 2, \ldots \,. \tag{10.2.1}$$

How about the eigenvalues of $N(a')$ for FD statistics? Here

$$\begin{aligned} N(a')\big(1 - N(a')\big) &= \psi(a')^\dagger \psi(a')\big(1 - \psi(a')^\dagger \psi(a')\big) \\ &= \psi(a')^\dagger \underbrace{\psi(a')\psi(a')}_{=0} \psi(a')^\dagger = 0 \,, \end{aligned} \tag{10.2.2}$$

so that

$$N(a')' = n(a') = 0, 1 \,. \tag{10.2.3}$$

If two particles occupied the same mode, that situation would be symmetrical; it cannot occur for FD statistics. This is Pauli's Exclusion Principle: one particle in a mode excludes any additional ones.

A degree of freedom with just *two* outcomes of measurement? That's also familiar: spin $\frac{1}{2}$. The construction

$$\psi = \tfrac{1}{2}(\sigma_1 - \mathrm{i}\sigma_2) \,, \qquad \psi^\dagger = \tfrac{1}{2}(\sigma_1 + \mathrm{i}\sigma_2) \tag{10.2.4}$$

gives

$$\sigma_1 = \psi + \psi^\dagger \,, \quad \sigma_2 = \mathrm{i}\psi - \mathrm{i}\psi^\dagger \,, \quad \sigma_3 = \frac{1}{\mathrm{i}}\sigma_1\sigma_2 = \psi^\dagger\psi - \psi\psi^\dagger \,, \tag{10.2.5}$$

for which the basic algebraic relations,

$$\begin{gathered} \sigma_1^2 = 1 \,, \quad \sigma_2^2 = 1 \,, \quad \sigma_3^2 = 1 \,, \\ \{\sigma_1, \sigma_2\} = 0 \,, \quad \{\sigma_2, \sigma_3\} = 0 \,, \quad \{\sigma_3, \sigma_1\} = 0 \,, \end{gathered} \tag{10.2.6}$$

are verified easily, and

$$N = \psi^\dagger\psi = \tfrac{1}{2}\{\psi^\dagger, \psi\} + \tfrac{1}{2}[\psi^\dagger, \psi] = \tfrac{1}{2} + \tfrac{1}{2}\sigma_3 \tag{10.2.7}$$

has eigenvalues $N' = \frac{1}{2} + \frac{1}{2} = 1$ and $N' = \frac{1}{2} - \frac{1}{2} = 0$, indeed.

But what about the fact that FD operators for different modes, which should be different degrees of freedom, anticommute rather than commute? This is considered in Problem 10-2, and also here, using the commuting σ's associated with different degrees of freedom. Let the modes be numbered, so that $a'' > a'$ means that a'' is later in the sequence than a'. Consider the constructions

$$\psi(a') = \left[\prod_{a'''(>a')} \sigma_3(a''')\right] \frac{\sigma_1 - i\sigma_2}{2}(a') ,$$

$$\psi(a')^\dagger = \left[\prod_{a'''(>a')} \sigma_3(a''')\right] \frac{\sigma_1 + i\sigma_2}{2}(a') . \tag{10.2.8}$$

Clearly if one multiplies operators for a common mode – that is $\psi(a')\psi(a')$ or $\psi(a')\psi(a')^\dagger$ or $\psi(a')^\dagger\psi(a')^\dagger$ – all is as before, because

$$\left[\prod_{a'''(>a')} \sigma_3(a''')\right]^2 = \prod_{a'''(>a')} \left[\sigma_3(a''')\right]^2 = 1 . \tag{10.2.9}$$

Now consider $a' < a''$, for instance, and look at

$$\psi(a')\psi(a'') = \left[\prod_{a'''(>a')} \sigma_3(a''')\right] \frac{\sigma_1 - i\sigma_2}{2}(a') \left[\prod_{a'''(>a'')} \sigma_3(a''')\right] \frac{\sigma_1 - i\sigma_2}{2}(a'') \tag{10.2.10}$$

versus

$$\psi(a'')\psi(a') = \left[\prod_{a'''(>a'')} \sigma_3(a''')\right] \frac{\sigma_1 - i\sigma_2}{2}(a'') \left[\prod_{a'''(>a')} \sigma_3(a''')\right] \frac{\sigma_1 - i\sigma_2}{2}(a') , \tag{10.2.11}$$

where one wants to move the left-hand $\frac{1}{2}(\sigma_1 - i\sigma_2)$ over to join its right-hand partner. In doing so, $\frac{1}{2}(\sigma_1 - i\sigma_2)(a')$ moves through the product $\prod \sigma_3(a''')$ with $a''' > a'' > a'$, so that a' is not in this product and nothing happens. On the other hand $\frac{1}{2}(\sigma_1 - i\sigma_2)(a'')$ passes through the product with $a''' > a'$ which, since $a'' > a'$, includes $\sigma_3(a'')$, and therefore there is a sign reversal:

$$\psi(a'')\psi(a') = -\psi(a')\psi(a'') \quad \text{or} \quad \{\psi(a'), \psi(a'')\} = 0 . \tag{10.2.12}$$

Similarly, the other anticommutators in (10.1.44) are verified.

All this gives any one-particle operator (10.1.4) the construction

$$F = \sum_{a',a''} \psi(a')^\dagger \langle a'|f|a''\rangle \psi(a'') . \tag{10.2.13}$$

Here is the origin of the term "second quantization". Suppose we consider a single system (particle) and write down the expectation value of f in the state described by the wave function $\psi(a')$. It is

$$\langle f\rangle = \langle\ |f|\ \rangle = \sum_{a',a''} \underbrace{\langle\ |a'\rangle}_{=\psi(a')^\dagger} \langle a'|f|a''\rangle \underbrace{\langle a''|\ \rangle}_{=\psi(a'')} . \tag{10.2.14}$$

F, which refers to any number of particles looks as though it has been produced by elevating the wave function (first quantization) into an operator (second quantization).

In Section 3.4, we have already met an example of this construction in another notation (dimensionless variables):

$$\boldsymbol{J} = y^\dagger \frac{1}{2}\boldsymbol{\sigma} y = \sum_{\sigma',\sigma''} y(\sigma')^\dagger \langle\sigma'|\tfrac{1}{2}\boldsymbol{\sigma}|\sigma''\rangle y(\sigma'') \tag{10.2.15}$$

with

$$\begin{gathered}\big[y(\sigma), y(\sigma')\big] = 0\,, \quad \big[y^\dagger(\sigma), y^\dagger(\sigma')\big] = 0\,,\\ \big[y(\sigma), y^\dagger(\sigma')\big] = \delta(\sigma,\sigma')\,,\end{gathered} \tag{10.2.16}$$

which, we now recognize, says that any angular momentum is a BE collection of spin-$\frac{1}{2}$'s. In this connection we noted then that $\boldsymbol{J}$ obeys the proper commutation relations because $\frac{1}{2}\boldsymbol{\sigma}$ does. That is an example of the statement

$$[F, G] = \sum_{a',a''} \psi(a')^\dagger \langle a'|[f,g]|a''\rangle \psi(a'')\,, \tag{10.2.17}$$

which follows from the commutation properties of the $\psi(a')^\dagger$, $\psi(a'')$, a restatement of the commutator

$$\left[\sum_k f_k, \sum_l g_l\right] = \sum_k [f_k, g_k]\,. \tag{10.2.18}$$

How about the individual product FG? On the one hand,

$$FG = \sum_k f_k \sum_l g_l = \sum_k f_k g_k + \sum_{k\neq l} f_k g_l\,, \tag{10.2.19}$$

a sum of a one-particle operator and a two-particle operator, and, on the other,

$$\begin{aligned} FG = &\sum_{a',a''} \psi(a')^\dagger \langle a'|f|a''\rangle \psi(a'')\\ &\times \sum_{a''',a^{\mathrm{iv}}} \psi(a''')^\dagger \langle a'''|g|a^{\mathrm{iv}}\rangle \psi(a^{\mathrm{iv}})\,. \end{aligned} \tag{10.2.20}$$

Here we encounter (10.1.35) [with $C(a',a''')C(a''',a'') = (\pm 1)^2 = 1$]

$$\begin{aligned}\psi(a')^\dagger \psi(a'')\psi(a''')^\dagger \psi(a^{\mathrm{iv}}) = &\,\delta(a'',a''')\psi(a')^\dagger \psi(a^{\mathrm{iv}})\\ &+ \psi(a')^\dagger \psi(a''')^\dagger \psi(a^{\mathrm{iv}})\psi(a'')\end{aligned} \tag{10.2.21}$$

so

$$FG = \sum_{a',a''} \psi(a')^\dagger \langle a'|fg|a''\rangle \psi(a'')$$
$$+ \sum_{a',\dots,a^{\prime v}} \psi(a')^\dagger \psi(a''')^\dagger \langle a'|f|a''\rangle \langle a'''|g|a^{\prime v}\rangle \psi(a^{\prime v})\psi(a'') \,. \quad (10.2.22)$$

A one-particle operator has a single ψ (and a single $\psi^\dagger$); only one particle is required for it to contribute. A two-particle operator has two ψ's (and two $\psi^\dagger$'s); at least two particles are required for it to function.

More generally, a two-particle operator is

$$F = \frac{1}{2}\sum_{k\neq l} f_{kl} \quad \text{with} \quad f_{kl} = f_{lk} \,. \quad (10.2.23)$$

Now

$$f_{kl} = \sum_{a',\dots,a^{\prime v}} |a'a'|_k |a'''a'''|_l f_{kl} |a^{\prime v}a^{\prime v}|_l |a''a''|_k$$
$$= \sum_{a',\dots,a^{\prime v}} \underbrace{\langle a_l'''|\langle a_k'|f_{kl}|a_k''\rangle|a_l^{\prime v}\rangle}_{\substack{=\langle a',a'''|f|a'',a^{\prime v}\rangle \\ =\langle a''',a'|f|a^{\prime v},a''\rangle}} |a'a''|_k |a'''a^{\prime v}|_l \quad (10.2.24)$$

so

$$F = \frac{1}{2} \sum_{a',\dots,a^{\prime v}} \langle a'a'''|f|a''a^{\prime v}\rangle \sum_{k\neq l} |a'a''|_k |a'''a^{\prime v}|_l \quad (10.2.25)$$

where

$$\sum_{k\neq l} |a'a''|_k |a'''a^{\prime v}|_l = |a'a''||a'''a^{\prime v}| - \delta(a'',a''')|a'a^{\prime v}|$$
$$= \psi(a')^\dagger \psi(a'')\psi(a''')^\dagger \psi(a^{\prime v}) - \delta(a'',a''')\psi(a')^\dagger \psi(a^{\prime v})$$
$$= \psi(a')^\dagger \psi(a''')^\dagger \psi(a^{\prime v})\psi(a'') \quad (10.2.26)$$

and the general form of a two-particle operator emerges as

$$F = \frac{1}{2} \sum_{a',\dots,a^{\prime v}} \psi(a')^\dagger \psi(a''')^\dagger \langle a'a'''|f|a''a^{\prime v}\rangle \psi(a^{\prime v})\psi(a'') \,. \quad (10.2.27)$$

Seeing this, it's easy to appreciate that a k-particle operator is

$$F = \frac{1}{k!} \sum_{a^{(1)},\dots,a^{(2k)}} \psi(a^{(1)})^\dagger \cdots \psi(a^{(2k-1)})^\dagger$$
$$\times \langle a^{(1)},\dots,a^{(2k-1)}|f|a^{(2)},\dots,a^{(2k)}\rangle$$
$$\times \psi(a^{(2k)}) \cdots \psi(a^{(2)}) \,. \quad (10.2.28)$$

10.3 Multi-particle states

What are the eigenvectors (10.1.19) of states with specified numbers in the various modes? As an obvious generalization of the BE angular momentum treatment, we have

$$\big|\{n\}\big\rangle = \prod_{a'} \frac{\big(\psi(a')^\dagger\big)^{n(a')}}{\sqrt{n(a')!}}\big|0\big\rangle \tag{10.3.1}$$

with

$$\psi(a')\big|0\big\rangle = 0 \quad \text{for all } a' ; \tag{10.3.2}$$

the vector $|0\rangle$ describes the state with no particles, the vacuum. We see that

$$\psi(a')^\dagger\big|\{n\}\big\rangle = \big|\{n+1(a')\}\big\rangle\sqrt{n(a')+1} \tag{10.3.3}$$

and that ($\psi \to \partial/\partial\psi^\dagger$)

$$\psi(a')\big|\{n\}\big\rangle = \big|\{n-1(a')\}\big\rangle\sqrt{n(a')}\,, \tag{10.3.4}$$

in words:

$$\text{application of}\left\{\begin{matrix}\psi(a')^\dagger\\ \psi(a')\end{matrix}\right\}\text{ to }\big|\{n\}\big\rangle\left\{\begin{matrix}\text{increases by one}\\ \text{decreases by one}\end{matrix}\right\}$$

$$\text{the number of particles in mode } a'. \tag{10.3.5}$$

These statements, of course, combine into

$$\begin{aligned}(\psi^\dagger\psi)(a')\big|\{n\}\big\rangle &= \big|\{n\}\big\rangle n(a')\,,\\ (\psi\psi^\dagger)(a')\big|\{n\}\big\rangle &= \big|\{n\}\big\rangle\big(n(a')+1\big)\,.\end{aligned} \tag{10.3.6}$$

The same construction applies to FD statistics, except that – with the restriction $n(a') = 0, 1$ – the factorial $n(a')! = 1$ is always unity. Also the order of multiplication is significant, some standard order must be adopted. In effect,

$$\big|\{n\}\big\rangle = \prod_{(n(a')=1)} \psi(a')^\dagger\big|0\big\rangle \quad \text{with} \quad \psi(a')\big|0\big\rangle = 0 \tag{10.3.7}$$

is simply a product over occupied modes. Now one has

$$\begin{aligned}\psi(a')\big|\{n\}\big\rangle &= \big|\{n-1(a')\}\big\rangle(-1)^{n(<a')}\sqrt{n(a')}\\ &= \begin{cases}\big|\{n-1(a')\}\big\rangle(-1)^{n(<a')} & \text{for } n(a')=1\,,\\ 0 & \text{for } n(a')=0\,,\end{cases}\end{aligned} \tag{10.3.8}$$

and

$$\psi(a')^\dagger|\{n\}\rangle = |\{n+1(a')\}\rangle(-1)^{n(<a')}\sqrt{1-n(a')}$$
$$= \begin{cases} 0 & \text{for } n(a') = 1\,, \\ |\{n+1(a')\}\rangle(-1)^{n(<a')} & \text{for } n(a') = 0\,. \end{cases} \tag{10.3.9}$$

The sign depends on $n(<a')$, the number of particles in modes $a'' < a'$, that is: $\psi(a'')^\dagger$ stand to the left of $\psi(a')^\dagger$ in (10.3.1),

$$n(<a') = \sum_{a''(<a')} n(a'')\,. \tag{10.3.10}$$

10.4 Dynamical basics

So far, this has been kinematics, the study of operators and vectors, without reference to time. Now to dynamics. The basic variables are $\psi(a',t)$ and $\psi(a',t)^\dagger$. How do they fit into the action principle? That is quite evident for the BE situation because the $\psi(a',t)$ and $\psi(a',t)^\dagger$ are just examples of non-Hermitian variables combining q,p type variables. We know that

$$L = \sum_{a'} \mathrm{i}\hbar\psi(a',t)^\dagger \frac{\mathrm{d}}{\mathrm{d}t}\psi(a',t) - H \tag{10.4.1}$$

from which we derive the equations of motion

$$\mathrm{i}\hbar\frac{\mathrm{d}}{\mathrm{d}t}\psi(a',t) = \frac{\partial H}{\partial\psi(a',t)^\dagger} \tag{10.4.2}$$

and

$$\mathrm{i}\hbar\frac{\mathrm{d}}{\mathrm{d}t}\psi(a',t)^\dagger = -\frac{\partial H}{\partial\psi(a',t)}\,, \tag{10.4.3}$$

which are adjoints of each other, and the generators

$$G = G_\psi + G_t = \sum_{a'} \mathrm{i}\hbar\psi(a',t)^\dagger\delta\psi(a',t) - H\delta t\,. \tag{10.4.4}$$

If the Hamiltonian is just a one-particle operator

$$H = \sum_{a',a''} \psi(a',t)^\dagger\langle a'|h|a''\rangle\psi(a'',t) \tag{10.4.5}$$

we get the equation of motion

$$i\hbar\frac{d}{dt}\psi(a',t) = \sum_{a''}\langle a'|h|a''\rangle\psi(a'',t)\,, \tag{10.4.6}$$

compactly written as

$$i\hbar\frac{d}{dt}\psi = h\psi\,, \tag{10.4.7}$$

a linear operator equation that is essentially a Schrödinger equation. Let's illustrate this with the example of angular momentum, considered as a BE collection of spin-$\frac{1}{2}$'s.

10.5 Example: General spin dynamics

As an example we treat the motion of an arbitrary spin $\boldsymbol{S}$ in a time varying magnetic field $\boldsymbol{B}(t)$ to which the magnetic moment $\gamma\boldsymbol{S}$ couples. The Hamilton operator

$$H = -\gamma\boldsymbol{S}\cdot\boldsymbol{B}(t) \tag{10.5.1}$$

has a parametric time dependence through $\boldsymbol{B}(t)$. Recalling that for any angular momentum $\delta\boldsymbol{J} = (i\hbar)^{-1}[\boldsymbol{J},\delta\boldsymbol{\omega}\cdot\boldsymbol{J}] = \delta\boldsymbol{\omega}\times\boldsymbol{J}$, we find the equation of motion

$$\frac{d\boldsymbol{S}}{dt} = \frac{1}{i\hbar}[\boldsymbol{S},H] = \gamma\boldsymbol{S}\times\boldsymbol{B}\,; \tag{10.5.2}$$

the right-hand side is, of course, the torque on the magnetic moment $\gamma\boldsymbol{S}$ in the magnetic field $\boldsymbol{B}$.

The conservation of $\boldsymbol{S}^2$ suggests to solve (10.5.2) for the given value of s $[\boldsymbol{S}^{2'} = s(s+1)\hbar^2;\ s = 0, \frac{1}{2}, 1, \ldots]$, but it is much simpler, and more systematic, to deal with all s values at once. We regard arbitrary $\boldsymbol{S}$ as a BE collection of spin-$\frac{1}{2}$'s,

$$\frac{1}{\hbar}\boldsymbol{S} = y^\dagger\frac{1}{2}\boldsymbol{\sigma}y\,,\qquad y = \begin{pmatrix} y_+ \\ y_- \end{pmatrix}\,,\qquad y^\dagger = (y_+^\dagger, y_-^\dagger)\,, \tag{10.5.3}$$

which uses the notation of Section 3.4. The Hamiltonian (10.5.1) then appears as

$$H = -\hbar\gamma y^\dagger\frac{1}{2}\boldsymbol{\sigma}\cdot\boldsymbol{B}y\,; \tag{10.5.4}$$

it is of the one-particle form (10.4.5) with $\psi(a',t)\to y_\pm$. From the BE viewpoint, the equations of motion are

$$i\frac{d}{dt}y = \frac{1}{\hbar}\frac{\partial H}{\partial y^\dagger} = -\gamma\tfrac{1}{2}\boldsymbol{\sigma}\cdot\boldsymbol{B}y \tag{10.5.5}$$

which are just two ($\pm$) linear operator equations.

The linearity of the equations of motion says that

$$y(t_1) = \mathcal{U} y(t_2) \tag{10.5.6}$$

where $\mathcal{U}$ is a 2×2 matrix, and therefore

$$y(t_1)^\dagger = \left[\mathcal{U} y(t_2)\right]^\dagger = y(t_2)^\dagger \mathcal{U}^{\mathrm{T}*} \tag{10.5.7}$$

which involves the complex conjugate (∗) of the transposed (T) matrix. Explicitly these are

$$\begin{aligned} y_\sigma(t_1) &= \sum_{\sigma'=\pm} \mathcal{U}_{\sigma\sigma'} y_{\sigma'}(t_2)\,, \\ y_\sigma(t_1)^\dagger &= \sum_{\sigma'=\pm} y_{\sigma'}(t_2)^\dagger (\mathcal{U}^{\mathrm{T}})^*_{\sigma'\sigma} = \sum_{\sigma'=\pm} \mathcal{U}^*_{\sigma\sigma'} y_{\sigma'}(t_2)^\dagger\,. \end{aligned} \tag{10.5.8}$$

The commutation relations require that

$$\begin{aligned} \left[y_\sigma(t_1), y_{\sigma'}(t_1)^\dagger\right] = \delta_{\sigma\sigma'} &= \Bigl[\sum_{\sigma''} \mathcal{U}_{\sigma\sigma''} y_{\sigma''}(t_2), \sum_{\sigma'''} y^\dagger_{\sigma'''}(t_2) \bigl(\mathcal{U}^{\mathrm{T}*}\bigr)_{\sigma'''\sigma'}\Bigr] \\ &= \sum_{\sigma'',\sigma'''} \mathcal{U}_{\sigma\sigma''} \underbrace{\left[y_{\sigma''}(t_2), y^\dagger_{\sigma'''}(t_2)\right]}_{=\delta_{\sigma''\sigma'''}} \bigl(\mathcal{U}^{\mathrm{T}*}\bigr)_{\sigma'''\sigma'} \\ &= \bigl(\mathcal{U}\mathcal{U}^{\mathrm{T}*}\bigr)_{\sigma\sigma'} \end{aligned} \tag{10.5.9}$$

which says that $\mathcal{U}$ is a unitary matrix,

$$\mathcal{U}\mathcal{U}^{\mathrm{T}*} = 1\,. \tag{10.5.10}$$

The dynamical evolution of the system is described by the time transformation function $\langle y^{\dagger\prime}, t_1 | y'', t_2\rangle$. It suffices to consider the dependence on $y^{\dagger\prime}$, for example,

$$\delta\langle y^{\dagger\prime}, t_1| = \frac{\mathrm{i}}{\hbar}\langle y^{\dagger\prime}, t_1|G_{y^\dagger} \tag{10.5.11}$$

where

$$G_{y^\dagger} = -\mathrm{i}\hbar\,\delta y^{\dagger\prime} y \tag{10.5.12}$$

[see (7.1.5)], so

$$\delta\langle y^{\dagger\prime}, t_1 | y'', t_2\rangle = \langle y^{\dagger\prime}, t_1 | \delta y^{\dagger\prime} y(t_1) | y'', t_2\rangle = \langle y^{\dagger\prime}, t_1 | \delta y^{\dagger\prime} \mathcal{U} \underset{\substack{\downarrow \\ y''}}{y(t_2)} | y'', t_2\rangle \tag{10.5.13}$$

or

$$\delta \log\langle y^{\dagger\prime}, t_1 | y'', t_2\rangle = \delta\bigl[y^{\dagger\prime}\mathcal{U} y''\bigr]\,. \tag{10.5.14}$$

The result,

$$\langle y^{\dagger\prime}, t_1 | y'', t_2 \rangle = e^{y^{\dagger\prime} \mathcal{U} y''} \tag{10.5.15}$$

is correctly normalized because $\langle y^{\dagger\prime} = 0|$ is the state of zero angular momentum, $\langle y^{\dagger\prime} = 0|\boldsymbol{S} = 0$, and nothing happens.

All information is contained in the 2×2 matrix $\mathcal{U}$, produced by solving (10.5.5) and writing the solution in the form (10.5.6). As a specific example we choose the rotating field

$$\begin{aligned}\boldsymbol{B} = B_1 (\boldsymbol{e}_x \cos(\omega t) + \boldsymbol{e}_y \sin(\omega t)) \\ + B_0 \boldsymbol{e}_z\end{aligned} \tag{10.5.16}$$

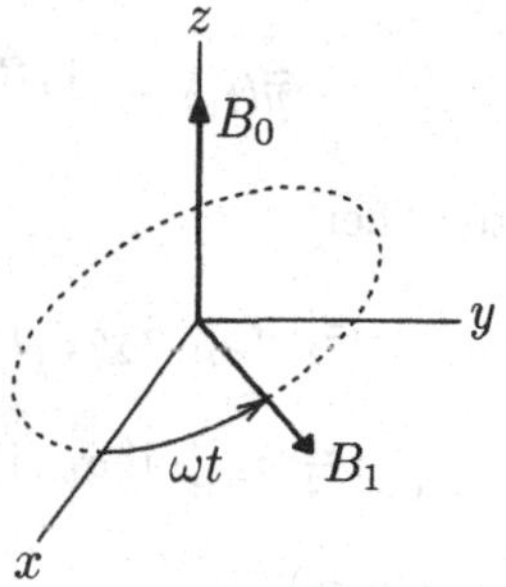

so that

$$\begin{aligned}\boldsymbol{\sigma} \cdot \boldsymbol{B} &= B_1 (\sigma_x \cos(\omega t) + \sigma_y \sin(\omega t)) + B_0 \sigma_z \\ &= B_1 \, e^{-i\sigma_z \frac{\omega t}{2}} \sigma_x \, e^{i\sigma_z \frac{\omega t}{2}} + B_0 \sigma_z \,.\end{aligned} \tag{10.5.17}$$

Accordingly we write

$$y(t) = e^{-i\frac{\omega t}{2}\sigma_z} \overline{y}(t) \tag{10.5.18}$$

and get

$$\left(i\frac{d}{dt} + \frac{\omega}{2}\sigma_z \right) \overline{y} = -\gamma \frac{1}{2} (B_1 \sigma_x + B_0 \sigma_z) \overline{y} \tag{10.5.19}$$

or

$$i\frac{d}{dt}\overline{y} = -\gamma \frac{1}{2} [B_1 \sigma_x + (B_0 + \omega/\gamma) \sigma_z] \overline{y} \,. \tag{10.5.20}$$

Now, with B and θ defined by

$$\begin{aligned}B &= \sqrt{B_1^2 + (B_0 + \omega/\gamma)^2} \,, \\ B \sin\theta &= B_1 \,, \quad B \cos\theta = B_0 + \omega/\gamma\end{aligned} \tag{10.5.21}$$

we have

$$B_1\sigma_x + (B_0 + \omega/\gamma)\sigma_z = e^{-i\frac{\theta}{2}\sigma_y} B\sigma_z\, e^{i\frac{\theta}{2}\sigma_y} \,. \tag{10.5.22}$$

So with

$$\overline{y} = e^{-i\frac{\theta}{2}\sigma_y}\overline{\overline{y}} \tag{10.5.23}$$

we get

$$i\frac{d}{dt}\overline{\overline{y}} = -\gamma B\frac{1}{2}\sigma_z\overline{\overline{y}} \tag{10.5.24}$$

and then

$$\overline{\overline{y}}(t_1) = e^{i\gamma B\frac{1}{2}\sigma_z(t_1 - t_2)}\overline{\overline{y}}(t_2)\,. \tag{10.5.25}$$

To put it together,

$$\begin{aligned} y(t_1) &= e^{-i\frac{\omega t_1}{2}\sigma_z}\, e^{-i\frac{\theta}{2}\sigma_y}\, e^{i\gamma B\frac{1}{2}\sigma_z(t_1 - t_2)}\overline{\overline{y}}(t_2) \\ &= e^{-i\frac{\omega t_1}{2}\sigma_z}\, e^{-i\frac{\theta}{2}\sigma_y}\, e^{i\gamma B\frac{1}{2}\sigma_z(t_1 - t_2)}\, e^{i\frac{\theta}{2}\sigma_y}\, e^{i\frac{\omega t_2}{2}\sigma_z} y(t_2) \\ &= \mathcal{U}y(t_2) \end{aligned} \tag{10.5.26}$$

with

$$\mathcal{U} = e^{-i\frac{\omega t_1}{2}\sigma_z}\overline{\mathcal{U}}(t_1 - t_2)\, e^{i\frac{\omega t_2}{2}\sigma_z}\,, \tag{10.5.27}$$

where

$$\begin{aligned} \overline{\mathcal{U}}(T) &= e^{-i\frac{\theta}{2}\sigma_y}\, e^{i\gamma B\frac{1}{2}\sigma_z T}\, e^{i\frac{\theta}{2}\sigma_y} \\ &= \cos(\tfrac{1}{2}\gamma BT) + i\sin(\tfrac{1}{2}\gamma BT)\big(\sigma_z\cos\theta + \sigma_x\sin\theta\big)\,. \end{aligned} \tag{10.5.28}$$

The explicit matrix elements of

$$\mathcal{U} = \begin{pmatrix} \mathcal{U}_{++} & \mathcal{U}_{+-} \\ \mathcal{U}_{-+} & \mathcal{U}_{--} \end{pmatrix} \tag{10.5.29}$$

are

$$\begin{aligned} \mathcal{U}_{++} &= e^{-i\frac{1}{2}\omega T}\big[\cos(\tfrac{1}{2}\gamma BT) + i\cos\theta\,\sin(\tfrac{1}{2}\gamma BT)\big]\,, \\ \mathcal{U}_{--} &= e^{i\frac{1}{2}\omega T}\big[\cos(\tfrac{1}{2}\gamma BT) - i\cos\theta\,\sin(\tfrac{1}{2}\gamma BT)\big] = \mathcal{U}_{++}^*\,, \\ \mathcal{U}_{+-} &= e^{-i\frac{1}{2}\omega(t_1 + t_2)} i\sin\theta\,\sin(\tfrac{1}{2}\gamma BT)\,, \\ \mathcal{U}_{-+} &= e^{i\frac{1}{2}\omega(t_1 + t_2)} i\sin\theta\,\sin(\tfrac{1}{2}\gamma BT) = -\mathcal{U}_{+-}^* \end{aligned} \tag{10.5.30}$$

with $T = t_1 - t_2$.

Now, applying the lessons of Section 3.4, the expansion

$$\langle y^{\dagger\prime}, t_1 | y'', t_2 \rangle = e^{y^{\dagger\prime} \mathcal{U} y''} = \sum_{n=0}^{\infty} \frac{1}{n!} (y^{\dagger\prime} \mathcal{U} y'')^n \tag{10.5.31}$$

introduces the wave functions for all the spin states

$$\langle y^{\dagger\prime} | s, m \rangle = \frac{(y_+^{\dagger\,\prime})^{s+m} (y_-^{\dagger\,\prime})^{s-m}}{\sqrt{(s+m)!(s-m)!}}$$

$$\langle s, m | y'' \rangle = \frac{(y_+'')^{s+m} (y_-'')^{s-m}}{\sqrt{(s+m)!(s-m)!}} \tag{10.5.32}$$

with

$$s + m = n_+ \quad \text{and} \quad s - m = n_- \, , \tag{10.5.33}$$

that is

$$n = 2s = 0, 1, 2, \ldots \, . \tag{10.5.34}$$

So, for a spin s, we have

$$\frac{1}{n!} (y^{\dagger\prime} \mathcal{U} y'')^{2s} = \sum_m \langle y^{\dagger\prime} | s, m \rangle \langle s, m; t_1 | s, m'; t_2 \rangle^{\boldsymbol{B}} \langle s, m' | y'' \rangle \tag{10.5.35}$$

which gives all the required propability amplitudes in terms of the four matrix elements of $\mathcal{U}$. For $s = \frac{1}{2}$, where

$$\langle y^{\dagger\prime} | \tfrac{1}{2}, \tfrac{1}{2} \rangle = y_+^{\dagger\,\prime} \, , \quad \langle y^{\dagger\prime} | \tfrac{1}{2}, -\tfrac{1}{2} \rangle = y_-^{\dagger\,\prime} \, , \; \ldots \tag{10.5.36}$$

we have immediately

$$\langle \tfrac{1}{2}, \tfrac{1}{2}\sigma; t_1 | \tfrac{1}{2}, \tfrac{1}{2}\sigma'; t_2 \rangle^{\boldsymbol{B}} = \mathcal{U}_{\sigma\sigma'} \, . \tag{10.5.37}$$

In particular, the probability of the transition $\frac{1}{2} \leftrightarrow -\frac{1}{2}$ in time T is

$$p(\tfrac{1}{2}, -\tfrac{1}{2}) = |\mathcal{U}_{+-}|^2 = |\mathcal{U}_{-+}|^2 = \sin^2\theta \, \sin^2(\tfrac{1}{2}\gamma B T) \, . \tag{10.5.38}$$

Notice that it oscillates in time, with $\sin^2(\frac{1}{2}\gamma BT)$ ranging between 0 and 1. The maximum value is given by

$$p(\tfrac{1}{2}, -\tfrac{1}{2}) \le \sin^2\theta = \left(\frac{B_1}{B}\right)^2 = \frac{(\gamma B_1)^2}{(\omega + \gamma B_0)^2 + (\gamma B_1)^2} \tag{10.5.39}$$

which reaches unity for

$$\omega = -\gamma B_0 \, , \tag{10.5.40}$$

the condition of ***resonance*** between the rotating field and the rotation of the angular momentum produced by field B_0. Also, at resonance, the time required to change the probability from 0 to 1 is

$$\Delta T = \frac{\pi}{\gamma B_1} \, . \tag{10.5.41}$$

So, the smaller B_1, the sharper the resonance, half-width: $\gamma B_1 = \Delta\omega$, but the longer it takes to build up the transition: $\Delta T = \pi/\Delta\omega$. See Problems 10-4 for $s > \frac{1}{2}$.

10.6 General dynamics

For BE systems, with commutation relations

$$\begin{gathered} \left[\psi(a'), \psi(a'')\right] = 0 \, , \qquad \left[\psi(a')^\dagger, \psi(a'')^\dagger\right] = 0 \, , \\ \left[\psi(a'), \psi(a'')^\dagger\right] = \delta(a', a'') \, , \end{gathered} \tag{10.6.1}$$

one uses infinitesimal variations $\delta\psi(a')$, $\delta\psi(a')^\dagger$ that *commute* with all operators ψ, $\psi^\dagger$. This maintains the commutation relations, as

$$\left[\delta\psi(a'), \psi(a'')^\dagger\right] = 0 \, , \tag{10.6.2}$$

for example. Therefore, analogously, for FD systems, with *anti*-commutation relations, one uses operator variations that *anticommute* with all ψ, $\psi^\dagger$:

$$\{\delta\psi(a'), \psi(a'')\} = 0 \, , \quad \{\delta\psi(a'), \psi(a'')^\dagger\} = 0 \, , \; \ldots \, . \tag{10.6.3}$$

Do such completely anticommuting quantities exist? Recall that

$$\begin{aligned} \psi(a') &= \left[\prod_{a'''(>a')} \sigma_3(a''')\right] \frac{1}{2}(\sigma_1 - i\sigma_2)(a') \, , \\ \psi(a')^\dagger &= \left[\prod_{a'''(>a')} \sigma_3(a''')\right] \frac{1}{2}(\sigma_1 + i\sigma_2)(a') \, . \end{aligned} \tag{10.6.4}$$

What anticommutes with each and every one of these operators? The product

$$\prod_{a''} \sigma_3(a'') \tag{10.6.5}$$

of *all* σ_3's.

We now want to recognize that dynamics, as described by the Lagrangian

$$L = \sum_{a'} \mathrm{i}\hbar\psi(a',t)^\dagger \frac{\mathrm{d}}{\mathrm{d}t}\psi(a',t) - H , \tag{10.6.6}$$

for BE systems, also includes FD systems. The distinction is implicit in the nature of the $\delta\psi$, $\delta\psi^\dagger$. Begin with

$$W_{12} = \int_2^1 \mathrm{d}t\, L = \int \Big(\sum_{a'} \mathrm{i}\hbar\psi^\dagger \mathrm{d}\psi - H\mathrm{d}t\Big) \tag{10.6.7}$$

and get

$$\begin{aligned}\delta W_{12} = &\int \mathrm{d}\Big(\sum_{a'} \mathrm{i}\hbar\psi^\dagger\delta\psi - H\delta t\Big)\\ &+ \int \Big(\sum_{a'} \mathrm{i}\hbar\delta\psi^\dagger \mathrm{d}\psi - \sum_{a'} \mathrm{i}\hbar \mathrm{d}\psi^\dagger\delta\psi - \delta H \mathrm{d}t + \mathrm{d}H\delta t\Big)\\ =& G_1 - G_2 .\end{aligned} \tag{10.6.8}$$

So $G = G_\psi + G_t$ with

$$G_\psi = \mathrm{i}\hbar \sum_{a'} \psi(a',t)^\dagger\delta\psi(a',t) \quad \text{and} \quad G_t = -H\delta t . \tag{10.6.9}$$

The significance of G_ψ as the generator of variations of the ψ is conveyed by

$$\begin{aligned}\delta\psi(a',t) &= \frac{1}{\mathrm{i}\hbar}[\psi(y't), G_\psi] = \sum_{a''}[\psi(a',t), \psi(a'',t)^\dagger\delta\psi(a'',t)] ,\\ 0 &= \frac{1}{\mathrm{i}\hbar}[\psi(y't)^\dagger, G_\psi] = \sum_{a''}[\psi(a',t)^\dagger, \psi(a'',t)^\dagger\delta\psi(a'',t)] .\end{aligned} \tag{10.6.10}$$

For BE statistics, the $\delta\psi$ commute with all ψ, $\psi^\dagger$, so that

$$[\psi, \psi^\dagger\delta\psi] = [\psi, \psi^\dagger]\delta\psi , \tag{10.6.11}$$

for instance, and we conclude that

$$[\psi(a',t), \psi(a'',t)^\dagger] = \delta(a',a'') , \quad [\psi(a',t)^\dagger, \psi(a'',t)^\dagger] = 0 . \tag{10.6.12}$$

With FD statistics the $\delta\psi$ anticommute with all ψ, $\psi^\dagger$, as in

$$\begin{aligned}[\psi, \psi^\dagger\delta\psi] &= \psi\psi^\dagger\delta\psi - \psi^\dagger\delta\psi\psi = \psi\psi^\dagger\delta\psi + \psi^\dagger\psi\delta\psi\\ &= \{\psi, \psi^\dagger\}\delta\psi ,\end{aligned} \tag{10.6.13}$$

for example, and

$$\{\psi(a',t),\psi(a'',t)^\dagger\} = \delta(a',a'')\,,$$
$$\{\psi(a',t)^\dagger,\psi(a'',t)^\dagger\} = 0\,, \quad \{\psi(a',t),\psi(a'',t)\} = 0 \tag{10.6.14}$$

follow.

The other inference from the action principle is

$$\delta H(\psi,\psi^\dagger,t) = \frac{\mathrm{d}H}{\mathrm{d}t}\delta t + \mathrm{i}\hbar\sum_{a'}\left(\delta\psi(a',t)^\dagger\frac{\mathrm{d}\psi(a',t)}{\mathrm{d}t} - \frac{\mathrm{d}\psi(a',t)^\dagger}{\mathrm{d}t}\delta\psi(a',t)\right) \tag{10.6.15}$$

which is set against the significance of $G_t = -H\delta t$ in producing the time derivative of any $F(\psi,\psi^\dagger,t)$,

$$\frac{\mathrm{d}F}{\mathrm{d}t} = \frac{\partial F}{\partial t} + \frac{1}{\mathrm{i}\hbar}[F,H]\,. \tag{10.6.16}$$

Again we get from both

$$\frac{\mathrm{d}H}{\mathrm{d}t} = \frac{\partial H}{\partial t}\,. \tag{10.6.17}$$

Now let $F = G_\psi$:

$$\frac{\mathrm{d}}{\mathrm{d}t}G_\psi - \frac{\partial}{\partial t}G_\psi = \frac{1}{\mathrm{i}\hbar}[G_\psi,H] = -\delta_\psi H\,, \tag{10.6.18}$$

so

$$\delta_\psi H = -\mathrm{i}\hbar\sum_{a'}\frac{\mathrm{d}\psi(a',t)^\dagger}{\mathrm{d}t}\delta\psi(a',t) - \mathrm{i}\hbar\sum_{a'}\psi(a',t)^\dagger\left(\frac{\mathrm{d}}{\mathrm{d}t} - \frac{\partial}{\partial t}\right)\delta\psi(a',t)\,. \tag{10.6.19}$$

Thus consistency requires that

$$\left(\frac{\mathrm{d}}{\mathrm{d}t} - \frac{\partial}{\partial t}\right)\delta\psi(a',t) = \frac{1}{\mathrm{i}\hbar}\left[\delta\psi(a',t),H\right] = 0\,, \tag{10.6.20}$$

which is trivial for BE systems, but for FD systems requires that the Hamilton operator H be an even function of ψ and $\psi^\dagger$. Of course, that's the kind of H we've been talking about, made of one-particle operators $[\psi^\dagger\cdots\psi]$ and two-particle operators $[\psi^\dagger\psi^\dagger\cdots\psi\psi]$.

Using an obvious notation for left and right derivatives, the equations of motion for both types are

$$\mathrm{i}\hbar\frac{\mathrm{d}}{\mathrm{d}t}\psi(a',t) = \frac{\partial_\mathrm{l} H}{\partial\psi(a',t)^\dagger}\,,$$
$$-\mathrm{i}\hbar\frac{\mathrm{d}}{\mathrm{d}t}\psi(a',t)^\dagger = \frac{\partial_\mathrm{r} H}{\partial\psi(a',t)}\,, \tag{10.6.21}$$

which are mutually adjoint.

10.7 Operator fields

Discrete indices are nice, but we want to describe particles that move in three-dimensional space and are specified by position $\boldsymbol{r}$ (as well as, e. g., spin). So, understanding the possible presence of, but not writing until needed, discrete spin indices, we replace $\psi(a',t) \to \psi(\boldsymbol{r},t)$. We are now dealing with operator-valued functions of space and time – operator fields. With

$$\sum_{a'} \to \int (\mathrm{d}\boldsymbol{r}) \tag{10.7.1}$$

the Lagrangian becomes

$$L = \int (\mathrm{d}\boldsymbol{r})\, \mathrm{i}\hbar\psi(\boldsymbol{r},t)^\dagger \frac{\partial}{\partial t}\psi(\boldsymbol{r},t) - H\,, \tag{10.7.2}$$

leading to the generator

$$G_\psi = \int (\mathrm{d}\boldsymbol{r})\, \mathrm{i}\hbar\psi(\boldsymbol{r},t)^\dagger \delta\psi(\boldsymbol{r},t)\,, \tag{10.7.3}$$

and

$$\begin{aligned} \delta\psi(\boldsymbol{r},t) &= \frac{1}{\mathrm{i}\hbar}\left[\psi(\boldsymbol{r},t), G_\psi\right] = \int (\mathrm{d}\boldsymbol{r}')\left[\psi(\boldsymbol{r},t), \psi(\boldsymbol{r}',t)^\dagger\delta\psi(\boldsymbol{r}',t)\right]\,,\\ 0 &= \frac{1}{\mathrm{i}\hbar}\left[\psi(\boldsymbol{r},t)^\dagger, G_\psi\right] = \int (\mathrm{d}\boldsymbol{r}')\left[\psi(\boldsymbol{r},t)^\dagger, \psi(\boldsymbol{r}',t)^\dagger\delta\psi(\boldsymbol{r}',t)\right]\,, \end{aligned} \tag{10.7.4}$$

or, using the $[\{\ \}]$ notation introduced in (10.1.45), for the two statistics:

$$\begin{aligned} \delta\psi(\boldsymbol{r},t) &= \int (\mathrm{d}\boldsymbol{r}')\, [\{\psi(\boldsymbol{r},t), \psi(\boldsymbol{r}',t)^\dagger\}]\, \delta\psi(\boldsymbol{r}',t)\,,\\ 0 &= \int (\mathrm{d}\boldsymbol{r}')\, [\{\psi(\boldsymbol{r},t)^\dagger, \psi(\boldsymbol{r}',t)^\dagger\}]\, \delta\psi(\boldsymbol{r}',t)\,, \end{aligned} \tag{10.7.5}$$

which yields the commutation relations

$$\begin{aligned} &[\{\psi(\boldsymbol{r},t), \psi(\boldsymbol{r}',t)^\dagger\}] = \delta(\boldsymbol{r}-\boldsymbol{r}')\,,\\ &[\{\psi(\boldsymbol{r},t)^\dagger, \psi(\boldsymbol{r}',t)^\dagger\}] = 0\,, \qquad [\{\psi(\boldsymbol{r},t), \psi(\boldsymbol{r}',t)\}] = 0\,, \end{aligned} \tag{10.7.6}$$

the continuous analogs of (10.1.46).

As for the Hamilton operator H, here is a one-particle term:

$$H^{(1)} = \int (\mathrm{d}\boldsymbol{r}')(\mathrm{d}\boldsymbol{r}'')\, \psi(\boldsymbol{r}',t)^\dagger \langle\boldsymbol{r}'|h(\boldsymbol{r},\boldsymbol{p},t)|\boldsymbol{r}''\rangle\psi(\boldsymbol{r}'',t)\,, \tag{10.7.7}$$

or with

$$\langle \boldsymbol{r}'|h(\boldsymbol{r},\boldsymbol{p},t)|\boldsymbol{r}''\rangle = h\Big(\boldsymbol{r}',\frac{\hbar}{\mathrm{i}}\boldsymbol{\nabla}',t\Big)\delta(\boldsymbol{r}'-\boldsymbol{r}'')\,, \tag{10.7.8}$$

somewhat more simply

$$H^{(1)} = \int (\mathrm{d}\boldsymbol{r})\,\psi(\boldsymbol{r},t)^\dagger h\Big(\boldsymbol{r},\frac{\hbar}{\mathrm{i}}\boldsymbol{\nabla},t\Big)\psi(\boldsymbol{r},t)\,. \tag{10.7.9}$$

A two-particle term (the model is potential energy pairs) is

$$\begin{aligned} H^{(2)} = \frac{1}{2}\int (\mathrm{d}\boldsymbol{r}')\cdots(\mathrm{d}\boldsymbol{r}^{\prime\mathrm{v}})\,&\psi(\boldsymbol{r}',t)^\dagger\psi(\boldsymbol{r}''',t)^\dagger \\ &\times\langle \boldsymbol{r}',\boldsymbol{r}'''|v(\boldsymbol{r}_1-\boldsymbol{r}_2)|\boldsymbol{r}'',\boldsymbol{r}^{\prime\mathrm{v}}\rangle \\ &\times\psi(\boldsymbol{r}^{\prime\mathrm{v}},t)\psi(\boldsymbol{r}'',t)\,, \end{aligned} \tag{10.7.10}$$

or with

$$\langle \boldsymbol{r}',\boldsymbol{r}'''|v(\boldsymbol{r}_1-\boldsymbol{r}_2)|\boldsymbol{r}'',\boldsymbol{r}^{\prime\mathrm{v}}\rangle = \delta(\boldsymbol{r}'-\boldsymbol{r}'')\delta(\boldsymbol{r}'''-\boldsymbol{r}^{\prime\mathrm{v}})v(\boldsymbol{r}'-\boldsymbol{r}''')\,, \tag{10.7.11}$$

more compactly

$$H^{(2)} = \frac{1}{2}\int (\mathrm{d}\boldsymbol{r})(\mathrm{d}\boldsymbol{r}')\,\psi(\boldsymbol{r},t)^\dagger\psi(\boldsymbol{r}',t)^\dagger v(\boldsymbol{r}-\boldsymbol{r}')\psi(\boldsymbol{r}',t)\psi(\boldsymbol{r},t)\,, \tag{10.7.12}$$

and the symmetry $v(\boldsymbol{r}-\boldsymbol{r}') = v(\boldsymbol{r}'-\boldsymbol{r})$ replaces $f_{kl} = f_{kl}$ of (10.2.23). The action principle gives (omitting the δt contribution)

$$\delta H = \mathrm{i}\hbar\int (\mathrm{d}\boldsymbol{r})\left[\delta\psi(\boldsymbol{r},t)^\dagger\frac{\partial\psi(\boldsymbol{r},t)}{\partial t} - \frac{\partial\psi(\boldsymbol{r},t)^\dagger}{\partial t}\delta\psi(\boldsymbol{r},t)\right]\,. \tag{10.7.13}$$

Define, in

$$\delta H(\psi^\dagger,\psi) = \int (\mathrm{d}\boldsymbol{r})\left[\delta\psi(\boldsymbol{r},t)^\dagger\frac{\delta_{\mathrm{l}}H}{\delta\psi(\boldsymbol{r},t)^\dagger} + \frac{\delta_{\mathrm{r}}H}{\delta\psi(\boldsymbol{r},t)}\delta\psi(\boldsymbol{r},t)\right] \tag{10.7.14}$$

the left functional derivative, with respect to $\psi^\dagger$, and the right functional derivative, with respect to ψ. That gives the equations of motion in the form

$$\begin{aligned} \mathrm{i}\hbar\frac{\partial}{\partial t}\psi(\boldsymbol{r},t) &= \frac{\delta_{\mathrm{l}}H}{\delta\psi(\boldsymbol{r},t)^\dagger}\,, \\ -\mathrm{i}\hbar\frac{\partial}{\partial t}\psi(\boldsymbol{r},t)^\dagger &= \frac{\delta_{\mathrm{r}}H}{\delta\psi(\boldsymbol{r},t)}\,, \end{aligned} \tag{10.7.15}$$

the continuous analogs of (10.6.21). For $H = H^{(1)} + H^{(2)}$, it is a matter of inspection that

$$\begin{aligned} \frac{\delta_{\mathrm{l}}H}{\delta\psi(\boldsymbol{r},t)^\dagger} &= h\Big(\boldsymbol{r},\frac{\hbar}{\mathrm{i}}\boldsymbol{\nabla},t\Big)\psi(\boldsymbol{r},t) \\ &\quad+\int (\mathrm{d}\boldsymbol{r}')\,\psi(\boldsymbol{r}',t)^\dagger v(\boldsymbol{r}-\boldsymbol{r}')\psi(\boldsymbol{r}',t)\,\psi(\boldsymbol{r},t)\,, \end{aligned} \tag{10.7.16}$$

giving the ψ equations of motion

$$i\hbar\frac{\partial}{\partial t}\psi(\boldsymbol{r},t) = \left[h\left(\boldsymbol{r},\frac{\hbar}{i}\boldsymbol{\nabla},t\right) + \int (\mathrm{d}\boldsymbol{r}')\,\psi(\boldsymbol{r}',t)^{\dagger}v(\boldsymbol{r}-\boldsymbol{r}')\psi(\boldsymbol{r}',t)\right]\psi(\boldsymbol{r},t)\,. \tag{10.7.17}$$

10.8 Non-interacting particles

Let's look first at the simplest problem: $v = 0$, $h = \boldsymbol{p}^2/(2M)$, a collection of non-interacting particles:

$$i\hbar\frac{\partial}{\partial t}\psi(\boldsymbol{r},t) = \frac{1}{2M}\left(\frac{\hbar}{i}\boldsymbol{\nabla}\right)^2\psi(\boldsymbol{r},t)\,. \tag{10.8.1}$$

What is the relation between $\psi(\boldsymbol{r},t_1)$ and $\psi(\boldsymbol{r}',t_2)$? We know all about that; the fact that ψ is now an operator and not a numerical wave function changes nothing. So,

$$\psi(\boldsymbol{r},t_1) = \int (\mathrm{d}\boldsymbol{r}')\,\langle\boldsymbol{r},t_1|\boldsymbol{r}',t_2\rangle\psi(\boldsymbol{r}',t_2)\,, \tag{10.8.2}$$

where

$$\langle\boldsymbol{r},t_1|\boldsymbol{r}',t_2\rangle = \int\frac{(\mathrm{d}\boldsymbol{p})}{(2\pi\hbar)^3}\,\mathrm{e}^{\frac{i}{\hbar}\boldsymbol{p}\cdot(\boldsymbol{r}-\boldsymbol{r}')}\,\mathrm{e}^{-\frac{i}{\hbar}E_{\boldsymbol{p}}(t_1-t_2)} \tag{10.8.3}$$

with

$$E_{\boldsymbol{p}} = \frac{\boldsymbol{p}^2}{2M} \tag{10.8.4}$$

is the time transformation function of (5.4.14).

We want to study the dependence of the time transformation function $\langle\psi^{\dagger\prime},t_1|\psi'',t_2\rangle$ on the quantum numbers $\psi^{\dagger\prime}$. Recall (10.7.3),

$$G_{\psi} = i\hbar\int(\mathrm{d}\boldsymbol{r})\,\psi(\boldsymbol{r},t)^{\dagger}\delta\psi(\boldsymbol{r},t)\,. \tag{10.8.5}$$

Then $G_{\psi^\dagger}$ is produced as

$$\begin{aligned} G_{\psi^\dagger} &= G_{\psi} - \delta\left(i\hbar\int(\mathrm{d}\boldsymbol{r})\,\psi(\boldsymbol{r},t)^{\dagger}\psi(\boldsymbol{r},t)\right) \\ &= -i\hbar\int(\mathrm{d}\boldsymbol{r})\,\delta\psi(\boldsymbol{r},t)^{\dagger}\psi(\boldsymbol{r},t)\,. \end{aligned} \tag{10.8.6}$$

So

$$\delta_{\psi^{\dagger\prime}}\langle\psi^{\dagger\prime},t_1|\psi'',t_2\rangle = \frac{\mathrm{i}}{\hbar}\langle\psi^{\dagger\prime}|G_{\psi^\dagger}(t_1)|\psi'',t_2\rangle$$
$$= \langle\psi^{\dagger\prime},t_1\Big|\int(\mathrm{d}\boldsymbol{r})\,\delta\psi^\dagger(\boldsymbol{r})'\psi(\boldsymbol{r},t_1)\Big|\psi'',t_2\rangle \qquad (10.8.7)$$

where, using (10.8.2),

$$\psi(\boldsymbol{r},t_1)|\psi'',t_2\rangle \to |\psi'',t_2\rangle\int(\mathrm{d}\boldsymbol{r}')\,\langle\boldsymbol{r},t_1|\boldsymbol{r}',t_2\rangle\psi(\boldsymbol{r}')'' \qquad (10.8.8)$$

gives

$$\delta_{\psi^{\dagger\prime}}\log\langle\psi^{\dagger\prime},t_1|\psi'',t_2\rangle = \delta_{\psi^\dagger}\left(\int(\mathrm{d}\boldsymbol{r})(\mathrm{d}\boldsymbol{r}')\,\psi(\boldsymbol{r})^{\dagger\prime}\langle\boldsymbol{r},t_1|\boldsymbol{r}',t_2\rangle\psi(\boldsymbol{r}')''\right) \qquad (10.8.9)$$

and then

$$\langle\psi^{\dagger\prime},t_1|\psi'',t_2\rangle = \mathrm{e}^{\int(\mathrm{d}\boldsymbol{r})(\mathrm{d}\boldsymbol{r}')\,\psi(\boldsymbol{r})^{\dagger\prime}\langle\boldsymbol{r},t_1|\boldsymbol{r}',t_2\rangle\psi(\boldsymbol{r}')''}\,. \qquad (10.8.10)$$

Again, this is properly normalized because $\langle 0,t_1|$ is the state of no particles, the vacuum, which stays the vacuum.

To draw the physical consequences of this expression it helps to exhibit $\langle\boldsymbol{r},t_1|\boldsymbol{r}',t_2\rangle$ in discrete form, as produced by breaking the $\int(\mathrm{d}\boldsymbol{p})$ integral into a sum over small $(\mathrm{d}\boldsymbol{p})$ cells (we keep the same notation), so now

$$\langle\boldsymbol{r},t_1|\boldsymbol{r}',t_2\rangle = \sum_{\boldsymbol{p}}\frac{(\mathrm{d}\boldsymbol{p})}{(2\pi\hbar)^3}\,\mathrm{e}^{\frac{\mathrm{i}}{\hbar}\boldsymbol{p}\cdot(\boldsymbol{r}-\boldsymbol{r}')}\,\mathrm{e}^{-\frac{\mathrm{i}}{\hbar}E_{\boldsymbol{p}}(t_1-t_2)}$$
$$= \sum_{\boldsymbol{p}}\phi_{\boldsymbol{p}}(\boldsymbol{r})\,\mathrm{e}^{-\frac{\mathrm{i}}{\hbar}E_{\boldsymbol{p}}(t_1-t_2)}\phi_{\boldsymbol{p}}(\boldsymbol{r}')^* \qquad (10.8.11)$$

where

$$\phi_{\boldsymbol{p}}(\boldsymbol{r}) = \sqrt{\frac{(\mathrm{d}\boldsymbol{p})}{(2\pi\hbar)^3}}\,\mathrm{e}^{\frac{\mathrm{i}}{\hbar}\boldsymbol{p}\cdot\boldsymbol{r}}\,. \qquad (10.8.12)$$

Then define

$$\int(\mathrm{d}\boldsymbol{r})\,\psi(\boldsymbol{r})^{\dagger\prime}\phi_{\boldsymbol{p}}(\boldsymbol{r}) = \psi_{\boldsymbol{p}}^{\dagger\prime}\,, \qquad \int(\mathrm{d}\boldsymbol{r}')\,\phi_{\boldsymbol{p}}(\boldsymbol{r}')^*\psi(\boldsymbol{r}')'' = \psi_{\boldsymbol{p}}'' \qquad (10.8.13)$$

and get, with $T = t_1 - t_2$,

$$\begin{aligned}\langle\psi^{\dagger\prime},t_1|\psi'',t_2\rangle &= \mathrm{e}^{\sum_p \psi_p^{\dagger\prime}\,\mathrm{e}^{-\frac{\mathrm{i}}{\hbar}E_pT}\,\psi_p''}\\ &= \prod_p \mathrm{e}^{\psi_p^{\dagger\prime}\,\mathrm{e}^{-\frac{\mathrm{i}}{\hbar}E_pT}\,\psi_p''}\\ &= \prod_p \sum_{n_p=0}^{\infty} \frac{1}{n_p!} \underbrace{\left(\psi_p^{\dagger\prime}\,\mathrm{e}^{-\frac{\mathrm{i}}{\hbar}E_pT}\,\psi_p''\right)^{n_p}}_{=\left(\psi_p^{\dagger\prime}\right)^{n_p}\,\mathrm{e}^{-\frac{\mathrm{i}}{\hbar}n_pE_pT}\,\left(\psi_p''\right)^{n_p}} .\end{aligned} \tag{10.8.14}$$

Consciously thinking of BE statistics, we get

$$\begin{aligned}\langle\psi^{\dagger\prime},t_1|\psi'',t_2\rangle &= \prod_p \sum_{n_p=0}^{\infty} \frac{\left(\psi_p^{\dagger\prime}\right)^{n_p}}{\sqrt{n_p!}}\,\mathrm{e}^{-\frac{\mathrm{i}}{\hbar}n_pE_pT}\,\frac{\left(\psi_p{''}\right)^{n_p}}{\sqrt{n_p!}}\\ &= \sum_{\{n\}} \langle\psi^{\dagger\prime}|\{n\}\rangle\,\mathrm{e}^{-\frac{\mathrm{i}}{\hbar}E(\{n\})T}\,\langle\{n\}|\psi''\rangle ,\end{aligned} \tag{10.8.15}$$

where

$$E(\{n\}) = \sum_p n_pE_p , \qquad n_p = 0,1,2,\ldots \tag{10.8.16}$$

is the energy of the multi-particle state specified by $\{n\}$, and

$$\langle\psi^{\dagger\prime}|\{n\}\rangle = \prod_p \frac{\left(\psi_p^{\dagger\prime}\right)^{n_p}}{\sqrt{n_p!}} , \qquad \langle\{n\}|\psi''\rangle = \prod_p \frac{\left(\psi_p{''}\right)^{n_p}}{\sqrt{n_p!}} \tag{10.8.17}$$

are its wave functions. These are the evident energy eigenvalues and the familiar oscillator wave functions, now for (infinitely) many degrees of freedom.

How does it work out for FD statistics? To this point we have not specified the statistics and taken for granted that one can work with eigenvectors and eigenvalues of the ψ and $\psi^\dagger$ in either situation. But what does, say,

$$\psi_p|\psi'\rangle = |\psi'\rangle\psi_p' \tag{10.8.18}$$

mean for FD statistics? We have

$$\psi_p\psi_{p'}|\psi'\rangle = \psi_p|\psi'\rangle\psi_{p'}' = |\psi'\rangle\psi_p'\psi_{p'}' \tag{10.8.19}$$

and therefore the algebraic properties of the ψ's must be obeyed by the eigenvalues ψ':

$$\{\psi_p', \psi_{p'}'\} = 0 . \tag{10.8.20}$$

The ψ_p', and the $\psi_p^{\dagger\prime}$, are a set of *totally anticommuting* numbers, analogs of the totally commuting numbers of BE statistics. [Totally anticommuting entities were foreshadowed by Grassmann* about 1840.] It is this total anticommutativity, which includes

$$(\psi_p'')^2 = 0 \quad \text{and} \quad (\psi_p^{\dagger\prime})^2 = 0\,, \tag{10.8.21}$$

that assures the FD property: $n_p = 0, 1$.

Now we return to (10.8.14) and note that, for FD statistics, the expansion of the exponential terminates with the linear term,

$$\begin{aligned}\langle \psi^{\dagger\prime}, t_1 | \psi'', t_2 \rangle &= \prod_p \mathrm{e}^{\psi_p^{\dagger\prime}\, \mathrm{e}^{-\frac{\mathrm{i}}{\hbar}E_p T}\, \psi_p''} \\ &= \prod_p \left[1 + \psi_p^{\dagger\prime}\, \mathrm{e}^{-\frac{\mathrm{i}}{\hbar}E_p T}\, \psi_p''\right] \\ &= \sum_{\{n\}} \langle \psi^{\dagger\prime} | \{n\} \rangle\, \mathrm{e}^{-\frac{\mathrm{i}}{\hbar}E(\{n\})T} \langle \{n\} | \psi'' \rangle\,, \end{aligned} \tag{10.8.22}$$

and we see that

$$\langle \psi^{\dagger\prime} | \{n\} \rangle \langle \{n\} | \psi'' \rangle = \prod_{(n_p=1)} \left(\psi_p^{\dagger\prime} \psi_p''\right)\,. \tag{10.8.23}$$

As an example take $n_p = 1$ for modes $1, 3, 7$ and $n_p = 0$ otherwise, then

$$\begin{aligned}\prod_{(n_p=1)} \left(\psi_p^{\dagger\prime} \psi_p''\right) &= \psi_1^{\dagger\prime} \psi_1'' \,\psi_3^{\dagger\prime} \psi_3'' \,\psi_7^{\dagger\prime} \psi_7'' \\ &= (-1)^2 \psi_3^{\dagger\prime} \psi_1^{\dagger\prime} \psi_1'' \,\psi_3'' \,\psi_7^{\dagger\prime} \psi_7'' \\ &= (-1)^{2+4} \psi_7^{\dagger\prime} \psi_3^{\dagger\prime} \psi_1^{\dagger\prime} \psi_1'' \,\psi_3'' \,\psi_7'' \\ &= \left(\psi_7^{\dagger\prime} \psi_3^{\dagger\prime} \psi_1^{\dagger\prime}\right)\left(\psi_1'' \psi_3'' \psi_7''\right)\,, \end{aligned} \tag{10.8.24}$$

which involves an even number of sign changes. In general then, adopting a standard multiplication order: $\prod$, and its reverse $\prod^{\mathrm{T}}$, we have

$$\langle \psi^{\dagger\prime} | \{n\} \rangle = \prod_p^{\mathrm{T}} \left(\psi_p^{\dagger\prime}\right)^{n_p}\,, \qquad \langle \{n\} | \psi'' \rangle = \prod_p (\psi_p'')^{n_p}\,, \tag{10.8.25}$$

where one could, for uniformity, include $1/\sqrt{n_p!}$, which is one.

The consistency of these results with the interpretation of the ψ and $\psi^\dagger$ as creation and annihilation operators can be checked. Note that for the vacuum state

$$\langle \psi^{\dagger\prime} | 0 \rangle = 1\,, \qquad \langle 0 | \psi'' \rangle = 1\,. \tag{10.8.26}$$

*Hermann Günther GRASSMANN (1809–1877)

So, for both statistics,

$$\langle\psi^{\dagger\prime}|\{n\}\rangle = \prod_{p}^{\mathrm{T}} \frac{\left(\psi_p^{\dagger\prime}\right)^{n_p}}{\sqrt{n_p!}}\langle\psi^{\dagger\prime}|0\rangle = \langle\psi^{\dagger\prime}|\prod_{p}^{\mathrm{T}}\frac{\left(\psi_p^{\dagger}\right)^{n_p}}{\sqrt{n_p!}}|0\rangle \tag{10.8.27}$$

and therefore

$$|\{n\}\rangle = \prod_{p}^{\mathrm{T}}\frac{\left(\psi_p^{\dagger}\right)^{n_p}}{\sqrt{n_p!}}|0\rangle \tag{10.8.28}$$

as required by the creation operators significance of the $\psi^{\dagger}$. Similarly

$$\langle\{n\}|\psi''\rangle = \langle 0|\psi''\rangle \prod_{p}\frac{\left(\psi_p''\right)^{n_p}}{\sqrt{n_p!}} = \langle 0|\prod_{p}\frac{\left(\psi_p\right)^{n_p}}{\sqrt{n_p!}}|\psi''\rangle \tag{10.8.29}$$

and

$$\langle\{n\}| = \langle 0|\prod_{p}\frac{\left(\psi_p\right)^{n_p}}{\sqrt{n_p!}}\,, \tag{10.8.30}$$

which left vector is indeed the adjoint of the preceding right vector, inasmuch as $\dagger$ reverses the multiplication order.

The $\phi_{\boldsymbol{p}}(\boldsymbol{r})$ are the eigenfunctions of $h = \boldsymbol{p}^2/(2M)$

$$\frac{1}{2M}\left(\frac{\hbar}{\mathrm{i}}\boldsymbol{\nabla}\right)^2\phi_{\boldsymbol{p}}(\boldsymbol{r}) = E_{\boldsymbol{p}}\phi_{\boldsymbol{p}}(\boldsymbol{r})\,. \tag{10.8.31}$$

Equally well, for any single-particle energy $h(\boldsymbol{r},\boldsymbol{p})$, say $h = \boldsymbol{p}^2/(2M) + V(\boldsymbol{r})$, one can introduce eigenfunctions $\langle\boldsymbol{r}|\underbrace{E,\ldots}_{a}\rangle = \phi_a(\boldsymbol{r})$ in accordance with

$$h\left(\boldsymbol{r},\frac{\hbar}{\mathrm{i}}\boldsymbol{\nabla}\right)\phi_a(\boldsymbol{r}) = E_a\phi_a(\boldsymbol{r}) \tag{10.8.32}$$

such that

$$\begin{aligned}\langle\boldsymbol{r}|\,\mathrm{e}^{-\frac{\mathrm{i}}{\hbar}ht_1}\,\mathrm{e}^{\frac{\mathrm{i}}{\hbar}ht_2}|\boldsymbol{r}'\rangle &= \langle\boldsymbol{r},t_1|\boldsymbol{r}',t_2\rangle \\ &= \sum_a \underbrace{\phi_a(\boldsymbol{r})}_{=\langle\boldsymbol{r}|a\rangle}\,\mathrm{e}^{-\frac{\mathrm{i}}{\hbar}E_a(t_1-t_2)}\,\underbrace{\phi_a(\boldsymbol{r}')^*}_{=\langle a|\boldsymbol{r}'\rangle}\end{aligned} \tag{10.8.33}$$

and all goes as before. In particular, for FD statistics,

$$|\{n\}\rangle = \prod_{(n_a=1)}^{\mathrm{T}}\psi_a^{\dagger}|0\rangle \tag{10.8.34}$$

with

$$\psi_a^\dagger = \int (\mathrm{d}\boldsymbol{r}')\,\psi(\boldsymbol{r}')^\dagger \phi_a(\boldsymbol{r}') \tag{10.8.35}$$

and

$$\psi_a|0\rangle = \int (\mathrm{d}\boldsymbol{r})\,\phi_a(\boldsymbol{r}')^*\,\underbrace{\psi(\boldsymbol{r})|0\rangle}_{=0} = 0\,, \tag{10.8.36}$$

for which the 1-particle state

$$|1_a\rangle = \int (\mathrm{d}\boldsymbol{r}')\,\psi(\boldsymbol{r}')^\dagger \phi_a(\boldsymbol{r}')|0\rangle\,, \tag{10.8.37}$$

and the 2-particle state

$$|1_a, 1_b\rangle = \int (\mathrm{d}r'')\,\psi(\boldsymbol{r}'')^\dagger \phi_b(\boldsymbol{r}'') \int (\mathrm{d}\boldsymbol{r}')\,\psi(\boldsymbol{r}')^\dagger \phi_a(\boldsymbol{r}')|0\rangle \tag{10.8.38}$$

are examples.

What is the effect of $\psi(\boldsymbol{r})$ on these states? First

$$\begin{aligned}\psi(\boldsymbol{r})|1_a\rangle &= \int (\mathrm{d}\boldsymbol{r}')\,\underbrace{\psi(\boldsymbol{r})\psi(\boldsymbol{r}')^\dagger}_{=\delta(\boldsymbol{r}-\boldsymbol{r}')-\psi(\boldsymbol{r}')^\dagger\psi(\boldsymbol{r})\to\delta(\boldsymbol{r}-\boldsymbol{r}')}\,\phi_a(\boldsymbol{r}')|0\rangle \\ &= \phi_a(\boldsymbol{r})|0\rangle \end{aligned} \tag{10.8.39}$$

which is most reasonable: the wave function $\phi_a(\boldsymbol{r})$ represents $\psi(\boldsymbol{r})$ for a 1-particle state. Next

$$\begin{aligned}\psi(\boldsymbol{r})|1_a, 1_b\rangle &= \int (\mathrm{d}r'')\,\underbrace{\psi(\boldsymbol{r})\psi(\boldsymbol{r}'')^\dagger}_{=\delta(\boldsymbol{r}-\boldsymbol{r}'')-\psi(\boldsymbol{r}'')^\dagger\psi(\boldsymbol{r})}\,\phi_b(\boldsymbol{r}'') \\ &\qquad\times \underbrace{\int (\mathrm{d}\boldsymbol{r}')\psi(\boldsymbol{r}')^\dagger \phi_a(\boldsymbol{r}')|0\rangle}_{=|1_a\rangle} \\ &= \phi_b(\boldsymbol{r})|1_a\rangle - \int (\mathrm{d}r'')\psi(\boldsymbol{r}'')^\dagger \phi_b(\boldsymbol{r}'')\,\underbrace{\psi(\boldsymbol{r})|1_a\rangle}_{=|0\rangle\phi_a(\boldsymbol{r})} \\ &= \phi_b(\boldsymbol{r})|1_a\rangle - \phi_a(\boldsymbol{r})|1_b\rangle\,. \end{aligned} \tag{10.8.40}$$

Here are two ways of annihilating one particle, with both sides *antisymmetrical* in the a, b labels. Clearly this is general as illustrated by $\psi(\boldsymbol{r})|1_a, 1_b, 1_c\rangle$ which is antisymmetrical in any pair of indices and unchanged by cyclic (even number) permutations. So

$$\psi(\boldsymbol{r})|1_a,1_b,1_c\rangle = \phi_c(\boldsymbol{r})|1_a,1_b\rangle + \phi_b(\boldsymbol{r})|1_c,1_a\rangle + \phi_a(\boldsymbol{r})|1_b,1_c\rangle\,. \tag{10.8.41}$$

Now try two annihilations:

$$\begin{aligned}\psi(\boldsymbol{r})\psi(\boldsymbol{r}')|1_a,1_b\rangle &= \psi(\boldsymbol{r})\left[\phi_b(\boldsymbol{r}')|1_a\rangle - \phi_a(\boldsymbol{r}')|1_b\rangle\right] \\ &= \left(\phi_a(\boldsymbol{r})\phi_b(\boldsymbol{r}') - \phi_b(\boldsymbol{r})\phi_a(\boldsymbol{r}')\right)|0\rangle\,, \end{aligned}\tag{10.8.42}$$

properly antisymmetrical in a, b and in $\boldsymbol{r}, \boldsymbol{r}'$. Here we see a 2-particle wave function. Similarly for three particles, using cyclic symmetry:

$$\begin{aligned}\psi(\boldsymbol{r})\psi(\boldsymbol{r}')|1_a,1_b,1_c\rangle &= \left(\phi_b(\boldsymbol{r})\phi_c(\boldsymbol{r}') - \phi_c(\boldsymbol{r})\phi_b(\boldsymbol{r}')\right)|1_a\rangle \\ &\quad + \left(\phi_c(\boldsymbol{r})\phi_a(\boldsymbol{r}') - \phi_a(\boldsymbol{r})\phi_c(\boldsymbol{r}')\right)|1_b\rangle \\ &\quad + \left(\phi_a(\boldsymbol{r})\phi_b(\boldsymbol{r}') - \phi_b(\boldsymbol{r})\phi_a(\boldsymbol{r}')\right)|1_c\rangle\,, \end{aligned}\tag{10.8.43}$$

and so forth for states with 4, 5, ... particles.

Problems

10-1a BE statistics: Evaluate

$$\left(\psi(a')\right)^2|\{n\}\rangle\,, \qquad \left(\psi(a')^\dagger\right)^2|\{n\}\rangle\,,$$

and for $a' \neq a''$

$$\psi(a')\psi(a'')|\{n\}\rangle\,, \quad \psi(a')^\dagger\psi(a'')^\dagger|\{n\}\rangle\,, \quad \psi(a')^\dagger\psi(a'')|\{n\}\rangle\,.$$

10-1b FD statistics: Same questions (only $a' \neq a''$, of course).

10-2 FD statistics: Show that, for $a' \neq a''$, the operators

$${}'\psi(a') = (-1)^{N_{>a'}}\psi(a')\,, \quad {}'\psi(a'') = (-1)^{N_{>a''}}\psi(a'')$$

are commutative. Here, the operator $N_{>a'}$, for example, counts the number of particles in all modes after a' in some standard ordering. What is the connection with the construction given in lecture?

10-3 Verify (10.3.8) and (10.3.9).

10-4a Spin s in the rotating magnetic field: What is the probability that, in time T, the transition $m = -s \to m = s$ happens? What is it for $m = s \to m = -s$?

10-4b Same set-up, for integer s. What is the probability that $m = -s \to m = 0$; that $m = s \to m = 0$; that $m = 0 \to m = \pm s$?

10-4c Again, for $s = 1$. Use the information available from Problems 10-4a and 10-4b and find the probability that $m = 0 \to m = 0$ in time T.

10-5 Concerning the spin dynamics of Section 10.5: Consider a magnetic field that changes *slowly* from $\boldsymbol{B}(t < 0) = B_0 \boldsymbol{e}_z$ to $\boldsymbol{B}(t > T) = -B_0 \boldsymbol{e}_z$. The initial state of given s has $m = s$. Find the final state for

$$\boldsymbol{B}(t) = B_0 \frac{T - 2t}{T} \boldsymbol{e}_z$$

and for

$$\boldsymbol{B}(t) = B_0 \big(\boldsymbol{e}_z \cos(\pi t/T) + \boldsymbol{e}_x \sin(\pi t/T)\big) .$$

10-6 For both statistics,

$$G_\psi = \mathrm{i}\hbar \sum_{a'} \psi(a',t)^\dagger \, \delta\psi(a',t) ,$$

$$G_{\psi^\dagger} = \mathrm{i}\hbar \sum_{a'} \delta\psi(a',t)^\dagger \, \psi(a',t) .$$

Use the known FD commutation relations to check that $G_{\psi^\dagger}$ does the expected things.

10-7 Operator F is of degree n in ψ and $\psi^\dagger$. Show that

$$\delta_\psi F = \frac{1}{\mathrm{i}\hbar} [F, G_\psi]$$

for FD systems, implies that

$$n \text{ even:} \quad \big[F, \psi(a',t)^\dagger\big] = \frac{\partial_{\mathrm{r}} F}{\partial\psi(a',t)} = -\frac{\partial_{\mathrm{l}} F}{\partial\psi(a',t)} ,$$

$$n \text{ odd:} \quad \big\{F, \psi(a',t)^\dagger\big\} = \frac{\partial_{\mathrm{r}} F}{\partial\psi(a',t)} = \frac{\partial_{\mathrm{l}} F}{\partial\psi(a',t)} .$$

10-8 Given the Hamiltonian

$$H = \sum_{a',a''} \psi(a',t)^\dagger \langle a'|h|a''\rangle \psi(a'',t)$$
$$+ \frac{1}{2} \sum_{a',\dots,a^{\mathrm{iv}}} \psi(a',t)^\dagger \psi(a''',t)^\dagger \langle a',a'''|v|a'',a^{\mathrm{iv}}\rangle \psi(a^{\mathrm{iv}},t)\psi(a'',t) ,$$

what are the equations of motion of a BE system, of a FD system?

10-9 Use (10.8.2) and the equal-time commutation relations (10.7.6) to find, for non-interacting particles, the commutation relations at unequal times.

11. Many-Electron Atoms

11.1 Hartree–Fock method

We now apply the methods of Chapter 10 to a study of neutral atoms with Z electrons. As basic physical approximations we take into account only the electrostatic nucleus–electron interactions, and we treat the nucleus as an infinitely massive point charge (of strength Ze).

In the Hamilton operator

$$\begin{aligned} H &= H^{(1)} + H^{(2)} \\ &= \int (\mathrm{d}\boldsymbol{r})\, \psi(\boldsymbol{r})^\dagger \left(-\frac{\hbar^2}{2m_{\mathrm{el}}} \boldsymbol{\nabla}^2 - \frac{Ze^2}{r} \right) \psi(\boldsymbol{r}) \\ &\quad + \frac{1}{2} \int (\mathrm{d}\boldsymbol{r})(\mathrm{d}\boldsymbol{r}')\, \psi(\boldsymbol{r})^\dagger \psi(\boldsymbol{r}')^\dagger \frac{e^2}{|\boldsymbol{r}-\boldsymbol{r}'|} \psi(\boldsymbol{r}')\psi(\boldsymbol{r}) \end{aligned} \tag{11.1.1}$$

the one-particle term $H^{(1)}$ represents the kinetic energy of the electrons and the nucleus–electron Coulomb interaction energy, and the two-particle term $H^{(2)}$ is the energy of the electrostatic electron–electron interaction. They are of the general forms (10.7.9) and (10.7.12) with

$$h\!\left(\boldsymbol{r}, \frac{\hbar}{\mathrm{i}}\boldsymbol{\nabla}\right) = -\frac{\hbar^2}{2m_{\mathrm{el}}}\boldsymbol{\nabla}^2 - \frac{Ze^2}{r} \quad \text{and} \quad v(\boldsymbol{r}-\boldsymbol{r}') = \frac{e^2}{|\boldsymbol{r}-\boldsymbol{r}'|}\,, \tag{11.1.2}$$

respectively. All conceivable multi-electron states are described by eigenvectors of the total number operator with eigenvalue Z,

$$\int (\mathrm{d}\boldsymbol{r})\, \psi(\boldsymbol{r})^\dagger \psi(\boldsymbol{r}) |\,\rangle = |\,\rangle Z\,. \tag{11.1.3}$$

We are interested in the ground state, treated approximately, under the assumption that all electrons are occupying *different* modes – electrons are FD particles, fermions. So the approximate ground state is

$$|\mathrm{g}\rangle = \prod_a^{\mathrm{T}} \psi_a^\dagger |0\rangle \tag{11.1.4}$$

where the a's label Z different modes. In general, the expectation value of H in any state is larger than the true ground-state energy E_g,

$$\langle\ |H|\ \rangle = \sum_{E',\ldots} E' \left|\langle E',\ldots|\ \rangle\right|^2 \geq \sum_{E',\ldots} E_{\rm g} \left|\langle E',\ldots|\ \rangle\right|^2 = E_{\rm g}\,. \tag{11.1.5}$$

So the best choice of approximate state is the one that minimizes $\langle H\rangle$. We are, of course, heading toward an application of the Rayleigh–Ritz variational method of Section 6.11.

To begin, we note that

$$\begin{aligned}\langle{\rm g}|\psi(\boldsymbol{r})^\dagger &= \sum_a (\pm)_a \phi_a(\boldsymbol{r})^* \langle{\rm g}-1_a|\,,\\ \psi(\boldsymbol{r}')|{\rm g}\rangle &= \sum_a |{\rm g}-1_a\rangle(\pm)_a\phi_a(\boldsymbol{r}')\end{aligned} \tag{11.1.6}$$

with the sign depending on the conventional order implicit in $\prod^{\rm T}$. So

$$\langle{\rm g}|\psi(\boldsymbol{r})^\dagger\psi(\boldsymbol{r}')|{\rm g}\rangle = \sum_a \phi_a(\boldsymbol{r})^*\phi_a(\boldsymbol{r}')\,, \tag{11.1.7}$$

and therefore the expectation value of the one-particle term is

$$\langle{\rm g}|H^{(1)}|{\rm g}\rangle = \sum_a \int (d\boldsymbol{r})\,\phi_a(\boldsymbol{r})^*\left(-\frac{\hbar^2}{2m_{\rm el}}\boldsymbol{\nabla}^2 - \frac{Ze^2}{r}\right)\phi_a(\boldsymbol{r})\,. \tag{11.1.8}$$

Similarly we have

$$\psi(\boldsymbol{r}')\psi(\boldsymbol{r})|{\rm g}\rangle = \sum_{a<b}|{\rm g}-1_a-1_b\rangle(\pm)_{ab}\left[\phi_b(\boldsymbol{r}')\phi_a(\boldsymbol{r}) - \phi_a(\boldsymbol{r}')\phi_b(\boldsymbol{r})\right] \tag{11.1.9}$$

and then

$$\begin{aligned}\langle{\rm g}|\psi(\boldsymbol{r})^\dagger\psi(\boldsymbol{r}')^\dagger\psi(\boldsymbol{r}')\psi(\boldsymbol{r})|{\rm g}\rangle &= \sum_{a<b}\left[\phi_a(\boldsymbol{r})\phi_b(\boldsymbol{r}') - \phi_b(\boldsymbol{r})\phi_a(\boldsymbol{r}')\right]^*\\ &\quad\times\left[\phi_a(\boldsymbol{r})\phi_b(\boldsymbol{r}') - \phi_b(\boldsymbol{r})\phi_a(\boldsymbol{r}')\right]\end{aligned} \tag{11.1.10}$$

or, with $\sum_{a<b} \to \frac{1}{2}\sum_{a\neq b}$,

$$\begin{aligned}&\langle{\rm g}|\psi(\boldsymbol{r})^\dagger\psi(\boldsymbol{r}')^\dagger\psi(\boldsymbol{r}')\psi(\boldsymbol{r})|{\rm g}\rangle\\ &\qquad= \sum_{a\neq b}\left[\phi_a(\boldsymbol{r})\phi_b(\boldsymbol{r}')\right]^*\left[\phi_a(\boldsymbol{r})\phi_b(\boldsymbol{r}') - \phi_b(\boldsymbol{r})\phi_a(\boldsymbol{r}')\right]\end{aligned} \tag{11.1.11}$$

and so we get

$$\begin{aligned}\langle{\rm g}|H^{(2)}|{\rm g}\rangle = \frac{1}{2}\sum_{a\neq b}&\int(d\boldsymbol{r})(d\boldsymbol{r}')\,\phi_a(\boldsymbol{r})^*\phi_b(\boldsymbol{r}')^*\\ &\times\frac{e^2}{|\boldsymbol{r}-\boldsymbol{r}'|}\left[\phi_a(\boldsymbol{r})\phi_b(\boldsymbol{r}') - \phi_b(\boldsymbol{r})\phi_a(\boldsymbol{r}')\right]\,.\end{aligned} \tag{11.1.12}$$

Since $\phi_a(\boldsymbol{r})\phi_b(\boldsymbol{r}') - \phi_b(\boldsymbol{r})\phi_a(\boldsymbol{r}') = 0$ for $a = b$, the replacement $\sum_{a\neq b} \to \sum_{a,b}$ is permissible; it does not change the value of $\langle \mathrm{g}|H^{(2)}|\mathrm{g}\rangle$. Accordingly, the energy estimate is (remember: it's an upper bound on the true ground-state energy)

$$\begin{aligned} E = &\sum_a \int (\mathrm{d}\boldsymbol{r})\, \phi_a^* \, h \, \phi_a \\ &+ \frac{1}{2}\sum_{a,b} \int (\mathrm{d}\boldsymbol{r})(\mathrm{d}\boldsymbol{r}') \left[\phi_a(\boldsymbol{r})^* \phi_b(\boldsymbol{r}')^* v(\boldsymbol{r}-\boldsymbol{r}')\phi_b(\boldsymbol{r}')\phi_a(\boldsymbol{r})\right. \\ &\qquad\qquad \left. - \phi_a(\boldsymbol{r})^* \phi_b(\boldsymbol{r}')^* v(\boldsymbol{r}-\boldsymbol{r}')\phi_a(\boldsymbol{r}')\phi_b(\boldsymbol{r})\right] \end{aligned} \tag{11.1.13}$$

with h and v of (11.1.2). Of course, the single-electron wave functions are orthonormal,

$$\int (\mathrm{d}\boldsymbol{r})\, \phi_a(\boldsymbol{r})^* \phi_b(\boldsymbol{r}) = \delta_{ab} \,. \tag{11.1.14}$$

Now look for the minimum of E by varying the ϕ_a^* and the ϕ_a:

$$\begin{aligned} \delta E = &\sum_a \int (\mathrm{d}\boldsymbol{r})\, \delta\phi_a(\boldsymbol{r})^* \Big(h\phi_a(\boldsymbol{r}) + \sum_b \int (\mathrm{d}\boldsymbol{r}') \big[\phi_b(\boldsymbol{r}')^* v(\boldsymbol{r}-\boldsymbol{r}')\phi_b(\boldsymbol{r}')\phi_a(\boldsymbol{r}) \\ &\qquad\qquad - \phi_b(\boldsymbol{r}')^* v(\boldsymbol{r}-\boldsymbol{r}')\phi_a(\boldsymbol{r}')\phi_b(\boldsymbol{r})\big] \Big) \\ &+ \{\text{its complex conjugate}\} \,, \end{aligned} \tag{11.1.15}$$

subject to the constraint (11.1.14), that is

$$\int (\mathrm{d}\boldsymbol{r}) \left[\delta\phi_a^* \, \phi_b + \phi_a^* \, \delta\phi_b\right] = 0 \,, \tag{11.1.16}$$

from which we conclude that

$$\begin{aligned} &\left[h + \sum_b \int (\mathrm{d}\boldsymbol{r}')\phi_b(\boldsymbol{r}')^* v(\boldsymbol{r}-\boldsymbol{r}')\phi_b(\boldsymbol{r}') \right] \phi_a(\boldsymbol{r}) \\ &\quad \underbrace{- \sum_b \int (\mathrm{d}\boldsymbol{r}')\, \phi_b(\boldsymbol{r}')^* v(\boldsymbol{r}-\boldsymbol{r}')\phi_a(\boldsymbol{r}')\, \phi_b(\boldsymbol{r})}_{\text{exchange terms}} = \sum_b \mathcal{E}_{a,b}\phi_b(\boldsymbol{r}) \end{aligned} \tag{11.1.17}$$

with energy parameters $\mathcal{E}_{a,b}$ to be determined self-consistently. These are the Hartree*–Fock equations.

Things are simpler if, as an approximation, we omit the so-called exchange terms. Note that, because they involve $\phi_b(\boldsymbol{r}')^*\phi_a(\boldsymbol{r}')$ the a and b wave functions must overlap which limits their contribution, particularly for $Z \gg 1$.

*Douglas Rayner HARTREE (1897–1958)

Then the ϕ_a can be chosen as the eigenfunctions of a single effective potential $V(\boldsymbol{r})$:

$$\left(-\frac{\hbar^2}{2m_{\rm el}}\boldsymbol{\nabla}^2 + V\right)\phi_a = \mathcal{E}_a\phi_a \tag{11.1.18}$$

with

$$V(\boldsymbol{r}) = -\frac{Ze^2}{r} + \sum_b \int (\mathrm{d}\boldsymbol{r}') \, \frac{e^2|\phi_b(\boldsymbol{r}')|^2}{|\boldsymbol{r}-\boldsymbol{r}'|} \; ; \tag{11.1.19}$$

these are called Hartree equations.

Clearly, here we meet the average electron density

$$n(\boldsymbol{r}) = \sum_b |\phi_b(\boldsymbol{r})|^2 \, , \qquad \int (\mathrm{d}\boldsymbol{r}) \, n(\boldsymbol{r}) = \sum_b 1 = Z \; ; \tag{11.1.20}$$

the latter statement repeats (11.1.3) in the present context. The total energy is

$$\begin{aligned} E = \sum_a \int (\mathrm{d}\boldsymbol{r})\phi_a^* \left(-\frac{\hbar^2}{2m_{\rm el}}\boldsymbol{\nabla}^2 - \frac{Ze^2}{r}\right)\phi_a \\ + \frac{e^2}{2}\int (\mathrm{d}\boldsymbol{r})(\mathrm{d}\boldsymbol{r}')\frac{n(\boldsymbol{r})n(\boldsymbol{r}')}{|\boldsymbol{r}-\boldsymbol{r}'|} \, , \end{aligned} \tag{11.1.21}$$

or

$$E = \underbrace{\sum_a \int (\mathrm{d}\boldsymbol{r})\phi_a^* \left(-\frac{\hbar^2}{2m_{\rm el}}\boldsymbol{\nabla}^2 + V\right)\phi_a}_{=\sum_a \mathcal{E}_a} - \frac{1}{2}\int (\mathrm{d}\boldsymbol{r})(\mathrm{d}\boldsymbol{r}')e^2\frac{n(\boldsymbol{r})n(\boldsymbol{r}')}{|\boldsymbol{r}-\boldsymbol{r}'|} \, . \tag{11.1.22}$$

The first term sums the independent-particle energies $\mathcal{E}_a$ and counts the interaction energy twice, and the second, negative, term subtracts the doubly counted interaction energy once.

Here we have looked for the best choice of the ϕ_a. Another procedure is to accept that the ϕ_a are the wave functions of a common effective potential and look for the best choice of V. This can be done including exchange, but we omit it here. So we start with the energy expression (11.1.21) and choose the ϕ_a's such that (11.1.18) holds, which gives

$$E = \sum_a \mathcal{E}_a - \int (\mathrm{d}\boldsymbol{r})\left(V + \frac{Ze^2}{r}\right)n(\boldsymbol{r}) + \frac{e^2}{2}\int (\mathrm{d}\boldsymbol{r})(\mathrm{d}\boldsymbol{r}')\frac{n(\boldsymbol{r})n(\boldsymbol{r}')}{|\boldsymbol{r}-\boldsymbol{r}'|} \, . \tag{11.1.23}$$

To vary V we recall that

$$\delta\mathcal{E}_a = \langle\delta V\rangle = \int(\mathrm{d}\boldsymbol{r})\,\delta V\,|\phi_a|^2 \tag{11.1.24}$$

so that

$$\delta\sum_a \mathcal{E}_a = \int(\mathrm{d}\boldsymbol{r})\,\delta V \sum_a |\phi_a|^2 = \int(\mathrm{d}\boldsymbol{r})\,\delta V\, n\,. \tag{11.1.25}$$

Then

$$\begin{aligned}\delta E = \int(\mathrm{d}\boldsymbol{r})\,\delta V\, n - \int(\mathrm{d}\boldsymbol{r})\left[\delta V\, n + \left(V + \frac{Ze^2}{r}\right)\delta n\right] \\ + \int(\mathrm{d}\boldsymbol{r})\,\delta n\, e^2 \int(\mathrm{d}\boldsymbol{r}')\frac{n(\boldsymbol{r}')}{|\boldsymbol{r}-\boldsymbol{r}'|}\,,\end{aligned} \tag{11.1.26}$$

or

$$\delta E = \int(\mathrm{d}\boldsymbol{r})\,\delta n\left[-V - \frac{Ze^2}{r} + e^2\int(\mathrm{d}\boldsymbol{r}')\,\frac{n(\boldsymbol{r}')}{|\boldsymbol{r}-\boldsymbol{r}'|}\right], \tag{11.1.27}$$

where we recognize that E is actually a functional of the electron density n. In view of the constraint

$$\int(\mathrm{d}\boldsymbol{r})\,\delta n = 0\,, \tag{11.1.28}$$

$\delta E = 0$ then implies that $[\cdots]$ in (11.1.27) is constant. Now, since $V(\boldsymbol{r}) \to V(\boldsymbol{r}) + V_0$ requires $\mathcal{E}_a \to \mathcal{E}_a + V_0$ for consistency and does not lead to a change of the energy (11.1.23), we can agree on the natural convention that $V(\boldsymbol{r}) \to 0$ as $r \to \infty$, and then this constant vanishes. Therefore, the best choice for V is such that

$$V(\boldsymbol{r}) = -\frac{Ze^2}{r} + e^2\int(\mathrm{d}\boldsymbol{r}')\frac{n(\boldsymbol{r}')}{|\boldsymbol{r}-\boldsymbol{r}'|}\,, \tag{11.1.29}$$

which is (11.1.19), indeed. The set of equations (11.1.18), (11.1.20), and (11.1.29) must be solved jointly, and iterative methods suggest themselves for this purpose.

Before continuing notice the implication of Poisson's equation:

$$-\boldsymbol{\nabla}^2\left(V + \frac{Ze^2}{r}\right) = 4\pi e^2 n\,, \tag{11.1.30}$$

namely that

$$-\int (\mathrm{d}\boldsymbol{r}) \left(V + \frac{Ze^2}{r}\right) n + \frac{1}{2}\int (\mathrm{d}\boldsymbol{r})\, n(\boldsymbol{r}) \underbrace{\int (\mathrm{d}\boldsymbol{r}')\, e^2 \frac{n(\boldsymbol{r}')}{|\boldsymbol{r}-\boldsymbol{r}'|}}_{=V+Ze^2/r}$$

$$= -\frac{1}{2}\int (\mathrm{d}\boldsymbol{r}) \left(V + \frac{Ze^2}{r}\right) \frac{1}{4\pi e^2} (-\boldsymbol{\nabla}^2) \left(V + \frac{Ze^2}{r}\right)$$

$$= -\frac{1}{8\pi e^2}\int (\mathrm{d}\boldsymbol{r}) \left[\boldsymbol{\nabla}\left(V + \frac{Ze^2}{r}\right)\right]^2 , \qquad (11.1.31)$$

which is just the negative electrostatic energy of the electrons.

Indeed

$$E = \sum_a \mathcal{E}_a - \frac{1}{8\pi e^2}\int (\mathrm{d}\boldsymbol{r}) \left[\boldsymbol{\nabla}\left(V + \frac{Ze^2}{r}\right)\right]^2 , \qquad (11.1.32)$$

which is a functional of the effective potential V, gives back the condition to determine V:

$$\delta E = \int (\mathrm{d}\boldsymbol{r})\, \delta V\, n - \frac{1}{4\pi e^2}\int (\mathrm{d}\boldsymbol{r})\, \boldsymbol{\nabla}\delta V \cdot \boldsymbol{\nabla}\left(V + \frac{Ze^2}{r}\right)$$

$$= \int (\mathrm{d}\boldsymbol{r})\, \delta V \left[n + \frac{1}{4\pi e^2}\boldsymbol{\nabla}^2 \left(V + \frac{Ze^2}{r}\right)\right] = 0 , \qquad (11.1.33)$$

in the form of Poisson's equation.

11.2 Semiclassical treatment: Thomas–Fermi model

Hartree's program involves making an initial choice of $V(\boldsymbol{r})$, solving Schrödinger's equation to find the wave functions, then constructing the electron density, to find a new $V(\boldsymbol{r})$, and so on. Noting that the effective potential is spherically symmetric, $V(\boldsymbol{r}) = V(r)$, since nothing distinguishes one spatial direction from the others, we can give an approximate version of this by using the WKB approximation of Problem 8-9a

$$\frac{1}{\pi\hbar}\int_{r_1}^{r_2} \mathrm{d}r \sqrt{2m_{\mathrm{el}}[\mathcal{E} - V(r)] - \frac{\hbar^2 \left(l+\frac{1}{2}\right)^2}{r^2}} = n_r + \frac{1}{2} , \qquad (11.2.1)$$

in which the integration is over the classically allowed region where the square root has a positive argument. Individual energies are labeled $\mathcal{E}_{n_r+\frac{1}{2},l+\frac{1}{2}}$ and we get the Z electrons by filling up the energy states from the bottom, up to the energy $-\zeta$:

$$Z = \sum_a \eta(-\mathcal{E}_a - \zeta) \qquad (11.2.2)$$

with

$$\sum_a \cdots = 2\sum_{n_r=0}^{\infty}\sum_{l=0}^{\infty}(2l+1)\cdots\,, \tag{11.2.3}$$

where $2l+1$ is the multiplicity of orbital angular momentum l, and the factor of 2 is the spin multiplicity. Accordingly, it is convenient to write E as

$$E = \underbrace{\sum_a (\mathcal{E}_a+\zeta)\eta(-\mathcal{E}_a-\zeta)}_{\equiv E_1} - \zeta Z - \frac{1}{8\pi e^2}\int(\mathrm{d}\boldsymbol{r})\left[\boldsymbol{\nabla}\left(V+\frac{Ze^2}{r}\right)\right]^2\,, \tag{11.2.4}$$

where $E_1-\zeta Z$ is the sum term of (11.1.32). The purpose of the step function $\eta(-\mathcal{E}_a-\zeta)$ is, of course, to select the occupied states, the ones with $\mathcal{E}_a \le -\zeta$, the actual value of ζ being determined by the count of states (11.2.2).

Inasmuch as we do not know $V(r)$, yet, we can hardly go at this directly, that is: beginning with V, find the $\mathcal{E}_{n_r+\frac12,l+\frac12}$ and work out E. As a step toward an approximate way of finding V, we replace these summations by equivalent integrations. First recall that

$$\frac{1}{2\pi}\sum_{m=-\infty}^{\infty}\mathrm{e}^{im(\phi-\phi')} = \delta(\phi-\phi') \quad\text{for}\quad -\pi<\phi,\phi'<\pi\,, \tag{11.2.5}$$

which is generalized to ($\phi-\phi'\to 2\pi x$, $-\infty<x<\infty$)

$$\sum_{m=-\infty}^{\infty}\mathrm{e}^{2\pi imx} = \sum_{p=-\infty}^{\infty}\delta(x-p)\,, \tag{11.2.6}$$

both sides being periodic in x with period 1. We use two examples of this so-called Poisson sum formula

$$\sum_{k=-\infty}^{\infty}\mathrm{e}^{2\pi ik(\lambda-\frac12)} = \sum_{l=-\infty}^{\infty}\delta\left(\lambda-\tfrac12-l\right)\,, \tag{11.2.7}$$

$$\sum_{j=-\infty}^{\infty}\mathrm{e}^{2\pi ij(\nu-\frac12)} = \sum_{n_r=-\infty}^{\infty}\delta\left(\nu-\tfrac12-n_r\right)\,. \tag{11.2.8}$$

Then we can write

$$\begin{aligned}
&\sum_{n_r,l=0}^{\infty} f\left(n_r+\tfrac12, l+\tfrac12\right)\\
&\quad= \int_0^{\infty}\mathrm{d}\nu\,\mathrm{d}\lambda\, f(\nu,\lambda)\sum_{n_r,l=-\infty}^{\infty}\delta\left(\nu-\tfrac12-n_r\right)\delta\left(\lambda-\tfrac12-l\right)\\
&\quad= \int_0^{\infty}\mathrm{d}\nu\,\mathrm{d}\lambda\, f(\nu,\lambda)\sum_{j,k=-\infty}^{\infty}\mathrm{e}^{2\pi i[j(\nu-\frac12)+k(\lambda-\frac12)]}\,.
\end{aligned} \tag{11.2.9}$$

In particular, for the contribution E_1 in (11.2.4),

$$E_1 = 2 \sum_{n_r,l=0}^{\infty} 2\left(l+\tfrac{1}{2}\right)\left(\mathcal{E}_{n_r+\frac{1}{2},l+\frac{1}{2}}+\zeta\right)\eta\left(-\mathcal{E}_{n_r+\frac{1}{2},l+\frac{1}{2}}-\zeta\right)$$
$$= 4\int_0^\infty \mathrm{d}\nu\,\mathrm{d}\lambda\,\lambda(\mathcal{E}_{\nu,\lambda}+\zeta)\eta(-\mathcal{E}_{\nu,\lambda}-\zeta) \sum_{j,k=-\infty}^{\infty} \mathrm{e}^{2\pi\mathrm{i}\,[j\,(\nu-\frac{1}{2})+k\,(\lambda-\frac{1}{2})]} \,. \tag{11.2.10}$$

All the terms here are integrals of oscillatory functions, *except* $j = k = 0$. We pick this out and for historical reasons call this highly semiclassical approximation the Thomas*–Fermi (TF) approximation,

$$E_1^{\mathrm{TF}} = 4\int_0^\infty \mathrm{d}\nu\,\lambda\mathrm{d}\lambda\,(\mathcal{E}_{\nu,\lambda}+\zeta)\eta(-\mathcal{E}_{\nu,\lambda}-\zeta) \tag{11.2.11}$$

where, this is (11.2.1),

$$\nu = \frac{1}{\pi}\int_{r_1}^{r_2} \mathrm{d}r\,\sqrt{\frac{2m_{\mathrm{el}}}{\hbar^2}(\mathcal{E}_{\nu,\lambda}-V)-\frac{\lambda^2}{r^2}} \,. \tag{11.2.12}$$

In essence, the sum over discrete quantum numbers n_r, l in (11.2.4) is approximated by an integral over continuous quantum numbers ν, λ.

As written, (11.2.12) gives an implicit definition of $\mathcal{E}_{\nu,\lambda}$. It is more convenient to read it as function $\nu(\mathcal{E},\lambda)$,

$$\nu(\mathcal{E},\lambda) = \frac{1}{\pi}\int_{r_1}^{r_2} \mathrm{d}r\,\sqrt{\frac{2m_{\mathrm{el}}}{\hbar^2}(\mathcal{E}-V)-\frac{\lambda^2}{r^2}} \,, \tag{11.2.13}$$

and to switch the integration variable from ν to $\mathcal{E}$. This is done with the aid of a partial integration. Note that

$$\mathrm{d}\nu\,(\mathcal{E}+\zeta)\eta(-\mathcal{E}-\zeta) = \mathrm{d}\big[\underbrace{(\mathcal{E}+\zeta)\eta(-\mathcal{E}-\zeta)\nu}_{\substack{\text{vanishes at both}\\ \text{integration limits}}}\big] - \nu\mathrm{d}\big[(\mathcal{E}+\zeta)\eta(-\mathcal{E}-\zeta)\big] \tag{11.2.14}$$

and therefore, with $\mathrm{d}\big[(\mathcal{E}+\zeta)\eta(-\mathcal{E}-\zeta)\big] = \mathrm{d}\mathcal{E}\,\eta(-\mathcal{E}-\zeta)$,

$$\int_0^\infty \mathrm{d}\nu\,(\mathcal{E}+\zeta)\eta(-\mathcal{E}-\zeta) = -\int_{-\infty}^{-\zeta} \mathrm{d}\mathcal{E}\,\nu(\mathcal{E},\lambda) \,, \tag{11.2.15}$$

which uses that $\nu = 0$ obtains for $\mathcal{E} \to -\infty$, so that

$$E_1^{\mathrm{TF}} = -4\int_0^\infty \mathrm{d}\lambda\,\lambda\int_{-\infty}^{-\zeta} \mathrm{d}\mathcal{E}\,\nu(\mathcal{E},\lambda) \,. \tag{11.2.16}$$

*Llewellyn Hilleth THOMAS (1903–1992)

In view of

$$\begin{aligned}\lambda\nu(\mathcal{E},\lambda) &= \lambda\frac{\partial}{\partial\mathcal{E}}\frac{1}{\pi}\int \mathrm{d}r\,\frac{2}{3}\frac{\hbar^2}{2m_{\mathrm{el}}}\left(\frac{2m_{\mathrm{el}}}{\hbar^2}[\mathcal{E}-V(r)]-\frac{\lambda^2}{r^2}\right)^{\frac{3}{2}}\\ &= \frac{\partial}{\partial\lambda}\frac{\partial}{\partial\mathcal{E}}\frac{1}{4\pi^2}\int \underbrace{4\pi\mathrm{d}r\,r^2}_{\to(\mathrm{d}\boldsymbol{r})}\left(-\frac{2}{15}\right)\frac{\hbar^2}{2m_{\mathrm{el}}}\left(\frac{2m_{\mathrm{el}}}{\hbar^2}[\mathcal{E}-V(r)]-\frac{\lambda^2}{r^2}\right)^{\frac{5}{2}}\end{aligned} \tag{11.2.17}$$

we then get

$$E_1^{\mathrm{TF}} = -\frac{1}{15\pi^2\,m_{\mathrm{el}}\hbar^3}\int(\mathrm{d}\boldsymbol{r})\left(2m_{\mathrm{el}}[-V(r)-\zeta]\right)^{\frac{5}{2}}. \tag{11.2.18}$$

Recalling (11.2.2) and the definition of E_1 in (11.2.4), we have

$$\frac{\partial E_1}{\partial\zeta} = Z \tag{11.2.19}$$

in general, and in the particular TF context

$$Z = \frac{1}{3\pi^2\hbar^3}\int(\mathrm{d}\boldsymbol{r})\left[2m_{\mathrm{el}}(-V-\zeta)\right]^{\frac{3}{2}}. \tag{11.2.20}$$

Next we look at variations of V as in (11.1.25), here:

$$\int(\mathrm{d}\boldsymbol{r})\,\delta V\,n = \delta_V\sum_a \mathcal{E}_a = \delta_V\,(E_1-\zeta Z) = \delta_V E_1\Big|_{\zeta\ \text{constant}} \tag{11.2.21}$$

since the induced changes of ζ do not contribute,

$$\delta_\zeta(E_1-\zeta Z) = 0\,. \tag{11.2.22}$$

So, the TF approximation for the electron density is

$$n(\boldsymbol{r}) = \frac{1}{3\pi^2\hbar^3}\left[2m_{\mathrm{el}}(-V(\boldsymbol{r})-\zeta)\right]^{\frac{3}{2}} \tag{11.2.23}$$

which is consistent with (11.2.20),

$$\int(\mathrm{d}\boldsymbol{r})\,n(\boldsymbol{r}) = Z\,. \tag{11.2.24}$$

Now Poisson's equation (11.1.30) says that

$$-\boldsymbol{\nabla}^2\left(V+\frac{Ze^2}{r}\right) = 4\pi e^2\frac{1}{3\pi^2\hbar^3}\left[2m_{\mathrm{el}}(-V-\zeta)\right]^{\frac{3}{2}}, \tag{11.2.25}$$

a non-linear differential equation to find $V(r)$, subject to the boundary conditions

$$rV \to -Ze^2 \quad \text{for} \quad r \to 0 \quad \text{and} \quad V \to 0 \quad \text{for} \quad r \to \infty\,. \tag{11.2.26}$$

We introduce an auxiliary function f by

$$-V - \zeta = \frac{Ze^2}{r} f \quad \text{with} \quad f \to 1 \quad \text{as} \quad r \to 0 \tag{11.2.27}$$

and note [recall the Laplacian differential operator in spherical coordinates; cf. (7.5.6)]

$$\boldsymbol{\nabla}^2 \left(-V - \zeta - \frac{Ze^2}{r} \right) = \boldsymbol{\nabla}^2 \left(\frac{Ze^2}{r}(f-1) \right) = \frac{Ze^2}{r} \frac{\mathrm{d}^2}{\mathrm{d}r^2} f\,. \tag{11.2.28}$$

Then (11.2.25) appears as

$$\frac{\mathrm{d}^2 f}{\mathrm{d}r^2} = \frac{4\pi}{Z} \frac{r}{3\pi^2\hbar^3} \left(\frac{2m_{\mathrm{el}} Ze^2}{r} \right)^{\frac{3}{2}} f^{\frac{3}{2}} = \frac{4}{3\pi} \left(\frac{Z}{r} \right)^{\frac{1}{2}} \left(\frac{2f}{a_0} \right)^{\frac{3}{2}}\,, \tag{11.2.29}$$

[$a_0 = \hbar^2/(m_{\mathrm{el}} e^2)$ is Bohr's radius once more] and with

$$r = aZ^{-\frac{1}{3}} x \tag{11.2.30}$$

this reads

$$\frac{\mathrm{d}^2 f(x)}{\mathrm{d}x^2} = \frac{[f(x)]^{\frac{3}{2}}}{x^{\frac{1}{2}}} \quad \text{with} \quad f(0) = 1 \tag{11.2.31}$$

if

$$a = \frac{1}{2} \left(\frac{3\pi}{4} \right)^{\frac{2}{3}} a_0 = 0.88534\, a_0\,. \tag{11.2.32}$$

The important point of this *TF differential equation* is that it is universal, independent of Z; its solution is the TF function $f(x)$.

The boundary conditions for large x need some discussion (this is for neutral atoms). At the edge of the atom, the electron density must drop to zero, so (11.2.23) requires

$$V + \zeta = 0 \tag{11.2.33}$$

and the total potential must vanish (no net charge):

$$V = 0 \quad \text{and therefore} \quad \zeta = 0\,. \tag{11.2.34}$$

Also, according to Gauss's theorem,

$$Z = \frac{1}{4\pi^2 e^2} 4\pi r^2 \left(-\frac{\mathrm{d}}{\mathrm{d}r} \right) \left(V + \frac{Ze^2}{r} \right) \bigg|_{\mathrm{edge}} = Z - \frac{r^2}{e^2} \frac{\mathrm{d}V}{\mathrm{d}r} \bigg|_{\mathrm{edge}} \tag{11.2.35}$$

so

$$\left.\frac{dV}{dr}\right|_{\text{edge}} = 0\,. \tag{11.2.36}$$

All this says that we must have, 'at the edge',

$$f = 0 \quad \text{and} \quad f' \equiv \frac{df}{dx} = 0\,. \tag{11.2.37}$$

For neutral atoms then, the edge is at $x = \infty$.

We begin at $x = 0$, where $f(0) = 1$, start with some downward slope, $f'(0) < 0$ and proceed according to $f'' = f\sqrt{f/x} > 0$, which says that the function is always curving upward. There are three qualitative possibilities according as $-f'(0) >, =, < B$ where B is the *correct* value of $-f'(0)$. The plot

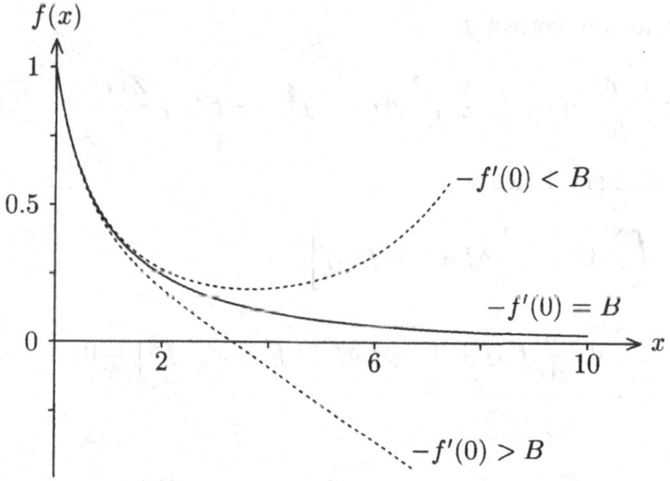

shows that $f(0) = 0$ occurs at finite x, where $-f' \neq 0$, if $-f(0) > B$; and that $f' = 0$ occurs at *finite* x, where $f \neq 0$, if $-f(0) < B$.

Of course, the value of B, the initial slope of the TF function, is found by numerical integration. But we want to a get a feeling for it, as one can by seeing its connection with the all important energy, which is

$$\begin{aligned} E^{\text{TF}} = &-\frac{1}{15\pi^2}\frac{1}{m_{\text{el}}\hbar^3}\int (d\boldsymbol{r})\,[2m_{\text{el}}(-V-\zeta)]^{\frac{5}{2}} - \zeta Z \\ &- \frac{1}{8\pi e^2}\int (d\boldsymbol{r}) \left[\boldsymbol{\nabla}\left(V + \frac{Ze^2}{r}\right)\right]^2 \end{aligned} \tag{11.2.38}$$

in the TF approximation. With $\zeta = 0$ and $V(r) = -(Ze^2/r)f(Z^{\frac{1}{3}}r/a) = -(Z^{\frac{4}{3}}e^2/a)\,f(x)/x$ this is

$$E^{\mathrm{TF}} = -\frac{Z^{\frac{7}{3}}e^2}{a}\left[\frac{1}{2}\int_0^\infty \mathrm{d}x \left(\frac{\mathrm{d}f}{\mathrm{d}x}\right)^2 + \frac{2}{5}\int_0^\infty \mathrm{d}x\, x^{-\frac{1}{2}} f^{\frac{5}{2}}\right] \tag{11.2.39}$$

since

$$\begin{aligned}\frac{aZ^{-\frac{7}{3}}}{4\pi e^2}\int(\mathrm{d}\boldsymbol{r})\left[\boldsymbol{\nabla}\left(V + \frac{Ze^2}{r}\right)\right]^2 &= \int_0^\infty \mathrm{d}x\, x^2\left(\frac{\mathrm{d}}{\mathrm{d}x}\frac{f-1}{x}\right)^2 \\ &= \int_0^\infty \mathrm{d}x \left[\left(\frac{\mathrm{d}f}{\mathrm{d}x}\right)^2 - \frac{\mathrm{d}}{\mathrm{d}x}\frac{(f-1)^2}{x}\right] \\ &= \int_0^\infty \mathrm{d}x \left(\frac{\mathrm{d}f}{\mathrm{d}x}\right)^2 - \underbrace{\left.\frac{(f-1)^2}{x}\right|_{x=0}^{\infty}}_{=0} .\end{aligned} \tag{11.2.40}$$

Clearly, for the correct f,

$$S \equiv \frac{1}{2}\int_0^\infty \mathrm{d}x\, f'^2 + \frac{2}{5}\int_0^\infty \mathrm{d}x\, x^{-\frac{1}{2}} f^{\frac{5}{2}} = -E^{\mathrm{TF}}\Big/\frac{Z^{\frac{7}{3}}e^2}{a} \tag{11.2.41}$$

must be stationary. Indeed

$$\begin{aligned}\delta S &= \int_0^\infty \mathrm{d}x \left[f'\frac{\mathrm{d}}{\mathrm{d}x}\delta f + x^{-\frac{1}{2}} f^{\frac{3}{2}}\delta f\right] \\ &= \underbrace{\int_0^\infty \mathrm{d}x \frac{\mathrm{d}}{\mathrm{d}x} f'\delta f}_{=f'\delta f|_0^\infty} + \int_0^\infty \mathrm{d}x\, \delta f\left[-f'' + x^{-\frac{1}{2}} f^{\frac{3}{2}}\right] = 0 ,\end{aligned} \tag{11.2.42}$$

when $f'' = f\sqrt{f/x}$ and $f(0) = 1$, so that $f(\infty) = 0$, $\delta f = 0$ at $x = 0$ and $x = \infty$.

We learn more by considering a scale change: $f(x) \to f(\lambda x)$,

$$S(\lambda) = \lambda\frac{1}{2}\int_0^\infty \mathrm{d}x\, f'^2 + \lambda^{-\frac{1}{2}}\frac{2}{5}\int_0^\infty \mathrm{d}x\, x^{-\frac{1}{2}} f^{\frac{5}{2}} \tag{11.2.43}$$

where we must have

$$\left.\frac{\mathrm{d}}{\mathrm{d}\lambda}S(\lambda)\right|_{\lambda=1} = 0 \quad \text{or} \quad \frac{1}{2}\int_0^\infty \mathrm{d}x\, f'^2 - \frac{1}{2}\frac{2}{5}\int_0^\infty \mathrm{d}x\, x^{-\frac{1}{2}} f^{\frac{5}{2}} = 0 , \tag{11.2.44}$$

so

$$\int_0^\infty \mathrm{d}x\, f'^2 = \frac{2}{5}\int_0^\infty \mathrm{d}x\, x^{-\frac{1}{2}} f^{\frac{5}{2}} \qquad \text{for correct } f . \tag{11.2.45}$$

Also

$$\int_0^\infty \mathrm{d}x\, f'^2 = \int_0^\infty \mathrm{d}x \frac{\mathrm{d}}{\mathrm{d}x}(ff') - \int_0^\infty \mathrm{d}x\, ff''$$
$$= \underbrace{-f(0)}_{=B} - \int_0^\infty \mathrm{d}x\, x^{-\frac{1}{2}} f^{\frac{5}{2}} \qquad \text{for correct } f\,. \tag{11.2.46}$$

For the correct f then,

$$\int_0^\infty \mathrm{d}x\, f'^2 = \frac{2}{7}B\,, \qquad \int_0^\infty \mathrm{d}x\, x^{-\frac{1}{2}} f^{\frac{5}{2}} = \frac{5}{7}B \tag{11.2.47}$$

and

$$S = \left(\frac{1}{7} + \frac{2}{7}\right) B = \frac{3}{7}B\,. \tag{11.2.48}$$

This means that

$$-E^{\mathrm{TF}} = Z^{\frac{7}{3}} \frac{e^2}{a} \frac{3}{7} B \tag{11.2.49}$$

where $\frac{3}{7}B$ is the stationary value of S. Before using the latter connection to get an approximate value for B let's see another way of understanding this result. We return to (11.2.39) and, inasmuch as we are, in the TF limit, no longer restricted to integer values of Z, consider $\partial/\partial Z$. Because the functional (11.2.39) is stationary, induced changes of V and ζ do not contribute and only the *explicit* dependence on Z counts:

$$\frac{\partial E^{\mathrm{TF}}}{\partial Z} = -\frac{1}{4\pi}\int (\mathrm{d}\boldsymbol{r}) \boldsymbol{\nabla}\left(V + \frac{Ze^2}{r}\right) \boldsymbol{\nabla}\frac{1}{r} - \zeta$$
$$= \frac{1}{4\pi}\int (\mathrm{d}\boldsymbol{r}) \left(V + \frac{Ze^2}{r}\right) \boldsymbol{\nabla}^2\frac{1}{r} - \zeta \tag{11.2.50}$$

or with

$$\boldsymbol{\nabla}^2 \frac{1}{r} = -4\pi\delta(\boldsymbol{r}) \tag{11.2.51}$$

and $\zeta = 0$

$$\frac{\partial E^{\mathrm{TF}}}{\partial Z} = -\left(V + \frac{Ze^2}{r}\right)(0)\,. \tag{11.2.52}$$

There is a simple way of understanding this: $\partial E/\partial Z$ is the change in E produced by placing an additional unit positive charge at the nucleus. That is the *negative* of the change in E produced by an additional *negative* unit charge at the nucleus. But the latter is just the interaction energy of an electron at $r = 0$ with the rest of the electrons, which is $\left(V + Ze^2/r\right)(0)$.

Now ($\zeta = 0$)

$$V + \frac{Ze^2}{r}\Big|_{r\to 0} = Z^{\frac{4}{3}} \frac{e^2}{a}\frac{1-f(x)}{x}\Big|_{x\to 0} = Z^{\frac{4}{3}}\frac{e^2}{a}B \tag{11.2.53}$$

from which follows

$$\frac{\partial E^{\mathrm{TF}}}{\partial Z} = -Z^{\frac{4}{3}}\frac{e^2}{a}B \tag{11.2.54}$$

and then (11.2.49).

Now let's return to the calculation of B according to (11.2.48) and (11.2.41). In Problem 11-5 we learn that the stationary value of S is a minimum, so that $\frac{3}{7}B \le S$ and the equal sign holds only if $f(x)$ solves the TF equation. All trial functions are, of course, subject to the boundary conditions $f(0) = 1$ and $f(\infty) = 0$.

As a simple example we consider

$$f(x) = \frac{1}{(1+\lambda x)^{\alpha}}\,, \tag{11.2.55}$$

so that, with (11.2.43),

$$\frac{3}{7}B \lesssim \lambda S_1 + \lambda^{-\frac{1}{2}} S_2 \tag{11.2.56}$$

where

$$S_1 = \frac{1}{2}\alpha^2 \int_0^{\infty} \frac{\mathrm{d}x}{(1+x)^{2\alpha+2}} = \frac{1}{2}\frac{\alpha^2}{2\alpha+1} \tag{11.2.57}$$

and [substitute $x = 1/t - 1$ and recognize Euler's beta function integral (9.5.54)]

$$\begin{aligned} S_2 &= \frac{2}{5}\int_0^{\infty} \mathrm{d}x\, x^{-\frac{1}{2}}(1+x)^{-\frac{5}{2}\alpha} = \frac{2}{5}\int_0^1 \mathrm{d}t\, t^{\frac{5}{2}\alpha-\frac{3}{2}}(1-t)^{-\frac{1}{2}} \\ &= \frac{2}{5}\sqrt{\pi}\frac{(\frac{5}{2}\alpha-\frac{3}{2})!}{(\frac{5}{2}\alpha-1)!}\,. \end{aligned} \tag{11.2.58}$$

Upon differentiating (11.2.56) we find that the optimal choice for λ is $\lambda = (\frac{1}{2}S_2/S_1)^{\frac{2}{3}}$ which gives the estimate

$$\frac{3}{7}B \lesssim \frac{3}{2}(2S_1 S_2^2)^{\frac{1}{3}} \tag{11.2.59}$$

or

$$B \lesssim \frac{7}{5}\left(\frac{5}{3}\pi\right)^{\frac{1}{3}}\left[\frac{\alpha}{\sqrt{2\alpha+1}}\frac{(\frac{5}{2}\alpha-\frac{3}{2})!}{(\frac{5}{2}\alpha-1)!}\right]^{\frac{2}{3}}\,. \tag{11.2.60}$$

For $\alpha = 1, \frac{7}{5}, \frac{3}{2}, 2$ this gives $B \lesssim 1.5960, 1.5908, 1.5910, 1.5940$, respectively. Certainly, then $B \leq 1.5908$. Numerical integration gives

$$B = 1.588\,071 \tag{11.2.61}$$

for the (negative) initial slope of the TF function. That is: to three significant figures, $B = 1.59$ is correct, and the variational estimate for $a = \frac{7}{5}$ is less than 0.2% in excess.

Now to the binding energy:

$$-E^{\mathrm{TF}} = \frac{3}{7} B Z^{\frac{7}{3}} \frac{e^2}{a} = \frac{3}{7} \frac{B}{a/a_0} B Z^{\frac{7}{3}} \frac{e^2}{a_0} = 0.7687\, Z^{\frac{7}{3}} \frac{e^2}{a_0}\,, \tag{11.2.62}$$

or measuring the energy as a multiple of the atomic energy unit e^2/a_0,

$$\frac{-E^{\mathrm{TF}}}{\frac{1}{2}Z^2} = 1.537\, Z^{\frac{1}{3}} \qquad \text{(atomic units)}\,. \tag{11.2.63}$$

The division by $\frac{1}{2}Z^2$ is convenient because values then range, over the periodic table, from 1 to about 6. The points (or rather little circles) in

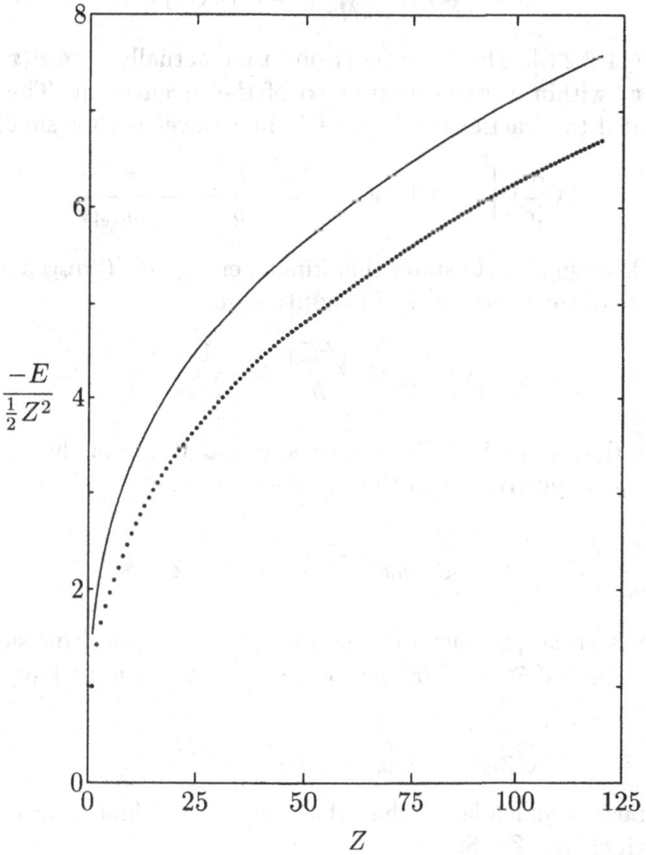

report binding energies (in atomic units) as calculated by Hartree–Fock methods, and in agreement with experiment where that is available (up to about $Z = 20$). This collection of points cries out for a smooth connecting curve, and as a first step toward this goal we have superimposed on the 'experimental' points the smooth curve that corresponds to the TF energy (11.2.63). The general pattern is right, but one may get the impression that the two curves are diverging for large Z; in fact they are not, and more important, the fractional error is decreasing: It begins at about 50% for $Z = 1$, and has dropped to about 15% for $Z = 100$; it is about 25% for $Z = 27$, $Z^{1/3} = 3$.

11.3 Correction for strongly bound electrons

Can we improve E^{TF} without getting involved with the oscillating terms of E_1 in (11.2.10) [because there is no sign of oscillations in the data just plotted]? For that it helps to understand where the TF approximation must break down. The clue is in the local nature of

$$n^{\mathrm{TF}}(\boldsymbol{r}) = \frac{1}{3\pi^2\hbar^3}\left[-2m_{\mathrm{el}}V(\boldsymbol{r})\right]^{\frac{3}{2}} \tag{11.3.1}$$

[$\zeta = 0$ in (11.2.23)]. The wave functions that actually give $n(\boldsymbol{r})$ certainly involve $V(\boldsymbol{r})$ within a wavelength or so of the given point. This would be unimportant if the fractional change of V in a wavelength is small:

$$\left|\lambda\frac{\mathrm{d}}{\mathrm{d}r}V\right| \ll |V| \quad \text{with} \quad \lambda = \frac{\hbar}{p} \cong \frac{\hbar}{\sqrt{-2m_{\mathrm{el}}V}} \tag{11.3.2}$$

[the virial theorem (8.3.10) states that kinetic energy $\boldsymbol{p}^2/(2m_{\mathrm{el}})$ and potential energy V are of the same order of magnitude] or

$$\frac{\mathrm{d}}{\mathrm{d}r}\frac{1}{\sqrt{-V}} \ll \frac{\sqrt{m_{\mathrm{el}}}}{\hbar} = \frac{1}{\sqrt{a_0e^2}}\,. \tag{11.3.3}$$

Observe that at typical TF distances, r and V are of the order $Z^{-\frac{1}{3}}a_0$ and $Z^{\frac{4}{3}}e^2/a_0$, respectively, and this criterion becomes

$$\frac{Z^{\frac{1}{3}}}{a_0}\left(Z^{\frac{4}{3}}\frac{e^2}{a_0}\right)^{-\frac{1}{2}} \ll \left(a_0e^2\right)^{-\frac{1}{2}} \quad \text{or} \quad Z^{\frac{1}{3}} \gg 1\,. \tag{11.3.4}$$

Here is the essential parameter of the TF approximation. Now suppose that $r \ll Z^{-\frac{1}{3}}a_0$; then $V \cong -Ze^2/r$ and for TF validity we must have

$$\frac{1}{\sqrt{Ze^2r}} \ll \frac{1}{\sqrt{a_0e^2}} \quad \text{or} \quad r \gg \frac{a_0}{Z}\,, \tag{11.3.5}$$

that is: r must be much larger than the radius of the first Bohr orbit for the nucleus with charge Ze. So

$$\frac{a_0}{Z} \ll r \ll \frac{a_0}{Z^{\frac{1}{3}}} \tag{11.3.6}$$

which must be considered satisfied for r as small as $Z^{-2/3}a_0$ (assuming $Z^{\frac{1}{3}} \gg 1$, of course) or energies of the order $Ze^2/(Z^{-2/3}a_0) = Z^{\frac{5}{3}}e^2/a_0$ but certainly fails for distances of the order $Z^{-1}a_0$ or (binding) energies of the order Z^2e^2/a_0. Strongly bound electrons, those for which $\mathcal{E} \gg Z^{\frac{5}{3}}e^2/a_0$ are a source of error for the TF approximation.

To correct this we regard E_1 as a function of ζ and write ($\zeta = 0$ eventually)

$$E_1(\zeta) = \left[E_1(\zeta) - E_1(\zeta_s)\right] + E_1(\zeta_s) \tag{11.3.7}$$

with ζ_s of the order $Z^{\frac{5}{3}}e^2/a_0$, thereby separating the contribution from the strongly bound electrons, $E_1(\zeta_s)$, for which the TF treatment is in error, from the rest, $E_1(\zeta) - E_1(\zeta_s)$, whose TF value can be trusted. Since $r \ll a_0/Z^{\frac{1}{3}}$ for the strongly bound electrons, they move in the potential

$$V \cong -\frac{Ze^2}{r}\left(1 - B\frac{r}{a}Z^{\frac{1}{3}}\right) = -\frac{Ze^2}{r} + B\frac{e^2}{a}Z^{\frac{4}{3}}\,, \tag{11.3.8}$$

which is essentially the Coulomb potential of the nuclear charge, and therefore the calculation of $E_1(\zeta_s)$ can be done exactly. Of course, for consistency – contributions of oscillatory terms are omitted from $E_1(\zeta)$ – we want only the non-oscillatory part of $E_1(\zeta_s)$.

First, we find the TF value of $E_1(\zeta) - E_1(\zeta_s)$ as

$$\begin{aligned}
&\left[E_1(\zeta = 0) - E_1(\zeta_s)\right]^{\mathrm{TF}} \\
&\quad = -\frac{1}{15\pi^2}\frac{1}{m_{\mathrm{el}}\hbar^3}\int (\mathrm{d}\boldsymbol{r})\,\left[2m_{\mathrm{el}}(-V)\right]^{\frac{5}{2}} \\
&\quad\quad + \frac{1}{15\pi^2}\frac{1}{m_{\mathrm{el}}\hbar^3}\int (\mathrm{d}\boldsymbol{r})\left[2m_{\mathrm{el}}\left(-\zeta_s - B\frac{e^2}{a}Z^{\frac{4}{3}} + \frac{Ze^2}{r}\right)\right]^{\frac{5}{2}} .
\end{aligned} \tag{11.3.9}$$

The latter integral covers $r \le r_s$ with

$$\zeta_s + B\frac{e^2}{a}Z^{\frac{4}{3}} = \frac{Ze^2}{r_s} \tag{11.3.10}$$

and has the value

$$\begin{aligned}
&\frac{4\pi}{15\pi^2}\frac{1}{m_{\mathrm{el}}\hbar^3}(2m_{\mathrm{el}}Ze^2)^{\frac{5}{2}}\int_0^{r_s}\mathrm{d}r\,r^2\left(\frac{1}{r} - \frac{1}{r_s}\right)^{\frac{5}{2}} \\
&\quad = \frac{32}{15\pi}\frac{Z^2e^2}{a_0}\underbrace{\left(\frac{m_{\mathrm{el}}e^2}{\hbar^2}a_0\right)^{\frac{3}{2}}}_{=1}\left(\frac{Zr_s}{2a_0}\right)^{\frac{1}{2}}\underbrace{\int_0^1\mathrm{d}t\,t^{-\frac{1}{2}}(1-t)^{\frac{5}{2}}}_{=5\pi/16} \\
&\quad = \frac{2}{3}\frac{Z^2e^2}{a_0}\left(\frac{Zr_s}{2a_0}\right)^{\frac{1}{2}} = \frac{2}{3}\frac{Z^2e^2}{a_0}n_s\,,
\end{aligned} \tag{11.3.11}$$

where we introduce an effective continuous quantum number n_s,

$$r_\mathrm{s} = \frac{2a_0}{Z}n_\mathrm{s}^2 \quad \text{or} \quad \zeta_\mathrm{s} + B\frac{e^2}{a}Z^{\frac{4}{3}} = \frac{Z^2e^2}{2a_0n_\mathrm{s}^2} \,. \tag{11.3.12}$$

Then, summing the Bohr energies with multiplicities $2n^2$, $E_1(\zeta_\mathrm{s})$ is

$$E_1(\zeta_\mathrm{s}) = \sum_{n=1}^{\infty} 2n^2 \underbrace{\left(-\frac{Z^2e^2}{2a_0n^2} + B\frac{e^2}{a}Z^{\frac{4}{3}} + \zeta_\mathrm{s}\right)}_{=\frac{Z^2e^2}{2a_0}\left(-\frac{1}{n^2}+\frac{1}{n_\mathrm{s}^2}\right)} \eta\left(\frac{1}{n^2} - \frac{1}{n_\mathrm{s}^2}\right)$$

$$= \frac{Z^2e^2}{a_0}\sum_{n=1}^{\lfloor n_\mathrm{s}\rfloor}\left(-1 + \frac{n^2}{n_\mathrm{s}^2}\right) \tag{11.3.13}$$

where $\lfloor n_\mathrm{s}\rfloor$ is the largest integer that does not exceed n_s. So, the correction for the strongly bound electrons adds

$$\frac{Z^2e^2}{a_0}\left[\frac{2}{3}n_\mathrm{s} + \sum_{n=1}^{\lfloor n_\mathrm{s}\rfloor}\left(\frac{n^2}{n_\mathrm{s}^2} - 1\right)\right] \tag{11.3.14}$$

to the energy, but this still contains oscillatory contributions which we must identify and then discard.

The challenge here is to separate the summations into smooth and oscillatory parts. Begin with the Poisson sum formula (11.2.6) in the form

$$\sum_{n=-\infty}^{\infty} \delta(x-n) = 1 + 2\sum_{m=1}^{\infty}\cos(2\pi m x) \tag{11.3.15}$$

and integrate x from ϵ to n_s with $0 < \epsilon < 1$,

$$\begin{aligned}\int_{\epsilon}^{n_\mathrm{s}} \mathrm{d}x \sum_{n=-\infty}^{\infty}\delta(x-n) &= \sum_{n=1}^{\lfloor n_\mathrm{s}\rfloor} 1 = \lfloor n_\mathrm{s}\rfloor \\ &= n_\mathrm{s} + \frac{1}{\pi}\sum_{m=1}^{\infty}\frac{\sin(2\pi m n_\mathrm{s})}{m} \\ &\quad - \underbrace{\left[\epsilon + \frac{1}{\pi}\sum_{m=1}^{\infty}\frac{\sin(2\pi m \epsilon)}{m}\right]}_{=\frac{1}{2}} \,.\end{aligned} \tag{11.3.16}$$

The left-hand side does not depend on ϵ, and so the ϵ terms on the right-hand side must be constant. Putting $\epsilon = \frac{1}{2}$ gives the value stated. But, wouldn't we get zero for $\epsilon = +0$? No; it is essential to realize that the summation is not zero for infinitesimal *positive* ϵ. Indeed, it has contributions for sufficiently large m that an integration can be used:

$$\frac{1}{\pi}\sum_{m=1}^{\infty}\frac{\sin\bigl(2\pi m(+0)\bigr)}{m} = \frac{1}{\pi}\int^{\infty}\frac{\mathrm{d}m}{m}\sin\bigl(2\pi m(+0)\bigr) = \frac{1}{\pi}\int_0^{\infty}\frac{\mathrm{d}t}{t}\sin t = \frac{1}{2}\,. \tag{11.3.17}$$

So

$$\lfloor n_\mathrm{s}\rfloor = \underbrace{n_\mathrm{s}-\tfrac{1}{2}}_{\text{smooth}} + \underbrace{\frac{1}{\pi}\sum_{m=1}^{\infty}\frac{\sin(2\pi m n_\mathrm{s})}{m}}_{\text{oscillatory}} \tag{11.3.18}$$

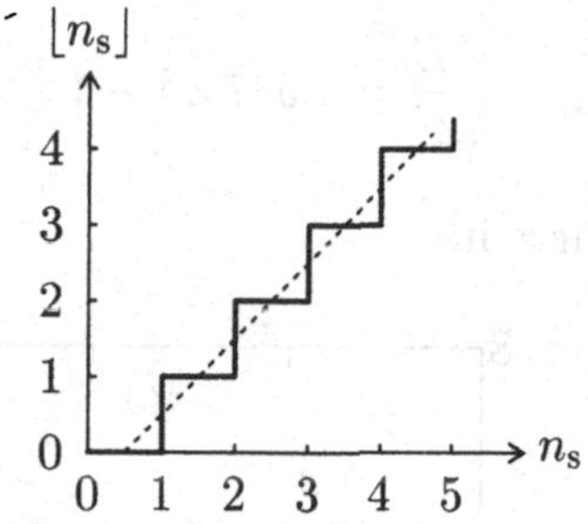

identifies the smooth and oscillatory parts of $\lfloor n_\mathrm{s}\rfloor$.

More generally, we multiply (11.3.15) by x^k, $k = 1, 2, \ldots$ and use successive partial integrations to get

$$\begin{aligned}\sum_{n=1}^{\lfloor n_\mathrm{s}\rfloor} n^k &= \int_{\frac{1}{2}}^{n_\mathrm{s}} \mathrm{d}x\, x^k \sum_{n=-\infty}^{\infty}\delta(x-n)\\ &= \left[\frac{x^{k+1}}{k+1} + 2\mathrm{Re}\sum_{j=0}^{k}(-1)^{k-j}\frac{j!}{k!}x^j\sum_{m=1}^{\infty}\frac{\mathrm{e}^{\mathrm{i}2\pi m x}}{(\mathrm{i}2\pi m)^{1+k-j}}\right]\Bigg|_{x\,=\,\frac{1}{2}}^{n_\mathrm{s}}\,.\end{aligned} \tag{11.3.19}$$

With

$$2\sum_{m=1}^{\infty}\frac{\cos(\pi m)}{(2\pi m)^2} = \frac{1}{2\pi^2}\sum_{m=1}^{\infty}\frac{(-1)^m}{m^2} = -\frac{1}{24} \tag{11.3.20}$$

this gives for $k = 1$

$$\sum_{n=1}^{\lfloor n_\mathrm{s}\rfloor} n = \underbrace{\frac{1}{2}{n_\mathrm{s}}^2 - \frac{1}{12}}_{\text{smooth}} \underbrace{+2\sum_{m=1}^{\infty}\left[n_\mathrm{s}\frac{\sin(2\pi m n_\mathrm{s})}{2\pi m} + \frac{\cos(2\pi m n_\mathrm{s})}{(2\pi m)^2}\right]}_{\text{oscillatory}} \tag{11.3.21}$$

and for $k = 2$

$$\sum_{n=1}^{\lfloor n_\mathrm{s}\rfloor} n^2 = \underbrace{\frac{1}{3}{n_\mathrm{s}}^3}_{\text{smooth}} \underbrace{+2\sum_{m=1}^{\infty}\left[{n_\mathrm{s}}^2\frac{\sin(2\pi m n_\mathrm{s})}{2\pi m} + 2n_\mathrm{s}\frac{\cos(2\pi m n_\mathrm{s})}{(2\pi m)^2} - 2\frac{\sin(2\pi m n_\mathrm{s})}{(2\pi m)^3}\right]}_{\text{oscillatory}}. \tag{11.3.22}$$

Accordingly, the smooth part of (11.3.14) is

$$\frac{Z^2e^2}{a_0}\left[\frac{2}{3}n_s + \frac{\frac{1}{3}{n_s}^3}{{n_s}^2} - \left(n_s - \frac{1}{2}\right)\right] = \frac{1}{2}Z^2\frac{e^2}{a_0}\,. \tag{11.3.23}$$

The net result is independent of n_s, as it should be, and tells us that the correction for strongly bound electrons adds $\frac{1}{2}Z^2$ atomic units to the energy. Therefore (11.2.63) is changed to

$$\frac{-E}{\frac{1}{2}Z^2} = 1.537\,Z^{\frac{1}{3}} - 1 \qquad \text{(atomic units)}\,. \tag{11.3.24}$$

The second curve in

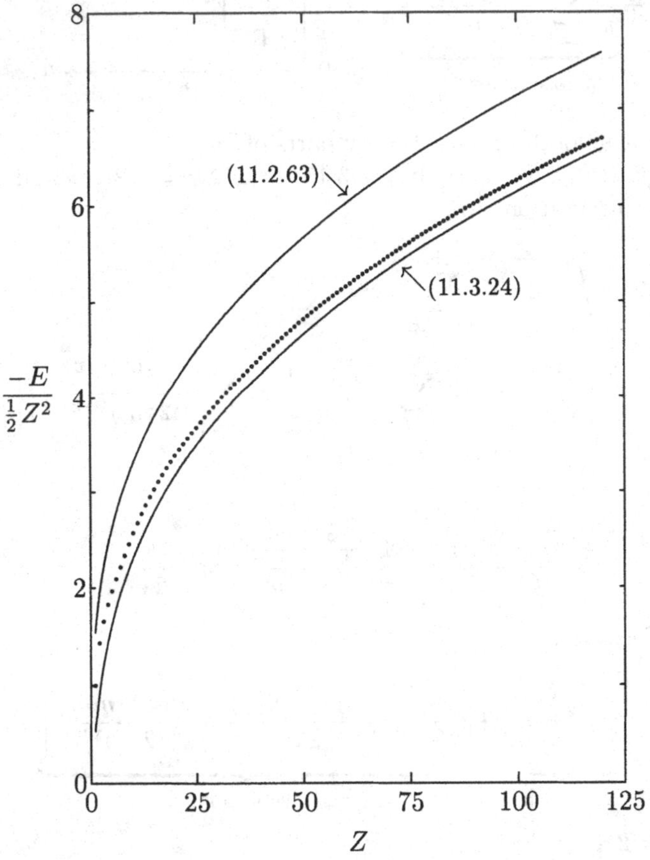

shows the remarkable improvement that these corrections for the innermost electrons bring about.

11.4 Quantum corrections and exchange energy

How about the remaining discrepancies? Where to look? Problem 11-3 shows that TF is a classical limit, with E^{TF} given by a classical phase space integral (including the spin multiplicity of 2),

$$E_1^{\mathrm{TF}}(\zeta) = 2\int \frac{(\mathrm{d}\boldsymbol{r})(\mathrm{d}\boldsymbol{p})}{(2\pi\hbar)^3}\left(H(\boldsymbol{r},\boldsymbol{p})+\zeta\right)\eta\left(-H(\boldsymbol{r},\boldsymbol{p})-\zeta\right), \tag{11.4.1}$$

the leading semiclassical approximation to

$$E_1 = \mathrm{tr}\,\{(H+\zeta)\eta(-H-\zeta)\} \tag{11.4.2}$$

with the single-particle Hamilton operator

$$H(\boldsymbol{r},\boldsymbol{p}) = \frac{\boldsymbol{p}^2}{2m_{\mathrm{el}}} + V(\boldsymbol{r})\,. \tag{11.4.3}$$

That suggests looking at the first quantum corrections to E_1^{TF}.

All we need is the three-dimensional generalization of the one-dimensional result of Problem 6-17b, which requires that we recognize

$$\omega^2 = \frac{1}{M}\frac{\mathrm{d}^2}{\mathrm{d}x^2}V(x) \longrightarrow \frac{1}{m_{\mathrm{el}}}\boldsymbol{\nabla}^2 V(\boldsymbol{r}) \tag{11.4.4}$$

as the essential ingredient. So, the leading quantum correction to E_1^{TF} is

$$\begin{aligned} E_1^{\mathrm{qm}} &= -2\int \frac{(\mathrm{d}\boldsymbol{r})\,(\mathrm{d}\boldsymbol{p})}{(2\pi\hbar)^3}\,\frac{\hbar^2}{24m_{\mathrm{el}}}\,\boldsymbol{\nabla}^2 V\,\frac{\partial^2}{\partial h^2}h\eta(-h)\Big|_{h\,=\,H(\boldsymbol{r},\boldsymbol{p})+\zeta} \\ &= 2\int \frac{(\mathrm{d}\boldsymbol{r})\,(\mathrm{d}\boldsymbol{p})}{(2\pi\hbar)^3}\,\frac{\hbar^2}{24m_{\mathrm{el}}}\,\boldsymbol{\nabla}^2 V\,\delta\Big(-\frac{\boldsymbol{p}^2}{2m_{\mathrm{el}}}-V(\boldsymbol{r})-\zeta\Big) \end{aligned} \tag{11.4.5}$$

or, after evaluating the momentum integral,

$$E_1^{\mathrm{qm}} = \frac{1}{24\pi^2\hbar}\int (\mathrm{d}\boldsymbol{r})\,\boldsymbol{\nabla}^2 V\,\left[2m_{\mathrm{el}}(-V-\zeta)\right]^{\frac{1}{2}}. \tag{11.4.6}$$

Notice the appearance of $\boldsymbol{\nabla}^2 V$, not $\boldsymbol{\nabla}^2(V+Ze^2/r)$ where

$$\boldsymbol{\nabla}^2 V(\boldsymbol{r}) = 4\pi Ze^2\delta(\boldsymbol{r}) - 4\pi e^2 n(\boldsymbol{r})\,. \tag{11.4.7}$$

But we must not forget the special treatment of strongly bound electrons. What we want is

$$\left[E_1(\zeta)-E_1(\zeta_{\mathrm{s}})\right]^{\mathrm{qm}} \tag{11.4.8}$$

where $E_1(\zeta_{\mathrm{s}})$ is computed from the potential $V_{\mathrm{s}} = -Ze^2/r + (B/a)Z^{\frac{4}{3}}$ such that $\boldsymbol{\nabla}^2 V_{\mathrm{s}} = 4\pi Ze^2\delta(\boldsymbol{r})$. So, the necessary correction for the strongly bound electrons removes the virtual $\delta(\boldsymbol{r})$ contribution, and we get the quantum correction

$$\Delta_{\rm qm}E = \frac{1}{24\pi^2\hbar}\int(\mathrm{d}\boldsymbol{r})\,\boldsymbol{\nabla}^2\left(V+\frac{Ze^2}{r}\right)\left[2m_{\rm el}(-V-\zeta)\right]^{\frac{1}{2}} . \tag{11.4.9}$$

Since this is a small correction to the TF energy, we evaluate it consistently for the TF potential that obeys the differential equation (11.2.25),

$$\begin{aligned}\Delta_{\rm qm}E &= -\frac{1}{18\pi^3}\frac{e^2}{\hbar^4}\int(\mathrm{d}\boldsymbol{r})\left[2m_{\rm el}(-V-\zeta)\right]^2 \\ &= -\frac{2}{9\pi^3}\frac{1}{a_0^2e^2}\int(\mathrm{d}\boldsymbol{r})\left[-V(\boldsymbol{r})-\zeta\right]^2 .\end{aligned} \tag{11.4.10}$$

With the TF parameterization $r = Z^{-\frac{1}{3}}ax$, $V+\zeta = -Z^{\frac{4}{3}}(e^2/a)f(x)/x$ this becomes

$$\Delta_{\rm qm}E = -\underbrace{\frac{8}{9\pi^2}\frac{a}{a_0}}_{=\frac{1}{4}\left(\frac{3\pi}{4}\right)^{-\frac{4}{3}}}Z^{\frac{5}{3}}\frac{e^2}{a_0}\int_0^\infty \mathrm{d}x\,\left[f(x)\right]^2 . \tag{11.4.11}$$

For the neutral-atom TF function $f(x)$, the integral is 0.615 435, and

$$\Delta_{\rm qm}E = -0.04907\,Z^{\frac{5}{3}}\frac{e^2}{a_0} . \tag{11.4.12}$$

Next we look at exchange, for which purpose we return to (11.1.13), pick out the exchange term, and make the spin summation explicit:

$$E_{\rm ex} = -\frac{1}{2}\sum_{a,b}\int(\mathrm{d}\boldsymbol{r})(\mathrm{d}\boldsymbol{r}')\sum_{\sigma,\sigma'}\phi_a(\boldsymbol{r},\sigma)^*\phi_b(\boldsymbol{r}',\sigma')^*v(\boldsymbol{r}-\boldsymbol{r}')\phi_a(\boldsymbol{r}',\sigma')\phi_b(\boldsymbol{r},\sigma) . \tag{11.4.13}$$

Since, for example,

$$\sum_\sigma \phi_a(\boldsymbol{r},\sigma)^*\phi_b(\boldsymbol{r}',\sigma) = 0 \quad \text{if modes } a, b \text{ have different spin} \tag{11.4.14}$$

only the a, b pairs of the same spin (net factor of 2) contribute, so that

$$E_{\rm ex} = -e^2\int(\mathrm{d}\boldsymbol{r})(\mathrm{d}\boldsymbol{r}')\frac{\sum_a\phi_a(\boldsymbol{r})^*\phi_a(\boldsymbol{r}')\sum_b\phi_b(\boldsymbol{r}')^*\phi_b(\boldsymbol{r})}{|\boldsymbol{r}-\boldsymbol{r}'|} \tag{11.4.15}$$

with the Coulomb interaction potential $v(\boldsymbol{r}-\boldsymbol{r}')$ of (11.1.2). In the semiclassical TF limit the wave functions are those of (10.8.12),

$$\phi_a(\boldsymbol{r}) = \sqrt{\frac{(\mathrm{d}\boldsymbol{p})}{(2\pi\hbar)^3}}\,\mathrm{e}^{\frac{\mathrm{i}}{\hbar}\boldsymbol{p}\cdot\boldsymbol{r}} \quad \text{with} \quad |\boldsymbol{p}| < \sqrt{2m_{\mathrm{el}}(-V-\zeta)} \equiv P \tag{11.4.16}$$

and we have

$$E_{\mathrm{ex}} = -e^2 \int (\mathrm{d}\boldsymbol{r})(\mathrm{d}\boldsymbol{r}') \frac{(\mathrm{d}\boldsymbol{p})}{(2\pi\hbar)^3} \frac{(\mathrm{d}\boldsymbol{p}')}{(2\pi\hbar)^3} \frac{\mathrm{e}^{-\frac{\mathrm{i}}{\hbar}(\boldsymbol{p}-\boldsymbol{p}')\cdot(\boldsymbol{r}-\boldsymbol{r}')}}{|\boldsymbol{r}-\boldsymbol{r}'|} \,. \tag{11.4.17}$$

We first integrate over $\boldsymbol{r}'$, where only the $\hbar/P$ vicinity of $\boldsymbol{r}$ contributes, and meet the Fourier transform of the Coulomb potential,

$$\begin{aligned}
&\int (\mathrm{d}\boldsymbol{r}') \frac{\mathrm{e}^{-\frac{\mathrm{i}}{\hbar}(\boldsymbol{p}-\boldsymbol{p}')\cdot(\boldsymbol{r}-\boldsymbol{r}')}}{|\boldsymbol{r}-\boldsymbol{r}'|} \\
&= \frac{\hbar^2}{(\boldsymbol{p}-\boldsymbol{p}')^2} \int (\mathrm{d}\boldsymbol{r}') \frac{1}{|\boldsymbol{r}-\boldsymbol{r}'|} \left(-\boldsymbol{\nabla}'^2\right) \mathrm{e}^{-\frac{\mathrm{i}}{\hbar}(\boldsymbol{p}-\boldsymbol{p}')\cdot(\boldsymbol{r}-\boldsymbol{r}')} \\
&= \frac{\hbar^2}{(\boldsymbol{p}-\boldsymbol{p}')^2} \int (\mathrm{d}\boldsymbol{r}')\, \mathrm{e}^{-\frac{\mathrm{i}}{\hbar}(\boldsymbol{p}-\boldsymbol{p}')\cdot(\boldsymbol{r}-\boldsymbol{r}')} \underbrace{\left(-\boldsymbol{\nabla}'^2\right) \frac{1}{|\boldsymbol{r}-\boldsymbol{r}'|}}_{=4\pi\delta(\boldsymbol{r}'-\boldsymbol{r})} \\
&= \frac{4\pi\hbar^2}{(\boldsymbol{p}-\boldsymbol{p}')^2} \,.
\end{aligned} \tag{11.4.18}$$

Then (11.4.17) becomes

$$E_{\mathrm{ex}} = -\frac{4\pi e^2}{(2\pi)^6\hbar^4} \int (\mathrm{d}\boldsymbol{r}) \int \frac{(\mathrm{d}\boldsymbol{p})(\mathrm{d}\boldsymbol{p}')}{(\boldsymbol{p}-\boldsymbol{p}')^2} \tag{11.4.19}$$

where $|\boldsymbol{p}|, |\boldsymbol{p}'| < P(\boldsymbol{r}) = \sqrt{2m_{\mathrm{el}}(-V(\boldsymbol{r})-\zeta)}$ specifies the range covered by the two momentum integrations, so that [see Problem 11-11]

$$\begin{aligned}
E_{\mathrm{ex}} &= -\frac{4\pi}{(2\pi)^4} \frac{e^2}{\hbar^4} \int (\mathrm{d}\boldsymbol{r}) \left[2m_{\mathrm{el}}(-V(\boldsymbol{r})-\zeta)\right]^2 \\
&= -\frac{1}{\pi^3} \frac{1}{e^2 a_0^2} \int (\mathrm{d}\boldsymbol{r}) \left[-V(\boldsymbol{r})-\zeta\right]^2
\end{aligned} \tag{11.4.20}$$

and, recalling (11.4.10), we have

$$E_{\mathrm{ex}} = \frac{9}{2} \Delta_{\mathrm{qm}} E \,. \tag{11.4.21}$$

Therefore

$$E_{\mathrm{ex}} + \Delta_{\mathrm{qm}} E = \frac{11}{2} \Delta_{\mathrm{qm}} E = -\underbrace{\frac{11}{8} \left(\frac{3\pi}{4}\right)^{-\frac{4}{3}} \int_0^\infty \mathrm{d}x \left[f(x)\right]^2}_{=0.2699} Z^{\frac{5}{3}} \frac{e^2}{a_0} \,, \tag{11.4.22}$$

and together with the terms of (11.3.24) we arrive at

$$\frac{-E_{\text{smooth}}}{\frac{1}{2}Z^2} = 1.537\,Z^{\frac{1}{3}} - 1 + 0.540\,Z^{-\frac{1}{3}} \qquad \text{(atomic units)}\,. \qquad (11.4.23)$$

Now we have a third curve in the comparison with the 'experimental' data,

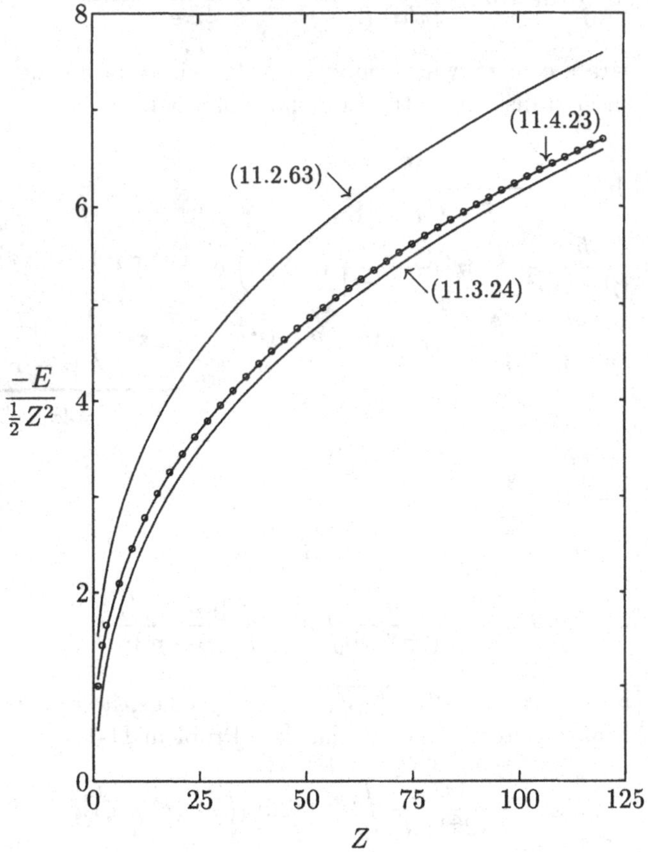

a curve that runs right through the little circles! (Except that it stops 7.7% above $Z = 1$.) That the third curve be visible at all, two thirds of the circles are removed, only those for $Z = 1, 2, 3, 6, 9, \ldots, 120$ are displayed, with a larger diameter than in the figures on pages 419 and 424. In three steps we've reached the goal of understanding the smooth Z dependence of the Hartree–Fock energies, first seen in the figure on page 419.

11.5 Energy oscillations

Do we stop with this success? Not quite. We have got to see the contributions of oscillatory terms. When should they show up? They are there to enforce the integer nature of n_r and l, or if you like, the discreteness of the individual electron. In contrast, the TF treatment regards the electrons as infinitely divisible, as if their total number were continuous. The implied error must be roughly the ratio of the actual unit of charge, e, relative to the total charge (for neutral atoms) Ze: $1/Z$. Relative to the TF energy (11.2.62), $E^{\mathrm{TF}} \propto Z^{7/3}$, this is $Z^{\frac{4}{3}}$, one order below where we are.

Can we anticipate anything about the structure of these oscillatory terms? Recall (11.2.10),

$$E_1 = 4\int \mathrm{d}\nu\,\mathrm{d}\lambda\,\lambda\,(\mathcal{E}_{\nu,\lambda}+\zeta)\,\eta(-\mathcal{E}_{\nu,\lambda}-\zeta)\sum_{j,k} \mathrm{e}^{2\pi\mathrm{i}\,[j(\nu-\frac{1}{2})+k(\lambda-\frac{1}{2})]}\,. \tag{11.5.1}$$

We might expect that a dominant contribution comes from regions where λ approaches its maximum value ($\equiv \Lambda$), a region of constructive interference. Now, in (11.2.12),

$$\nu = \frac{1}{\pi}\int\frac{\mathrm{d}r}{r}\sqrt{\frac{2m_{\mathrm{el}}}{\hbar^2}r^2(\mathcal{E}-V)-\lambda^2}\,, \tag{11.5.2}$$

we have

$$\nu \geq 0 \quad\text{and}\quad \mathcal{E} \leq -\zeta\,, \tag{11.5.3}$$

so that the largest λ occurs for the smallest ν, that is: $\nu = 0$, and the largest $\mathcal{E}$, that is: $\mathcal{E} = 0$ for neutral atoms ($\zeta = 0$). In fact

$$\begin{aligned}\Lambda^2 &= \operatorname*{Max}_r\left\{\frac{2m_{\mathrm{el}}}{\hbar^2}r^2\left[-V(r)\right]\right\}\\ &= Z^{2/3}\frac{2a}{a_0}\operatorname*{Max}_x\{xf(x)\}\\ &= \left[Z^{1/3}\left(\frac{3\pi}{4}\right)^{\frac{1}{3}}\sqrt{x_{\max}f(x_{\max})}\right]^2 .\end{aligned} \tag{11.5.4}$$

From the numerically known $f(x)$ it turns out that $xf(x)$ is maximum at $x_{\max} = 2.104$, and from the value of f there, $f(x_{\max}) = 0.2312$, one gets

$$\Lambda = 0.928 Z^{\frac{1}{3}}\,. \tag{11.5.5}$$

This suggests looking at

$$\frac{-E_{\mathrm{osc}}}{Z^{\frac{4}{3}}} \equiv \frac{(-E)-(-E_{\mathrm{smooth}})}{Z^{\frac{4}{3}}} \tag{11.5.6}$$

as a function of $Z^{\frac{1}{3}}$. That is done in this figure (atomic energy units used once more):

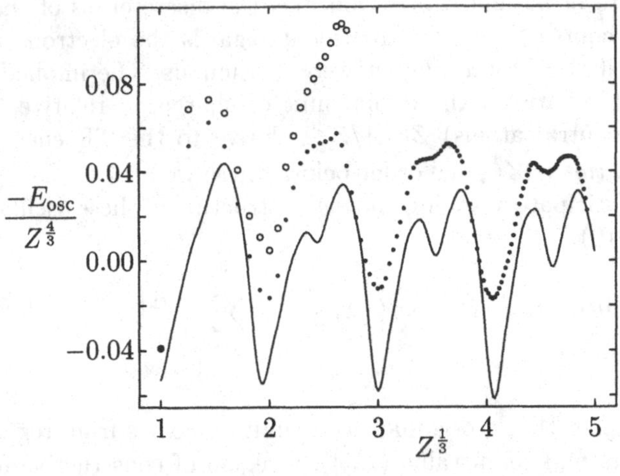

where Hartree–Fock data is marked by little circles, and the larger circles display real experimental data (including a smooth relativistic contribution). There are the oscillations! And their period is indeed close to $Z^{\frac{1}{3}}$. The result of calculation, with leading terms only, is the continuous curve. It refers only to oscillatory terms – a smooth term of order $Z^{\frac{4}{3}}$ in the energy is *not* included [nor is the relativistic contribution whose dominant term is $\frac{1}{8}Z^2(Z\alpha)^2$, $\alpha \cong 1/137$ being Sommerfeld's fine structure constant]. With that in mind, the agreement in magnitude and structure is impressive.

Problems

11-1a For a variational estimate of the ground-state energy of two-electron atoms, Hamiltonian:

$$H = \underbrace{\sum_{j=1}^{2} \frac{\boldsymbol{p}_j^2}{2m_{\text{el}}}}_{\equiv H_{\text{kin.}}} - \underbrace{\sum_{j=1}^{2} \frac{Ze^2}{|\boldsymbol{r}_j|}}_{\equiv H_{\text{n-e}}} + \underbrace{\frac{e^2}{|\boldsymbol{r}_1 - \boldsymbol{r}_2|}}_{\equiv H_{\text{e-e}}},$$

consider states with zero total angular momentum (antisymmetric spin state) and symmetric spatial wave functions of the form

$$\psi(\boldsymbol{r}_1', \boldsymbol{r}_2') = C\left(\mathrm{e}^{-Z(\alpha r_1' + \beta r_2')/a_0} + \mathrm{e}^{-Z(\beta r_1' + \alpha r_2')/a_0}\right)$$

with $r_1' = |\boldsymbol{r}_1'|$, $r_2' = |\boldsymbol{r}_2'|$ and $\alpha, \beta > 0$. Determine the (positive) normalization constant C. Then verify that the expectation values of the three pieces of H are given by

$$\langle H_{\text{kin.}}\rangle \Big/ \frac{e^2}{a_0} = \frac{1}{2}Z^2 \left[\frac{\alpha^2+\beta^2}{(\alpha\beta)^3} + 2\alpha\beta\left(\frac{2}{\alpha+\beta}\right)^6\right] \Big/ \left[\frac{1}{(\alpha\beta)^3} + \left(\frac{2}{\alpha+\beta}\right)^6\right],$$

$$\langle H_{\text{n-e}}\rangle \Big/ \frac{e^2}{a_0} = -Z^2(\alpha+\beta)\,,$$

$$\langle H_{\text{e-e}}\rangle \Big/ \frac{e^2}{a_0} = \frac{1}{8}Z\left(\frac{2}{\alpha+\beta}\right)^3 \left[\frac{1}{\alpha^2} + \frac{1}{\beta^2} + \frac{3}{\alpha\beta} + 5\left(\frac{2}{\alpha+\beta}\right)^2\right] \times \left[\frac{1}{(\alpha\beta)^3} + \left(\frac{2}{\alpha+\beta}\right)^6\right]^{-1}.$$

Why could you have anticipated that $\alpha \leftrightarrow \beta$ changes nothing?

11-1b Now, for simplicity, put $\alpha = \beta$ and find the α value for which $\langle H\rangle$ is minimal. Is this estimate good enough to explain the existence of the stable negative hydrogen ion H^-?

11-1c To facilitate the optimization of both α and β, introduce new parameters x, y in accordance with

$$x = \frac{\frac{1}{2}(\alpha+\beta)}{\sqrt{\alpha\beta}}\,, \qquad y = \sqrt{\alpha\beta}\,.$$

You should find that $\langle H\rangle$ is a quadratic function of y, so that – for any given $x \geq 1$ (why this restriction?) – the optimal y can be found immediately. Then optimize x. Is the estimate now good enough?

11-1d Compare both estimates with experimental values:

$$-E = 0.52776,\ 2.9038,\ 7.2804,\ 13.657\,,22.036 \times e^2/a_0 \quad \text{for} \quad Z = 1,\ldots,5\,.$$

11-2a Non-interacting electrons filling ν closed Bohr shells. Energy:

$$E = \sum_{n=1}^{\nu} 2n^2\left(-\frac{Z^2}{2n^2}\right) = -Z^2\nu \qquad \text{(atomic energy units).}$$

Show that, for N electrons,

$$N = \sum_{n=1}^{\nu} 2n^2 = \tfrac{2}{3}(\nu+\tfrac{1}{2})^3 - \tfrac{1}{6}(\nu+\tfrac{1}{2})\,.$$

Verify that, for sufficiently large N,

$$\nu = (\tfrac{3}{2}N)^{\frac{1}{3}} - \tfrac{1}{2} + \tfrac{1}{12}(\tfrac{3}{2}N)^{-\frac{1}{3}} + \cdots$$

and

$$-E = Z^2\left[\left(\tfrac{3}{2}N\right)^{\frac{1}{3}} - \tfrac{1}{2} + \tfrac{1}{12}\left(\tfrac{3}{2}N\right)^{\frac{1}{3}} + \cdots\right]$$

$$\underset{\overbrace{N=Z}}{\overset{\uparrow}{=}} \left(\tfrac{3}{2}\right)^{\frac{1}{3}} Z^{\frac{7}{3}} - \tfrac{1}{2}Z^2 + \tfrac{1}{12}\left(\tfrac{2}{3}\right)^{\frac{1}{3}} Z^{\frac{5}{3}} + \cdots .$$

11-2b The asymptotic expansion of ν as a function of N obviously holds only for large N. Nevertheless, put $N = 2$ (!) and compare the numerical value produced by the first three terms with the actual value of ν. Similarly for $N = 10$.

11-3 The classical limit of tr $\{f(q,p)\}$, n degrees of freedom, is

$$\int \frac{(\mathrm{d}q)\,(\mathrm{d}p)}{(2\pi\hbar)^n} f(q,p) .$$

Consider $f = \big(H(\boldsymbol{r},\boldsymbol{p}) + \zeta\big)\eta\big(-H(\boldsymbol{r},\boldsymbol{p}) - \zeta\big)$ so that

$$\mathrm{tr}\,\{f\} = \sum_a (\mathcal{E}_a + \zeta)\eta(-\mathcal{E}_a - \zeta) = E_1 .$$

Then ($n = 3$ plus spin-$\frac{1}{2}$)

$$E_{1,\mathrm{class.}} = 2\int \frac{(\mathrm{d}\boldsymbol{r})\,(\mathrm{d}\boldsymbol{p})}{(2\pi\hbar)^3} \big(H(\boldsymbol{r},\boldsymbol{p}) + \zeta\big)\eta\big(-H(\boldsymbol{r},\boldsymbol{p}) - \zeta\big) .$$

Verify, for $H = \boldsymbol{p}^2/(2m_{\mathrm{el}}) + V(\boldsymbol{r})$, that

$$E_{1,\mathrm{class.}} = \int (\mathrm{d}\boldsymbol{r}) \left(-\frac{1}{15\pi^2}\right) \frac{1}{m_{\mathrm{el}}\hbar^3} \left[2m_{\mathrm{el}}(-V - \zeta)\right]^{\frac{5}{2}} = E_1^{\mathrm{TF}} .$$

11-4a Show that a solution of the TF differential equation (it is the asymptotic form for $x \gg 1$) is

$$f(x) = \frac{(12)^2}{x^3} = \lambda^3 \frac{(12)^2}{(\lambda x)^3} .$$

Then demonstrate that if $f(x)$ is any solution of the differential equation, so also is $\lambda^3 f(\lambda x)$. [Of course, the boundary condition $f(0) = 1$ cannot be maintained.]

11-4b Prove that δf, a small difference between two solutions, obeys

$$\frac{\mathrm{d}^2}{\mathrm{d}x^2}\delta f(x) = \frac{3}{2}\left(\frac{f(x)}{x}\right)^{\frac{1}{2}} \delta f(x) .$$

Show that this δf equation is satisfied by $3f + x\frac{\mathrm{d}f}{\mathrm{d}x}$. How could you have anticipated that?

11-5 With the convention that $f^{\frac{1}{2}} = 0$ for $f < 0$ show that

$$f_1^{\frac{5}{2}} = f_2^{\frac{5}{2}} + \tfrac{5}{2} f_2^{\frac{3}{2}}(f_1 - f_2) + \int_{f_2}^{f_1} \mathrm{d}f\,(f_1 - f)\tfrac{15}{4} f^{\frac{1}{2}} .$$

Use this to prove that the stationary value of the f functional S of (11.2.41) is a global minimum.

11-6a The variational estimate (11.2.60) is an upper bound on B, the initial slope of the TF function. To get a lower bound, take

$$g(x) = \frac{\mathrm{d}}{\mathrm{d}x} f(x) = f'(x)$$

as the looked-for function. Then show that

$$\overline{S} = -\frac{3}{5}\int_0^\infty \mathrm{d}x\, x^{\frac{1}{3}} \left[g'(x)\right]^{\frac{5}{3}} - g(0) - \frac{1}{2}\int_0^\infty \mathrm{d}x\, \left[g(x)\right]^2$$

equals $\frac{3}{7}B$ for the correct $g(x)$; that $\overline{S}$ is stationary at the correct g, if trial functions are restricted by $g'(x) \geq 0$ and $g(x \to \infty) = 0$; and that the stationary value is a global maximum.

11-6b Now try, for example,

$$g(x) = -\lambda_1\, \mathrm{e}^{-\lambda_2 \sqrt{x}} \quad \text{with} \quad \lambda_{1,2} > 0 ,$$

and optimize λ_1 and λ_2. Compare the resulting lower bound on B with the actual value (11.2.61).

11-7 Use the asymptotic form given in Problem 11-4a to test for the failure of the TF approximation at large r.

11-8a Positive TF ions: $N < Z$. Modify the arguments presented for the neutral atom to show that $f(x)$ still obeys the TF differential equation but with the following boundary conditions at $r_0 = aZ^{-\frac{1}{3}}x_0$, the edge of the atom:

$$f(x_0) = 0 , \qquad -x_0 \frac{\mathrm{d}f}{\mathrm{d}x}(x_0) = 1 - \frac{N}{Z} .$$

Of course, $f(0) = 1$, as for $N = Z$. What can you say qualitatively about $-f'(0)$ in relation to $B = 1.588$, the value for neutral atoms?

11-8b Show that

$$N = \int (\mathrm{d}\boldsymbol{r})\, n = Z \int_0^{x_0} \mathrm{d}x\, x^{\frac{1}{2}} \left[f(x)\right]^{\frac{3}{2}} .$$

Then use the differential equation and boundary conditions to prove that the last form is indeed equal to N.

11-9a TF atoms: energy and scaling. Consider $E = E_1 - \zeta N + E_2$ with

$$E_1 = -\frac{1}{15\pi^2}\frac{1}{m_{\mathrm{el}}\hbar^3}\int (\mathrm{d}\boldsymbol{r})\left[2m_{\mathrm{el}}(-V-\zeta)\right]^{\frac{5}{2}}$$

and

$$E_2 = -\frac{1}{8\pi e^2}\int (\mathrm{d}\boldsymbol{r})\left[\boldsymbol{\nabla}\left(V + \frac{Ze^2}{r}\right)\right]^2$$

and recall the physical boundary condition

$$\lim_{r\to 0} rV(r) = -Ze^2 .$$

Use the scaling transformation

$$V(r) \to \lambda V(\lambda r) , \quad \zeta \to \lambda\zeta , \quad Z \to Z$$

to show that

$$-\tfrac{1}{2}E_1 - \zeta N + E_2 = 0 \quad \text{or} \quad E = \tfrac{3}{2}E_1 .$$

Then use the different transformation

$$V(r) \to \lambda^4 V(\lambda r) , \quad \zeta \to \lambda^4\zeta , \quad Z \to \lambda^3 Z$$

to prove that

$$7(E_1 + E_2) - 4\zeta N = 3Z\frac{\partial}{\partial Z}E(Z,N)$$

or

$$E = \frac{3}{7}\left[Z\frac{\partial}{\partial Z}E(Z,N) - \zeta N\right] .$$

11-9b Prove that

$$\frac{\partial}{\partial N}E(Z,N) = -\zeta ,$$

and so arrive at

$$\left(Z\frac{\partial}{\partial Z}+N\frac{\partial}{\partial N}\right)E(Z,N)=\frac{7}{3}E(Z,N)\,.$$

First consider neutral atoms, $N = Z$, and use the above to prove again that $E(Z,Z) \propto Z^{\frac{7}{3}}$. Then demonstrate that, in general,

$$E(Z,N)=Z^{\frac{7}{3}}\varepsilon\left(\frac{N}{Z}\right)\,.$$

What can you say about the function $\varepsilon(N/Z)$ for $N = Z$ and for $N \ll Z$?

11-10 The TF density is proportional to $r^{-\frac{3}{2}}$ for very small r. With the method that is used in Section 11.3 derive $n(\boldsymbol{r} = 0)$, the electron density at the site of the nucleus, with corrections for the strongly bound electrons taken into account. You will need the density of a closed Bohr shell at $\boldsymbol{r} = 0$. How do you get this from (8.3.30)?

11-11 In the following integral, $\boldsymbol{p}$ and $\boldsymbol{p}'$ range over the interior of a sphere with radius P. Prove that

$$\int (\mathrm{d}\boldsymbol{p})\,(\mathrm{d}\boldsymbol{p}')\,\frac{1}{|\boldsymbol{p}-\boldsymbol{p}'|^2}=(2\pi)^2P^4\,.$$

12. Electromagnetic Radiation

12.1 Lagrangian, modes, equations of motion

We now turn to another application of the methods of Chapter 10: the electromagnetic field interacting with charged particles. Aiming at applications in atomic physics, where particle dynamics is predominantly non-relativistic, we continue to use the non-relativistic description of the particles developed in Chapters 4–11.

The Lagrangian consists of a sum over the particles (masses m_a, charges e_a, positions $\boldsymbol{r}_a$, momenta $\boldsymbol{p}_a$, velocities $\boldsymbol{v}_a$) and an integral over the Lagrangian density of the field:

$$L = \sum_a \left[\boldsymbol{p} \cdot \left(\frac{\mathrm{d}\boldsymbol{r}}{\mathrm{d}t} - \boldsymbol{v} \right) + \frac{1}{2} m v^2 - e\Phi(\boldsymbol{r},t) + \frac{e}{c} \boldsymbol{v} \cdot \boldsymbol{A}(\boldsymbol{r},t) \right]_a$$
$$+ \int (\mathrm{d}\boldsymbol{r}) \frac{1}{4\pi} \left[-\boldsymbol{E} \cdot \left(\frac{1}{c}\frac{\partial}{\partial t} \boldsymbol{A} + \boldsymbol{\nabla}\Phi \right) - \boldsymbol{B} \cdot \boldsymbol{\nabla} \times \boldsymbol{A} + \frac{1}{2} \left(\boldsymbol{B}^2 - \boldsymbol{E}^2 \right) \right] \tag{12.1.1}$$

(symmetrized products understood). For a single particle, we have already seen the particle part in Section 8.6. We check the field part from its consequences: varying the electric field $\boldsymbol{E}$ and the vector potential $\boldsymbol{A}$, which are paired in the time-derivative term, gives two equations of motion,

$$\delta\boldsymbol{E}: \qquad \boldsymbol{E} = -\frac{1}{c}\frac{\partial}{\partial t}\boldsymbol{A} - \boldsymbol{\nabla}\Phi\,,$$
$$\delta\boldsymbol{A}: \qquad \boldsymbol{\nabla} \times \boldsymbol{B} - \frac{1}{c}\frac{\partial}{\partial t}\boldsymbol{E} = \frac{4\pi}{c}\sum_a e_a \boldsymbol{v}_a \delta(\boldsymbol{r} - \boldsymbol{r}_a) = \frac{4\pi}{c}\boldsymbol{j}\,, \tag{12.1.2}$$

where $\boldsymbol{j}$ is the electric current density of the moving charges; whereas varying the magnetic field $\boldsymbol{B}$ and the scalar potential Φ, which have no time-derivative terms in the Lagrangian, gives two constraints,

$$\delta\boldsymbol{B}: \qquad \boldsymbol{B} = \boldsymbol{\nabla} \times \boldsymbol{A}\,,$$
$$\delta\Phi: \qquad \boldsymbol{\nabla} \cdot \boldsymbol{E} = 4\pi \sum_a e_a \delta(\boldsymbol{r} - \boldsymbol{r}_a) = 4\pi\rho\,, \tag{12.1.3}$$

where ρ is the electric charge density.

Using these constraints, we can eliminate the fields $\boldsymbol{B}$, $\boldsymbol{\Phi}$ that do not obey equations of motion. For $\boldsymbol{B}$, this is easy: accept $\boldsymbol{B} = \boldsymbol{\nabla} \times \boldsymbol{A}$ as a definition. For $\boldsymbol{\Phi}$, we turn to the $\delta\boldsymbol{\Phi}$ equation in (12.1.3) and split $\boldsymbol{E}$ into a longitudinal and a transverse part,

$$\boldsymbol{E} = \boldsymbol{E}_{\parallel} + \boldsymbol{E}_{\perp} \quad \text{with} \quad \boldsymbol{\nabla} \times \boldsymbol{E}_{\parallel} = 0 \quad \text{and} \quad \boldsymbol{\nabla} \cdot \boldsymbol{E}_{\perp} = 0\,, \tag{12.1.4}$$

where we might as well write

$$\boldsymbol{E}_{\parallel} = -\boldsymbol{\nabla}\boldsymbol{\Phi}\,, \qquad \text{ensuring } \boldsymbol{\nabla} \times \boldsymbol{E}_{\parallel} = 0\,, \tag{12.1.5}$$

and get Poisson's equation for $\boldsymbol{\Phi}$,

$$-\boldsymbol{\nabla}^2\boldsymbol{\Phi} = 4\pi\rho\,, \qquad \boldsymbol{\Phi}(\boldsymbol{r},t) = \int(\mathrm{d}\boldsymbol{r}')\,\frac{\rho(\boldsymbol{r}',t)}{|\boldsymbol{r}-\boldsymbol{r}'|}\,. \tag{12.1.6}$$

Similarly

$$\boldsymbol{A} = \boldsymbol{A}_{\parallel} + \boldsymbol{A}_{\perp} \quad \text{with} \quad \boldsymbol{\nabla} \times \boldsymbol{A}_{\parallel} = 0 \quad \text{and} \quad \boldsymbol{\nabla} \cdot \boldsymbol{A}_{\perp} = 0\,, \tag{12.1.7}$$

but here the freedom of gauge transformations, $\boldsymbol{A} \to \boldsymbol{A} + \boldsymbol{\nabla}\lambda$ (see Section 8.6 and Problem 12-1), says that the longitudinal part $\boldsymbol{A}_{\parallel}$ is arbitrary and can be chosen to vanish:

$$\boldsymbol{A}_{\parallel} = 0\,, \quad \boldsymbol{A} = \boldsymbol{A}_{\perp}\,, \qquad \boldsymbol{\nabla} \cdot \boldsymbol{A} = 0\,. \tag{12.1.8}$$

This is the radiation gauge. It retains only the gauge-invariant transverse part of $\boldsymbol{A}$. Incidentally, (12.1.5) in the $\boldsymbol{E}$ equation of (12.1.2) requires $\partial\boldsymbol{A}_{\parallel}/\partial t = 0$ for consistency.

In the radiation gauge then, note that the time-derivative term involves transverse fields only,

$$\int(\mathrm{d}\boldsymbol{r})\,\frac{1}{4\pi}\left[-\boldsymbol{E}\cdot\frac{1}{c}\frac{\partial}{\partial t}\boldsymbol{A}\right] = \int(\mathrm{d}\boldsymbol{r})\,\frac{1}{4\pi}\left[-\boldsymbol{E}_{\perp}\cdot\frac{1}{c}\frac{\partial}{\partial t}\boldsymbol{A}\right], \tag{12.1.9}$$

because a partial integration removes the virtual $\boldsymbol{E}_{\parallel}$ contribution; that the integral of $\boldsymbol{E}^2$ splits in two,

$$\int(\mathrm{d}\boldsymbol{r})\,\frac{\boldsymbol{E}^2}{8\pi} = \int(\mathrm{d}\boldsymbol{r})\,\frac{\boldsymbol{E}_{\perp}^2}{8\pi} + \int(\mathrm{d}\boldsymbol{r})\,\frac{\boldsymbol{E}_{\parallel}^2}{8\pi}\,; \tag{12.1.10}$$

that this longitudinal contribution is the electrostatic interaction energy of the charges,

$$\begin{aligned}\int(\mathrm{d}\boldsymbol{r})\,\frac{\boldsymbol{E}_{\parallel}^2}{8\pi} &= \int(\mathrm{d}\boldsymbol{r})\,\frac{(\boldsymbol{\nabla}\boldsymbol{\Phi})^2}{8\pi} = \frac{1}{2}\int(\mathrm{d}\boldsymbol{r})\,\rho\boldsymbol{\Phi}\\ &= \frac{1}{2}\int(\mathrm{d}\boldsymbol{r})(\mathrm{d}\boldsymbol{r}')\,\frac{\rho(\boldsymbol{r},t)\rho(\boldsymbol{r}',t)}{|\boldsymbol{r}-\boldsymbol{r}'|} = \frac{1}{2}\sum_{a,b}\frac{e_a e_b}{|\boldsymbol{r}_a-\boldsymbol{r}_b|}\,,\end{aligned} \tag{12.1.11}$$

where the $a = b$ term will be omitted; and that

$$\sum_a \left[e\Phi(\boldsymbol{r},t)\right]_a = \int (\mathrm{d}\boldsymbol{r})\, \rho\Phi = \sum_{a,b} \frac{e_a e_b}{|\boldsymbol{r}_a - \boldsymbol{r}_b|} \tag{12.1.12}$$

is twice this interaction energy.

So the Lagrangian reduces to

$$L = \sum_a \left[\boldsymbol{p}\cdot\left(\frac{\mathrm{d}\boldsymbol{r}}{\mathrm{d}t} - \boldsymbol{v}\right) + \frac{1}{2}mv^2 + \frac{e}{c}\boldsymbol{v}\cdot\boldsymbol{A}(\boldsymbol{r},t)\right]_a - \frac{1}{2}\sum_{a\neq b}\frac{e_a e_b}{|\boldsymbol{r}_a - \boldsymbol{r}_b|}$$
$$+ \int (\mathrm{d}\boldsymbol{r}) \left[-\frac{1}{4\pi c}\boldsymbol{E}_\perp \cdot \frac{\partial}{\partial t}\boldsymbol{A} - \frac{(\boldsymbol{\nabla}\times\boldsymbol{A})^2 + \boldsymbol{E}_\perp^2}{8\pi}\right] . \tag{12.1.13}$$

In addition to terms that refer solely to the particles,

$$L_{\text{part.}} = \sum_a \left[\boldsymbol{p}\cdot\left(\frac{\mathrm{d}\boldsymbol{r}}{\mathrm{d}t} - \boldsymbol{v}\right) + \frac{1}{2}mv^2\right]_a - \frac{1}{2}\sum_{a\neq b}\frac{e_a e_b}{|\boldsymbol{r}_a - \boldsymbol{r}_b|} ; \tag{12.1.14}$$

or solely to the radiation field,

$$L_{\text{field}} = \int (\mathrm{d}\boldsymbol{r}) \left[-\frac{1}{4\pi c}\boldsymbol{E}_\perp \cdot \frac{\partial}{\partial t}\boldsymbol{A} - \frac{(\boldsymbol{\nabla}\times\boldsymbol{A})^2 + \boldsymbol{E}_\perp^2}{8\pi}\right] , \tag{12.1.15}$$

there is a single interaction term,

$$L_{\text{int.}} = \sum_a \frac{e_a}{c}\boldsymbol{v}_a(t)\cdot\boldsymbol{A}(\boldsymbol{r}_a(t),t) = \int (\mathrm{d}\boldsymbol{r})\, \frac{1}{c}\boldsymbol{j}(\boldsymbol{r},t)\cdot\boldsymbol{A}(\boldsymbol{r},t) . \tag{12.1.16}$$

The two equivalent ways of writing $L_{\text{int.}}$ emphasize the particles or the radiation, respectively.

Now note that, owing to Stokes's* theorem and the transverse nature of $\boldsymbol{A}$,

$$0 < \int (\mathrm{d}\boldsymbol{r})\, (\boldsymbol{\nabla}\times\boldsymbol{A})^2 = \int (\mathrm{d}\boldsymbol{r})\, [\boldsymbol{A}\cdot(\boldsymbol{\nabla}\times(\boldsymbol{\nabla}\times\boldsymbol{A}))]$$
$$= \int (\mathrm{d}\boldsymbol{r})\, \boldsymbol{A}\cdot(-\boldsymbol{\nabla}^2)\boldsymbol{A} , \tag{12.1.17}$$

so that the negative Laplacian $-\boldsymbol{\nabla}^2$ is a positive differential operator. It is expedient to introduce transverse vector eigenfunctions of $-\boldsymbol{\nabla}^2$:

$$-\boldsymbol{\nabla}^2\boldsymbol{A}_\alpha(\boldsymbol{r}) = \left(\frac{\omega_\alpha}{c}\right)^2 \boldsymbol{A}_\alpha(\boldsymbol{r}) , \qquad \boldsymbol{\nabla}\cdot\boldsymbol{A}_\alpha(\boldsymbol{r}) = 0 \tag{12.1.18}$$

(with $\omega_\alpha > 0$ by convention) that are complete (as transverse functions, see Problem 12-3 for details), and orthonormal:

*Sir George Gabriel Stokes (1819–1903)

$$\int (\mathrm{d}\boldsymbol{r})\, \boldsymbol{A}_\alpha(\boldsymbol{r})^* \cdot \boldsymbol{A}_\beta(\boldsymbol{r}) = \delta_{\alpha\beta} \,. \tag{12.1.19}$$

Another important property is

$$\int (\mathrm{d}\boldsymbol{r})\, \boldsymbol{A}_\alpha(\boldsymbol{r}) \cdot \boldsymbol{A}_\beta(\boldsymbol{r}) = 0 \quad \text{if } \omega_\alpha \neq \omega_\beta \,, \tag{12.1.20}$$

which holds since $\boldsymbol{A}_\alpha^*$ obeys the eigenfunction equation (12.1.18) with the same ω value as $\boldsymbol{A}_\alpha$.

To handle the infinitely many degrees of freedom of the radiation field, we express the transverse fields $\boldsymbol{E}_\perp$ and $\boldsymbol{A}$ in terms of these mode functions as Hermitian operators,

$$\begin{aligned} \frac{1}{c}\boldsymbol{A}(\boldsymbol{r},t) &= \sum_\alpha \sqrt{\frac{2\pi\hbar}{\omega_\alpha}} \left(y_\alpha(t)\boldsymbol{A}_\alpha(\boldsymbol{r}) + y_\alpha(t)^\dagger \boldsymbol{A}_\alpha(\boldsymbol{r})^* \right) , \\ \boldsymbol{E}_\perp(\boldsymbol{r},t) &= \sum_\alpha \sqrt{2\pi\hbar\omega_\alpha} \left(\mathrm{i}y_\alpha(t)\boldsymbol{A}_\alpha(\boldsymbol{r}) - \mathrm{i}y_\alpha(t)^\dagger \boldsymbol{A}_\alpha(\boldsymbol{r})^* \right) . \end{aligned} \tag{12.1.21}$$

Then, to within additional time derivatives,

$$\int (\mathrm{d}\boldsymbol{r}) \left(-\frac{1}{4\pi c} \right) \boldsymbol{E}_\perp \cdot \frac{\partial}{\partial t}\boldsymbol{A} \longrightarrow \sum_\alpha \mathrm{i}\hbar y_\alpha^\dagger \frac{\mathrm{d}}{\mathrm{d}t} y_\alpha \tag{12.1.22}$$

and, to within an additive constant,

$$\int (\mathrm{d}\boldsymbol{r})\, \frac{(\boldsymbol{\nabla}\times\boldsymbol{A})^2 + \boldsymbol{E}_\perp^2}{8\pi} \longrightarrow \sum_\alpha \hbar\omega_\alpha y_\alpha^\dagger y_\alpha \,. \tag{12.1.23}$$

So we arrive at

$$\begin{aligned} L = \sum_a \Biggl[\boldsymbol{p}\cdot\left(\frac{\mathrm{d}\boldsymbol{r}}{\mathrm{d}t} - \boldsymbol{v}\right) + \frac{1}{2}m\boldsymbol{v}^2 + e\boldsymbol{v}\cdot\sum_\alpha \sqrt{\frac{2\pi\hbar}{\omega_\alpha}} \left(y_\alpha \boldsymbol{A}_\alpha(\boldsymbol{r}) + y_\alpha^\dagger \boldsymbol{A}_\alpha(\boldsymbol{r})^* \right) \Biggr]_a \\ - \frac{1}{2}\sum_{a\neq b} \frac{e_a e_b}{|\boldsymbol{r}_a - \boldsymbol{r}_b|} + \underbrace{\sum_\alpha \left[\mathrm{i}\hbar y_\alpha^\dagger \frac{\mathrm{d}}{\mathrm{d}t} y_\alpha - \hbar\omega_\alpha y_\alpha^\dagger y_\alpha \right]}_{= L_{\text{field}}} . \end{aligned} \tag{12.1.24}$$

From the structure of this Lagrangian one reads off the commutation relations and equations of motion. Of course, operators referring to different degrees of freedom commute, and within each degree of freedom, the only non-vanishing ones are

$$\frac{1}{\mathrm{i}\hbar}[r_k, p_k] = 1 \,, \qquad \frac{1}{\mathrm{i}\hbar}[y, \mathrm{i}\hbar y^\dagger] = [y, y^\dagger] = 1 \,. \tag{12.1.25}$$

Varying the $y_\alpha^\dagger$'s produces the equations of motion of the y_α's,

$$\left(\mathrm{i}\frac{\mathrm{d}}{\mathrm{d}t} - \omega_\alpha\right) y_\alpha = -\sqrt{\frac{2\pi}{\hbar\omega_\alpha}} \sum_a e_a \boldsymbol{v}_a \cdot \boldsymbol{A}_\alpha(\boldsymbol{r}_a)^* , \tag{12.1.26}$$

and variation of the y_α's gives the adjoint equations,

$$\left(-\mathrm{i}\frac{\mathrm{d}}{\mathrm{d}t} - \omega_\alpha\right) y_\alpha^\dagger = -\sqrt{\frac{2\pi}{\hbar\omega_\alpha}} \sum_a e_a \boldsymbol{v}_a \cdot \boldsymbol{A}_\alpha(\boldsymbol{r}_a) . \tag{12.1.27}$$

Upon introducing current components,

$$\begin{aligned} j_\alpha(t) &= \sum_a e_a \boldsymbol{v}_a(t) \cdot \boldsymbol{A}_\alpha\big(\boldsymbol{r}_a(t)\big)^* = \int (\mathrm{d}\boldsymbol{r})\, \boldsymbol{j}(\boldsymbol{r},t) \cdot \boldsymbol{A}_\alpha(\boldsymbol{r})^* , \\ j_\alpha(t)^\dagger &= \sum_a e_a \boldsymbol{v}_a(t) \cdot \boldsymbol{A}_\alpha\big(\boldsymbol{r}_a(t)\big) = \int (\mathrm{d}\boldsymbol{r})\, \boldsymbol{j}(\boldsymbol{r},t) \cdot \boldsymbol{A}_\alpha(\boldsymbol{r}) , \end{aligned} \tag{12.1.28}$$

(remember: products are symmetrized if necessary), the interaction Lagrangian has the compact appearance,

$$L_{\text{int.}} = \sum_\alpha \sqrt{\frac{2\pi\hbar}{\omega_\alpha}} \left(j_\alpha^\dagger y_\alpha + y_\alpha^\dagger j_\alpha\right) , \tag{12.1.29}$$

and the equations of motion read

$$\begin{aligned} \frac{\mathrm{d}}{\mathrm{d}t}\left(\mathrm{e}^{\mathrm{i}\omega_\alpha t} y_\alpha(t)\right) &= \mathrm{i}\sqrt{\frac{2\pi}{\hbar\omega_\alpha}}\, \mathrm{e}^{\mathrm{i}\omega_\alpha t} j_\alpha(t) , \\ \frac{\mathrm{d}}{\mathrm{d}t}\left(\mathrm{e}^{-\mathrm{i}\omega_\alpha t} y_\alpha(t)^\dagger\right) &= -\mathrm{i}\sqrt{\frac{2\pi}{\hbar\omega_\alpha}}\, \mathrm{e}^{-\mathrm{i}\omega_\alpha t} j_\alpha(t)^\dagger . \end{aligned} \tag{12.1.30}$$

The electromagnetic modes are described by driven oscillators – driven by the electric current of the charges.

12.2 Effective action

First consider the situation in which the radiation field is *not* driven because no charges are present. Then we just have a collection of independent oscillators, and from Section 10.8 we know that

$$\begin{aligned} \langle y^{\dagger\prime}, t_1 | y'', t_2\rangle &= \prod_\alpha \underbrace{\langle y_\alpha^{\dagger\prime}, t_1 | y_\alpha'', t_2\rangle}_{= \mathrm{e}^{y_\alpha^{\dagger\prime}\, \mathrm{e}^{-\mathrm{i}\omega_\alpha(t_1 - t_2)} y_\alpha''}} \\ &= \sum_{\{n\}} \langle y^{\dagger\prime} | \{n\}\rangle\, \mathrm{e}^{-\frac{\mathrm{i}}{\hbar} E(\{n\})(t_1 - t_2)} \langle \{n\} | y''\rangle , \end{aligned} \tag{12.2.1}$$

where $|\{n\}\rangle$ is the vector describing the state in which there are n_α light quanta – photons – in each mode α, with the multi-photon energy

$$E(\{n\}) = \sum_\alpha \hbar\omega_\alpha n_\alpha \tag{12.2.2}$$

and the multi-photon wave functions

$$\langle y^{\dagger\prime}|\{n\}\rangle = \prod_\alpha \frac{\left(y_\alpha^{\dagger\prime}\right)^{n_\alpha}}{\sqrt{n_\alpha!}} \,, \qquad \langle\{n\}|y''\rangle = \prod_\alpha \frac{(y''_\alpha)^{n_\alpha}}{\sqrt{n_\alpha!}} \,. \tag{12.2.3}$$

When the charged particles *are* present we examine the transformation function

$$\langle y^{\dagger\prime},\ldots,t_1|y'',\ldots,t_2\rangle \equiv \langle 1|2\rangle \,, \tag{12.2.4}$$

where the dots represent some choice of description for the particles. Keeping in mind the structure of the Lagrangian

$$L = L_{\text{part.}} + L_{\text{int.}} + L_{\text{field}} \,, \tag{12.2.5}$$

we see that, if we vary particle variables only, which appear in $L_{\text{part.}}$ and $L_{\text{int.}}$,

$$\delta_{\text{part.}}\langle 1|2\rangle = \frac{\mathrm{i}}{\hbar}\langle 1|\delta_{\text{part.}}\left[\int_2^1 \mathrm{d}t\left(L_{\text{part.}} + L_{\text{int.}}\right)\right]|2\rangle \tag{12.2.6}$$

where

$$\delta_{\text{part.}}L_{\text{int.}} = \sum_\alpha \sqrt{\frac{2\pi\hbar}{\omega_\alpha}}\left[\delta_{\text{part.}}j_\alpha^\dagger\, y_\alpha + y_\alpha^\dagger\,\delta_{\text{part.}}j_\alpha\right] . \tag{12.2.7}$$

Now, the solution of the driven oscillator equations of motion (12.1.30) are [compare with (7.2.9) and (7.2.10)]

$$\begin{aligned} y_\alpha(t) &= \mathrm{e}^{-\mathrm{i}\omega_\alpha(t-t_2)}y_\alpha(t_2) + \mathrm{i}\sqrt{\frac{2\pi}{\hbar\omega_\alpha}}\int_2^1 \mathrm{d}t'\,\eta(t-t')\,\mathrm{e}^{-\mathrm{i}\omega_\alpha(t-t')}j_\alpha(t') \,, \\ y_\alpha(t)^\dagger &= y_\alpha(t_1)^\dagger\,\mathrm{e}^{-\mathrm{i}\omega_\alpha(t_1-t)} + \mathrm{i}\sqrt{\frac{2\pi}{\hbar\omega_\alpha}}\int_2^1 \mathrm{d}t'\,j_\alpha(t')^\dagger\,\mathrm{e}^{-\mathrm{i}\omega_\alpha(t'-t)}\eta(t'-t) \,. \end{aligned} \tag{12.2.8}$$

As a consequence of the initial specification of the y_α, $y_\alpha(t_2) \to y''_\alpha$, and the final specification of the $y_\alpha^\dagger$, $y_\alpha(t_1)^\dagger \to y_\alpha^{\dagger\prime}$, one piece of the action variation $\delta_{\text{part.}}\int_2^1 \mathrm{d}t\,L_{\text{int.}}$ is

$$\int_2^1 \mathrm{d}t \sum_\alpha \sqrt{\frac{2\pi\hbar}{\omega_\alpha}} \Big[\delta_{\text{part.}} j_\alpha(t)^\dagger \,\mathrm{e}^{-\mathrm{i}\omega_\alpha(t-t_2)} y''_\alpha + y^{\dagger\,\prime}_\alpha \,\mathrm{e}^{-\mathrm{i}\omega_\alpha(t_1-t)} \delta_{\text{part.}} j_\alpha(t)\Big]$$

$$= \delta_{\text{part.}} \left[\int_2^1 \mathrm{d}t \sum_a \frac{e_a}{c} \boldsymbol{v}_a \cdot \boldsymbol{A}'\big(\boldsymbol{r}_a(t), t\big)\right] \tag{12.2.9}$$

where

$$\frac{1}{c}\boldsymbol{A}'(\boldsymbol{r},t) = \sum_\alpha \sqrt{\frac{2\pi\hbar}{\omega_\alpha}} \left[\boldsymbol{A}_\alpha(\boldsymbol{r})\,\mathrm{e}^{-\mathrm{i}\omega_\alpha(t-t_2)} y''_\alpha + y^{\dagger\,\prime}_\alpha \,\mathrm{e}^{-\mathrm{i}\omega_\alpha(t_1-t)} \boldsymbol{A}_\alpha(\boldsymbol{r})^*\right] . \tag{12.2.10}$$

The second piece of $\delta_{\text{part.}} \int_2^1 \mathrm{d}t\, L_{\text{int.}}$ is

$$\int_2^1 \mathrm{d}t\,\mathrm{d}t'\,\mathrm{i} \sum_\alpha \frac{2\pi}{\omega_\alpha} \Big[\delta_{\text{part.}} j_\alpha(t)^\dagger \eta(t-t')\,\mathrm{e}^{-\mathrm{i}\omega_\alpha(t-t')} j_\alpha(t) + j_\alpha(t)^\dagger \eta(t-t')\,\mathrm{e}^{-\mathrm{i}\omega_\alpha(t-t')} \delta_{\text{part.}} j_\alpha(t)\Big]$$

$$= \delta_{\text{part.}} \left[\int_2^1 \mathrm{d}t\,\mathrm{d}t'\,\mathrm{i} \sum_\alpha \frac{2\pi}{\omega_\alpha} j_\alpha(t)^\dagger \eta(t-t')\,\mathrm{e}^{-\mathrm{i}\omega_\alpha(t-t')} j_\alpha(t')\right] . \tag{12.2.11}$$

So, for a given effective field $\boldsymbol{A}'(\boldsymbol{r},t)$, we have an effective particle action

$$W_{12}^{\boldsymbol{A}'} = \int_2^1 \mathrm{d}t \Big[L_{\text{part.}} + \sum_a \frac{e_a}{c} \boldsymbol{v}_a \cdot \boldsymbol{A}'\big(\boldsymbol{r}_a(t),t\big)\Big] + \int_2^1 \mathrm{d}t\,\mathrm{d}t'\,\mathrm{i} \sum_\alpha \frac{2\pi}{\omega_\alpha} j_\alpha(t)^\dagger \eta(t-t')\,\mathrm{e}^{-\mathrm{i}\omega_\alpha(t-t')} j_\alpha(t') , \tag{12.2.12}$$

which determines the particle part of the time transformation function,

$$\delta\langle \ldots, t_1 | \ldots, t_2\rangle^{\boldsymbol{A}'} = \frac{\mathrm{i}}{\hbar} \langle \ldots, t_1 | \delta W_{12}^{\boldsymbol{A}'} | \ldots, t_2\rangle \tag{12.2.13}$$

where $\boldsymbol{A}'$, that is: y''_α and $y^{\dagger\,\prime}_\alpha$, held fixed. The complete time transformation function is given just by

$$\langle y^{\dagger\,\prime}, \ldots, t_1 | y'', \ldots, t_2\rangle = \underbrace{\langle y^{\dagger\,\prime}, t_1 | y'', t_2\rangle}_{\substack{\text{photons}\\ \text{only}}} \underbrace{\langle \ldots, t_1 | \ldots, t_2\rangle^{\boldsymbol{A}'}}_{\substack{\text{particles under}\\ \text{the influence}\\ \text{of photons}}} . \tag{12.2.14}$$

12.3 Consistency check

Of course we should check what must be true: that this also contains the influence of the charged particles on the photons. So consider

$$\frac{\partial}{\partial y_\alpha^{\dagger\,\prime}}\langle 1|2\rangle = \langle 1|y_\alpha(t_1)|2\rangle$$

$$= \langle 1|\Big[\mathrm{e}^{-i\omega_\alpha(t_1 - t_2)}\underset{\substack{\downarrow \\ y''_\alpha}}{y_\alpha(t_2)} + \mathrm{i}\sqrt{\frac{2\pi}{\hbar\omega_\alpha}}\int_2^1 \mathrm{d}t\ \mathrm{e}^{-\mathrm{i}\omega_\alpha(t_1 - t)} j_\alpha(t)\Big]|2\rangle$$

$$= \mathrm{e}^{-i\omega_\alpha(t_1 - t_2)} y''_\alpha\langle 1|2\rangle + \mathrm{i}\sqrt{\frac{2\pi}{\hbar\omega_\alpha}}\int_2^1 \mathrm{d}t\ \mathrm{e}^{-\mathrm{i}\omega_\alpha(t_1 - t)}\langle 1|j_\alpha(t)|2\rangle \tag{12.3.1}$$

and compare it with

$$\frac{\partial}{\partial y_\alpha^{\dagger\,\prime}}\Big[\langle y^{\dagger\prime}, t_1|y'', t_2\rangle\langle\ldots, t_1|\ldots, t_2\rangle^{\boldsymbol{A}'}\Big]$$

$$= \left(\frac{\partial}{\partial y_\alpha^{\dagger\,\prime}}\langle y^{\dagger\prime}, t_1|y'', t_2\rangle\right)\langle\ldots, t_1|\ldots, t_2\rangle^{\boldsymbol{A}'}$$

$$+ \langle y^{\dagger\prime}, t_1|y'', t_2\rangle\frac{\partial}{\partial y_\alpha^{\dagger\,\prime}}\langle\ldots, t_1|\ldots, t_2\rangle^{\boldsymbol{A}'} . \tag{12.3.2}$$

As for the first part, we have

$$\frac{\partial}{\partial y_\alpha^{\dagger\,\prime}}\langle y^{\dagger\prime}, t_1|y'', t_2\rangle = \frac{\partial}{\partial y_\alpha^{\dagger\,\prime}}\mathrm{e}^{\sum_\beta y_\beta^{\dagger\,\prime}\,\mathrm{e}^{-i\omega_\beta(t_1 - t_2)}y''_\beta}$$

$$= \mathrm{e}^{-i\omega_\alpha(t_1 - t_2)}y''_\alpha\langle y^{\dagger\prime}, t_1|y'', t_2\rangle . \tag{12.3.3}$$

For the second part we use the action principle to vary $\boldsymbol{A}'(\boldsymbol{r}, t)$, specifically by an infinitesimal change of $y_\alpha^{\dagger\,\prime}$:

$$\delta_{\boldsymbol{A}'}\langle\ldots, t_1|\ldots, t_2\rangle^{\boldsymbol{A}'} = \frac{\mathrm{i}}{\hbar}\langle\ldots, t_1|\int_2^1 \mathrm{d}t\sum_a \frac{e_a}{c}\boldsymbol{v}_a\cdot\delta\boldsymbol{A}'(\boldsymbol{r}_a(t), t)|\ldots, t_2\rangle^{\boldsymbol{A}'} , \tag{12.3.4}$$

where, for the present purposes,

$$\frac{1}{c}\delta\boldsymbol{A}' = \sqrt{\frac{2\pi\hbar}{\omega_\alpha}}\,\delta y_\alpha^{\dagger\,\prime}\,\mathrm{e}^{-\mathrm{i}\omega_\alpha(t_1 - t)}\boldsymbol{A}_\alpha(\boldsymbol{r})^* . \tag{12.3.5}$$

So,

$$\frac{\partial}{\partial y_\alpha^{\dagger\prime}}\langle \dots, t_1 | \dots, t_2 \rangle^{A'}$$
$$= \mathrm{i}\sqrt{\frac{2\pi\hbar}{\omega_\alpha}}\int_2^1 \mathrm{d}t\, \mathrm{e}^{-\mathrm{i}\omega_\alpha(t_1 - t)}\langle \dots, t_1 | j_\alpha(t) | \dots, t_2 \rangle^{A'} , \tag{12.3.6}$$

and the two versions are the same, according to the clearly consistent interpretation of a particle matrix element,

$$\langle 1 | j_\alpha(t) | 2 \rangle = \langle y^{\dagger\prime}, t_1 | y'', t_2 \rangle \langle \dots, t_1 | j_\alpha(t) | \dots, t_2 \rangle^{A'} . \tag{12.3.7}$$

We have carried out variations of particle variables and photon variables. But what about t_1 and t_2? Consider, for example, the t_1 Schrödinger equation

$$\mathrm{i}\hbar\frac{\partial}{\partial t_1}\langle 1 | 2 \rangle = \langle 1 | H | 2 \rangle . \tag{12.3.8}$$

Is this exactly reproduced by

$$\frac{\partial \langle y^{\dagger\prime}, t_1 | y'', t_2 \rangle}{\partial t_1}\langle \dots, t_1 | \dots, t_2 \rangle^{A'} + \langle y^{\dagger\prime}, t_1 | y'', t_2 \rangle \frac{\partial \langle \dots, t_1 | \dots, t_2 \rangle^{A'}}{\partial t_1} ? \tag{12.3.9}$$

It simplifies our task here to adopt the Lagrangian viewpoint, in which

$$\boldsymbol{v}_a = \frac{\mathrm{d}\boldsymbol{r}_a}{\mathrm{d}t} , \tag{12.3.10}$$

so that

$$\mathrm{d}t\, L_{\text{part.}} = \sum_a \frac{m_a}{2}\frac{(\mathrm{d}\boldsymbol{r}_a)^2}{\mathrm{d}t} - \mathrm{d}t\,\frac{1}{2}\sum_{a \neq b}\frac{e_a e_b}{|\boldsymbol{r}_a - \boldsymbol{r}_b|} , \tag{12.3.11}$$

and [recall the definition of j_α in (12.1.28)]

$$\mathrm{d}t\, L_{\text{int.}} = \sum_\alpha \sqrt{\frac{2\pi\hbar}{\omega_\alpha}}\sum_a \left[y_\alpha e_a \mathrm{d}\boldsymbol{r}_a \cdot \boldsymbol{A}_\alpha(\boldsymbol{r}_a) + y_\alpha^\dagger e_a \mathrm{d}\boldsymbol{r}_a \cdot \boldsymbol{A}_\alpha(\boldsymbol{r}_a)^*\right] , \tag{12.3.12}$$

along with

$$\mathrm{d}t\, L_{\text{field}} = \sum_\alpha \left[\mathrm{i}\hbar y_\alpha^\dagger\, \mathrm{d}y_\alpha - \mathrm{d}t\, \hbar\omega_\alpha y_\alpha^\dagger y_\alpha\right] . \tag{12.3.13}$$

We identify H as the coefficient of $-\mathrm{d}\delta t$ in $\delta\big[\mathrm{d}t\, L\big]$. Thus, we see that

$$H = \underbrace{\sum_a \frac{m_a}{2}\left(\frac{\mathrm{d}\boldsymbol{r}_a}{\mathrm{d}t}\right)^2 + \frac{1}{2}\sum_{a\neq b}\frac{e_a e_b}{|\boldsymbol{r}_a - \boldsymbol{r}_b|}}_{= H_{\text{part.}}} + \sum_\alpha \hbar\omega_\alpha y_\alpha^\dagger y_\alpha\,. \qquad (12.3.14)$$

The t_1 Schrödinger equation now reads more explicitly

$$\frac{\partial}{\partial t_1}\langle y^{\dagger\prime},\dots,t_1|y'',\dots,t_2\rangle$$
$$= -\frac{\mathrm{i}}{\hbar}\langle y^{\dagger\prime},\dots,t_1\Big|\Big[H_{\text{part.}} + \sum_\alpha \hbar\omega_\alpha y_\alpha(t_1)^\dagger y_\alpha(t_1)\Big]\Big|y'',\dots,t_2\rangle\,, \qquad (12.3.15)$$

where we can introduce $y, y^\dagger$ eigenvalues:

$$y_\alpha(t_1)^\dagger \to y_\alpha^{\dagger\prime}\,, \qquad y_\alpha(t_1) \to \mathrm{e}^{-i\omega_\alpha(t_1 - t_2)}y_\alpha'' + \mathrm{i}\sqrt{\frac{2\pi}{\hbar\omega_\alpha}}\int_2^1 \mathrm{d}t\;\mathrm{e}^{-\mathrm{i}\omega_\alpha(t_1 - t)}j_\alpha(t)\,, \qquad (12.3.16)$$

and recognize immediately that the purely photonic part,

$$-\frac{\mathrm{i}}{\hbar}\sum_\alpha \hbar\omega_\alpha y_\alpha^{\dagger\prime}\,\mathrm{e}^{-i\omega_\alpha(t_1 - t_2)}y_\alpha''\langle 1|2\rangle\,, \qquad (12.3.17)$$

is just what emerges from the t_1 derivative of

$$\langle y^{\dagger\prime},t_1|y'',t_2\rangle = \mathrm{e}^{\sum_\alpha y_\alpha^{\dagger\prime}\,\mathrm{e}^{-i\omega_\alpha(t_1 - t_2)}y_\alpha''}\,. \qquad (12.3.18)$$

The only question is whether

$$-\frac{\mathrm{i}}{\hbar}\langle y^{\dagger\prime},\dots,t_1\Big|\Big[H_{\text{part.}} + \mathrm{i}\sum_\alpha \sqrt{2\pi\hbar\omega_\alpha}\,y_\alpha^{\dagger\prime}\int_2^1 \mathrm{d}t\;\mathrm{e}^{-\mathrm{i}\omega_\alpha(t_1 - t)}j_\alpha(t)\Big]\Big|y'',\dots,t_2\rangle \qquad (12.3.19)$$

equals

$$\langle y^{\dagger\prime},t_1|y'',t_2\rangle\frac{\partial}{\partial t_1}\langle\dots,t_1|\dots,t_2\rangle^{\boldsymbol{A}'}\,. \qquad (12.3.20)$$

This is also the question: Is the t_1 Hamiltonian associated with the particle function $\langle\dots,t_1|\dots,t_2\rangle^{\boldsymbol{A}'}$ given by

$$H_1^{\boldsymbol{A}'} = H_{\text{part.}} + \mathrm{i}\sum_\alpha \sqrt{2\pi\hbar\omega_\alpha}\,y_\alpha^{\dagger\prime}\int_2^1 \mathrm{d}t\;\mathrm{e}^{-\mathrm{i}\omega_\alpha(t_1 - t)}j_\alpha(t)\;? \qquad (12.3.21)$$

To answer this, we look back at the effective action $W_{12}^{\boldsymbol{A}'}$ in (12.2.12) where

$$\mathrm{d}t \sum_a \frac{e_a}{c} \boldsymbol{v}_a \cdot \boldsymbol{A}' = \sum_a e_a \, \mathrm{d}\boldsymbol{r}_a \sum_\alpha \sqrt{\frac{2\pi\hbar}{\omega_\alpha}} \Big[\boldsymbol{A}_\alpha(\boldsymbol{r}_a) \, \mathrm{e}^{-\mathrm{i}\omega_\alpha(t - t_2)} y_\alpha'' + y_\alpha^{\dagger\,\prime} \, \mathrm{e}^{-\mathrm{i}\omega_\alpha(t_1 - t)} \boldsymbol{A}_\alpha(\boldsymbol{r}_a)^* \Big] \tag{12.3.22}$$

and, for example,

$$\mathrm{d}t \, j_\alpha^\dagger = \sum_a e_a \, \mathrm{d}\boldsymbol{r}_a \cdot \boldsymbol{A}_\alpha(\boldsymbol{r}_a) \,. \tag{12.3.23}$$

Observe first that the t variation of the $j^\dagger \cdots j$ term in $W_{12}^{\boldsymbol{A}'}$ contains overt references to δt but not to $\frac{\mathrm{d}}{\mathrm{d}t}\delta t$; *no* contribution to $H_1^{\boldsymbol{A}'}$ here. Of course, $L_{\text{part.}}$ gives $H_{\text{part.}}$, as before. That leaves the terms of (12.3.22). Again there is δt, but not $\frac{\mathrm{d}}{\mathrm{d}t}\delta t$. Problem? No. There is *explicit* dependence on the final time t_1 in the effective field $\boldsymbol{A}'$. The relevant part of $W_{12}^{\boldsymbol{A}'}$ is

$$\begin{aligned} &\int \mathrm{d}t \sum_a e_a \boldsymbol{v}_a \cdot \sum_\alpha \sqrt{\frac{2\pi\hbar}{\omega_\alpha}} (-\mathrm{i}\omega_\alpha \delta t_1) y_\alpha^{\dagger\,\prime} \, \mathrm{e}^{-\mathrm{i}\omega_\alpha(t_1 - t)} \boldsymbol{A}_\alpha(\boldsymbol{r}_a)^* \\ &= -\,\delta t_1 \left[\mathrm{i} \sum_\alpha \sqrt{2\pi\hbar\omega_\alpha} \, y_\alpha^{\dagger\,\prime} \int_2^1 \mathrm{d}t \; \mathrm{e}^{-\mathrm{i}\omega_\alpha(t_1 - t)} j_\alpha(t) \right] , \end{aligned} \tag{12.3.24}$$

which displays just the required contribution to $H_1^{\boldsymbol{A}'}$.

12.4 Free-space photon mode functions

Let's be more explicit about the $\boldsymbol{A}_\alpha(\boldsymbol{r})$, the photon mode functions, for the situation of unbounded space. Then, there are no boundary conditions to watch in addition to the defining properties stated in (12.1.18) and (12.1.19). The basic solutions are plane waves,

$$\mathrm{e}^{\mathrm{i}\boldsymbol{k} \cdot \boldsymbol{r}} \quad \text{with} \quad |\boldsymbol{k}| = \frac{\omega}{c} \,, \tag{12.4.1}$$

which are given a transverse-vector character by polarization vectors $\boldsymbol{e}_{\boldsymbol{k}\lambda}$ (with $\lambda = 1, 2$ for linear polarization or $\lambda = \pm$ for circular polarization, for instance) subject to

$$\boldsymbol{k} \cdot \boldsymbol{e}_{\boldsymbol{k}\lambda} = 0 \quad \text{and} \quad {\boldsymbol{e}_{\boldsymbol{k}\lambda}}^* \cdot \boldsymbol{e}_{\boldsymbol{k}\lambda'} = \delta_{\lambda,\lambda'} \,, \tag{12.4.2}$$

and are normalized by reference to small wave number cells,

$$A_{k\lambda}(r) = e_{k\lambda} \frac{1}{\sqrt{(2\pi)^3\,(\delta k)}} \int_{(\delta k)} (dk)\, e^{ik\cdot r} . \tag{12.4.3}$$

Indeed, in view of

$$\int_{(\delta k)} (dk) \int_{(\delta k')} (dk') \underbrace{\int (dr)\, e^{-i(k-k')\cdot r}}_{=(2\pi)^3\delta(k-k')} = (2\pi)^3\,(\delta k)\delta_{k,k'} \tag{12.4.4}$$

their orthonormality is easily verified,

$$\begin{aligned} \int (dr)\, A_{k\lambda}(r)^* \cdot A_{k'\lambda'}(r) &= e^*_{k\lambda} \cdot e_{k'\lambda'} \frac{1}{(2\pi)^3} \frac{1}{\sqrt{(\delta k)\,(\delta k')}} \\ &\quad \times \int_{(\delta k)} (dk) \int_{(\delta k')} (dk') \int (dr)\, e^{-i(k-k')\cdot r} \\ &= \delta_{k,k'}\delta_{\lambda,\lambda'} . \end{aligned} \tag{12.4.5}$$

As long as the range of r is restricted,

$$|\Delta k \cdot r| \ll 1 , \tag{12.4.6}$$

where Δk measures the size of a (δk) cell, it suffices to write [now using (dk) rather than (δk)]

$$A_{k\lambda}(r) = e_{k\lambda} \sqrt{\frac{(dk)}{(2\pi)^3}}\, e^{ik\cdot r} . \tag{12.4.7}$$

This is, of course, just the transverse-vector analog of the scalar mode function (10.8.12).

With this choice we have ($\omega = c|k|$)

$$\begin{aligned} \frac{1}{c} A'(r,t) = \sum_{k,\lambda} \frac{1}{2\pi} \sqrt{\frac{\hbar(dk)}{\omega}} \Big[& e_{k\lambda}\, e^{i[k\cdot r - \omega(t-t_2)]} y''_{k\lambda} \\ & + e^*_{k\lambda}\, e^{-i[k\cdot r - \omega(t-t_1)]} y^{\dagger\prime}_{k\lambda} \Big] , \end{aligned} \tag{12.4.8}$$

where it might be clearer to associate the explicit t_1, t_2 dependence with the corresponding eigenvalues, that is

$$e^{i[k\cdot r - \omega t]} y''_{k\lambda}\, e^{i\omega t_2} ; \qquad e^{-i[k\cdot r - \omega t]} y^{\dagger\prime}_{k\lambda}\, e^{-i\omega t_1} . \tag{12.4.9}$$

As for the non-local in time interaction term,

$$W_{12,\text{n-loc}} \equiv \int_2^1 dt\, dt'\, i \sum_\alpha \frac{2\pi}{\omega_\alpha} j_\alpha(t)^\dagger \eta(t-t')\, e^{-i\omega_\alpha(t-t')} j_\alpha(t'), \tag{12.4.10}$$

we have

$$j_a(t)^\dagger \to \sum_a e_a \boldsymbol{v}_a(t) \cdot \boldsymbol{e}_{\boldsymbol{k}\lambda} \sqrt{\frac{(\mathrm{d}\boldsymbol{k})}{(2\pi)^3}} \, \mathrm{e}^{\mathrm{i}\boldsymbol{k} \cdot \boldsymbol{r}_a(t)} ,$$

$$j_a(t') \to \sum_b e_b \boldsymbol{v}_b(t') \cdot \boldsymbol{e}^*_{\boldsymbol{k}\lambda} \sqrt{\frac{(\mathrm{d}\boldsymbol{k})}{(2\pi)^3}} \, \mathrm{e}^{-\mathrm{i}\boldsymbol{k} \cdot \boldsymbol{r}_b(t')} , \tag{12.4.11}$$

so that

$$W_{12,\text{n-loc}} = \int_2^1 \mathrm{d}t\,\mathrm{d}t' \sum_\lambda \int \frac{(\mathrm{d}\boldsymbol{k})}{(2\pi)^2 \omega} \Big(\sum_a e_a \boldsymbol{v}_a(t) \cdot \boldsymbol{e}_{\boldsymbol{k}\lambda} \, \mathrm{e}^{\mathrm{i}\boldsymbol{k} \cdot \boldsymbol{r}_a(t)} \Big)$$
$$\times \, \mathrm{i}\eta(t - t') \, \mathrm{e}^{-\mathrm{i}\omega(t - t')} \Big(\sum_b e_b \boldsymbol{v}_b(t') \cdot \boldsymbol{e}^*_{\boldsymbol{k}\lambda} \, \mathrm{e}^{-\mathrm{i}\boldsymbol{k} \cdot \boldsymbol{r}_b(t')} \Big) . \tag{12.4.12}$$

12.5 Physical mass

To begin our applications, let all $y'' = 0$ (no photons initially) and all $y^{\dagger\,\prime} = 0$ (nor finally), so that the effective field vanishes: $\boldsymbol{A}'(\boldsymbol{r}, t) = 0$, and consider just one particle (mass m_0, charge e, position $\boldsymbol{r}$, momentum $\boldsymbol{p}$, velocity $\boldsymbol{v}$). Then

$$W_{12} \to \int_2^1 \mathrm{d}t \left[\boldsymbol{p} \cdot \left(\frac{\mathrm{d}\boldsymbol{r}}{\mathrm{d}t} - \boldsymbol{v} \right) - \frac{1}{2} m_0 \boldsymbol{v}^2 \right]$$
$$+ \int_2^1 \mathrm{d}t\,\mathrm{d}t' \, e^2 \sum_\lambda \int \frac{(\mathrm{d}\boldsymbol{k})}{(2\pi)^2 \omega} \boldsymbol{v}(t) \cdot \boldsymbol{e}_{\boldsymbol{k}\lambda} \, \mathrm{e}^{\mathrm{i}\boldsymbol{k} \cdot \boldsymbol{r}(t)}$$
$$\times \, \mathrm{i}\eta(t - t') \, \mathrm{e}^{-\mathrm{i}\omega(t - t')} \boldsymbol{v}(t') \cdot \boldsymbol{e}^*_{\boldsymbol{k}\lambda} \, \mathrm{e}^{-\mathrm{i}\boldsymbol{k} \cdot \boldsymbol{r}(t')} , \tag{12.5.1}$$

where the time–non-local term accounts for the net effect that results from the emission of a photon (at the earlier times t') and its subsequent reabsorption (at the later times t).

In this situation we can well anticipate that the whole system moves with constant velocity, and constant momentum, as described non-relativistically by an effective Hamiltonian

$$H = \frac{\boldsymbol{p}^2}{2m} , \qquad \frac{\mathrm{d}}{\mathrm{d}t} \boldsymbol{p} = 0 , \tag{12.5.2}$$

so that $[\boldsymbol{r} \equiv \boldsymbol{r}(t = 0)]$

$$\boldsymbol{v} = \frac{\boldsymbol{p}}{m} , \qquad \boldsymbol{r}(t) = \boldsymbol{r} + \frac{\boldsymbol{p}}{m} t = \boldsymbol{r} + \boldsymbol{v}t . \tag{12.5.3}$$

Then we encounter the *operator* product

$$\boldsymbol{v}\cdot\boldsymbol{e}_{\boldsymbol{k}\lambda}\,\mathrm{e}^{\mathrm{i}\boldsymbol{k}\cdot(\boldsymbol{r}+\boldsymbol{v}t)}\,\mathrm{e}^{-\mathrm{i}\boldsymbol{k}\cdot(\boldsymbol{r}+\boldsymbol{v}t')}\boldsymbol{v}\cdot\boldsymbol{e}^{*}_{\boldsymbol{k}\lambda}\,, \tag{12.5.4}$$

where, for example, $\boldsymbol{v}\cdot\boldsymbol{e}_{\boldsymbol{k}\lambda}$ and $\mathrm{e}^{\mathrm{i}\boldsymbol{k}(\boldsymbol{r}+\boldsymbol{v}t)}$ commute because $\boldsymbol{e}_{\boldsymbol{k}\lambda}$ and $\boldsymbol{k}$ are perpendicular [$\boldsymbol{k}\cdot\boldsymbol{e}_{\boldsymbol{k}\lambda}$=0, cf. (12.4.2)]. Now, one knows, in various ways, that

$$\mathrm{e}^{A}\,\mathrm{e}^{B}=\mathrm{e}^{A+B+\frac{1}{2}[A,B]+\cdots}\,, \tag{12.5.5}$$

where (recall Problem 2-10b) the series terminates if, as here, $[A,B]$ commutes with A and B:

$$A=\mathrm{i}\boldsymbol{k}\cdot(\boldsymbol{r}+\boldsymbol{v}t)\,,\qquad B=-\mathrm{i}\boldsymbol{k}\cdot(\boldsymbol{r}+\boldsymbol{v}t')\,,$$

$$[A,B]=\underbrace{\Big[\boldsymbol{k}\cdot\boldsymbol{r},\boldsymbol{k}\cdot\frac{\boldsymbol{p}}{m}t'\Big]}_{=\mathrm{i}\hbar\frac{\boldsymbol{k}^2}{m}t'}+\underbrace{\Big[\boldsymbol{k}\cdot\frac{\boldsymbol{p}}{m}t,\boldsymbol{k}\cdot\boldsymbol{r}\Big]}_{=-\mathrm{i}\hbar\frac{\boldsymbol{k}^2}{m}t}=-\mathrm{i}\hbar\frac{\boldsymbol{k}^2}{m}(t-t')\,. \tag{12.5.6}$$

So

$$\begin{aligned}\mathrm{e}^{\mathrm{i}\boldsymbol{k}\cdot(\boldsymbol{r}+\boldsymbol{v}t)}\,\mathrm{e}^{-\mathrm{i}\boldsymbol{k}\cdot(\boldsymbol{r}+\boldsymbol{v}t')}&=\mathrm{e}^{\mathrm{i}\boldsymbol{k}\cdot\boldsymbol{v}(t-t')}\,\mathrm{e}^{-i\hbar\frac{\boldsymbol{k}^2}{2m}(t-t')}\\&=\mathrm{e}^{\mathrm{i}\boldsymbol{k}\cdot\boldsymbol{v}(t-t')}\,\mathrm{e}^{-i\frac{\hbar\omega^2}{2mc^2}(t-t')}\,,\end{aligned} \tag{12.5.7}$$

which presents us with the time integral

$$\int_2^1 \mathrm{d}t\,\mathrm{d}t'\;\mathrm{e}^{-\mathrm{i}[\omega-\boldsymbol{k}\cdot\boldsymbol{v}+\hbar\omega^2/(2mc^2)](t-t')}\mathrm{i}\eta(t-t')\,. \tag{12.5.8}$$

If we restrict attention to photons of non-relativistic energy,

$$\hbar\omega\ll mc^2\,, \tag{12.5.9}$$

the ratio of the third to the first two terms is

$$\frac{\hbar\omega}{2mc^2}\ll 1\,, \tag{12.5.10}$$

and we neglect the third term (while noting the potential for a relativistic treatment). For the second term we have

$$|\boldsymbol{k}\cdot\boldsymbol{v}|\leq\frac{\omega}{c}|\boldsymbol{v}|=\omega\frac{|\boldsymbol{v}|}{c}\,, \tag{12.5.11}$$

which is neglected relative to ω for non-relativistic motion, $|\boldsymbol{v}|\ll c$.

So, the exponent in the integrand of (12.5.8) is effectively equal to $-\mathrm{i}\omega(t-t')$, and we are left with

$$\int_2^1 \mathrm{d}t\,\mathrm{d}t'\ \mathrm{e}^{-\mathrm{i}\omega(t-t')}\mathrm{i}\eta(t-t') = \int_{t_2}^{t_1}\mathrm{d}t\int_{t_2}^{t}\mathrm{d}t'\,\mathrm{i}\,\mathrm{e}^{-\mathrm{i}\omega(t-t')}$$
$$= \int_{t_2}^{t_1}\mathrm{d}t\,\frac{1}{\omega}\left(1-\mathrm{e}^{-\mathrm{i}\omega(t-t_2)}\right)$$
$$= \int_{t_2}^{t_1}\frac{\mathrm{d}t}{\omega} - \frac{\mathrm{i}}{\omega^2}\left(\mathrm{e}^{-\mathrm{i}\omega(t_1-t_2)}-1\right). \tag{12.5.12}$$

Concentrate first on the secular term, the one growing linearly with the duration $T = t_1 - t_2$. That contribution to W_{12} is

$$\int \mathrm{d}t\, e^2 \sum_\lambda \int \frac{(\mathrm{d}\boldsymbol{k})}{(2\pi)^2\omega^2}\boldsymbol{v}\cdot\boldsymbol{e}_{\boldsymbol{k}\lambda}\,\boldsymbol{v}\cdot\boldsymbol{e}^*_{\boldsymbol{k}\lambda}\,. \tag{12.5.13}$$

Inasmuch as there is no explicit dependence on $\boldsymbol{k}$, the polarization sum is over two of three orthogonal directions, so on the average

$$\sum_\lambda \boldsymbol{v}\cdot\boldsymbol{e}_{\boldsymbol{k}\lambda}\,\boldsymbol{v}\cdot\boldsymbol{e}^*_{\boldsymbol{k}\lambda} \to \frac{2}{3}\boldsymbol{v}^2\,, \tag{12.5.14}$$

and

$$(\mathrm{d}\boldsymbol{k}) \to 4\pi k^2\mathrm{d}k = 4\pi\frac{\omega^2}{c^2}\mathrm{d}k\,. \tag{12.5.15}$$

This gives for the additional action term

$$\int_2^1 \mathrm{d}t\, e^2\frac{2}{3}\boldsymbol{v}^2\frac{1}{\pi c^2}\mathrm{d}k = \int_2^1 \mathrm{d}t\,\frac{1}{2}\delta m\,\boldsymbol{v}^2 \tag{12.5.16}$$

where

$$\delta m = \frac{4}{3\pi}\frac{e^2}{c^2}\int_0^{2\pi/\lambda_{\min}}\mathrm{d}k = \frac{8}{3}\frac{e^2/\lambda_{\min}}{c^2} \tag{12.5.17}$$

so that

$$\delta m \ll m \quad \text{if} \quad \lambda_{\min} \gg \frac{e^2}{mc^2}\,, \tag{12.5.18}$$

requiring a restriction to the non-relativistic situation to stop the linearly divergent integral [which, as Weisskopf* and Furry† noticed, is only logarithmic, relativistically, when ω gets replaced by $\omega + \hbar k^2/(2m)$]. This piece of the action adds directly to $\int_2^1 \mathrm{d}t\,\frac{1}{2}m_0\boldsymbol{v}^2$ to effectively change m_0 into

$$m = m_0 + \delta m, \tag{12.5.19}$$

the "renormalized mass".

---

*Victor Frederick WEISSKOPF (b. 1908) †Wendell Hinkle FURRY (1907–1984)

12.6 Infrared photons

Now we turn to the non-secular, transient term of (12.5.12), which is produced from (12.5.13) by the substitution

$$\int_2^1 dt\,[\cdots] \to [\cdots]\left(-\frac{i}{\omega}\right)\left(e^{-i\omega T}-1\right), \tag{12.6.1}$$

so, with (12.5.14) and (12.5.15),

$$\begin{aligned} W_{12,\text{n-sec}} &= \int dk\,\frac{2}{3}\frac{e^2}{\pi c^2}\boldsymbol{v}^2\left(-\frac{i}{kc}\right)\left(e^{-i\omega T}-1\right) \\ &= -i\frac{2}{3}\frac{e^2\boldsymbol{v}^2}{\pi c^3}\int_0^{\omega_{\max}}\frac{d\omega}{\omega}\left(e^{-i\omega T}-1\right). \end{aligned} \tag{12.6.2}$$

For a state of definite momentum (velocity), this is a numerical addition to the action, one that produces a change in the time transformation function by the multiplicative factor

$$\begin{aligned} e^{\frac{i}{\hbar}W_{12,\text{n-sec}}} &= e^{\frac{2}{3}\frac{e^2\boldsymbol{v}^2}{\pi\hbar c^3}\int\frac{d\omega}{\omega}\left(e^{-i\omega T}-1\right)} \\ &= e^{-\frac{2}{3}\frac{e^2\boldsymbol{v}^2}{\pi\hbar c^3}\int\frac{d\omega}{\omega}}\left[1+\frac{2}{3}\frac{e^2\boldsymbol{v}^2}{\pi\hbar c^3}\int\frac{d\omega}{\omega}e^{-i\omega T}+\cdots\right]. \end{aligned} \tag{12.6.3}$$

The factors $e^{-i\omega T}$, $e^{-i(\omega+\omega')T}$, ... clearly indicate the presence of one or more photons. What's going on?

The initial choice $y''_\alpha = 0$, "setting the initial field equal to zero", denies the existence of the magnetic field associated with the uniformly moving charge. In effect, we have set $\boldsymbol{v} = 0$ at time t_2. Then at time $t_2 + 0$, the magnetic field springs into being, as though the velocity $\boldsymbol{v}$ comes into being instantaneously. That, as we know, produces radiation. Indeed, for long wavelength, 'infrared' photons, it is known that the relative probability of emitting a photon in the range $d\omega$ is

$$\frac{2}{3\pi}\frac{e^2}{\hbar c}\frac{\boldsymbol{v}^2}{c^2}\frac{d\omega}{\omega}, \tag{12.6.4}$$

as one sees above. In fact

$$e^{-\frac{2}{3\pi}\frac{e^2\boldsymbol{v}^2}{\hbar c^3}},\quad e^{-\frac{2}{3\pi}\frac{e^2\boldsymbol{v}^2}{\hbar c^3}}\,\frac{2}{3\pi}\frac{e^2\boldsymbol{v}^2}{\hbar c^3},\quad \ldots \tag{12.6.5}$$

are the absolute probabilities for emitting no photons, one photon, ... ; that these probabilities add to unity is immediately apparent:

$$\lim_{T\to 0} e^{\frac{2}{3}\frac{e^2 v^2}{\pi\hbar c^3}\int \frac{d\omega}{\omega}\left(e^{-i\omega T}-1\right)} = 1\,. \tag{12.6.6}$$

One may wonder why these factors are probabilities, rather than probability amplitudes. That is because they contain both the probability amplitude for emission (at time t_2) and the complex conjugate probability amplitude for absorption (at time t_1). The absolute squared amplitude is the probability.

One can live with this description, but it would be more physical to make the magnetic field explicit from the beginning. For that purpose, go back to the Lagrangian (12.1.24), with $L_{\text{int.}}$ in the form (12.1.29), and the y_α equations of motion (12.1.30). If there were no time dependences in the current components j_α, the steady-state solution for y_α would be

$$y_\alpha \to \frac{1}{\omega_\alpha}\sqrt{\frac{2\pi}{\hbar\omega_\alpha}}\,j_\alpha\,. \tag{12.6.7}$$

Accordingly, let us, more generally, redefine y_α, $y_\alpha^\dagger$,

$$\begin{aligned} y_\alpha(t) &\to y_\alpha(t) + \frac{1}{\omega_\alpha}\sqrt{\frac{2\pi}{\hbar\omega_\alpha}}\,j_\alpha(t)\,, \\ y_\alpha(t)^\dagger &\to y_\alpha(t)^\dagger + \frac{1}{\omega_\alpha}\sqrt{\frac{2\pi}{\hbar\omega_\alpha}}\,j_\alpha(t)^\dagger\,. \end{aligned} \tag{12.6.8}$$

Then, for example,

$$\begin{aligned} -\sum_\alpha \hbar\omega_\alpha y_\alpha^\dagger y_\alpha \to &-\sum_\alpha \hbar\omega_\alpha y_\alpha^\dagger y_\alpha - \sum_\alpha y_\alpha^\dagger\sqrt{\frac{2\pi\hbar}{\omega_\alpha}}\,j_\alpha \\ &-\sum_\alpha \sqrt{\frac{2\pi\hbar}{\omega_\alpha}}\,j_\alpha^\dagger y_\alpha - \sum_\alpha \frac{2\pi}{\omega_\alpha^2} j_\alpha^\dagger j_\alpha\,, \end{aligned} \tag{12.6.9}$$

whereas $L_{\text{int.}}$ becomes

$$\sum_\alpha \sqrt{\frac{2\pi\hbar}{\omega_\alpha}}\left(j_\alpha^\dagger y_\alpha + y_\alpha^\dagger j_\alpha\right) + 2\sum_\alpha \frac{2\pi}{\omega_\alpha^2} j_\alpha^\dagger j_\alpha\,; \tag{12.6.10}$$

the sum of the two is just

$$-\sum_\alpha \hbar\omega_\alpha y_\alpha^\dagger y_\alpha + \sum_\alpha \frac{2\pi}{\omega_\alpha^2} j_\alpha^\dagger j_\alpha\,. \tag{12.6.11}$$

No interaction between photons and charges? It is there, in

$$\begin{aligned} \sum_\alpha i\hbar y_\alpha^\dagger \frac{d}{dt} y_\alpha \to &\sum_\alpha i\hbar y_\alpha^\dagger \frac{d}{dt} y_\alpha + \sum_\alpha \frac{i}{\omega_\alpha}\sqrt{\frac{2\pi}{\hbar\omega_\alpha}}\,j_\alpha^\dagger \frac{d}{dt} y_\alpha \\ &+\sum_\alpha \frac{i}{\omega_\alpha}\sqrt{\frac{2\pi}{\hbar\omega_\alpha}}\,y_\alpha^\dagger \frac{d}{dt} j_\alpha + \sum_\alpha \frac{i\,2\pi}{\omega_\alpha^3} j_\alpha^\dagger \frac{d}{dt} j_\alpha\,. \end{aligned} \tag{12.6.12}$$

Thus, we now have a new interaction Lagrangian $L_{\text{int.}}$ (shift the time derivative to $j^\dagger$),

$$L_{\text{int.}} \to \sum_\alpha \frac{1}{\omega_\alpha}\sqrt{\frac{2\pi}{\hbar\omega_\alpha}}\,\mathrm{i}\left(y_\alpha^\dagger \frac{\mathrm{d}}{\mathrm{d}t} j_\alpha - \frac{\mathrm{d}}{\mathrm{d}t} j_\alpha^\dagger\, y_\alpha\right) , \tag{12.6.13}$$

and a new particle Lagrangian $L_{\text{part.}}$,

$$L_{\text{part.}} \to \sum_a \left[\boldsymbol{p}\cdot\left(\frac{\mathrm{d}\boldsymbol{r}}{\mathrm{d}t} - \boldsymbol{v}\right) - \frac{1}{2} m_0 \boldsymbol{v}^2\right]_a - \frac{1}{2}\sum_{a\neq b} \frac{e_a e_b}{|\boldsymbol{r}_a - \boldsymbol{r}_b|} + \sum_\alpha \frac{2\pi}{\omega_\alpha^2} j_\alpha^\dagger j_\alpha , \tag{12.6.14}$$

setting aside, as of only marginal interest *at the moment*, the particle contribution involving the $\mathrm{i}(j^\dagger \frac{\mathrm{d}}{\mathrm{d}t} j - \frac{\mathrm{d}}{\mathrm{d}t} j^\dagger\, j)$ terms of the last sum in (12.6.12); we'll remember about them later, on page 458.

What is the additional particle term, for just a single particle? It is, with the single-particle versions of (12.4.11),

$$\int \frac{(\mathrm{d}\boldsymbol{k})}{(2\pi)^3} \frac{2\pi}{\omega^2} e^2 \sum_\lambda \boldsymbol{e}_{\boldsymbol{k}\lambda}\cdot\boldsymbol{v}\, \mathrm{e}^{\mathrm{i}\boldsymbol{k}\cdot\boldsymbol{r}}\, \mathrm{e}^{-\mathrm{i}\boldsymbol{k}\cdot\boldsymbol{r}} \boldsymbol{e}_{\boldsymbol{k}\lambda}^*\cdot\boldsymbol{v}$$
$$= \int \frac{(\mathrm{d}\boldsymbol{k})}{(2\pi)^2} \frac{e^2}{k^2c^2}\frac{2}{3}\boldsymbol{v}^2 = \frac{2}{3\pi}\frac{e^2}{c^2}\int_0^{2\pi/\lambda_{\min}} \mathrm{d}k\ \boldsymbol{v}^2 = \frac{1}{2}\delta m\, \boldsymbol{v}^2 , \tag{12.6.15}$$

producing, for each particle, the mass renormalization (12.5.19).

How about the interaction between different particles? That is contained in

$$\int \frac{(\mathrm{d}\boldsymbol{k})}{(2\pi)^3}\frac{2\pi}{\omega^2}\sum_\lambda\sum_{a,b} e_a \boldsymbol{e}_{\boldsymbol{k}\lambda}\cdot\boldsymbol{v}_a\, \mathrm{e}^{\mathrm{i}\boldsymbol{k}\cdot\boldsymbol{r}_a}\, \mathrm{e}^{-\mathrm{i}\boldsymbol{k}\cdot\boldsymbol{r}_b} e_b \boldsymbol{e}_{\boldsymbol{k}\lambda}^*\cdot\boldsymbol{v}_b$$
$$\to \sum_{a\neq b} e_a e_b \int \frac{(\mathrm{d}\boldsymbol{k})}{(2\pi)^2 k^2 c^2}\, \boldsymbol{v}_a\cdot\boldsymbol{e}_{\boldsymbol{k}\lambda}\boldsymbol{e}_{\boldsymbol{k}\lambda}^*\cdot\boldsymbol{v}_b\, \mathrm{e}^{\mathrm{i}\boldsymbol{k}\cdot(\boldsymbol{r}_a - \boldsymbol{r}_b)} , \tag{12.6.16}$$

where, with $\mathbf{1}$ denoting the unit dyadic,

$$\sum_{\lambda=1,2} \boldsymbol{e}_{\boldsymbol{k}\lambda}\boldsymbol{e}_{\boldsymbol{k}\lambda}^* = \mathbf{1} - \frac{\boldsymbol{k}\boldsymbol{k}}{k^2} , \tag{12.6.17}$$

which, of course, states the geometrical fact, exploited already in (12.5.14), that this dyadic sum is the projector to the plane perpendicular to $\boldsymbol{k}$. This then gives

$$\frac{1}{2}\sum_{a\neq b}\frac{e_a e_b}{c^2}\frac{1}{2\pi^2}\int\frac{(\mathrm{d}\boldsymbol{k})}{k^2}\left[\boldsymbol{v}_a\cdot\boldsymbol{v}_b + \frac{1}{k^2}\boldsymbol{v}_a\cdot\boldsymbol{\nabla}\,\boldsymbol{v}_b\cdot\boldsymbol{\nabla}\right]\mathrm{e}^{\mathrm{i}\boldsymbol{k}\cdot\boldsymbol{r}}\Big|_{\boldsymbol{r}=\boldsymbol{r}_a-\boldsymbol{r}_b} , \tag{12.6.18}$$

where

$$\int \frac{(\mathrm{d}\boldsymbol{k})}{k^2}\,\mathrm{e}^{\mathrm{i}\boldsymbol{k}\cdot\boldsymbol{r}} = 4\pi \int_0^\infty \mathrm{d}k\,\frac{\sin(kr)}{kr} = \frac{2\pi^2}{r} \tag{12.6.19}$$

[note that this is essentially the inverse Fourier transform of the Coulomb potential that we've seen in (11.4.18)], and

$$\begin{aligned}\int \frac{(\mathrm{d}\boldsymbol{k})}{(k^2)^2}\,\mathrm{e}^{\mathrm{i}\boldsymbol{k}\cdot\boldsymbol{r}} &\to \int \frac{(\mathrm{d}\boldsymbol{k})}{(k^2)^2}\Big[\mathrm{e}^{\mathrm{i}\boldsymbol{k}\cdot\boldsymbol{r}} - 1 - \mathrm{i}\boldsymbol{k}\cdot\boldsymbol{r}\Big] \\ &= 4\pi \int_0^\infty \frac{\mathrm{d}k}{k^2}\left[\frac{\sin(kr)}{kr} - 1\right] \\ &= \frac{4\pi}{r}\int_0^\infty \mathrm{d}k\,\big[\sin(kr) - kr\big]\frac{\mathrm{d}}{\mathrm{d}k}\left(-\frac{1}{2k^2}\right) \\ &= 2\pi\int_0^\infty \mathrm{d}k\,\frac{\cos(kr)-1}{k^2} = -\pi^2 r\,,\end{aligned} \tag{12.6.20}$$

where the equivalent replacement in the first step subtracts an $\boldsymbol{r}$ independent term, which is spurious, since this is differentiated eventually,

$$\boldsymbol{v}_a\cdot\boldsymbol{\nabla}\,\boldsymbol{v}_b\cdot\boldsymbol{\nabla}\,r = \boldsymbol{v}_a\cdot\boldsymbol{\nabla}\,\frac{\boldsymbol{v}_b\cdot\boldsymbol{r}}{r} = \frac{\boldsymbol{v}_a\cdot\boldsymbol{v}_b}{r} - \frac{\boldsymbol{v}_a\cdot\boldsymbol{r}\,\boldsymbol{v}_b\cdot\boldsymbol{r}}{r^3}\,. \tag{12.6.21}$$

Thus,

$$\begin{aligned}&\frac{1}{2\pi^2}\int\frac{(\mathrm{d}\boldsymbol{k})}{k^2}\left[\boldsymbol{v}_a\cdot\boldsymbol{v}_b + \frac{1}{k^2}\boldsymbol{v}_a\cdot\boldsymbol{\nabla}\,\boldsymbol{v}_b\cdot\boldsymbol{\nabla}\right]\mathrm{e}^{\mathrm{i}\boldsymbol{k}\cdot\boldsymbol{r}} \\ &= \frac{\boldsymbol{v}_a\cdot\boldsymbol{v}_b}{r} - \frac{1}{2}\left[\frac{\boldsymbol{v}_a\cdot\boldsymbol{v}_b}{r} - \frac{\boldsymbol{v}_a\cdot\boldsymbol{r}\,\boldsymbol{v}_b\cdot\boldsymbol{r}}{r^3}\right] = \frac{1}{2}\left[\frac{\boldsymbol{v}_a\cdot\boldsymbol{v}_b}{r} + \frac{\boldsymbol{v}_a\cdot\boldsymbol{r}\,\boldsymbol{v}_b\cdot\boldsymbol{r}}{r^3}\right],\end{aligned} \tag{12.6.22}$$

which gives us the magnetic-energy interaction contribution to $L_{\text{part.}}$,

$$\frac{1}{4}\sum_{a\neq b}\frac{e_a e_b}{c^2}\,\frac{\boldsymbol{v}_a\cdot\boldsymbol{v}_b + \boldsymbol{v}_a\cdot\frac{\boldsymbol{r}}{r}\,\boldsymbol{v}_b\cdot\frac{\boldsymbol{r}}{r}}{r}\Bigg|_{\boldsymbol{r}=\boldsymbol{r}_a-\boldsymbol{r}_b}\,, \tag{12.6.23}$$

known as Darwin* term.

12.7 Effective Hamiltonian

The new form of the interaction Lagrangian $L_{\text{int.}}$ in (12.6.13) is obtained from the old one in (12.1.29) by the substitution

*Charles Galton DARWIN (1887–1962)

$$j_\alpha \to \frac{\mathrm{i}}{\omega_\alpha}\frac{\mathrm{d}}{\mathrm{d}t} j_\alpha\,, \qquad j_\alpha^\dagger \to -\frac{\mathrm{i}}{\omega_\alpha}\frac{\mathrm{d}}{\mathrm{d}t} j_\alpha^\dagger\,. \tag{12.7.1}$$

This implies no essential change in the $\boldsymbol{A}'$ term because, for example,

$$\mathrm{e}^{\mathrm{i}\omega_\alpha t} j_\alpha \quad \text{and} \quad \mathrm{e}^{\mathrm{i}\omega_\alpha t}\frac{\mathrm{i}}{\omega_\alpha}\frac{\mathrm{d}}{\mathrm{d}t} j_\alpha \tag{12.7.2}$$

differ by a time derivative. But there is a significant change in the time–non-local interaction term of the effective action (12.2.12), which now reads

$$\begin{aligned} W_{12,\text{n-loc}} &= \int_2^1 \mathrm{d}t\,\mathrm{d}t'\,\mathrm{i}\sum_\alpha \frac{2\pi}{\omega_\alpha^3}\frac{\mathrm{d}j_\alpha(t)}{\mathrm{d}t}^\dagger \eta(t-t')\,\mathrm{e}^{-\mathrm{i}\omega_\alpha(t-t')}\frac{\mathrm{d}j_\alpha(t')}{\mathrm{d}t'} \\ &= \int_2^1 \mathrm{d}t\,\mathrm{d}t' \sum_\lambda \int \frac{(\mathrm{d}\boldsymbol{k})}{(2\pi)^3}\frac{2\pi}{\omega^3}\left[\frac{\mathrm{d}}{\mathrm{d}t}\sum_a e_a \boldsymbol{v}_a(t)\cdot \boldsymbol{e}_{\boldsymbol{k}\lambda}\,\mathrm{e}^{\mathrm{i}\boldsymbol{k}\cdot\boldsymbol{r}_a(t)}\right] \\ &\qquad \times \mathrm{i}\eta(t-t')\,\mathrm{e}^{-\mathrm{i}\omega(t-t')} \\ &\qquad \times \left[\frac{\mathrm{d}}{\mathrm{d}t'}\sum_b e_b \boldsymbol{v}_b(t')\cdot \boldsymbol{e}^*_{\boldsymbol{k}\lambda}\,\mathrm{e}^{-\mathrm{i}\boldsymbol{k}\cdot\boldsymbol{r}_b(t')}\right]. \end{aligned} \tag{12.7.3}$$

Note here that, for instance,

$$\frac{\mathrm{d}}{\mathrm{d}t}\boldsymbol{v}\,\mathrm{e}^{\mathrm{i}\boldsymbol{k}\cdot\boldsymbol{r}} = \frac{\mathrm{d}\boldsymbol{v}}{\mathrm{d}t}\,\mathrm{e}^{\mathrm{i}\boldsymbol{k}\cdot\boldsymbol{r}} + \boldsymbol{v}\mathrm{i}\boldsymbol{k}\cdot\boldsymbol{v}\,\mathrm{e}^{\mathrm{i}\boldsymbol{k}\cdot\boldsymbol{r}}\,, \tag{12.7.4}$$

so that the magnitude of the second term, compared to the first, is of the order

$$\frac{|\boldsymbol{v}|\frac{\omega}{c}|\boldsymbol{v}|}{\left|\frac{\mathrm{d}\boldsymbol{v}}{\mathrm{d}t}\right|} = \frac{|\boldsymbol{v}|}{c}\frac{\omega|\boldsymbol{v}|}{\left|\frac{\mathrm{d}\boldsymbol{v}}{\mathrm{d}t}\right|} \cong \frac{|\boldsymbol{v}|}{c} \ll 1 \tag{12.7.5}$$

because one expects the important values of ω to be of the order of $\left|\frac{\mathrm{d}\boldsymbol{v}}{\mathrm{d}t}\right|/|\boldsymbol{v}|$.

For the simplicity of writing in what follows let the system be a hydrogenic atom, so there is effectively only one moving charge. Then the non-local in time supplement to the particle action becomes

$$\begin{aligned} W_{12,\text{n-loc}} &= \int_2^1 \mathrm{d}t\,\mathrm{d}t'\,e^2 \sum_\lambda \int \frac{(\mathrm{d}\boldsymbol{k})}{(2\pi)^2\omega^3}\,\boldsymbol{e}_{\boldsymbol{k}\lambda}\cdot\frac{\mathrm{d}\boldsymbol{v}(t)}{\mathrm{d}t}\,\mathrm{e}^{\mathrm{i}\boldsymbol{k}\cdot\boldsymbol{r}(t)} \\ &\qquad \times \mathrm{i}\eta(t-t')\,\mathrm{e}^{-\mathrm{i}\omega(t-t')}\frac{\mathrm{d}\boldsymbol{v}(t')}{\mathrm{d}t'}\cdot \boldsymbol{e}^*_{\boldsymbol{k}\lambda}\,\mathrm{e}^{-\mathrm{i}\boldsymbol{k}\cdot\boldsymbol{r}(t')}\,. \end{aligned} \tag{12.7.6}$$

We should try to find an effective Hamiltonian H in a self-consistent way. It begins by writing

$$\frac{\mathrm{d}\boldsymbol{v}(t')}{\mathrm{d}t'}\,\mathrm{e}^{-\mathrm{i}\boldsymbol{k}\cdot\boldsymbol{r}(t')} = \mathrm{e}^{-\frac{\mathrm{i}}{\hbar}H(t-t')}\frac{\mathrm{d}\boldsymbol{v}(t)}{\mathrm{d}t}\,\mathrm{e}^{-\mathrm{i}\boldsymbol{k}\cdot\boldsymbol{r}(t)}\,\mathrm{e}^{\frac{\mathrm{i}}{\hbar}H(t-t')} \tag{12.7.7}$$

and by supposing that the system is initially in the E_0 eigenstate of H, so that on the right side, $H \to E_0$. Then the essential terms of the t' integral are

$$\int_{t_2}^{t} \mathrm{d}t' \,\mathrm{i}\,\mathrm{e}^{-\frac{\mathrm{i}}{\hbar}(\hbar\omega + H - E_0)(t-t')} = \int_0^{t-t_2} \mathrm{d}\tau \,\mathrm{i}\,\mathrm{e}^{-\frac{\mathrm{i}}{\hbar}(\hbar\omega + H - E_0)\tau}$$
$$= \frac{\hbar}{\hbar\omega + H - E_0}\left[1 - \mathrm{e}^{-\frac{\mathrm{i}}{\hbar}(\hbar\omega + H - E_0)(t-t_2)}\right], \tag{12.7.8}$$

which has the form

$$\frac{1}{x} - \frac{1}{x}\mathrm{e}^{-\mathrm{i}x\tau} \tag{12.7.9}$$

with $x = \omega + (H - E_0)/\hbar$ and $\tau = t - t_2$, and equals $\mathrm{i}\tau$ for $x = 0$.

Now we must recognize that τ, which is of the order of the total measurement time $T = t_1 - t_2$, is a very long time on the atomic scale, particularly if we are interested in energy measurements or the details of transition processes. This means that the rapidly oscillating x function $\mathrm{e}^{-\mathrm{i}x\tau}$ will generally not contribute to x integrals, except in the neighborhood of $x = 0$. Now, breaking (12.7.9) into real and imaginary parts,

$$\frac{1-\cos(x\tau)}{x} + \mathrm{i}\frac{\sin(x\tau)}{x}, \tag{12.7.10}$$

we observe that

$$\int_{-x_0}^{x_0} \mathrm{d}x\, \frac{\sin(x\tau)}{x} \cong \int_{-\infty}^{\infty} \mathrm{d}y\, \frac{\sin y}{y} = \pi \quad \text{for} \quad x_0\tau \gg 1, \tag{12.7.11}$$

so that

$$\frac{\sin(x\tau)}{x} \to \pi\delta(x), \tag{12.7.12}$$

whereas $[1 - \cos(x\tau)]/x$, which vanishes at $x = 0$, is $1/x$ with the singularity at $x = 0$ removed: the principal value ($\mathcal{P}$). In summary then,

$$\frac{1 - \mathrm{e}^{-\mathrm{i}x\tau}}{x} \to \mathcal{P}\frac{1}{x} + \mathrm{i}\pi\delta(x), \tag{12.7.13}$$

effectively.

With only a single integral left, we can present the change δW_{12} in the action in terms of an addition δH to the Hamiltonian,

$$\delta H = -e^2 \sum_\lambda \int \frac{(\mathrm{d}\boldsymbol{k})}{(2\pi)^2}\frac{\hbar}{\omega^3}\boldsymbol{e}_{\boldsymbol{k}\lambda}\cdot\frac{\mathrm{d}\boldsymbol{v}}{\mathrm{d}t}$$
$$\times\, \mathrm{e}^{\mathrm{i}\boldsymbol{k}\cdot\boldsymbol{r}}\left[\mathcal{P}\frac{1}{\hbar\omega + H - E_0} + \pi\mathrm{i}\delta(\hbar\omega + H - E_0)\right]\mathrm{e}^{-\mathrm{i}\boldsymbol{k}\cdot\boldsymbol{r}}$$
$$\times\, \boldsymbol{e}^*_{\boldsymbol{k}\lambda}\cdot\frac{\mathrm{d}\boldsymbol{v}}{\mathrm{d}t}. \tag{12.7.14}$$

Now, for any function of $\boldsymbol{p}$ and $\boldsymbol{r}$,

$$\mathrm{e}^{\mathrm{i}\boldsymbol{k}\cdot\boldsymbol{r}} f(\boldsymbol{p},\boldsymbol{r})\,\mathrm{e}^{-\mathrm{i}\boldsymbol{k}\cdot\boldsymbol{r}} = f(\boldsymbol{p}-\hbar\boldsymbol{k},\boldsymbol{r})\,. \tag{12.7.15}$$

Is this change of the electron momentum $\boldsymbol{p}$ by a photon momentum $\hbar\boldsymbol{k}$ relevant in the non-relativistic regime we are considering? No, it isn't because the important photon energies $\hbar\omega$ are of the order of atomic energies, so that $|\hbar\boldsymbol{k}| = \hbar\omega/c$ is very small compared to $|\boldsymbol{p}| \sim$ (atomic energy)$/v$. Thus, the exponential factors can be discarded in (12.7.14), and then the polarization sum can be done once more,

$$\sum_{\lambda=1,2} \frac{\mathrm{d}\boldsymbol{v}}{\mathrm{d}t}\cdot\boldsymbol{e}_{\boldsymbol{k}\lambda}[\ldots]\boldsymbol{e}^*_{\boldsymbol{k}\lambda}\cdot\frac{\mathrm{d}\boldsymbol{v}}{\mathrm{d}t} \to \frac{2}{3}\frac{\mathrm{d}\boldsymbol{v}}{\mathrm{d}t}\cdot[\ldots]\frac{\mathrm{d}\boldsymbol{v}}{\mathrm{d}t}\,, \tag{12.7.16}$$

and we arrive at

$$\delta H = -\frac{2}{3\pi}\frac{e^2\hbar}{c^3}\int\frac{\mathrm{d}\omega}{\omega}\frac{\mathrm{d}\boldsymbol{v}}{\mathrm{d}t}\cdot\left[\mathcal{P}\frac{1}{\hbar\omega+H-E_0}+\pi\mathrm{i}\delta(\hbar\omega+H-E_0)\right]\frac{\mathrm{d}\boldsymbol{v}}{\mathrm{d}t}\,. \tag{12.7.17}$$

This addition to H from $W_{12,\text{n-loc}}$ is not the whole story, however, because the terms put aside on page 454 must be added. They give an energy contribution

$$\begin{aligned}
&-\sum_\alpha \frac{\pi}{\omega_\alpha^3}\left(\mathrm{i}j_\alpha^\dagger\frac{\mathrm{d}j_\alpha}{\mathrm{d}t}-\mathrm{i}\frac{\mathrm{d}j_\alpha^\dagger}{\mathrm{d}t}j_\alpha\right)\\
&\quad= -\pi\sum_\lambda\int\frac{(\mathrm{d}\boldsymbol{k})}{(2\pi)^3}\frac{e^2}{\omega^3}\mathrm{i}\Big[\boldsymbol{v}\cdot\boldsymbol{e}_{\boldsymbol{k}\lambda}\,\mathrm{e}^{\mathrm{i}\boldsymbol{k}\cdot\boldsymbol{r}}\frac{\mathrm{d}}{\mathrm{d}t}\left(\mathrm{e}^{-\mathrm{i}\boldsymbol{k}\cdot\boldsymbol{r}}\boldsymbol{e}^*_{\boldsymbol{k}\lambda}\cdot\boldsymbol{v}\right)\\
&\qquad\qquad -\frac{\mathrm{d}}{\mathrm{d}t}\left(\boldsymbol{v}\cdot\boldsymbol{e}_{\boldsymbol{k}\lambda}\,\mathrm{e}^{\mathrm{i}\boldsymbol{k}\cdot\boldsymbol{r}}\right)\mathrm{e}^{-\mathrm{i}\boldsymbol{k}\cdot\boldsymbol{r}}\boldsymbol{e}^*_{\boldsymbol{k}\lambda}\cdot\boldsymbol{v}\Big]\\
&\quad\to -\frac{1}{3\pi}\frac{e^2}{c^3}\int\frac{\mathrm{d}\omega}{\omega}\mathrm{i}\left(\boldsymbol{v}\cdot\frac{\mathrm{d}\boldsymbol{v}}{\mathrm{d}t}-\frac{\mathrm{d}\boldsymbol{v}}{\mathrm{d}t}\cdot\boldsymbol{v}\right),
\end{aligned} \tag{12.7.18}$$

where the intermediate steps are analogous to those between (12.7.3) and (12.7.6) as well as (12.7.16).

Putting the pieces together, the first-order correction in the effective Hamilton operator,

$$\delta H_1 - \mathrm{i}\delta H_2, \tag{12.7.19}$$

is not a Hermitian operator but has an imaginary part,

$$\delta H_2 = \frac{2}{3}\frac{e^2\hbar}{c^3}\int\frac{\mathrm{d}\omega}{\omega}\frac{\mathrm{d}\boldsymbol{v}}{\mathrm{d}t}\cdot\delta(\hbar\omega+H-E_0)\frac{\mathrm{d}\boldsymbol{v}}{\mathrm{d}t}\,, \tag{12.7.20}$$

in addition to its real part,

$$\delta H_1 = -\frac{2}{3\pi}\frac{e^2}{c^3}\int\frac{d\omega}{\omega}\left[\frac{d\boldsymbol{v}}{dt}\cdot\left(\mathcal{P}\frac{\hbar}{\hbar\omega + H - E_0}\right)\frac{d\boldsymbol{v}}{dt} + \frac{1}{2}i\left(\boldsymbol{v}\cdot\frac{d\boldsymbol{v}}{dt} - \frac{d\boldsymbol{v}}{dt}\cdot\boldsymbol{v}\right)\right]. \quad (12.7.21)$$

12.8 Energy shift

Let's first look at the energy change $\langle E_0|\delta H_1|E_0\rangle$ induced by this real part. A simplification is achieved by using

$$\frac{d\boldsymbol{v}}{dt} = \frac{1}{i\hbar}[\boldsymbol{v}, H] = \frac{1}{i\hbar}[\boldsymbol{v}, \hbar\omega + H - E_0] \equiv \frac{1}{i}[\boldsymbol{v}, \Omega] = i[\Omega, \boldsymbol{v}] \quad (12.8.1)$$

to rewrite the integrand (principal values understood):

$$\begin{aligned}\frac{d\boldsymbol{v}}{dt}\cdot\frac{1}{\Omega}\frac{d\boldsymbol{v}}{dt} + \frac{1}{2}i\left(\boldsymbol{v}\cdot\frac{d\boldsymbol{v}}{dt} - \frac{d\boldsymbol{v}}{dt}\cdot\boldsymbol{v}\right) &= (\boldsymbol{v}\Omega - \Omega\boldsymbol{v})\cdot\Omega^{-1}(\Omega\boldsymbol{v} - \boldsymbol{v}\Omega) \\ &\quad + \tfrac{1}{2}\boldsymbol{v}\cdot(\boldsymbol{v}\Omega - \Omega\boldsymbol{v}) + \tfrac{1}{2}(\Omega\boldsymbol{v} - \boldsymbol{v}\Omega)\cdot\boldsymbol{v} \\ &= \underset{\omega}{\underset{\downarrow}{\Omega}}\,\boldsymbol{v}\cdot\Omega^{-1}\boldsymbol{v}\,\underset{\omega}{\underset{\downarrow}{\Omega}} - \tfrac{1}{2}\boldsymbol{v}^2\,\underset{\omega}{\underset{\downarrow}{\Omega}} - \tfrac{1}{2}\,\underset{\omega}{\underset{\downarrow}{\Omega}}\,\boldsymbol{v}^2 .\end{aligned} \quad (12.8.2)$$

As indicated, the Ω's to the very left and the very right will stand next to $\langle E_0|$ and $|E_0\rangle$, respectively, and therefore they can be replaced by their eigenvalue ω. So

$$\frac{d\boldsymbol{v}}{dt}\cdot\frac{1}{\Omega}\frac{d\boldsymbol{v}}{dt} + \frac{1}{2}i\left(\boldsymbol{v}\cdot\frac{d\boldsymbol{v}}{dt} - \frac{d\boldsymbol{v}}{dt}\cdot\boldsymbol{v}\right) \to \omega^2\boldsymbol{v}\cdot\left(\frac{1}{\Omega} - \frac{1}{\omega}\right)\boldsymbol{v} \quad (12.8.3)$$

and we have

$$\begin{aligned}\delta H_1 &= -\frac{2}{3\pi}\frac{e^2}{c^3}\int d\omega\,\omega\,\boldsymbol{v}\cdot\left(\mathcal{P}\frac{\hbar}{\hbar\omega + H - E_0} - \mathcal{P}\frac{1}{\omega}\right)\boldsymbol{v} \\ &= -\frac{2}{3\pi}\frac{e^2\hbar}{m_{el}^2c^3}\int d\omega\,\omega\,\boldsymbol{p}\cdot\left(\mathcal{P}\frac{1}{\hbar\omega + H - E_0} - \mathcal{P}\frac{1}{\hbar\omega}\right)\boldsymbol{p},\end{aligned} \quad (12.8.4)$$

with $\boldsymbol{p} = m_{el}\boldsymbol{v}$.

Owing to the occurring subtraction, δH_1 contains no contribution proportional to $\boldsymbol{v}^2$; and, indeed, it shouldn't since mass renormalization is already taken care of by the extra term in (12.6.14), as we've seen at (12.6.15). And so the mass appearing in (12.8.4) is the physical mass of the electron.

Mindful of the $\langle E_0|\cdots|E_0\rangle$ context, we note that (principal value understood),

$$\begin{aligned}\delta H_1 &= -\frac{2}{3\pi}\frac{e^2\hbar}{m_{\text{el}}^2c^3}\int \mathrm{d}\omega\,\omega\left[\boldsymbol{p},\frac{1}{H+\hbar\omega-E_0}\right]\cdot\boldsymbol{p}\\ &= -\frac{2}{3\pi}\frac{e^2\hbar}{m_{\text{el}}^2c^3}\int \mathrm{d}\omega\,\omega\,\boldsymbol{p}\cdot\left[\frac{1}{H+\hbar\omega-E_0},\boldsymbol{p}\right]\end{aligned} \tag{12.8.5}$$

are equally good, and so is half their sum. Now

$$\left[\boldsymbol{p},A^{-1}\right] = -A^{-1}\left[\boldsymbol{p},A\right]A^{-1} \tag{12.8.6}$$

for any operator A, and

$$\left[\boldsymbol{p},H\right] = \frac{\hbar}{\mathrm{i}}\boldsymbol{\nabla}V \tag{12.8.7}$$

for $H = \boldsymbol{p}^2/(2m_{\text{el}}) + V(\boldsymbol{r})$, so that

$$\left[\boldsymbol{p},\frac{1}{H+\hbar\omega-E_0}\right]\cdot\boldsymbol{p} \to -\frac{1}{\hbar\omega}\frac{\hbar}{\mathrm{i}}\boldsymbol{\nabla}V\frac{1}{H+\hbar\omega-E_0}\cdot\boldsymbol{p} \tag{12.8.8}$$

and (take the adjoint)

$$\boldsymbol{p}\cdot\left[\frac{1}{H+\hbar\omega-E_0},\boldsymbol{p}\right] \to -\boldsymbol{p}\cdot\frac{1}{H+\hbar\omega-E_0}\left(-\frac{\hbar}{\mathrm{i}}\right)\boldsymbol{\nabla}V\frac{1}{\hbar\omega}\,. \tag{12.8.9}$$

This gives

$$\begin{aligned}\delta H_1 &\to -\frac{e^2\hbar}{3\pi m_{\text{el}}^2c^3}\int_0^{\omega_{\max}}\mathrm{d}\omega\,\mathrm{i}\Big[\boldsymbol{\nabla}V\cdot\frac{1}{H+\hbar\omega-E_0}\boldsymbol{p}\\ &\qquad\qquad -\boldsymbol{p}\cdot\frac{1}{H+\hbar\omega-E_0}\boldsymbol{\nabla}V\Big]\\ &= -\frac{e^2}{3\pi m_{\text{el}}^2c^3}\,\mathrm{i}\left[\boldsymbol{\nabla}V\cdot\log\left(\frac{\hbar\omega_{\max}}{|H-E_0|}\right)\boldsymbol{p}-\boldsymbol{p}\cdot\log\left(\frac{\hbar\omega_{\max}}{|H-E_0|}\right)\boldsymbol{\nabla}V\right]\end{aligned} \tag{12.8.10}$$

which stops the non-relativistic integration where it certainly breaks down, at $\hbar\omega_{\max}$ of the order of $m_{\text{el}}c^2$, the relativistic energy associated with the electron mass m_{el}.

Suppose we replace the logarithmic operator $\log\bigl(\hbar\omega_{\max}/|H-E_0|\bigr)$ by some effective numerical value $\log\bigl(m_{\text{el}}c^2/\Delta E\bigr)$. Then we get

$$\delta H_1 \to \frac{e^2}{3\pi m_{\text{el}}^2c^3}\log\left(\frac{m_{\text{el}}c^2}{\Delta E}\right)(-\mathrm{i})\underbrace{\bigl(\boldsymbol{\nabla}V\cdot\boldsymbol{p}-\boldsymbol{p}\cdot\boldsymbol{\nabla}V\bigr)}_{=\mathrm{i}\hbar\boldsymbol{\nabla}^2V} \tag{12.8.11}$$

and, with

$$\boldsymbol{\nabla}^2V = \boldsymbol{\nabla}^2\frac{-Ze^2}{r} = 4\pi Ze^2\delta(\boldsymbol{r})\,, \tag{12.8.12}$$

we have

$$\delta H_1 \to \frac{e^2\hbar}{3\pi m_{\rm el}^2 c^3}\log\left(\frac{m_{\rm el}c^2}{\Delta E}\right)4\pi Z e^2\delta(\boldsymbol{r})$$
$$= \frac{4}{3}\frac{e^2}{\hbar c}\left(\frac{\hbar}{m_{\rm el}c}\right)^2 Z e^2 \log\left(\frac{m_{\rm el}c^2}{\Delta E}\right)\delta(\boldsymbol{r})\,, \tag{12.8.13}$$

so that

$$\langle E_0|\delta H|E_0\rangle = \frac{4}{3}\frac{e^2}{\hbar c}\left(\frac{\hbar}{m_{\rm el}c}\right)^2 Z e^2 \log\left(\frac{m_{\rm el}c^2}{\Delta E}\right)|\psi_{E_0}(0)|^2\,, \tag{12.8.14}$$

which involves the wave function $\psi_{E_0}(\boldsymbol{r})$ at $\boldsymbol{r} = 0$, that is: at the site of the nucleus.

We know that, for $l = 0$ and principal quantum number n,

$$|\psi_{E_0}(0)|^2 = \frac{1}{\pi}\left(\frac{Z}{na_0}\right)^3 \tag{12.8.15}$$

[(8.3.30) in conjunction with $|Y_{00}|^2 = (4\pi)^{-1}$], and with the recognition that the ratio of the electron's Compton* wavelength $\hbar/(m_{\rm el}c)$ and the Bohr radius $a_0 = \hbar^2/(m_{\rm el}e^2)$ is Sommerfeld's fine structure constant α,

$$\frac{\hbar}{m_{\rm el}c}\Big/ a_0 = \alpha = \frac{1}{137.036} \tag{12.8.16}$$

we get Bethe's† result

$$\delta E_{n,0} = \langle n,0|\delta H_1|n,0\rangle = \frac{4}{3\pi}\frac{\alpha^3 Z^4}{n^3}\log\left(\frac{m_{\rm el}c^2}{\Delta E}\right)\frac{e^2}{a_0}\,. \tag{12.8.17}$$

This upward displacement of the $l = 0$ states is known as the Lamb‡ shift. Together with relativistic effects of order α^2 it gives a complete account of hydrogenic fine structure.

12.9 Transition rates

Having dealt with the real part δH_1 of (12.7.21) we now turn to the imaginary part δH_2 of (12.7.20). What is its significance? Look at the $\langle E_0, t_1|E_0, t_2\rangle$ probability amplitude. The "energy shift" $-\mathrm{i}\,\langle\delta H_2\rangle$ contributes to the time factor $\mathrm{e}^{-\frac{\mathrm{i}}{\hbar}ET}$ the real factor

$$\mathrm{e}^{-\frac{1}{\hbar}\langle\delta H_2\rangle T} \tag{12.9.1}$$

---

*Arthur Holly COMPTON (1982–1962) †Hans Albrecht BETHE (b. 1906) ‡Willis Eugene LAMB (b. 1913)

implying that the probability for persistence of the state after the elapse of time $T = t_1 - t_2$ is

$$e^{-\frac{2}{\hbar}\langle\delta H_2\rangle T} = e^{-\gamma T}, \tag{12.9.2}$$

so that γ is the probability per unit time for the system to leave the state. In other words, γ is the decay constant of the unstable system associated with spontaneous emission.

We have thus

$$\gamma = \frac{2}{\hbar}\langle E_0|\delta H_1|E_0\rangle = \frac{4}{3}\frac{e^2}{c^3}\langle E_0|\frac{d\boldsymbol{v}}{dt}\cdot\int_0^\infty \frac{d\omega}{\omega}\,\delta(\hbar\omega + H - E_0)\frac{d\boldsymbol{v}}{dt}|E_0\rangle \tag{12.9.3}$$

with

$$\begin{aligned}\int_0^\infty \frac{d\omega}{\omega}\,\delta(\hbar\omega + H - E_0) &= \frac{1}{E_0 - H}\eta(E_0 - H)\\ &= \sum_{E(<E_0)} |E,\ldots\rangle\frac{1}{E_0 - E}\langle E,\ldots|,\end{aligned} \tag{12.9.4}$$

where the ellipsis indicates further quantum numbers in case of energetic degeneracy. So

$$\gamma = \frac{4}{3}\frac{e^2}{c^3}\sum_{E(<E_0)}\frac{1}{E_0 - E}\left|\langle E,\ldots|\frac{d\boldsymbol{v}}{dt}|E_0\rangle\right|^2, \tag{12.9.5}$$

which is a sum over all states of energy *below* E_0. Alternative versions are obtained from

$$\frac{d}{dt}\boldsymbol{v} = \frac{1}{i\hbar}[\boldsymbol{v}, H] = \left(\frac{1}{i\hbar}\right)^2\big[[\boldsymbol{r}, H], H\big] \tag{12.9.6}$$

and $\boldsymbol{v} = \boldsymbol{p}/m_{\mathrm{el}}$; they read

$$\begin{aligned}\gamma &= \frac{4}{3}\frac{e^2}{\hbar c}\sum_{E(<E_0)}\frac{E_0 - E}{\hbar}\left|\langle E,\ldots|\frac{\boldsymbol{v}}{c}|E_0\rangle\right|^2\\ &= \frac{4}{3}\frac{e^2}{\hbar c}\sum_{E(<E_0)}\frac{E_0 - E}{\hbar}\left|\langle E,\ldots|\frac{\boldsymbol{p}}{m_{\mathrm{el}}c}|E_0\rangle\right|^2\\ &= \frac{4}{3}\left(\frac{e^2}{\hbar c}\right)^3\sum_{E(<E_0)}\frac{E_0 - E}{\hbar}\left(\frac{E_0 - E}{e^2/a_0}\right)^2\left|\langle E,\ldots|\frac{\boldsymbol{r}}{a_0}|E_0\rangle\right|^2\end{aligned} \tag{12.9.7}$$

and illustrate that one can equally well use transition matrix elements of $\boldsymbol{r}$ or $\boldsymbol{p}$ for the evaluation of γ.

The individual terms in (12.9.5) obviously represent the rate at which transitions are made with the emission of a photon of (angular) frequency

$$\omega = \frac{1}{\hbar}(E_0 - E) \,. \tag{12.9.8}$$

This is the quantum analog of the classical Larmor* formula for the rate of radiation of *energy*:

$$\frac{2}{3}\frac{e^2}{c^3}\left(\frac{\mathrm{d}\boldsymbol{v}}{\mathrm{d}t}\right)^2 = P \,. \tag{12.9.9}$$

In fact, the quantum rate of radiation of energy is obtained by multiplying each transition rate by the appropriate $\hbar\omega = E_0 - E$, giving

$$P = \sum_{E(<E_0)} \frac{4}{3}\frac{e^2}{c^3}\left|\langle E, \ldots | \frac{\mathrm{d}\boldsymbol{v}}{\mathrm{d}t} | E_0 \rangle\right|^2 \,. \tag{12.9.10}$$

For sufficiently excited states, the sum of matrix elements to lower energies becomes equal to the sum to higher energies, so that the replacement

$$\sum_{E(<E_0)} \to \frac{1}{2}\sum_{\text{all } E} \tag{12.9.11}$$

is permissible, and

$$P \to \frac{2}{3}\frac{e^2}{c^3}\langle E, \ldots | \left(\frac{\mathrm{d}\boldsymbol{v}}{\mathrm{d}t}\right)^2 | E_0 \rangle = \frac{2}{3}\frac{e^2}{c^3}\left\langle \left(\frac{\mathrm{d}\boldsymbol{v}}{\mathrm{d}t}\right)^2 \right\rangle_{E_0} \,. \tag{12.9.12}$$

This recovery of the classical Larmor rate is an example of how classical physics is contained in quantum mechanics as a limit.

How about a direct derivation of the transition rate? Return to (12.2.14),

$$\langle y^{\dagger\prime}, \ldots, t_1 | y', \ldots, t_2 \rangle = \langle y^{\dagger\prime}, t_1 | y', t_2 \rangle \langle \ldots, t_1 | \ldots, t_2 \rangle^{\boldsymbol{A}'} \,. \tag{12.9.13}$$

If we want the probability amplitude for one photon finally, given none initially, we are looking for a term with but a single $y_\alpha^{\dagger\prime} = \langle y^{\dagger\prime} | 1_\alpha \rangle$. That can only come from $\langle \ldots | \ldots \rangle^{\boldsymbol{A}'}$ since $\langle y^{\dagger\prime}, t_1 | 0, t_2 \rangle = 1$. So we want

$$\frac{\partial}{\partial y_\alpha^{\dagger\prime}} \langle \ldots, t_1 | \ldots, t_2 \rangle^{\boldsymbol{A}'} \Big|_{y^{\dagger\prime}=0} = \frac{\mathrm{i}}{\hbar}\langle \ldots, t_1 | \frac{\partial}{\partial y_\alpha^{\dagger\prime}} W_{12}^{\boldsymbol{A}'} | \ldots, t_2 \rangle^{\boldsymbol{A}'=0} \,, \tag{12.9.14}$$

where [cf. (12.3.1); here one particle only, for simplicity]

*Sir Joseph LARMOR (1857–1942)

$$\frac{\partial}{\partial y_\alpha^{\dagger\prime}} W_{12}^{A'} = \int_2^1 \mathrm{d}t\, e\boldsymbol{v}(t) \cdot \sqrt{\frac{2\pi\hbar}{\omega_\alpha}}\, \mathrm{e}^{-\mathrm{i}\omega_\alpha(t_1 - t)} \boldsymbol{A}_\alpha(\boldsymbol{r})^* , \tag{12.9.15}$$

where the so-called dipole approximation,

$$\boldsymbol{A}_\alpha(\boldsymbol{r})^* = \sqrt{\frac{(\mathrm{d}\boldsymbol{k})}{(2\pi)^3}}\, \boldsymbol{e}_\alpha\, \mathrm{e}^{-\mathrm{i}\boldsymbol{k}\cdot\boldsymbol{r}} \cong \sqrt{\frac{(\mathrm{d}\boldsymbol{k})}{(2\pi)^3}}\, \boldsymbol{e}_\alpha , \tag{12.9.16}$$

is applicable since typical photon wavelengths $\lambda = |\boldsymbol{k}|^{-1}$ are much larger than internal atomic distances r, so that $|\boldsymbol{k}\cdot\boldsymbol{r}| \leq r/\lambda \ll 1$. Then the probability of the transition $E_0 \to E, \ldots$ is

$$\frac{1}{\hbar^2}\frac{(\mathrm{d}\boldsymbol{k})}{(2\pi)^3}\frac{2\pi\hbar}{\omega} e^2 \left| \int_2^1 \mathrm{d}t\, \langle E, \ldots | \boldsymbol{e}_\alpha \cdot \boldsymbol{v}(t) | E_0 \rangle\, \mathrm{e}^{\mathrm{i}\omega t} \right|^2 \tag{12.9.17}$$

where

$$\begin{aligned} \langle E, \ldots | \boldsymbol{v}(t) | E_0 \rangle &= \langle E, \ldots | \mathrm{e}^{\frac{\mathrm{i}}{\hbar}Ht} \boldsymbol{v}\, \mathrm{e}^{-\frac{\mathrm{i}}{\hbar}Ht} | E_0 \rangle \\ &= \langle E, \ldots | \boldsymbol{v} | E_0 \rangle\, \mathrm{e}^{\frac{\mathrm{i}}{\hbar}(E - E_0)t} , \end{aligned} \tag{12.9.18}$$

which gives, see Problem 9-9d,

$$\left| \int_2^1 \mathrm{d}t\ \mathrm{e}^{\frac{\mathrm{i}}{\hbar}(E - E_0 + \hbar\omega)t} \right|^2 = T\, 2\pi\hbar\, \delta(E - E_0 + \hbar\omega) . \tag{12.9.19}$$

This is proportional to the duration $T = t_1 - t_2$, and so the probability per unit time is

$$\frac{1}{\hbar^2}\frac{(\mathrm{d}\boldsymbol{k})}{(2\pi)^3}\frac{2\pi\hbar}{\omega}\, e^2\, 2\pi\hbar\, \delta(E - E_0 + \hbar\omega) \left| \langle E | \boldsymbol{e}_\alpha \cdot \boldsymbol{v} | E_0 \rangle \right|^2 . \tag{12.9.20}$$

We sum over the two polarizations and integrate over $\boldsymbol{k}$, and get the rate for transitions $E_0 \to E$,

$$\begin{aligned} \gamma_{E\leftarrow E_0} &= \int \frac{4\pi\omega \mathrm{d}\omega}{2\pi c^3}\, \delta(\hbar\omega + E - E_0)\, e^2\, \frac{2}{3} \left| \langle E, \ldots | \boldsymbol{v} | E_0 \rangle \right|^2 \\ &= \frac{4}{3}\frac{e^2}{\hbar c}\frac{E_0 - E}{\hbar} \left| \langle E, \ldots | \frac{\boldsymbol{v}}{c} | E_0 \rangle \right|^2 \eta(E_0 - E) . \end{aligned} \tag{12.9.21}$$

The sum of these partial rates is the total rate γ of (12.9.7), which shows the consistency of the two ways of calculating γ.

If there had been n photons of this kind present initially, and we ask for the probability of one more, the additional $y^{\dagger\prime}$ factor multiplying $\langle y^{\dagger\prime} | n \rangle$ gives $\sqrt{n+1}\langle y^{\dagger\prime} | n+1 \rangle$, so that the probability (per unit time) has the factor $n + 1$:

$$(n+1)\gamma_{E\leftarrow E_0} = n\gamma_{E\leftarrow E_0} + \gamma_{E\leftarrow E_0} . \tag{12.9.22}$$

In addition to spontaneous emission (rate $\gamma_{E\leftarrow E_0}$) there is stimulated emission (rate $n\gamma_{E\leftarrow E_0}$), stimulated by the n photons already present. A similar discussion for absorption gives a rate proportional to n, the number of photons available.

12.10 Thomson scattering

As the last application, we consider a situation in which one photon (wave vector $\boldsymbol{k}'$, polarization λ') is present initially, another photon ($\boldsymbol{k}$, λ) finally, and there is a single charge (electron) that is at rest, $\boldsymbol{p} = 0$, initially and finally: scattering of a photon by a charge at rest. To extract the amplitude

$$\psi_{\text{scatt.}} = \left\langle 1_{\boldsymbol{k}\lambda}, \boldsymbol{p}=0, t_1 \middle| 1_{\boldsymbol{k}'\lambda'}, \boldsymbol{p}=0, t_2 \right\rangle \tag{12.10.1}$$

we differentiate twice

$$\psi_{\text{scatt.}} = \frac{\partial}{\partial y_\alpha^{\dagger\,\prime}} \frac{\partial}{\partial y_\beta''} \langle y^{\dagger\,\prime}, \boldsymbol{p}=0, t_1 | y'', \boldsymbol{p}=0, t_2 \rangle \Big|_{y^{\dagger\,\prime}=0,\ y''=0}$$

$$\text{with} \quad \alpha \equiv \boldsymbol{k}\lambda\,, \quad \beta \equiv \boldsymbol{k}'\lambda'\,, \quad \alpha \neq \beta\,. \tag{12.10.2}$$

So,

$$\psi_{\text{scatt.}} = \left\langle 0, \boldsymbol{p}=0, t_1 \middle| \Psi \middle| 0, \boldsymbol{p}=0, t_2 \right\rangle \tag{12.10.3}$$

where

$$\Psi = \left(y_\alpha(t_1) y_\beta(t_2)^\dagger + \frac{\partial y_\beta(t_2)^\dagger}{\partial y_\alpha^{\dagger\,\prime}} \right) \Bigg|_{y^{\dagger\,\prime}=0,\ y''=0} \tag{12.10.4}$$

or, equivalently,

$$\Psi = \left(y_\alpha(t_1) y_\beta(t_2)^\dagger + \frac{\partial y_\alpha(t_1)}{\partial y_\beta''} \right) \Bigg|_{y^{\dagger\,\prime}=0,\ y''=0}\,, \tag{12.10.5}$$

corresponding to the choice of order in which we differentiate.

In the long-wavelength limit we have in mind, the dipole approximation (12.9.16) is appropriate. Then ($T = t_1 - t_2$ again)

$$\begin{aligned} y_\alpha(t_1) &= y_\alpha(t_2)\, \mathrm{e}^{-\mathrm{i}\omega_\alpha T} + \mathrm{i}\sqrt{\frac{2\pi}{\hbar\omega_\alpha}} \int_2^1 \mathrm{d}t\ \mathrm{e}^{\mathrm{i}\omega_\alpha(t_1 - t)} e\boldsymbol{v}(t) \cdot \boldsymbol{A}_\alpha(\boldsymbol{r}(t))^* \\ &\to y_\alpha''\, \mathrm{e}^{-\mathrm{i}\omega_\alpha T} + \mathrm{i}\sqrt{\frac{2\pi}{\hbar\omega_\alpha}} \sqrt{\frac{(\mathrm{d}\boldsymbol{k}_\alpha)}{(2\pi)^3}} \int_2^1 \mathrm{d}t\ \mathrm{e}^{-\mathrm{i}\omega_\alpha(t_1 - t)} e\boldsymbol{v}(t) \cdot \boldsymbol{e}_\alpha^*\,, \end{aligned} \tag{12.10.6}$$

and with

$$\begin{aligned} m_{\text{el}}\boldsymbol{v} = \boldsymbol{p} - \frac{e}{c}\boldsymbol{A}' \to -e \sum_{\alpha'} \sqrt{\frac{2\pi\hbar}{\omega_{\alpha'}}} \sqrt{\frac{(\mathrm{d}\boldsymbol{k}_{\alpha'})}{(2\pi)^3}} \Big[& \boldsymbol{e}_{\alpha'}\, \mathrm{e}^{-\mathrm{i}\omega_{\alpha'}(t - t_2)} y_{\alpha'}'' \\ & + y_{\alpha'}^{\dagger\,\prime}\, \mathrm{e}^{-\mathrm{i}\omega_{\alpha'}(t_1 - t)} \boldsymbol{e}_{\alpha'}^* \Big] \end{aligned} \tag{12.10.7}$$

we get

$$y_\alpha(t_1) \to 0\,, \qquad \frac{\partial y_\alpha(t_1)}{\partial y''_\beta} \to \mathrm{i}\sqrt{\frac{2\pi}{\hbar\omega_\alpha}}\sqrt{\frac{(\mathrm{d}\boldsymbol{k}_\alpha)}{(2\pi)^3}}\sqrt{\frac{2\pi\hbar}{\omega_\beta}}\sqrt{\frac{(\mathrm{d}\boldsymbol{k}_\beta)}{(2\pi)^3}}\left(-\frac{e^2}{m_{\mathrm{el}}}\right) \times \int_2^1 \mathrm{d}t\, \mathrm{e}^{-\mathrm{i}\omega_\alpha(t_1 - t)}\, \mathrm{e}^{-\mathrm{i}\omega_\beta(t - t_2)} \boldsymbol{e}^*_\alpha \cdot \boldsymbol{e}_\beta\,. \tag{12.10.8}$$

So the transition probability is

$$|\psi_{\mathrm{scatt.}}|^2 = \left(\frac{e^2}{m_{\mathrm{el}}}\right)^2 \frac{2\pi}{\omega_\alpha}\frac{2\pi}{\omega_\beta}\frac{(\mathrm{d}\boldsymbol{k}_\alpha)}{(2\pi)^3}\frac{(\mathrm{d}\boldsymbol{k}_\beta)}{(2\pi)^3}\left(\boldsymbol{e}^*_\alpha \cdot \boldsymbol{e}_\beta\right)^2 \times \underbrace{\left|\int_2^1 \mathrm{d}t\, \mathrm{e}^{\mathrm{i}(\omega_\alpha - \omega_\beta)t}\right|^2}_{=2\pi T\,\delta(\omega_\alpha - \omega_\beta)}\,, \tag{12.10.9}$$

and

$$\frac{1}{T}|\psi_{\mathrm{scatt.}}|^2 = \left(\frac{e^2}{m_{\mathrm{el}}}\right)^2 \frac{1}{(2\pi)^3}\delta(\omega - \omega')\frac{(\mathrm{d}\boldsymbol{k})}{\omega}\frac{(\mathrm{d}\boldsymbol{k}')}{\omega'}\left(\boldsymbol{e}^*_\lambda \cdot \boldsymbol{e}_{\lambda'}\right)^2 \tag{12.10.10}$$

is the transition rate (now writing $\boldsymbol{k}_\alpha, \omega_\alpha \equiv \boldsymbol{k}, \omega$ and $\boldsymbol{k}_\beta, \omega_\beta \equiv \boldsymbol{k}', \omega'$). The δ function states the expected: the scattering is elastic, the scattered photon has the frequency of the incident one.

We are interested in the differential cross section $\mathrm{d}\sigma$ for scattering into solid angle $\mathrm{d}\Omega$, so we put

$$(\mathrm{d}\boldsymbol{k}) = \frac{\omega^2 \mathrm{d}\omega}{c^3}\mathrm{d}\Omega\,, \tag{12.10.11}$$

integrate over the frequency ω of the scattered photon, and divide by the incident flux $c(\mathrm{d}\boldsymbol{k}')/(2\pi)^3$, giving

$$\mathrm{d}\sigma = \left(\frac{e^2}{m_{\mathrm{el}}c^2}\right)^2 \left(\boldsymbol{e}^*_\lambda \cdot \boldsymbol{e}_{\lambda'}\right)^2 \mathrm{d}\Omega\,. \tag{12.10.12}$$

If we have no knowledge of the polarization of the incident photon, we must take the average of the two λ' possibilities,

$$\left(\boldsymbol{e}^*_\lambda \cdot \boldsymbol{e}_{\lambda'}\right)^2 \to \frac{1}{2}\sum_{\lambda'=1,2}\left(\boldsymbol{e}^*_\lambda \cdot \boldsymbol{e}_{\lambda'}\right)^2 = \frac{1}{2} - \frac{1}{2}\frac{\boldsymbol{k}' \cdot \boldsymbol{e}_\lambda\, \boldsymbol{e}^*_\lambda \cdot \boldsymbol{k}'}{k'^2}\,, \tag{12.10.13}$$

and if we don't discriminate between the two polarizations of the scattered photon, we must sum over the two possible λ values,

$$(e_\lambda^* \cdot e_{\lambda'})^2 \to \sum_{\lambda=1,2} (e_\lambda^* \cdot e_{\lambda'})^2 = 1 - \frac{k \cdot e_{\lambda'}\, e_{\lambda'}^* \cdot k}{k^2} . \tag{12.10.14}$$

Taken together, they amount to

$$(e_\lambda^* \cdot e_{\lambda'})^2 \to \frac{1}{2} + \frac{1}{2}\frac{(k \cdot k')^2}{k^2\, k'^2} = \frac{1}{2} + \frac{1}{2}\cos^2\theta , \tag{12.10.15}$$

where θ is the scattering angle, $k \cdot k' = kk' \cos\theta$. So the polarization-insensitive version of (12.10.12) is

$$\frac{d\sigma}{d\Omega} = \frac{1}{2}\left(\frac{e^2}{m_{\mathrm{el}}c^2}\right)^2 (1 + \cos^2\theta) . \tag{12.10.16}$$

The total cross section,

$$\sigma = \int d\Omega \frac{d\sigma}{d\Omega} = \frac{1}{2}\left(\frac{e^2}{m_{\mathrm{el}}c^2}\right)^2 4\pi\left(1 + \frac{1}{3}\right) = \frac{8\pi}{3}\left(\frac{e^2}{m_{\mathrm{el}}c^2}\right)^2 , \tag{12.10.17}$$

is the classical Thomson* cross section for light scattering by small obstacles. The remark made above about the Larmor formula applies here too: classical physics is contained in quantum mechanics as a limit.

Problems

12-1 That there are not equations of motion for all field variables of Lagrangian (12.1.1) is implied by the ambiguity, or freedom, associated with gauge transformations,

$$A \to A + \nabla\lambda , \quad \Phi \to \Phi - \frac{1}{c}\frac{\partial}{\partial t}\lambda ,$$
$$E \to E , \quad B \to B ,$$

where $\lambda(r, t)$ is arbitrary. Show that an infinitesimal gauge transformation gives

$$\delta_\lambda L = \int (dr) \left[\frac{1}{c} j \cdot \nabla\delta\lambda + \rho\frac{1}{c}\frac{\partial}{\partial t}\delta\lambda\right] .$$

Then apply the principle of stationary action to obtain the continuity equation

$$\nabla \cdot j + \frac{\partial}{\partial t}\rho = 0 .$$

Explain why this states the *local* conservation of electric charge.

*Sir John Joseph THOMSON (1856–1940)

12-2 Use vector identities to show that

$$F_{\|}(r) = \int (dr') \frac{1}{4\pi|r - r'|} \left[-\nabla' \nabla' \cdot F(r')\right] ,$$

$$F_{\perp}(r) = \int (dr') \frac{1}{4\pi|r - r'|} \nabla' \times \left[\nabla' \times F(r')\right]$$

are the longitudinal and transverse parts of vector field $F(r)$.

12-3 The completeness relation of the transverse mode functions A_α of (12.1.18),

$$\sum_\alpha A_\alpha(r)\, A_\alpha(r')^* = \delta_\perp(r, r') ,$$

is in terms of the transverse delta function, a dyadic with the properties

$$\nabla \cdot \delta_\perp(r, r') = 0 , \qquad \nabla' \cdot \delta_\perp(r, r') = 0 ,$$

and

$$\int (dr')\, \delta_\perp(r, r') \cdot F(r') = F_\perp(r)$$

for any $F = F_{\|} + F_\perp$. Show that

$$\int \frac{(dk)}{(2\pi)^3}\, e^{ik \cdot (r - r')} \left(1 - \frac{k\,k}{k^2}\right) ,$$

where **1** is the unit dyadic, has these properties.

12-4 The mode expansion (12.1.21) constructs the transverse fields A and $E_\perp$ from the mode functions A_α and the non-Hermitian variables $y_\alpha, y_\alpha^\dagger$. Reverse it and express $y_\alpha, y_\alpha^\dagger$ in terms of A and $E_\perp$.

12-5 Justify (12.1.22) and (12.1.23).

12-6a Given a set of orthonormal transverse mode functions $A_\alpha(r)$, show that the set of functions defined by

$$B_\alpha(r) = \frac{c}{\omega_\alpha} \nabla \times A_\alpha(r)$$

are mode functions that could be used equally well. Can you express the $A_\alpha(r)$'s in terms of the $B_\alpha(r)$'s?

12-6b Recognize that

$$B(r, t) = \nabla \times A(r, t) = \sum_\alpha \sqrt{2\pi\hbar\omega_\alpha} \left(y_\alpha(t) B_\alpha(r) + y_\alpha(t)^\dagger B_\alpha(r)^*\right) .$$

Then verify that the radiation–energy–density operator is

$$U = \frac{1}{8\pi}\left(\boldsymbol{E}_\perp^2 + \boldsymbol{B}^2\right) = \frac{\hbar}{2}\sum_{\alpha,\beta}\sqrt{\omega_\alpha\omega_\beta}\Big[\left(\boldsymbol{A}_\alpha^*\cdot\boldsymbol{A}_\beta + \boldsymbol{B}_\alpha^*\cdot\boldsymbol{B}_\beta\right)y_\alpha^\dagger y_\beta - \frac{1}{2}\left(\boldsymbol{A}_\alpha\cdot\boldsymbol{A}_\beta - \boldsymbol{B}_\alpha\cdot\boldsymbol{B}_\beta\right)y_\alpha y_\beta - \frac{1}{2}\left(\boldsymbol{A}_\alpha^*\cdot\boldsymbol{A}_\beta^* - \boldsymbol{B}_\alpha^*\cdot\boldsymbol{B}_\beta^*\right)y_\alpha^\dagger y_\beta^\dagger\Big]\,,$$

where the same terms are omitted as in (12.1.23), the terms that would give rise to a non-zero energy density of the vacuum. Check that

$$\int(\mathrm{d}\boldsymbol{r})\,U = \sum_\alpha \hbar\omega_\alpha y_\alpha^\dagger y_\alpha\,.$$

12-6c The energy–flux–density operator, the analog of the classical Poynting* vector, is

$$\boldsymbol{S} = \frac{c}{4\pi}\boldsymbol{E}_\perp \times \boldsymbol{B}\,.$$

Give its mode expansion (usual omissions). Use it to prove that

$$\frac{\partial}{\partial t}U + \boldsymbol{\nabla}\cdot\boldsymbol{S} = -\boldsymbol{j}_\perp\cdot\boldsymbol{E}_\perp\,;$$

this is the quantum analog of Poynting's theorem.

12-7a Apply the principle of stationary action to the Lagrangian (12.1.13) and find the fundamental field commutation relations

$$\left[\boldsymbol{a}\cdot\boldsymbol{A}(\boldsymbol{r},t), \boldsymbol{b}\cdot\boldsymbol{E}_\perp(\boldsymbol{r}',t)\right]\,,$$

where $\boldsymbol{a}$ and $\boldsymbol{b}$ are arbitrary numerical vectors. Then verify that the mode expansion (12.1.21) is consistent, provided that $y, y^\dagger$ have their usual commutation relations.

12-7b State the commutation relations

$$\left[\boldsymbol{a}\cdot\boldsymbol{B}(\boldsymbol{r},t), \boldsymbol{b}\cdot\boldsymbol{E}_\perp(\boldsymbol{r}',t)\right]\,.$$

12-8a Suppose the current components $j_\alpha(t)$ in (12.1.30) derive from a classical electric current $\boldsymbol{j}$, so that they are numerical quantities, not operators. Assume further that $\boldsymbol{j} = 0$ for $t < t_2$ and $t > t_1$, and find the vacuum persistence amplitude $\langle 0, t_1 | 0, t_2\rangle^{\boldsymbol{j}}$.

*John Henry Poynting (1852–1914)

12-8b Now consider $\boldsymbol{j} = \boldsymbol{j}_< + \boldsymbol{j}_>$ with $\boldsymbol{j}_< = 0$ for $t > T$ and $\boldsymbol{j}_> = 0$ for $t < T$, where T is an intermediate instant, $t_1 > T > t_2$. Use the identity

$$\langle 0, t_1 | 0, t_2 \rangle^{\boldsymbol{j}_< + \boldsymbol{j}_>} = \sum_{\{n\}} \langle 0, t_1 | \{n\}, T \rangle^{\boldsymbol{j}_>} \langle \{n\}, T | 0, t_2 \rangle^{\boldsymbol{j}_<}$$

to find $\langle \{n\}, t_1 | 0, t_2 \rangle^{\boldsymbol{j}}$ and $\langle 0, t_1 | \{n\}, t_2 \rangle^{\boldsymbol{j}}$.

12-8c Give an independent derivation by first constructing $\langle y^{\dagger\prime}, t_1 | 0, t_2 \rangle^{\boldsymbol{j}}$ and $\langle 0, t_1 | y'', t_2 \rangle^{\boldsymbol{j}}$; then compare.

12-9a For given wave vector $\boldsymbol{k}$, adopt a coordinate system in which $\boldsymbol{k} = k\boldsymbol{e}_3$. Then the real unit vectors

$$\boldsymbol{e}_{\boldsymbol{k}1} = \boldsymbol{e}_1 \,\widehat{=}\, \begin{pmatrix} 1 \\ 0 \\ 0 \end{pmatrix} , \quad \boldsymbol{e}_{\boldsymbol{k}2} = \boldsymbol{e}_2 \,\widehat{=}\, \begin{pmatrix} 0 \\ 1 \\ 0 \end{pmatrix}$$

specify linear polarization, and

$$\boldsymbol{e}_{\boldsymbol{k}+} = 2^{-\frac{1}{2}} (\boldsymbol{e}_1 + \mathrm{i}\boldsymbol{e}_2) \,\widehat{=}\, \frac{1}{\sqrt{2}} \begin{pmatrix} 1 \\ \mathrm{i} \\ 0 \end{pmatrix} , \quad \boldsymbol{e}_{\boldsymbol{k}-} = 2^{-\frac{1}{2}} (\mathrm{i}\boldsymbol{e}_1 + \boldsymbol{e}_2) \,\widehat{=}\, \frac{1}{\sqrt{2}} \begin{pmatrix} \mathrm{i} \\ 1 \\ 0 \end{pmatrix}$$

specify circular polarization (or, better, helicity). Verify explicitly that

$$\sum_{\lambda=1,2} \boldsymbol{e}_{\boldsymbol{k}\lambda} \boldsymbol{e}^*_{\boldsymbol{k}\lambda} = \sum_{\lambda=\pm} \boldsymbol{e}_{\boldsymbol{k}\lambda} \boldsymbol{e}^*_{\boldsymbol{k}\lambda} = \boldsymbol{1} - \frac{\boldsymbol{k}\,\boldsymbol{k}}{k^2} \,\widehat{=}\, \begin{pmatrix} 1\,0\,0 \\ 0\,1\,0 \\ 0\,0\,0 \end{pmatrix} .$$

12-9b Repeat for the pair of polarization vectors

$$\boldsymbol{e}_{\boldsymbol{k}a} = \boldsymbol{e}_1 \cos\vartheta + \boldsymbol{e}_2\, \mathrm{e}^{\mathrm{i}\varphi} \sin\vartheta , \quad \boldsymbol{e}_{\boldsymbol{k}b} = \boldsymbol{e}_2 \cos\vartheta - \boldsymbol{e}_1\, \mathrm{e}^{-\mathrm{i}\varphi} \sin\vartheta .$$

How are they related to the $\boldsymbol{e}_{\boldsymbol{k}\pm}$ pair? Polarization of which kind is specified by $\boldsymbol{e}_{\boldsymbol{k}a}, \boldsymbol{e}_{\boldsymbol{k}b}$?

12-10 Charged particle in an isotropic three-dimensional oscillator potential; initial state $|n'_x, n'_y, n'_z\rangle \equiv |2\rangle$; final state $\langle n_x, n_y, n_z| \equiv \langle 1|$. Find the transition rate $\gamma_{1\leftarrow 2}$. Then sum over all possible transitions to get the decay rate of the initial state.

12-11 Hydrogenic atoms: Concerning matrix elements of the type

$$\langle n, l, m | \boldsymbol{r} | n', l', m' \rangle ,$$

find a selection rule that states which elements are *not* automatically zero. [Hint: Consider $\boldsymbol{r} \to -\boldsymbol{r}$.]

12-12a Hydrogenic atoms: Find the decay rate of the states with principal quantum number $n = 2$, angular quantum number $l = 1$, magnetic quantum number $m = 1$ or $m = 0$ or $m = -1$. Why is it sufficient to consider one m value?

12-12b Repeat for $n = 3$, $l = 0, 1, 2$.

12-12c Repeat for arbitrary n, $l = n-1$, $m = l$, the circular Rydberg states of Problem 9-16.

12-13 The Thomson cross section (12.10.17), an area, is a multiple of a squared length, $e^2/(m_{\rm el}c^2)$, the so-called classical electron radius. What is its relation to the electron's Compton wavelength and to the Bohr radius?

Index

Printed in the United States
By Bookmasters